Study and Solutions Guide
to Accompany
Calculus

Study and Solutions Guide
to Accompany

Calculus

LARSON and HOSTETLER
THIRD EDITION

DAVID E. HEYD

D. C. Heath and Company
Lexington, Massachusetts   Toronto

Preface

This guide is designed as a supplement to *Calculus with Analytic Geometry,* Third Edition, by Roland E. Larson and Robert P. Hostetler. All references to chapters, theorems, and exercises relate to the main text. Although this supplement is not a cure-all for poor study habits, it can be of great value when incorporated into a well-planned course of study. The following suggestions are given to assist you in the use of the text, your lecture notes, and this guide.

Read the section in the text for general content before class to get the overall picture. Do not get bogged down in the details in your first reading. Make sure you do this reading *before* you attend the lecture on the topic. You will be surprised how much more you get from the lecture if you have done some preliminary reading. If you are familiar with the topic, you will understand more of the lecture and you will be able to take fewer (and better) notes.

The exercise sets in the text are divided into groups of similar problems and are presented in roughly the same order as the section topics. Try to get an overall picture of the various types of problems in the set. *As you work your way through the exercise set, reread the portion of the section that covers each type of problem* and pay particular attention to the solved examples.

You cannot learn calculus merely by reading any more than you can learn to play the piano or to bowl merely by reading. Learning calculus takes a lot of practice. Only after you have practiced the techniques of a section and have discovered your weak points can you make good use of the supplementary solutions in this guide. Solutions in the guide are given in greater algebraic detail than the solutions in the text. Students who need additional help with algebra should find this especially helpful.

During many years of teaching calculus I have found that one major problem for calculus students is a lack of good study habits. My students have found the following additional suggestions to be helpful in making the best use of their study time.

1. *Write neatly in pencil.* A notebook filled with unorganized scribbling is of little value.

2. *Work at a deliberate methodical pace without skipping steps.* When you hurry through a problem you are more apt to make careless arithmetic or algebraic errors that, in the long run, waste time.

3. *Keep up with the work.* This suggestion is crucial because calculus is a very structured topic. If you cannot do the problems in one section, you are not likely to be able to do the problems in the next. The night before a quiz or test is not the time to start working problems. Cramming just does not work!

4. After working some of the assigned exercises with access to the examples and answers, *try at least one of each type of exercise with the book closed.* This will increase your confidence on quizzes and tests.

5. *Do not be overly concerned with finding the most efficient way to solve a problem.* Your first goal is to find one way that works. Short cuts and clever methods come later.

6. If you have trouble with the algebra of calculus, refer to the *Algebra review* at the beginning of this guide.

The author wishes to acknowledge several people whose help and encouragement were invaluable in the production of this supplementary guide. First, I am grateful to Roland E. Larson and Robert P. Hostetler for the privilege of working with them for several years on the main text and for their helpful suggestions in the preparation of this guide. I also wish to thank Linda M. Bollinger, Linda M. Matta, and Nancy K. Stout for their assistance in preparing the manuscript, Timothy R. Larson and A. David Salvia for the computer graphics, and Kim Rieck Fisher and the staff at D. C. Heath and Company. I am grateful to my wife Jean for her encouragement and understanding of the many hours required in the production of this guide. I also thank our children, Ed, Ruth, and Andy who, for the first time, were able to give assistance in the production of the manuscript.

David E. Heyd

Contents

Algebra review

Monomial factors

Factor as indicated:

(a) $3x^4 + 4x^3 - x^2 = x^2($)

(b) $2\sqrt{x} + 6x^{3/2} = 2\sqrt{x}($)

(c) $e^{-x} - xe^{-x} + 2x^2e^{-x} = e^{-x}($)

(d) $x^{-1} - 2 + x = x^{-1}($)

(e) $\dfrac{x}{2} - 6x^2 = \dfrac{x}{2}($)

(f) $\sin x + \tan x = \sin x($)

(g) $\dfrac{1}{2x^2 + 4x} = \dfrac{1}{2x}\left(\right)$

Solution

(a) $3x^4 + 4x^3 - x^2 = x^2(3x^2 + 4x - 1)$

(b) $2\sqrt{x} + 6x^{3/2} = 2x^{1/2}(1 + 3x) = 2\sqrt{x}(1 + 3x)$

(c) $e^{-x} - xe^{-x} + 2x^2e^{-x} = e^{-x}(1 - x + 2x^2)$

(d) $x^{-1} - 2 + x = x^{-1}(1 - 2x + x^2)$

(e) $\dfrac{x}{2} - 6x^2 = \dfrac{x}{2}(1 - 12x)$

(f) $\sin x + \tan x = \sin x + \dfrac{\sin x}{\cos x} = \sin x\left(1 + \dfrac{1}{\cos x}\right)$

$$= \sin x(1 + \sec x)$$

(g) $\dfrac{1}{2x^2 + 4x} = \dfrac{1}{2x}\left(\dfrac{1}{x + 2}\right)$

2.

Binomial factors

Factor as indicated:

(a) $(x - 1)^2(x) - (x - 1) = (x - 1)(\qquad)$

(b) $3(x^2 + 4)(x^2 + 1) + 6(x^2 + 4)^2 = 3(x^2 + 4)(\qquad)$

(c) $\sqrt{x^2 + 1} - \dfrac{x^2}{\sqrt{x^2 + 1}} = \dfrac{1}{\sqrt{x^2 + 1}}(\qquad)$

(d) $(x - 3)^3(x + 2)(x + 2) - 2(x - 3)^2(x + 2)^2$
$= (x - 3)^2(x + 2)(\qquad)$

(e) $(2x + 1)^{3/2}(x^{1/2}) + (2x + 1)^{5/2}(x^{-1/2})$
$= (2x + 1)^{3/2}(x^{-1/2})(\qquad)$

Solution

(a) $(x - 1)^2(x) - (x - 1) = (x - 1)[(x - 1)x - 1]$
$= (x - 1)(x^2 - x - 1)$

(b) $3(x^2 + 4)(x^2 + 1) + 6(x^2 + 4)^2$
$= 3(x^2 + 4)[(x^2 + 1) + 2(x^2 + 4)]$
$= 3(x^2 + 4)(3x^2 + 9)$

(c) $\sqrt{x^2 + 1} - \dfrac{x^2}{\sqrt{x^2 + 1}} = (x^2 + 1)^{1/2} - x^2(x^2 + 1)^{-1/2}$
$= (x^2 + 1)^{-1/2}[(x^2 + 1) - x^2]$
$= \dfrac{1}{\sqrt{x^2 + 1}}$

(d) $(x - 3)^3(x + 2)(x + 2) - 2(x - 3)^2(x + 2)^2$
$= (x - 3)^2(x + 2)[(x - 3) - 2(x + 2)]$
$= (x - 3)^2(x + 2)(-x - 7)$

(e) $(2x + 1)^{3/2}(x^{1/2}) + (2x + 1)^{5/2}(x^{-1/2})$
$= (2x + 1)^{3/2}(x^{-1/2})[x + (2x + 1)]$
$= (2x + 1)^{3/2}(x^{-1/2})(3x + 1)$

3.

Factoring quadratic expressions

Factor as indicated:

(a) $x^2 - 3x + 2 = (\qquad)(\qquad)$

(b) $x^2 - 9 = (\qquad)(\qquad)$

(c) $x^2 + 5x - 6 = ($ $)($ $)$

(d) $x^2 + 5x + 6 = ($ $)($ $)$

(e) $2x^2 + 5x - 3 = ($ $)($ $)$

(f) $e^{2x} + 2 + e^{-2x} = ($ $)^2$

(g) $x^4 - 7x^2 + 12 = ($ $)($ $)($ $)$

(h) $1 - \sin^2 x = ($ $)($ $)$

Solution

(a) $x^2 - 3x + 2 = (x - 2)(x - 1)$

(b) $x^2 - 9 = (x + 3)(x - 3)$

(c) $x^2 + 5x - 6 = (x + 6)(x - 1)$

(d) $x^2 + 5x + 6 = (x + 2)(x + 3)$

(e) $2x^2 + 5x - 3 = (2x - 1)(x + 3)$

(f) $e^{2x} + 2 + e^{-2x} = (e^x + e^{-x})^2$

(g) $x^4 - 7x^2 + 12 = (x^2 - 3)(x + 2)(x - 2)$

(h) $1 - \sin^2 x = (1 + \sin x)(1 - \sin x)$

4.
Cancellation

Reduce each expression to lowest terms:

(a) $\dfrac{3x + 9}{6x}$

(b) $\dfrac{x^2}{x^{1/2}}$

(c) $\dfrac{(x + 1)^3(x - 2) + 3(x + 1)^2}{(x + 1)^4}$

(d) $\dfrac{x^{1/2} - x^{1/3}}{x^{1/6}}$

(e) $\dfrac{\sqrt{x - 1} + (x - 1)^{3/2}}{\sqrt{x - 1}}$

(f) $\dfrac{1 - (\sin x + \cos x)^2}{2 \sin x}$

Solution

(a) $\dfrac{3x + 9}{6x} = \dfrac{3(x + 3)}{3(2x)} = \dfrac{x + 3}{2x}$

(b) $\dfrac{x^2}{x^{1/2}} = \dfrac{(x^{1/2})(x^{3/2})}{x^{1/2}} = x^{3/2}$

(c) $\dfrac{(x + 1)^3(x - 2) + 3(x + 1)^2}{(x + 1)^4} = \dfrac{(x + 1)^2[(x + 1)(x - 2) + 3]}{(x + 1)^4}$

$= \dfrac{x^2 - x + 1}{(x + 1)^2}$

(d) $\dfrac{x^{1/2} - x^{1/3}}{x^{1/6}} = \dfrac{x^{1/6}(x^{2/6} - x^{1/6})}{x^{1/6}} = x^{1/3} - x^{1/6}$

(e) $\dfrac{\sqrt{x - 1} + (x - 1)^{3/2}}{\sqrt{x - 1}} = \dfrac{\sqrt{x - 1}[1 + (x - 1)]}{\sqrt{x - 1}} = x$

(f) $\dfrac{1 - (\sin x + \cos x)^2}{2 \sin x} = \dfrac{1 - (\sin^2 x + 2 \sin x \cos x + \cos^2 x)}{2 \sin x}$

$$= \dfrac{1 - (\sin^2 x + \cos^2 x) - 2 \sin x \cos x}{2 \sin x}$$

$$= \dfrac{1 - 1 - 2 \sin x \cos x}{2 \sin x} = -\cos x$$

5.
Quadratic formula

	Equation	*Solve for*
(a)	$x^2 - 4x - 1 = 0$	x
(b)	$2x^2 + x - 3 = 0$	x
(c)	$\cos^2 x + 3 \cos x + 2 = 0$	$\cos x$
(d)	$x^2 - xy - (1 + y^2) = 0$	x
(e)	$x^4 - 4x^2 + 2 = 0$	x^2

Solution

(a) $x = \dfrac{4 \pm \sqrt{16 + 4}}{2} = \dfrac{4 \pm \sqrt{20}}{2} = \dfrac{4 \pm 2\sqrt{5}}{2} = 2 \pm \sqrt{5}$

(b) $x = \dfrac{-1 \pm \sqrt{1 + 24}}{4} = \dfrac{-1 \pm 5}{4}$. Therefore

$$x = \frac{4}{4} = 1 \quad \text{or} \quad x = -\frac{6}{4} = -\frac{3}{2}.$$

(c) $\cos x = \dfrac{-3 \pm \sqrt{9 - 8}}{2} = \dfrac{-3 \pm 1}{2}$. Therefore

$$\cos x = -\frac{2}{2} = -1 \quad \text{or} \quad \cos x = -\frac{4}{2} = -2.$$

(d) $x = \dfrac{y \pm \sqrt{y^2 + 4(1 + y^2)}}{2} = \dfrac{y \pm \sqrt{y^2 + 4 + 4y^2}}{2}$

$$= \dfrac{y \pm \sqrt{5y^2 + 4}}{2}$$

(e) $x^2 = \dfrac{4 \pm \sqrt{16 - 8}}{2} = \dfrac{4 \pm \sqrt{8}}{2} = \dfrac{4 \pm 2\sqrt{2}}{2} = 2 \pm \sqrt{2}$

6.

Synthetic division

Use synthetic division to factor as indicated:

(a) $x^3 - 4x^2 + 2x + 1 = (x - 1)($ $)$

(b) $2x^3 + 5x + 7 = (x + 1)($ $)$

(c) $x^4 - 3x^3 + x^2 + x + 2 = (x - 2)($ $)$

(d) $4x^4 + 3x^2 - 1 = (2x - 1)($ $)$

Solution

(a)

$$x^3 - 4x^2 + 2x \quad + 1$$

$$\begin{array}{c|cccc} 1 & 1 & -4 & 2 & 1 \\ & & 1 & -3 & -1 \\ \hline & 1 & -3 & -1 & 0 \end{array}$$

$$x^2 - 3x - 1$$

$$x^3 - 4x^2 + 2x + 1 = (x - 1)(x^2 - 3x - 1)$$

(b)

$$2x^3 + 5x + 7$$

$$\begin{array}{c|cccc} -1 & 2 & 0 & 5 & 7 \\ & & -2 & 2 & -7 \\ \hline & 2 & -2 & 7 & 0 \end{array}$$

$$2x^2 - 2x + 7$$

$$2x^3 + 5x + 7 = (x + 1)(2x^2 - 2x + 7)$$

(c)

$$x^4 - 3x^3 + x^2 + x + 2$$

$$\begin{array}{c|ccccc} 2 & 1 & -3 & 1 & 1 & 2 \\ & & 2 & -2 & -2 & -2 \\ \hline & 1 & -1 & -1 & -1 & 0 \end{array}$$

$$x^3 - x^2 - x - 1$$

$$x^4 - 3x^3 + x^2 + x + 2 = (x - 2)(x^3 - x^2 - x - 1)$$

(d)

$$4x^4 \quad + \quad 3x^2 \quad - \quad 1$$

$$\begin{array}{c|ccccc} \frac{1}{2} & 4 & 0 & 3 & 0 & -1 \\ & & 2 & 1 & 2 & 1 \\ \hline & 4 & 2 & 4 & 2 & 0 \end{array}$$

$$4x^3 + 2x^2 + 4x + 2$$

$$4x^4 + 3x^2 - 1 = \left(x - \frac{1}{2}\right)(4x^3 + 2x^2 + 4x + 2)$$

$$= (2x - 1)(2x^3 + x^2 + 2x + 1)$$

7.
Special products

Factor completely (into linear or irreducible quadratic factors):

(a) $x^3 - 27$

(b) $x^3 - 3x^2 + 3x - 1$

(c) $x^3 + 6x^2 + 12x + 8$

(d) $x^4 - 25$

(e) $x^4 - 8x^3 + 24x^2 - 32x + 16$

Solution

(a) $x^3 - 27 = (x - 3)(x^2 + 3x + 9)$

(b) $x^3 - 3x^2 + 3x - 1 = (x - 1)^3$

(c) $x^3 + 6x^2 + 12x + 8 = x^3 + 3(2)x^2 + 3(2^2)x + 2^3 = (x + 2)^3$

(d) $x^4 - 25 = (x^2 + 5)(x^2 - 5) = (x^2 + 5)(x + \sqrt{5})(x - \sqrt{5})$

(e) $x^4 - 8x^3 + 24x^2 - 32x + 16$
$= x^4 - 4(2)x^3 + 6(2^2)x^2 - 4(2^3)x + 2^4 = (x - 2)^4$

8.
Factoring by grouping

Factor completely (into linear or irreducible quadratic factors):

(a) $x^3 + 4x^2 - 2x - 8$

(b) $x^3 + 2x^2 + 3x + 6$

(c) $5 \cos^2 x - 5 \sin^2 x + \sin x + \cos x$

(d) $\cos^2 x + 4 \cos x + 4 - \tan^2 x$

Solution

(a) $x^3 + 4x^2 - 2x - 8 = x^2(x + 4) - 2(x + 4)$
$\qquad\qquad\qquad\qquad\quad = (x^2 - 2)(x + 4)$

(b) $x^3 + 2x^2 + 3x + 6 = x^2(x + 2) + 3(x + 2)$
$\qquad\qquad\qquad\qquad\quad = (x^2 + 3)(x + 2)$

(c) $5 \cos^2 x - 5 \sin^2 x + \sin x + \cos x$
$\quad = (5)(\cos^2 x - \sin^2 x) + (\sin x + \cos x)$
$\quad = (5)(\cos x - \sin x)(\cos x + \sin x) + (\sin x + \cos x)$
$\quad = (\cos x + \sin x)[5(\cos x - \sin x) + 1]$
$\quad = (\cos x + \sin x)(5 \cos x - 5 \sin x + 1)$

(d) $\cos^2 x + 4 \cos x + 4 - \tan^2 x = (\cos x + 2)^2 - \tan^2 x$
$$= (\cos x + 2 - \tan x)(\cos x + 2 + \tan x)$$

9.
Factoring polynomials

Factor completely (into linear or irreducible quadratic factors):

(a) $x^3 - 2x^2 - 5x + 6$ (b) $2x^3 - 13x^2 + 27x - 18$

(c) $9x^4 - 3x^3 + 7x^2 - 3x - 2$ (d) $2x^3 - 17x^2 + 9x - 8$

(e) $4x^4 - 7x^2 - 5x - 1$ (f) $3x^3 - x^2 - 22x + 24$

Solution

(a) $x^3 - 2x^2 - 5x + 6 = (x - 1)(x + 2)(x - 3)$

(b) $2x^3 - 13x^2 + 27x - 18 = (x - 2)(x - 3)(2x - 3)$

(c) $9x^4 - 3x^3 + 7x^2 - 3x - 2 = (3x - 2)(3x + 1)(x^2 + 1)$

(d) $2x^3 - 17x^2 + 9x - 8 = (x - 8)(2x^2 - x + 1)$

(e) $4x^4 - 7x^2 - 5x - 1 = (2x + 1)^2(x^2 - x - 1)$

(f) $3x^3 - x^2 - 22x + 24 = (x - 2)(x + 3)(3x - 4)$

10.
Simplifying

Rewrite each of the following in simplest form:

(a) $\dfrac{(x - 1)(x + 3) - (x + 1)^2}{x + 1}$ (b) $\dfrac{\sqrt{x^2 + 1} - (1/\sqrt{x^2 + 1})}{x^2 + 1}$

(c) $\dfrac{x^2 - 5x + 6}{x^2 - 4x + 4}$ (d) $\dfrac{1}{x + 1} - \dfrac{1}{x - 1} - \dfrac{2}{x^2 - 1}$

(e) $\dfrac{x(-2x)}{2\sqrt{1 - x^2}} + \sqrt{1 - x^2} + \dfrac{1}{\sqrt{1 - x^2}}$

Solution

(a) $\dfrac{(x - 1)(x + 3) - (x + 1)^2}{x + 1} = \dfrac{(x^2 + 2x - 3) - (x^2 + 2x + 1)}{x + 1}$

$$= \dfrac{x^2 + 2x - 3 - x^2 - 2x - 1}{x + 1}$$

$$= \dfrac{-4}{x + 1}$$

(b) $\dfrac{\sqrt{x^2 + 1} - (1/\sqrt{x^2 + 1})}{x^2 + 1} = \dfrac{(1/\sqrt{x^2 + 1})(x^2 + 1 - 1)}{x^2 + 1}$

$$= \dfrac{x^2 + 1 - 1}{\sqrt{x^2 + 1}(x^2 + 1)} = \dfrac{x^2}{(x^2 + 1)^{3/2}}$$

(c) $\dfrac{x^2 - 5x + 6}{x^2 - 4x + 4} = \dfrac{(x - 2)(x - 3)}{(x - 2)^2} = \dfrac{x - 3}{x - 2}$

(d) $\dfrac{1}{x + 1} - \dfrac{1}{x - 1} - \dfrac{2}{x^2 - 1} = \dfrac{(x - 1) - (x + 1) - 2}{x^2 - 1} = \dfrac{-4}{x^2 - 1}$

(e) $\dfrac{x(-2x)}{2\sqrt{1 - x^2}} + \sqrt{1 - x^2} + \dfrac{1}{\sqrt{1 - x^2}}$

$$= \dfrac{-x^2}{\sqrt{1 - x^2}} + \dfrac{1 - x^2}{\sqrt{1 - x^2}} + \dfrac{1}{\sqrt{1 - x^2}}$$

$$= \dfrac{2 - 2x^2}{\sqrt{1 - x^2}} = \dfrac{2(1 - x^2)}{\sqrt{1 - x^2}} = 2\sqrt{1 - x^2}$$

11.
Rationalizing

Remove the sum from the denominator by multiplying by the denominator's conjugate:

(a) $\dfrac{1}{1 - \cos x}$ (b) $\dfrac{x}{1 - \sqrt{x^2 + 1}}$ (c) $\dfrac{2}{x + \sqrt{x^2 + 1}}$

Solution

(a) $\dfrac{1}{1 - \cos x} = \left(\dfrac{1}{1 - \cos x}\right)\left(\dfrac{1 + \cos x}{1 + \cos x}\right) = \dfrac{1 + \cos x}{1 - \cos^2 x}$

$$= \dfrac{1 + \cos x}{\sin^2 x}$$

(b) $\dfrac{x}{1 - \sqrt{x^2 + 1}} = \left(\dfrac{x}{1 - \sqrt{x^2 + 1}}\right)\left(\dfrac{1 + \sqrt{x^2 + 1}}{1 + \sqrt{x^2 + 1}}\right)$

$$= \dfrac{x(1 + \sqrt{x^2 + 1})}{1 - (x^2 + 1)} = \dfrac{x(1 + \sqrt{x^2 + 1})}{-x^2} = \dfrac{1 + \sqrt{x^2 + 1}}{-x}$$

(c) $\left(\dfrac{2}{x + \sqrt{x^2 + 1}}\right)\left(\dfrac{x - \sqrt{x^2 + 1}}{x - \sqrt{x^2 + 1}}\right) = \dfrac{2(x - \sqrt{x^2 + 1})}{x^2 - (x^2 + 1)}$

$$= 2(x - \sqrt{x^2 + 1})$$

12.
Algebraic errors to avoid

	Error	Correct form	Comment
1. Parentheses	$a - (x - b) \neq a - x - b$	$a - (x - b) = a - x + b$	Change all signs when distributing negative through parentheses.
	$(a + b)^2 \neq a^2 + b^2$	$(a + b)^2 = a^2 + 2ab + b^2$	Don't forget middle term when squaring binomials.
	$\left(\dfrac{1}{2}a\right)\left(\dfrac{1}{2}b\right) \neq \dfrac{1}{2}(ab)$	$\left(\dfrac{1}{2}a\right)\left(\dfrac{1}{2}b\right) = \dfrac{1}{4}(ab) = \dfrac{ab}{4}$	1/2 occurs twice as a factor.
2. Fractions	$\dfrac{a}{x + b} \neq \dfrac{a}{x} + \dfrac{a}{b}$	Leave as $\dfrac{a}{x + b}$	Don't add denominators when adding fractions.
	$\dfrac{\left(\dfrac{x}{a}\right)}{b} \neq \dfrac{bx}{a}$	$\dfrac{\left(\dfrac{x}{a}\right)}{b} = \left(\dfrac{x}{a}\right)\left(\dfrac{1}{b}\right) = \dfrac{x}{ab}$	Multiply by reciprocal when dividing.
	$\dfrac{1}{a} + \dfrac{1}{b} \neq \dfrac{1}{a + b}$	$\dfrac{1}{a} + \dfrac{1}{b} = \dfrac{a + b}{ab}$	Use definition for adding fractions.
	$\dfrac{1}{3x} \neq \dfrac{1}{3}x$	$\dfrac{1}{3x} = \dfrac{1}{3} \cdot \dfrac{1}{x}$	Use definition for multiplying fractions.
	$\dfrac{1}{3}x \neq \dfrac{1}{3x}$	$\dfrac{1}{3}x = \dfrac{1}{3} \cdot x = \dfrac{x}{3}$	Be careful when using a slash to denote division.
	$\dfrac{1}{x} + 2 \neq \dfrac{1}{(x + 2)}$	$\dfrac{1}{x} + 2 = \dfrac{1}{x} + \dfrac{2x}{x}$ $= \dfrac{1 + 2x}{x}$	Be careful when using a slash to denote division.

9

	Error	Correct form	Comment
3. Exponents and radicals	$(x^2)^3 \neq x^5$	$(x^2)^3 = x^{2 \cdot 3} = x^6$	Multiply exponents when an exponential form is raised to a power.
	$x^2 \cdot x^3 \neq x^6$	$x^2 \cdot x^3 = x^{2+3} = x^5$	Add exponents when multiplying exponentials with like bases.
	$2x^3 \neq (2x)^3$	$2x^3 = 2(x^3)$	Exponents have priority over coefficients.
	$\dfrac{1}{x^{1/2} - x^{1/3}} \neq x^{-1/2} - x^{-1/3}$	Leave as $\dfrac{1}{x^{1/2} - x^{1/3}}$	Don't shift term-by-term from denominator to numerator.
	$\sqrt{5x} \neq 5\sqrt{x}$	$\sqrt{5x} = \sqrt{5}\sqrt{x}$	Radicals apply to every factor inside radical.
	$\sqrt{x^2 + a^2} \neq x + a$	Leave as $\sqrt{x^2 + a^2}$	Don't apply radicals term-by-term.
	$\sqrt{-x^2 + a^2} \neq -\sqrt{x^2 - a^2}$	Leave as $\sqrt{-x^2 + a^2}$ or write as $\sqrt{a^2 - x^2}$	Don't factor negatives out of square roots.
4. Cancellations	$\dfrac{a + bx}{a} \neq 1 + bx$	$\dfrac{a + bx}{a} = \dfrac{a}{a} + \dfrac{bx}{a}$ $= 1 + \dfrac{b}{a}x$	Cancel common factors, not common terms.
	$\dfrac{a + ax}{a} \neq a + x$	$\dfrac{a + ax}{a} = \dfrac{a(1 + x)}{a}$ $= 1 + x$	Factor before canceling.
	$1 + \dfrac{x}{2x} \neq 1 + \dfrac{1}{x}$	$1 + \dfrac{x}{2x} = 1 + \dfrac{1}{2} = \dfrac{3}{2}$	When canceling factors with like bases, subtract exponents, not coefficients.

13.
Some algebra of calculus

	Required simplest algebraic form	Desired unsimplified form for calculus	Comment
1. Unusual factoring	$\dfrac{5x^4}{8}$	$\dfrac{5}{8}(x^4)$	Factor out fractional coefficient.
	$\dfrac{x^2 + 3x}{-6}$	$-\dfrac{1}{6}(x^2 + 3x)$	Factor out fractional coefficient.
	$2x^2 - x - 3$	$2\left(x^2 - \dfrac{x}{2} - \dfrac{3}{2}\right)$	Factor out leading coefficient.
	$\dfrac{x}{2}(x+1)^{-1/2} + (x+1)^{1/2}$	$\dfrac{(x+1)^{-1/2}}{2}[x + 2(x+1)]$	Remove factor with negative exponent.
2. Inserting required factors	$(2x-1)^3$	$\dfrac{1}{2}(2x-1)^3(2)$	Multiply and divide by desired factor.
	$7x^2(4x^3 - 5)^{1/2}$	$\dfrac{7}{12}(4x^3 - 5)^{1/2}(12x^2)$	Multiply and divide by desired factor.
	$\dfrac{4x^2}{9} - 4y^2 = 1$	$\dfrac{x^2}{9/4} - \dfrac{y^2}{1/4} = 1$	Invert and *divide*.
3. Rewriting with negative exponents	$\dfrac{9}{-5x^3}$	$\dfrac{9}{-5}(x^{-3})$	Move factor to numerator and change sign of exponent.
	$\dfrac{7}{\sqrt{2x-3}}$	$7(2x-3)^{-1/2}$	Move factor to numerator and change sign of exponent.

	Required simplest algebraic form	Desired unsimplified form for calculus	Comment
4. Writing a fraction as a sum of terms	$\dfrac{x + 2x^2 + 1}{\sqrt{x}}$	$x^{1/2} + 2x^{3/2} + x^{-1/2}$	Divide each term by $x^{1/2}$.
	$\dfrac{1 + x}{x^2 + 1}$	$\dfrac{1}{x^2 + 1} + \dfrac{x}{x^2 + 1}$	Rewrite fraction as sum of fractions.
	$\dfrac{2x}{x^2 + 2x + 1}$	$\dfrac{2x + 2 - 2}{x^2 + 2x + 1} =$	Add and subtract terms in numerator.
		$\dfrac{2x + 2}{x^2 + 2x + 1} - \dfrac{2}{(x + 1)^2}$	Rewrite fraction as difference of fractions.
	$\dfrac{x^2 - 2}{x + 1}$	$x - 1 - \dfrac{1}{x + 1}$	Long division.
	$\dfrac{x + 7}{x^2 - x - 6}$	$\dfrac{2}{x - 3} - \dfrac{1}{x + 2}$	Use method of *partial fractions*.

1 *The Cartesian plane and functions*

The real line

7. Determine whether the real number $\sqrt[3]{64}$ is rational or irrational.

    ### Solution

    Since $4^3 = 64$, it follows that $\sqrt[3]{64} = 4$ and is therefore rational.

13. Express $0.29\overline{7297}$ as a ratio of two integers.

    ### Solution

    Let $x = 0.297297\ldots$. Since the repeating pattern occurs every three decimal places, we multiply both sides of the equation by 1000 and obtain

    $$1000x = 297.297297\ldots$$
    $$\underline{x = 0.297297\ldots} \qquad \text{Subtract}$$
    $$999x = 297$$
    $$x = \frac{297}{999} = \frac{11}{37}$$

19. Solve the inequality $x - 5 \geq 7$ and graph the solution on the real line.

    ### Solution

    $$x - 5 \geq 7$$
    $$x - 5 + 5 \geq 7 + 5$$
    $$x \geq 12$$

7 8 9 10 11 12 13 14 15

27. Solve the inequality $-4 < 2x - 3 < 4$ and graph the solution on the real line.

Solution

$$-4 < 2x - 3 < 4$$

$$-4 + 3 < 2x - 3 + 3 < 4 + 3$$

$$-1 < 2x < 7$$

$$-\frac{1}{2} < x < \frac{7}{2}$$

35. Solve the inequality $\left|\dfrac{x-3}{2}\right| \geq 5$ and graph the solution on the real line.

Solution

Replace $\left|\dfrac{x-3}{2}\right| \geq 5$ by $-\dfrac{x-3}{2} \geq 5$ or $\dfrac{x-3}{2} \geq 5$.

$$-\frac{x-3}{2} \geq 5 \qquad \text{or} \qquad \frac{x-3}{2} \geq 5$$

$$-\frac{x-3}{2}(-2) \leq 5(-2) \qquad \text{or} \qquad \frac{x-3}{2}(2) \geq 5(2)$$

$$x - 3 \leq -10 \qquad \text{or} \qquad x - 3 \geq 10$$

$$x - 3 + 3 \leq -10 + 3 \qquad \text{or} \qquad x - 3 + 3 \geq 10 + 3$$

$$x \leq -7 \qquad \text{or} \qquad x \geq 13$$

57. Use absolute values to define the following intervals:

(a)

(b)

Solution

(a) $2 \leq x \leq 6$

$$-2 \leq x - 4 \leq 2 \qquad \text{Centered at 4}$$

$$|x - 4| \leq 2$$

(b) $-7 < x < -1$

$$-3 < x - (-4) < 3 \qquad \text{Centered at } -4$$

$$|x + 4| < 3$$

64. The heights, h, of two-thirds of the members of a certain popu-
     lation satisfy the inequality

$$\left| \frac{h - 68.5}{2.7} \right| \le 1$$

where h is measured in inches. Determine the solution interval
for these heights.

Solution $\left| \dfrac{h - 68.5}{2.7} \right| \le 1$

$$-1 \le \frac{h - 68.5}{2.7} \le 1$$

$$-2.7 \le h - 68.5 \le 2.7$$

$$65.8 \le h \le 71.2$$

69. Prove that $|ab| = |a||b|$.

Solution

Case I: $a > 0$, $b > 0$, and $ab > 0$.

$$|ab| = ab = |a||b|$$

Case II: $a < 0$, $b < 0$, and $ab > 0$.

$$|ab| = ab = (-a)(-b) = |a||b|$$

Case III: $a > 0$, $b < 0$, and $ab < 0$.

$$|ab| = -(ab) = a(-b) = |a||b|$$

Case IV: $a < 0$, $b > 0$, and $ab < 0$.

$$|ab| = -(ab) = (-a)(b) = |a||b|$$

1.2
The Cartesian plane, the Distance Formula, and circles

1. Plot the points $(2, 1)$ and $(4, 5)$. Find the distance between the
    points and find the midpoint of the line segment joining the
    points.

    Solution

    Let $(2, 1) = (x_1, y_1)$ and $(4, 5) = (x_2, y_2)$.

$$d = \sqrt{(x_2 - x_1)^2 + (y_2 - y_1)^2} = \sqrt{(4 - 2)^2 + (5 - 1)^2}$$
$$= \sqrt{2^2 + 4^2} = \sqrt{20} = \sqrt{(4)(5)} = 2\sqrt{5}$$

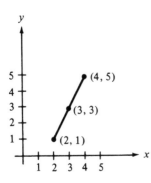

$$\text{midpoint} = \left(\frac{x_1 + x_2}{2}, \frac{y_1 + y_2}{2}\right) = \left(\frac{2 + 4}{2}, \frac{1 + 5}{2}\right) = (3, 3)$$

9. Show that the points $(4, 0)$, $(2, 1)$, $(-1, -5)$ are vertices of a right triangle.

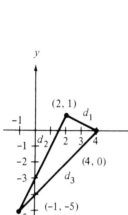

Solution

Let $d_1 =$ distance between $(4, 0)$ and $(2, 1)$. Then

$$d_1{}^2 = (2 - 4)^2 + (1 - 0)^2 = (-2)^2 + 1^2 = 5$$

Let $d_2 =$ distance between $(2, 1)$ and $(-1, -5)$. Then

$$d_2{}^2 = (-1 - 2)^2 + (-5 - 1)^2 = (-3)^2 + (-6)^2 = 45$$

Let $d_3 =$ distance between $(4, 0)$ and $(-1, -5)$. Then

$$d_3{}^2 = (-1 - 4)^2 + (-5 - 0)^2 = (-5)^2 + (-5)^2 = 50$$

Since $d_1{}^2 + d_2{}^2 = 5 + 45 = 50 = d_3{}^2$, the triangle must be a right triangle.

15. Use the Distance Formula to determine if the points $(-2, 1)$, $(-1, 0)$, $(2, -2)$ lie on a straight line.

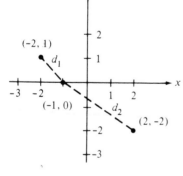

Solution

Let $d_1 =$ distance between $(-2, 1)$ and $(-1, 0)$. Then

$$d_1 = \sqrt{[-1 - (-2)]^2 + (0 - 1)^2} = \sqrt{1^2 + (-1)^2} = \sqrt{2}$$

Let $d_2 =$ distance between $(-1, 0)$ and $(2, -2)$. Then

$$d_2 = \sqrt{[2 - (-1)]^2 + (-2 - 0)^2} = \sqrt{3^2 + (-2)^2} = \sqrt{13}$$

Let $d_3 =$ distance between $(-2, 1)$ and $(2, -2)$. Then

$$d_3 = \sqrt{[2 - (-2)]^2 + (-2 - 1)^2} = \sqrt{4^2 + (-3)^2} = \sqrt{25} = 5$$

The points $(-2, 1)$, $(-1, 0)$, and $(2, -2)$ lie on a line only if $d_1 + d_2 = d_3$. Finally, since $\sqrt{2} + \sqrt{13} \approx 5.02 \neq 5$, the points are *not* collinear.

19. Find the relationship between x and y so that the point (x, y) is equidistant from $(4, -1)$ and $(-2, 3)$.

Solution

Let $d_1 =$ distance between $(4, -1)$ and (x, y). Then

$$d_1 = \sqrt{(x - 4)^2 + [y - (-1)]^2}$$

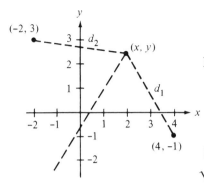

$$= \sqrt{x^2 - 8x + 16 + y^2 + 2y + 1}$$
$$= \sqrt{x^2 - 8x + y^2 + 2y + 17}$$

Let $d_2 =$ distance between $(-2, 3)$ and (x, y). Then

$$d_2 = \sqrt{[x - (-2)]^2 + (y - 3)^2}$$
$$= \sqrt{x^2 + 4x + 4 + y^2 - 6y + 9}$$
$$= \sqrt{x^2 + 4x + y^2 - 6y + 13}$$

Setting d_1 equal to d_2, we have

$$\sqrt{x^2 - 8x + y^2 + 2y + 17} = \sqrt{x^2 + 4x + y^2 - 6y + 13}$$
$$x^2 - 8x + y^2 + 2y + 17 = x^2 + 4x + y^2 - 6y + 13$$
$$-8x - 4x + 2y + 6y = 13 - 17$$
$$-12x + 8y = -4$$
$$3x - 2y = 1$$

21.   Use the Midpoint Rule successively to find the three points that divide the line segment joining (x_1, y_1) and (x_2, y_2) into four parts.

Solution

The midpoint of the given line segment is $\left(\dfrac{x_1 + x_2}{2}, \dfrac{y_1 + y_2}{2}\right)$.

The midpoint between (x_1, y_1) and $\left(\dfrac{x_1 + x_2}{2}, \dfrac{y_1 + y_2}{2}\right)$ is .

$$\left(\frac{x_1 + (x_1 + x_2)/2}{2}, \frac{y_1 + (y_1 + y_2)/2}{2}\right)$$
$$= \left(\frac{1}{2}\left(\frac{2x_1 + x_1 + x_2}{2}\right), \frac{1}{2}\left(\frac{2y_1 + y_1 + y_2}{2}\right)\right)$$
$$= \left(\frac{3x_1 + x_2}{4}, \frac{3y_1 + y_2}{4}\right)$$

The midpoint between $\left(\dfrac{x_1 + x_2}{2}, \dfrac{y_1 + y_2}{2}\right)$ and (x_2, y_2) is

$$\left(\frac{(x_1 + x_2)/2 + x_2}{2}, \frac{(y_1 + y_2)/2 + y_2}{2}\right)$$
$$= \left(\frac{1}{2}\left(\frac{x_1 + x_2 + 2x_3}{2}\right), \frac{1}{2}\left(\frac{y_1 + y_2 + 2y_2}{2}\right)\right)$$
$$= \left(\frac{x_1 + 3x_2}{4}, \frac{y_1 + 3y_2}{4}\right)$$

Thus the three points are

$$\left(\frac{3x_1 + x_2}{4}, \frac{3y_1 + y_2}{4}\right), \qquad \left(\frac{x_1 + x_2}{2}, \frac{y_1 + y_2}{2}\right),$$

$$\left(\frac{x_1 + 3x_2}{4}, \frac{y_1 + 3y_2}{4}\right)$$

33. Write the general equation of the circle with center at $(2, -1)$ and radius 4.

Solution

Let $(2, -1) = (h, k)$ and $r = 4$. Then using the standard equation of a circle, we have

$$(x - h)^2 + (y - k)^2 = r^2$$
$$(x - 2)^2 + [y - (-1)]^2 = 4^2$$
$$x^2 - 4x + 4 + y^2 + 2y + 1 = 16$$
$$x^2 + y^2 - 4x + 2y - 11 = 0$$

39. Write the general equation of the circle passing through the points $(0, 0)$, $(0, 8)$, and $(6, 0)$.

Solution

Since the general form of the equation of the circle is

$$x^2 + y^2 + Dx + Ey + F = 0$$

we must find the coefficients D, E, and F such that the given points are solution points.

Solution point	Resulting equation	Coefficient
$(0, 0)$	$(0)^2 + (0)^2 + D(0) + E(0) + F = 0$	$F = 0$
$(0, 8)$	$(0)^2 + (8)^2 + D(0) + E(8) + F = 0$	$E = -8$
$(6, 0)$	$(6)^2 + (0)^2 + D(6) + E(0) + F = 0$	$D = -6$

Therefore, the general equation is $x^2 + y^2 - 6x - 8y = 0$.

45. Write the equation $2x^2 + 2y^2 - 2x - 2y - 3 = 0$ in standard form and sketch its graph.

Solution

$$2x^2 + 2y^2 - 2x - 2y - 3 = 0$$
$$2x^2 - 2x + 2y^2 - 2y = 3$$

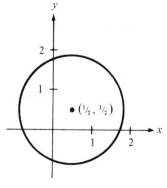

$$(x^2 - x + \quad) + (y^2 - y + \quad) = \frac{3}{2}$$

$$\left(x^2 - x + \frac{1}{4}\right) + \left(y^2 - y + \frac{1}{4}\right) = \frac{3}{2} + \frac{1}{4} + \frac{1}{4}$$

$$\left(x - \frac{1}{2}\right)^2 + \left(y - \frac{1}{2}\right)^2 = 2$$

Thus the circle is centered at $(\frac{1}{2}, \frac{1}{2})$ with a radius of $\sqrt{2}$.

51. Find the equation of the circle passing through the points (4, 3), (−2, −5), and (5, 2).

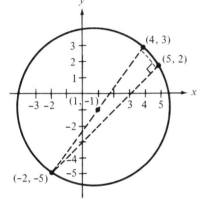

Solution

First, we observe that the points (4, 3), (−2, −5), and (5, 2) form the vertices of a right triangle. Then we can use a result from geometry that states that if a right triangle is inscribed in a circle, the hypotenuse of the triangle must lie on a diameter of the circle. Thus the center of the circle must be the midpoint of the line segment joining (4, 3) and (−2, −5).

$$(h, k) = \left(\frac{4 - 2}{2}, \frac{3 - 5}{2}\right) = (1, -1)$$

Furthermore, the distance between (4, 3) and (1, −1) must be the radius of the circle

$$r = \sqrt{(4 - 1)^2 + [3 - (-1)]^2} = \sqrt{9 + 16} = 5$$

Therefore, the equation of the circle is

$$(x - 1)^2 + [y - (-1)]^2 = 5^2$$
$$(x - 1)^2 + (y + 1)^2 = 25$$

53. Sketch the set of all points satisfying the inequality

$$x^2 + y^2 - 4x + 2y + 1 \leq 0$$

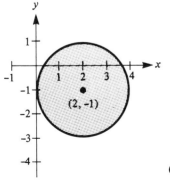

Solution

$$x^2 + y^2 - 4x + 2y + 1 \leq 0$$
$$(x^2 - 4x + 4) + (y^2 + 2y + 1) \leq -1 + 4 + 1$$
$$(x - 2)^2 + (y + 1)^2 \leq 4$$

Therefore, the inequality is satisfied by the set of all points lying on the boundary and in the interior of the circle with center (2, −1) and radius 2.

61. Prove that an angle inscribed in a semicircle is a right angle.

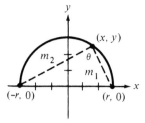

Solution

For simplicity, we assume the semicircle is centered at the origin with a radius of r. Since (x, y) lies on the semicircle, it must satisfy the equation

$$x^2 + y^2 = r^2$$
$$y^2 = r^2 - x^2$$
$$y = \sqrt{r^2 - x^2}$$

The slope of the line containing (x, y) and $(r, 0)$ is

$$m_1 = \frac{0 - y}{r - x} = \frac{-\sqrt{r^2 - x^2}}{r - x} = \frac{-\sqrt{(r - x)(r + x)}}{r - x} = \frac{-\sqrt{r + x}}{\sqrt{r - x}}$$

The slope of the line containing (x, y) and $(-r, 0)$ is

$$m_2 = \frac{0 - y}{-r - x} = \frac{-\sqrt{r^2 - x^2}}{-(r + x)} = \frac{\sqrt{(r - x)(r + x)}}{r + x} = \frac{\sqrt{r - x}}{\sqrt{r + x}}$$

Since $m_1 = -1/m_2$, the lines joining $(r, 0)$ and $(-r, 0)$ to (x, y) must be perpendicular. Hence the angle θ is a right angle.

1.3
Graphs of equations

11. Check $y^2 = x^3 - 4x$ for symmetry about both axes and the origin.

Solution

No symmetry about the y-axis since

$$y^2 = (-x)^3 - 4(-x) = -x^3 + 4x$$

is *not* equivalent to the original equation.
   Symmetry about the x-axis since

$$(-y)^2 = x^3 - 4x \qquad \text{or} \qquad y^2 = x^3 - 4x$$

is equivalent to the original equation.
   No symmetry about the origin since

$$(-y)^2 = (-x)^3 - 4(-x) \qquad \text{or} \qquad y^2 = -x^3 + 4x$$

is *not* equivalent to the original equation.

15. Check $y = x/(x^2 + 1)$ for symmetry about both axes and the origin.

Solution

No symmetry about the y-axis since

$$y = \frac{-x}{(-x)^2 + 1} = \frac{-x}{x^2 + 1}$$

is *not* equivalent to the original equation.
   No symmetry about the x-axis since

$$-y = \frac{x}{x^2 + 1}$$

is *not* equivalent to the original equation.
   Symmetry about the origin since

$$-y = \frac{-x}{(-x)^2 + 1} \qquad \text{or} \qquad y = \frac{x}{x^2 + 1}$$

is equivalent to the original equation.

19. Find the intercepts for the graph of $y = x^2 + x - 2$.

Solution

To find the x-intercepts, let $y = 0$. Then

$$x^2 + x - 2 = 0$$

and by factoring (or by the quadratic formula),

$$(x + 2)(x - 1) = 0$$

which implies that $x = -2$ or $x = 1$. Therefore the x-intercepts are $(-2, 0)$ and $(1, 0)$. To find the y-intercept, let $x = 0$. Then $y = 0^2 + 0 - 2 = -2$, and the y-intercept is $(0, -2)$.

25. Find the intercepts for the graph of $x^2 y - x^2 + 4y = 0$.

Solution

Let $y = 0$. Then $x^2(0) - x^2 + 4(0) = -x^2 = 0$, which implies that $x = 0$. Letting $x = 0$ yields the same result and the only intercept is $(0, 0)$.

41. Use the methods of this section to sketch the graph of the equation $x^2 + 4y^2 = 4$. Identify the intercepts and test for symmetry.

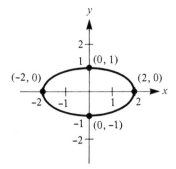

Solution

To find the x-intercepts, let $y = 0$. Then $x^2 = 4$ which implies that the x-intercepts are $(-2, 0)$ and $(2, 0)$. To find the y-intercepts, let $x = 0$. Then $4y^2 = 4$, which implies that the y-intercepts are $(0, -1)$ and $(0, 1)$.

There is symmetry with respect to the x-axis, the y-axis, and the origin since each of the variables is raised only to an even power.

Some first-quadrant solution points are

x	$\dfrac{1}{2}$	1	$\dfrac{3}{2}$
y	$\dfrac{\sqrt{15}}{4}$	$\dfrac{\sqrt{3}}{2}$	$\dfrac{\sqrt{7}}{4}$

49. Find the points of intersection of the graphs of $x + y = 7$ and $3x - 2y = 11$.

Solution

To solve the two equations

$$x + y = 7$$
$$3x - 2y = 11$$

simultaneously, we multiply the first equation by 2 and add. Thus

$$2x + 2y = 14$$
$$(+) \quad 3x - 2y = 11$$
$$\overline{5x = 25}$$
$$x = 5$$

Substituting $x = 5$ into the first equation, we have

$$5 + y = 7 \quad \text{or} \quad y = 2$$

Thus the point of intersection is $(5, 2)$.

51. Find the points of intersection of the graphs of $x^2 + y^2 = 5$ and $x - y = 1$.

Solution

To solve the two equations

$$x^2 + y^2 = 5$$
$$x - y = 1$$

simultaneously, we solve the second equation for x and obtain

$$x = y + 1$$

Substituting into the first equation, we have

$$(y + 1)^2 + y^2 = 5$$
$$(y^2 + 2y + 1) + y^2 = 5$$
$$2y^2 + 2y - 4 = 0$$
$$2(y + 2)(y - 1) = 0$$

which implies that $y = -2$ or $y = 1$. Therefore,

$$x = (-2) + 1 = -1 \qquad \text{or} \qquad x = 1 + 1 = 2$$

and the points of intersection are $(-1, -2)$ and $(2, 1)$.

55. Find the points of intersection of the graphs of $y = x^3 - 2x^2 + x - 1$ and $y = -x^2 + 3x - 1$.

Solution

By equating the two expressions for y, we have

$$x^3 - 2x^2 + x - 1 = -x^2 + 3x - 1$$
$$x^3 - 2x^2 + x^2 + x - 3x - 1 + 1 = 0$$
$$x^3 - x^2 - 2x = 0$$
$$x(x^2 - x - 2) = 0$$
$$x(x - 2)(x + 1) = 0$$

Thus $x = 0$, 2, or -1, and by substituting these values into the second (or first) equation, we have $y = -(0)^2 + 3(0) - 1 = -1$, $y = -(2)^2 + 3(2) - 1 = -4 + 6 - 1 = 1$, and $y = -(-1)^2 + 3(-1) - 1 = -1 - 3 - 1 = -5$. Therefore the points of intersection are $(0, -1)$, $(2, 1)$, and $(-1, -5)$.

63. For what values of k does the graph of $y = kx^3$ pass through these points?

(a) $(1, 4)$ (b) $(-2, 1)$ (c) $(0, 0)$ (d) $(-1, -1)$

Solution

(a) If $x = 1$ and $y = 4$, then $4 = k(1)^3$ and we obtain $k = 4$.

(b) If $x = -2$ and $y = 1$, then $1 = k(-2)^3$ and we obtain $k = -\frac{1}{8}$.

(c) If $x = y = 0$, then $0 = k(0)^3$ and k may have *any value*.

(d) If $x = y = -1$, then $-1 = k(-1)^3$ and we obtain $k = 1$.

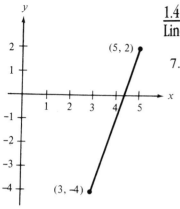

1.4
Lines in the plane

7. Plot the points $(3, -4)$ and $(5, 2)$ and find the slope of the line passing through them.

Solution

Let $(3, -4) = (x_1, y_1)$ and $(5, 2) = (x_2, y_2)$. The slope of the line passing through (x_1, y_1) and (x_2, y_2) is

$$m = \frac{y_2 - y_1}{x_2 - x_1} = \frac{2 - (-4)}{5 - 3} = \frac{6}{2} = 3$$

15. Find an equation of the line passing through $(2, 1)$ and $(0, -3)$ and sketch its graph.

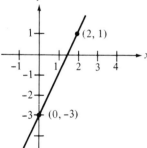

Solution

Let $(2, 1) = (x_1, y_1)$ and $(0, -3) = (x_2, y_2)$. The slope of the line passing through (x_1, y_1) and (x_2, y_2) is

$$m = \frac{y_2 - y_1}{x_2 - x_1} = \frac{-3 - 1}{0 - 2} = 2$$

Using the point-slope form of the equation of a line, we have

$$y - y_1 = m(x - x_1)$$
$$y - 1 = 2(x - 2)$$
$$y - 1 = 2x - 4$$
$$0 = 2x - y - 3 \qquad \text{General form}$$

23. Find an equation of the line passing through $(0, 3)$ with a slope of $m = \frac{3}{4}$ and sketch its graph.

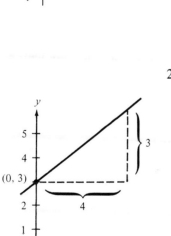

Solution

Using the point–slope equation, we have

$$y - y_1 = m(x - x_1)$$
$$y - 3 = \frac{3}{4}(x - 0)$$
$$4y - 12 = 3x$$
$$-3x + 4y - 12 = 0 \qquad \text{or} \qquad 3x - 4y + 12 = 0$$

35. Write an equation of the line whose x-intercept is 2 and whose y-intercept is 3.

Solution

From Exercise 34, the two-intercept equation of a line is $(x/a) + (y/b) = 1$, where a is the x-intercept and b is the y-intercept. Since $a = 2$ and $b = 3$, we have

$$\frac{x}{2} + \frac{y}{3} = 1$$
$$3x + 2y = 6$$
$$3x + 2y - 6 = 0$$

43. Write an equation of the line passing through $(2, 1)$ (a) parallel and (b) perpendicular to the line $4x - 2y = 3$.

Solution

The line given by $4x - 2y = 3$ has a slope of 2 since

$$4x - 2y = 3$$
$$-2y = -4x + 3$$
$$y = 2x - \frac{3}{2} = mx + b$$

(a) The line through $(2, 1)$ parallel to $4x - 2y = 3$ must also have a slope of 2. Thus its equation must be

$$y - y_1 = m(x - x_1)$$
$$y - 1 = 2(x - 2)$$
$$y - 1 = 2x - 4$$
$$-2x + y + 3 = 0$$

(b) The line through $(2, 1)$ perpendicular to $4x - 2y = 3$ must have a slope of $m = -\frac{1}{2}$. Thus its equation must be

$$y - y_1 = m(x - x_1)$$
$$y - 1 = -\frac{1}{2}(x - 2)$$
$$2y - 2 = -x + 2$$
$$x + 2y - 4 = 0$$

51. Use slope to determine if the points $(0, -4)$, $(2, 0)$, and $(3, 2)$ are collinear.

Solution

Let $(0, -4) = (x_1, y_1)$, $(2, 0) = (x_2, y_2)$, and $(3, 2) = (x, y)$. The point (x, y) lies on the line passing through (x_1, y_1) and (x_2, y_2) if and only if

$$\frac{y - y_1}{x - x_1} = m = \frac{y_2 - y_1}{x_2 - x_1}$$

Since $\dfrac{2 - (-4)}{3 - 0} = \dfrac{6}{3} = 2 = \dfrac{4}{2} = \dfrac{0 - (-4)}{2 - 0}$

the three points must be collinear.

61. A small business purchases a piece of equipment for $875. After five years the equipment will be obsolete and have no value. Write a linear equation giving the value V of the equipment during the five years it will be used.

Solution

Two solution points to the linear equation are $(t_1, V_1) = (0, 875)$ and $(t_2, V_2) = (5, 0)$. Therefore, the slope (rate of depreciation per year) is

$$m = \frac{0 - 875}{5 - 0} = -\$175 \text{ per year}$$

and the equation of the line is

$$V - 0 = -175(t - 5)$$
$$V = 875 - 175t \qquad \text{where} \qquad 0 \le t \le 5$$

67. Find the distance from the point $(2, 3)$ to the line $4x + 3y = 10$.

Solution

The distance from the point (x_1, y_1) to the line $Ax + By + C = 0$ is

$$d = \frac{|Ax_1 + By_1 + C|}{\sqrt{A^2 + B^2}}$$

Letting $(2, 3) = (x_1, y_1)$ and $4x + 3y - 10 = Ax + By + C = 0$, we have $A = 4$, $B = 3$, and $C = -10$. (Note that $C = -10$, not 10.) Thus

$$d = \frac{|4(2) + 3(3) - 10|}{\sqrt{4^2 + 3^2}} = \frac{|8 + 9 - 10|}{\sqrt{25}} = \frac{7}{5}$$

71. Find the distance between the parallel lines $3x - 4y = 1$ and $3x - 4y = 10$.

Solution

A point on the line $3x - 4y = 1$ is $(-1, -1)$. The distance between the given lines is equal to the distance from $(-1, -1)$ to the line $3x - 4y = 10$. Letting $(-1, -1) = (x_1, y_1)$ and $3x - 4y - 10 = Ax + By + C = 0$, we have

$$d = \frac{|Ax_1 + By_1 + C|}{\sqrt{A^2 + B^2}} = \frac{|3(-1) + (-4)(-1) + (-10)|}{\sqrt{3^2 + (-4)^2}}$$

$$= \frac{|-9|}{\sqrt{25}} = \frac{9}{5}$$

1.5
Functions

3. Given $f(x) = \sqrt{x + 3}$, find:

(a) $f(-2)$ (b) $f(6)$ (c) $f(c)$ (d) $f(x + \Delta x)$

Solution

(a) $f(-2) = \sqrt{-2 + 3}$ (b) $f(6) = \sqrt{6 + 3} = \sqrt{9} = 3$
$\qquad\quad = \sqrt{1} = 1$

(c) $f(c) = \sqrt{c + 3}$ (d) $f(x + \Delta x) = \sqrt{x + \Delta x + 3}$

9. Given $f(x) = x^3$, find $\dfrac{f(x + \Delta x) - f(x)}{\Delta x}$.

Solution

$$\frac{f(x + \Delta x) - f(x)}{\Delta x} = \frac{(x + \Delta x)^3 - x^3}{\Delta x}$$

$$= \frac{x^3 + 3x^2 \Delta x + 3x(\Delta x)^2 + (\Delta x)^3 - x^3}{\Delta x}$$

$$= \frac{\Delta x[3x^2 + 3x\Delta x + (\Delta x)^2]}{\Delta x}$$

$$= 3x^2 + 3x\Delta x + (\Delta x)^2$$

14. Given $f(x) = 1/x$ and $g(x) = x^2 - 1$, find:

(a) $f(g(2))$ (b) $g(f(2))$ (c) $f(g(1/\sqrt{2}))$

(d) $g(f(1\sqrt{2}))$ (e) $g(f(x))$ (f) $f(g(x))$

Solution

(a) $f(g(2)) = f(2^2 - 1) = f(3) = \dfrac{1}{3}$

(b) $g(f(2)) = g\left(\dfrac{1}{2}\right) = \left(\dfrac{1}{2}\right)^2 - 1 = \dfrac{1}{4} - 1 = -\dfrac{3}{4}$

(c) $f\left(g\left(\dfrac{1}{\sqrt{2}}\right)\right) = f\left(\left(\dfrac{1}{\sqrt{2}}\right)^2 - 1\right) = f\left(-\dfrac{1}{2}\right) = \dfrac{1}{-\frac{1}{2}} = -2$

(d) $g\left(f\left(\dfrac{1}{\sqrt{2}}\right)\right) = g\left(\dfrac{1}{1/\sqrt{2}}\right) = g(\sqrt{2}) = (\sqrt{2})^2 - 1 = 1$

(e) $g(f(x)) = g\left(\dfrac{1}{x}\right) = \left(\dfrac{1}{x}\right)^2 - 1 = \dfrac{1}{x^2} - 1 = \dfrac{1 - x^2}{x^2}$

(f) $f(g(x)) = f(x^2 - 1) = \dfrac{1}{x^2 - 1}$

23. Sketch the graph of the function $f(x) = \sqrt{9 - x^2}$ and give its domain and range.

Solution

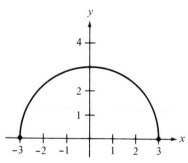

Since $9 - x^2$ must be nonnegative ($9 - x^2 \geq 0$), the domain is $[-3, 3]$. The range is $[0, 3]$. There is symmetry with respect to the y-axis since

$$y = \sqrt{9 - (-x)^2} = \sqrt{9 - x^2}$$

is equivalent to the original equation. Squaring both members of the equation, we have

$$y^2 = 9 - x^2 \quad \text{or} \quad x^2 + y^2 = 3^2$$

This is the standard form of an equation of a circle with center $(0, 0)$ and radius 3. Therefore, the graph of $f(x) = \sqrt{9 - x^2}$ is a semicircle in the first and second quadrants with center $(0, 0)$ and radius 3.

46. Determine if y is a function of x for the equation $x^2y - x^2 + 4y = 0$.

Solution

Solving the equation for y, we obtain

$$x^2y - x^2 + 4y = 0$$
$$(x^2 + 4)y = x^2$$
$$= \frac{x^2}{x^2 + 4}$$

For each value of the independent variable x, there corresponds exactly one value of the dependent variable y. Therefore, y is a function of x.

53. Determine if the function $f(x) = x(4 - x^2)$ is even, odd, or neither.

Solution

This function is odd since

$$f(-x) = (-x)[4 - (-x)^2] = -x(4 - x^2) = -f(x)$$

69. A man is in a boat 2 miles from the nearest point on the coast. He is to go to a point Q, 3 miles down the coast and 1 mile inland as shown in the accompanying figure. He can row at 2 mi/h and walk at 4 mi/h. Express the total time T of the trip as a function of x.

Solution

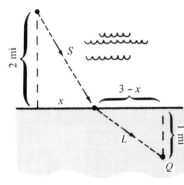

$S =$ distance on water $= \sqrt{x^2 + 4}$

$L =$ distance on land $= \sqrt{1 + (3 - x)^2} = \sqrt{x^2 - 6x + 10}$

Total time = (time on water) + (time on land)

$$T = \frac{S}{(\text{rate on water})} + \frac{L}{(\text{rate on land})}$$

$$= \frac{\sqrt{x^2 + 4}}{2} + \frac{\sqrt{x^2 - 6x + 10}}{4}$$

1.6
Review of trigonometric functions

7. Express the following angles in degree measure:

(a) $3\pi/2$ (b) $7\pi/6$

(c) $-7\pi/12$ (d) $\pi/9$

Solution

Since $180° = \pi$ radians, we conclude that 1 radian $= 180°/\pi$.

(a) $\dfrac{3\pi}{2}$ radians $= \left(\dfrac{3\pi}{2}\right)\left(\dfrac{180°}{\pi}\right) = 270°$

(b) $\dfrac{7\pi}{6}$ radians $= \left(\dfrac{7\pi}{6}\right)\left(\dfrac{180°}{\pi}\right) = 210°$

(c) $\dfrac{-7\pi}{12}$ radians $= \left(\dfrac{-7\pi}{12}\right)\left(\dfrac{180°}{\pi}\right) = -150°$

(d) $\dfrac{\pi}{9}$ radians $= \left(\dfrac{\pi}{9}\right)\left(\dfrac{180°}{\pi}\right) = 20°$

19.   Find $\cot \theta$ given $\cos \theta = \frac{4}{5}$.

Solution

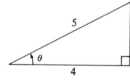

Using the fact that

$$\cos \theta = \frac{4}{5}$$

we construct the accompanying figure and obtain

$$\cot \theta = \frac{4}{y} = \frac{4}{\sqrt{25 - 16}} = \frac{4}{3}$$

25.   Evaluate the sine, cosine, and tangent of the following angles *without* using a calculator:

(a)   225°   (b)   −225°   (c)   300°   (d)   330°

Solution

(a)   The angle is in quadrant III and the reference angle is
225° − 180° = 45°. Therefore,

$$\sin 225° = -\sin 45° = -\frac{\sqrt{2}}{2}$$

$$\cos 225° = -\cos 45° = -\frac{\sqrt{2}}{2}$$

$$\tan 225° = \tan 45° = -1$$

(b)   The angle is in quadrant II and the reference angle is
225° − 180° = 45°. Therefore,

$$\sin (-225°) = \sin 45° = \frac{\sqrt{2}}{2}$$

$$\cos (-225°) = -\cos 45° = -\frac{\sqrt{2}}{2}$$

$$\tan (-225°) = -\tan 45° = -1$$

(c)   The angle is in quadrant IV and the reference angle is $360° - 300° = 60°$. Therefore,

$$\sin 300° = -\sin 60° = -\frac{\sqrt{3}}{2}$$

$$\cos 300° = \cos 60° = \frac{1}{2}$$

$$\tan 300° = -\tan 60° = -\sqrt{3}$$

(d)   The angle is in quadrant IV and the reference angle is $360° - 330° = 30°$. Therefore,

$$\sin 330° = -\sin 30° = -\frac{1}{2}$$

$$\cos 330° = \cos 30° = \frac{\sqrt{3}}{2}$$

$$\tan 330° = -\tan 30° = -\frac{\sqrt{3}}{3}$$

37.   Solve for x in $\tan^2 x = \tan x$, where $0 \le x \le 2\pi$.

Solution

$$\tan^2 x - \tan x = 0$$

$$\tan x(\tan x - 1) = 0$$

If $\tan x = 0$, then $x = \pi$ or $x = 0$. If $\tan x - 1 = 0$, we have $\tan x = 1$ and $x = \pi/4$ or $5\pi/4$. Thus for $0 \le x \le 2\pi$, there are four solutions:

$$x = 0, \pi, \frac{\pi}{4}, \frac{5\pi}{4}$$

43.   Solve for y and r in the accompanying triangle.

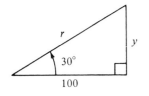

Solution

$$\tan 30° = \frac{\sqrt{3}}{3} = \frac{y}{100}$$

Therefore, $y = 100\sqrt{3}/3$.

$$\sec 30° = \frac{2\sqrt{3}}{3} = \frac{r}{100}$$

Therefore, $r = 200\sqrt{3}/3$.

49.   From a 150-foot observation tower on the coast, a Coast Guard

officer sights a boat in difficulty. The angle of depression of the boat is 4° as shown in the accompanying figure. How far is the boat from the shoreline?

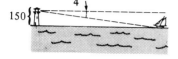

Solution

If d is the required distance, then

$$\cot 4° = \frac{d}{150}$$

$$d = 150 \cot 4° \approx 2145 \text{ ft}$$

69. Sketch the graph of the function $y = \csc (x/2)$.

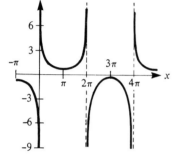

Solution

The graph of $y = \csc (x/2)$ has the following characteristics:

$$\text{period:} \quad \frac{2\pi}{\frac{1}{2}} = 4\pi$$

vertical asymptotes: $x = 2n\pi$, n an integer

Using the basic shape of the graph of the cosecant function, we sketch one period of the function on the interval $[0, 4\pi]$, following the pattern

minimum: $(\pi, 1)$ maximum: $(3\pi, -1)$

Review Exercises for Chapter 1

3. Sketch the intervals defined by $4 < (x + 3)^2$.

Solution

By finding the square root of each member of the inequality $4 < (x + 3)^2$ we obtain

$$2 < x + 3 \qquad \text{or} \qquad -2 > x + 3$$
$$-1 < x \qquad \text{or} \qquad -5 > x$$

15. Determine the value of c so that the given circle $x^2 - 6x + y^2 + 8y = c$ has a radius of 2.

Solution

$$x^2 - 6x + y^2 + 8y = c$$
$$(x^2 - 6x + 9) + (y^2 + 8y + 16) = c + 9 + 16$$

$$(x - 3)^2 + (y + 4)^2 = c + 25$$

If the radius of the circle is 2, then $c + 25 = 2^2$ or $c = -21$.

35. Express the value v of a farm at \$850 per acre, with buildings, livestock, and equipment worth \$300,000, as a function of the number of acres a, and give its domain.

Solution

The value of the farm is \$300,000 plus the value of the land. If the farm has one acre, its value is

$$v = 300,000 + 850$$

If the farm has two acres, its value is

$$v = 300,000 + 2(850)$$

Thus, in general, for a acres its value is

$$v = 300,000 + a(850) = 300,000 + 850a \qquad \text{(for } a > 0)$$

41. Express R and r, as defined in the accompanying figure, as functions of x.

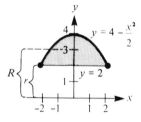

Solution

For a given value of x in the interval $[-2, 2]$, R is defined to be the distance from the x-axis to the graph of the parabola $y = 4 - (x^2/2)$. Therefore

$$R = 4 - \frac{x^2}{2}$$

For a given value of x in the interval $[-2, 2]$, r is defined to be the distance from the x-axis to the graph of the line $y = 2$. Therefore, r is the constant function

$$r = 2$$

50. Sketch the graph of $x^3 - y^2 + 1 = 0$ and determine if y is a function of x.

Solution

There is symmetry with respect to the x-axis since

$$x^3 - (-y)^2 + 1 = 0$$
$$x^3 - y^2 + 1 = 0$$

is equivalent to the original equation.

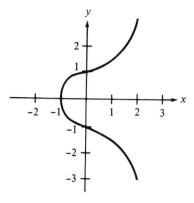

To find the y-intercepts, let $x = 0$. Then

$$-y^2 + 1 = 0$$
$$y^2 = 1$$
$$y = \pm 1$$

and the y-intercepts are $(0, 1)$ and $(0, -1)$. Therefore, y is not a function of x since for $x = 0$ there correspond two values of y. Now solving the equation for y, we obtain

$$y^2 = x^3 + 1$$
$$y = \pm\sqrt{x^3 + 1}$$

(Note: The domain is $x \geq -1$.)

x	-1	$-\dfrac{1}{2}$	0	2
y	0	$\pm\sqrt{\dfrac{7}{8}} = \pm 0.935$	± 1	± 3

69.   A six-foot person standing 12 ft from a streetlight casts an 8-foot shadow as shown in the accompanying figure. What is the height of the streetlight?

Solution

We begin be letting h be the height of the streetlight. Since the lengths of corresponding sides of similar triangles are proportional, we have

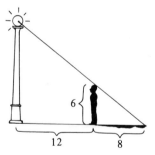

$$\frac{h}{20} = \frac{6}{8}$$

$$h = 20\left(\frac{6}{8}\right) = 15 \text{ ft}$$

73.   Graph $f(x) = \cos(2x - \pi/3)$ through 2 periods.

Solution

$$f(x) = \cos\left(2x - \frac{\pi}{3}\right) = \cos\left[2\left(x - \frac{\pi}{6}\right)\right]$$

Therefore, the graph of f has these characteristics:

amplitude: 1   period: $\dfrac{2\pi}{2} = \pi$

right shift of $\dfrac{\pi}{6}$

Using the basic shape of the graph of the cosine function, we sketch two periods of the function on the interval $[\pi/6, \, 13\pi/6]$, following the pattern

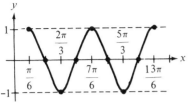

maximum: $\left(\dfrac{\pi}{6}, \, 1\right)$ minimum: $\left(\dfrac{2\pi}{3}, \, -1\right)$

maximum: $\left(\dfrac{7\pi}{6}, \, 1\right)$ minimum: $\left(\dfrac{5\pi}{3}, \, -1\right)$

2 Limits and their properties

2.1
An introduction to limits

5. Complete the following table and use the result to estimate the limit

$$\lim_{x \to 3} \frac{[1/(x + 1)] - (\frac{1}{4})}{x - 3}$$

x	2.9	2.99	2.999	3	3.001	3.01	3.1
$f(x)$				?			

Solution

The table lists values of $f(x)$ at several x-values near 3.

x	2.9	2.99	2.999	3	3.001	3.01	3.1
$f(x)$	-0.0641	-0.0627	-0.0625	?	-0.0625	-0.0623	-0.0610

As x approaches 3 from the left and from the right $f(x)$ approaches -0.0625. Therefore we estimate the limit to be $-\frac{1}{16}$.

17. Use the accompanying graph to find the limit (if it exists).

$$\lim_{x \to \pi/2} \tan x$$

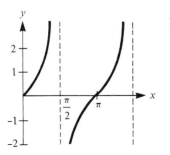

Solution

From the graph we see that as x approaches $\pi/2$ from the left, $f(x)$ increases without bound, and as x approaches $\pi/2$ from the right, $f(x)$ decreases without bound. Since $f(x)$ is not approaching a real number L as x approaches $\pi/2$, we say that the limit does not exist.

21. Find the limit $\lim_{x \to 2} (3x + 2) = L$ and then find $\delta > 0$ such that $|f(x) - L| < 0.01$ whenever $0 < |x - c| < \delta$.

Solution

We use the definition of limit to verify that

$$\lim_{x \to 2} (3x + 2) = 8$$

We are required to show that there exists a δ such that $|f(x) - L| < 0.01$ whenever $0 < |x - 2| < \delta$.

$$|f(x) - L| < 0.01$$
$$|(3x + 2) - 8| < 0.01$$
$$|3x - 6| < 0.01$$
$$3|x - 2| < 0.01$$
$$0 < |x - 2| < \frac{0.01}{3}$$

Therefore, $\delta = \dfrac{0.01}{3}$.

25. Find the limit $\lim_{x \to 2} (x + 3) = L$. Then for $\epsilon > 0$, find $\delta > 0$ such that $|f(x) - L| < \epsilon$ whenever $0 < |x - c| < \delta$.

Solution

We use the definition of limit to verify that

$$\lim_{x \to 2} (x + 3) = 5$$

Given $\epsilon > 0$,

$$|(x + 3) - 5| < \epsilon$$
$$|x - 2| < \epsilon = \delta$$

29. Find the limit $\lim_{x \to 0} \sqrt[3]{x} = L$. Then for $\epsilon > 0$, find δ such that $|f(x) - L| < \epsilon$ whenever $0 < |x - c| < \delta$.

Solution

We use the definition of limit to verify that

$$\lim_{x \to 0} \sqrt[3]{x} = 0$$

Given $\epsilon > 0$,

$$|\sqrt[3]{x}| < \epsilon$$
$$(|\sqrt[3]{x}|)^3 < \epsilon^3$$
$$|x| < \epsilon^3 = \delta$$

2.2
Properties of limits

7.   Find the limit $\lim_{x \to 3} \sqrt{x + 1}$.

   Solution

$$\lim_{x \to 3} \sqrt{x + 1} = \sqrt{3 + 1} = \sqrt{4} = 2$$

23.   Find the limit $\lim_{x \to 3} \tan(\pi x/4)$.

   Solution

$$\lim_{x \to 3} \tan \frac{\pi x}{4} = \tan \frac{3\pi}{4} = -1$$

26.   If $\lim_{x \to c} f(x) = \frac{3}{2}$ and $\lim_{x \to c} g(x) = \frac{1}{2}$, find:

   (a) $\lim_{x \to c} [f(x) + g(x)]$ (b) $\lim_{x \to c} [f(x)g(x)]$ (c) $\lim_{x \to c} \dfrac{f(x)}{g(x)}$

   Solution

   (a) $\lim_{x \to c} [f(x) + g(x)] = \lim_{x \to c} f(x) + \lim_{x \to c} g(x) = \frac{3}{2} + \frac{1}{2} = 2$

   (b) $\lim_{x \to c} [f(x)g(x)] = [\lim_{x \to c} f(x)][\lim_{x \to c} g(x)] = \frac{3}{2} \cdot \frac{1}{2} = \frac{3}{4}$

   (c)   Since $g(c) \neq 0$,

$$\lim_{x \to c} \frac{f(x)}{g(x)} = \frac{\lim_{x \to c} f(x)}{\lim_{x \to c} g(x)} = \frac{\frac{3}{2}}{\frac{1}{2}} = 3$$

2.3
Techniques for evaluating limits

9.   Find (if it exists) the limit $\lim_{x \to -1^-} \dfrac{x^2 - 1}{x + 1}$.

   Solution

$$\lim_{x \to -1^-} \frac{x^2 - 1}{x + 1} = \lim_{x \to -1^-} \frac{(x + 1)(x - 1)}{x + 1} = \lim_{x \to -1^-} (x - 1) = -2$$

17.   Find (if it exists) the limit

$$\lim_{\Delta x \to 0} \frac{(x + \Delta x)^2 - 2(x + \Delta x) + 1 - (x^2 - 2x + 1)}{\Delta x}$$

Solution

$$\lim_{\Delta x \to 0} \frac{(x + \Delta x)^2 - 2(x + \Delta x) + 1 - (x^2 - 2x + 1)}{\Delta x}$$

$$= \lim_{\Delta x \to 0} \frac{\cancel{x^2} + 2x\Delta x + (\Delta x)^2 - \cancel{2x} - 2\Delta x + \cancel{1} - \cancel{x^2} + \cancel{2x} - \cancel{1}}{\Delta x}$$

$$= \lim_{\Delta x \to 0} \frac{2x\Delta x + (\Delta x)^2 - 2\Delta x}{\Delta x} = \lim_{\Delta x \to 0} \frac{\cancel{\Delta x}(2x + \Delta x - 2)}{\cancel{\Delta x}}$$

$$= \lim_{\Delta x \to 0} (2x + \Delta x - 2) = 2x - 2$$

23. Find (if it exists) the limit $\displaystyle \lim_{x \to 0} \frac{\sqrt{3 + x} - \sqrt{3}}{x}$.

Solution

$$\lim_{x \to 0} \frac{\sqrt{3 + x} - \sqrt{3}}{x} = \lim_{x \to 0} \left(\frac{\sqrt{3 + x} - \sqrt{3}}{x} \right) \left(\frac{\sqrt{3 + x} + \sqrt{3}}{\sqrt{3 + x} + \sqrt{3}} \right)$$

$$= \lim_{x \to 0} \frac{\cancel{3} + x - \cancel{3}}{x(\sqrt{3 + x} + \sqrt{3})}$$

$$= \lim_{x \to 0} \frac{\cancel{x}}{\cancel{x}(\sqrt{3 + x} + \sqrt{3})}$$

$$= \lim_{x \to 0} \frac{1}{\sqrt{3 + x} + \sqrt{3}} = \frac{1}{2\sqrt{3}} = \frac{\sqrt{3}}{6}$$

25. Find (if it exists) the limit

$$\lim_{x \to 0} \frac{1/(2 + x) - \frac{1}{2}}{x}$$

Solution

$$\lim_{x \to 0} \frac{1/(2 + x) - \frac{1}{2}}{x} = \lim_{x \to 0} \frac{[2 - (2 + x)]/[2(2 + x)]}{x}$$

$$= \lim_{x \to 0} \frac{-x}{x(2)(2 + x)}$$

$$= \lim_{x \to 0} \frac{-1}{2(2 + x)} = -\frac{1}{4}$$

37. Find (if it exists) the limit

$$\lim_{x \to \pi/2} \frac{\cos x}{\cot x}$$

Solution

$$\lim_{x \to \pi/2} \frac{\cos x}{\cot x} = \lim_{x \to \pi/2} \frac{\cos x}{(\cos x)/(\sin x)}$$

$$= \lim_{x \to \pi/2} \sin x$$

$$= 1$$

2.4
Continuity and one-sided limits

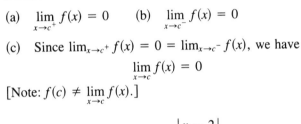

Use the graph to visually determine:

(a) $\lim_{x \to c^+} f(x)$ (b) $\lim_{x \to c^-} f(x)$ (c) $\lim_{x \to c} f(x)$

Solution

(a) $\lim_{x \to c^+} f(x) = 0$ (b) $\lim_{x \to c^-} f(x) = 0$

(c)   Since $\lim_{x \to c^+} f(x) = 0 = \lim_{x \to c^-} f(x)$, we have

$$\lim_{x \to c} f(x) = 0$$

[Note: $f(c) \neq \lim_{x \to c} f(x)$.]

14.   Find (if it exists) the limit $\lim_{x \to 2} \dfrac{|x - 2|}{x - 2}$.

Solution

$$\lim_{x \to 2^+} \frac{|x - 2|}{x - 2} = \lim_{x \to 2^+} \frac{x - 2}{x - 2} = 1$$

$$\lim_{x \to 2^-} \frac{|x - 2|}{x - 2} = \lim_{x \to 2^-} \frac{-(x - 2)}{x - 2} = -1$$

Since the limit from the left is not equal to the limit from the right, the limit does *not* exist.

15.   Find (if it exists) the limit $\lim_{x \to 3} f(x)$, where

$$f(x) = \begin{cases} \dfrac{x + 2}{2}, & x \leq 3 \\[2mm] \dfrac{12 - 2x}{2}, & x > 3 \end{cases}$$

Solution

$$\lim_{x \to 3^-} f(x) = \lim_{x \to 3^-} \frac{x + 2}{2} = \frac{5}{2} \qquad \lim_{x \to 3^+} f(x) = \lim_{x \to 3^+} \frac{12 - 2x}{3} = 2$$

Since the limit from the left is not equal to the limit from the right, the limit does *not* exist.

17. Find (if it exists) the limit $\lim_{x \to 1} f(x)$, where

$$f(x) = \begin{cases} x^3 + 1, & x < 1 \\ x + 1, & x \geq 1 \end{cases}$$

Solution

$$\lim_{x \to 1^-} f(x) = \lim_{x \to 1^-} (x^3 + 1) = 2, \qquad \lim_{x \to 1^+} f(x) = \lim_{x \to 1^+} (x + 1) = 2$$

Since these two limits are equal, we have $\lim_{x \to 1} f(x) = 2$.

31. Find the discontinuities (if any) for $f(x) = 1/(x - 1)$. Which of the discontinuities are removable?

Solution

From Theorem 2.11 we know that f is continuous for all x other than $x = 1$. At $x = 1$ the function is discontinuous and the discontinuity is nonremovable since the limit

$$\lim_{x \to 1} \frac{1}{x - 1}$$

does not exist.

35. Find the discontinuities (if any) for $f(x) = (x + 2)/(x^2 - 3x - 10)$. Which of the discontinuities are removable?

Solution

Since $x^2 - 3x - 10 = (x - 5)(x + 2)$, $x = 5$ and $x = -2$ are not in the domain of f. By Theorem 2.11, f is continuous for all x other than $x = 5$ or $x = -2$. At $x = 5$ the function is discontinuous and the discontinuity is nonremovable since the limit

$$\lim_{x \to 5} \frac{x + 2}{x^2 - 3x - 10}$$

does not exist. At $x = -2$ the function is discontinuous but the discontinuity is removable since

$$\lim_{x \to -2} \frac{x + 2}{x^2 - 3x - 10} = \lim_{x \to -2} \frac{x + 2}{(x + 2)(x - 5)}$$

$$= \lim_{x \to -2} \frac{1}{x - 5} = -\frac{1}{7}$$

39. Find the discontinuities (if any) for

$$f(x) = \begin{cases} \dfrac{x}{2} + 1, & x \le 2 \\[2mm] 3 - x, & x > 2 \end{cases}$$

Which of the discontinuities are removable?

Solution

Since f is linear to the right and left of $x = 2$, it is continuous for all x other than possibly at $x = 2$. At $x = 2$,

$$\lim_{x \to 2^-} f(x) = \lim_{x \to 2^-} \left(\frac{x}{2} + 1 \right) = 2$$

$$\lim_{x \to 2^+} f(x) = \lim_{x \to 2^+} (3 - x) = 1$$

Thus f is discontinuous at $x = 2$ and this discontinuity is non-removable since the limit from the left is not equal to the limit from the right.

44. Find the discontinuities (if any) for

$$f(x) = \begin{cases} 3 + x, & x \le 2 \\ x^2 + 1, & x > 2 \end{cases}$$

Which of the discontinuities are removable?

Solution

Since f is represented by a polynomial to the right and to the left of $x = 2$, it is continuous for all x other than possibly at $x = 2$. At $x = 2$,

$$\lim_{x \to 2^-} f(x) = \lim_{x \to 2^-} (3 + x) = 5$$

$$\lim_{x \to 2^+} f(x) = \lim_{x \to 2^+} (x^2 + 1) = 5$$

Thus f is continuous at $x = 2$ since

$$\lim_{x \to 2^-} f(x) = f(2) = \lim_{x \to 2^+} f(x)$$

Consequently, f is continuous over the entire real line.

67. Use the Intermediate Value Theorem to approximate the zero of the function $f(x) = x^3 + x - 1$ in the interval $[0, 1]$.
    (a) Begin by locating the zero in a subinterval of length 0.1.
    (b) Refine your approximation by locating the zero in a subinterval of length 0.01.

Solution

Since f is a polynomial, it is continuous. Therefore, by the Intermediate Value Theorem, if f changes signs on the interval $[a, b]$, a zero of f is in that interval. Since $f(0) = -1$ and $f(1) = 1$, a zero of f is in $[0, 1,]$. We now evaluate f for values of x in $[0, 1]$ to determine where f changes sign and therefore refine our estimate of the subinterval containing the zero of f. In each case the selected value of x is approximately the midpoint of each subinterval.

x	$f(x)$	*Interval containing the zero of f*
0.5	−0.375	$[0.5, 1]$
0.7	0.043	$[0.5, 0.7]$
0.6	−0.184	$[0.6, 0.7]$ Ans. part (a)
0.65	−0.075	$[0.65, 0.7]$
0.68	−0.006	$[0.68, 0.7]$
0.69	0.019	$[0.68, 0.69]$

Therefore, the subinterval of length 0.01 containing the zero of f is $[0.68, 0.69]$.

72. Verify the applicability of the Intermediate Value Theorem and find the value of c in $\left[\frac{5}{2}, 4\right]$ guaranteed by the theorem if $f(x) = (x^2 + x)/(x - 1)$ and $k = 6$.

Solution

By Theorem 2.11 $f(x) = (x^2 + x)/(x - 1)$ is continuous for every real number except $x = 1$. Therefore, it is continuous on $\left[\frac{5}{2}, 4\right]$. Also $f(a) < k < f(b)$ since

$$f\left(\frac{5}{2}\right) = \frac{35}{6} < 6 < \frac{20}{3} = f(4)$$

Hence the Intermediate Value Theorem applies.
  To find c, solve the equation

$$\frac{x^2 + x}{x - 1} = 6$$

$$\left(\frac{x^2 + x}{x - 1}\right)(x - 1) = 6(x - 1)$$

$$x^2 - 5x + 6 = 0$$

$$(x - 3)(x - 2) = 0$$

The solution of this equation is $x = 2$ or $x = 3$. Since 2 is not in $[\frac{5}{2}, 4]$, $c = 3$.

79.   Prove that if f and g are both continuous at c, then their product is continuous at c.

Solution

Since both f and g are continuous at $x = c$, we know that

$$\lim_{x \to c} f(x) = f(c) \qquad \text{and} \qquad \lim_{x \to c} g(x) = g(c)$$

Let $h(x) = f(x)g(x)$. Then

$$\lim_{x \to c} h(x) = \lim_{x \to c} [f(x)g(x)] = [\lim_{x \to c} f(x)][\lim_{x \to c} g(x)]$$
$$= f(c)g(c) = h(c)$$

Thus the product of f and g is continuous at $x = c$.

2.5
Infinite limits

15.   Find the vertical asymptotes (if any) for $f(x) = 1 - \dfrac{4}{x^2}$.

Solution

By rewriting the equation for this function as the ratio of two polynomials, we have

$$f(x) = 1 - \frac{4}{x^2} = \frac{x^2 - 4}{x^2}$$

Since the denominator of this function is zero only when x is zero, the only vertical asymptote is $x = 0$.

17.   Find the vertical asymptotes (if any) for $f(x) = \dfrac{x}{x^2 + x - 2}$.

Solution

Since the denominator factors as

$$x^2 + x - 2 = (x - 1)(x + 2)$$

we see that this function has two vertical asymptotes, one at $x = 1$ and the other at $x = -2$.

22. Determine whether the function $f(x) = \sin (x + 1)/(x + 1)$ has a vertical asymptote or a removable discontinuity at $x = -1$.

Solution

Since, by Theorem 2.9,

$$\lim_{x \to -1} \frac{\sin (x + 1)}{x + 1} = 1$$

there is a removable discontinuity at $x = -1$.

29. Find (if it exists) the limit $\lim\limits_{x \to 0^+} \dfrac{2}{\sin x}$.

Solution

Since

$$\lim_{x \to 0^+} 2 = 2 \qquad \text{and} \qquad \lim_{x \to 0^+} \sin x = 0$$

we conclude that

$$\lim_{x \to 0^+} \frac{2}{\sin x} = \infty$$

Review Exercises for Chapter 2

10. Evaluate $\lim\limits_{s \to 0} \dfrac{(1/\sqrt{1 + s}) - 1}{s}$.

Solution

$$\lim_{s \to 0} \frac{(1/\sqrt{1 + s}) - 1}{s} = \lim_{s \to 0} \frac{(1 - \sqrt{1 + s})/\sqrt{1 + s}}{s}$$

$$= \lim_{s \to 0} \frac{1 - \sqrt{1 + s}}{s\sqrt{1 + s}}$$

$$= \lim_{s \to 0} \left(\frac{1 - \sqrt{1 + s}}{s\sqrt{1 + s}} \right)\left(\frac{1 + \sqrt{1 + s}}{1 + \sqrt{1 + s}} \right)$$

$$= \lim_{s \to 0} \frac{1 - (1 + s)}{s\sqrt{1 + s} \, (1 + \sqrt{1 + s})}$$

$$= \lim_{s \to 0} \frac{-s}{s\sqrt{1+s}\,(1+\sqrt{1+s})}$$

$$= \lim_{s \to 0} \frac{-1}{\sqrt{1+s}\,(1+\sqrt{1+s})} = -\frac{1}{2}$$

12. Evaluate $\lim\limits_{x \to -2} \dfrac{x^2 - 4}{x^3 + 8}$.

   Solution

   $$\lim_{x \to -2} \frac{x^2 - 4}{x^3 + 8} = \lim_{x \to -2} \frac{(x+2)(x-2)}{(x+2)(x^2 - 2x + 4)}$$

   $$= \lim_{x \to -2} \frac{x-2}{x^2 - 2x + 4} = -\frac{1}{3}$$

25. Evaluate $\lim\limits_{\Delta x \to 0} \dfrac{\sin(\pi/6 + \Delta x) - 1/2}{\Delta x}$.

   Solution

   $$\lim_{\Delta x \to 0} \frac{\sin(\pi/6 + \Delta x) - 1/2}{\Delta x}$$

   $$= \lim_{\Delta x \to 0} \frac{(1/2)\cos \Delta x + (\sqrt{3}/2)\sin \Delta x - (1/2)}{\Delta x}$$

   $$= \lim_{\Delta x \to 0} \left[\frac{\sqrt{3}\sin \Delta x}{2\Delta x} + \frac{1}{2}\left(\frac{\cos \Delta x - 1}{\Delta x} \right) \right]$$

   $$= \frac{\sqrt{3}}{2}$$

31. Determine if $\lim\limits_{x \to 0} \dfrac{|x|}{x} = 1$ is true or false.

   Solution

   $$\lim_{x \to 0^+} \frac{|x|}{x} = \lim_{x \to 0^+} \frac{x}{x} = \lim_{x \to 0^+} 1 = 1$$

   $$\lim_{x \to 0^-} \frac{|x|}{x} = \lim_{x \to 0^-} \frac{-x}{x} = \lim_{x \to 0^-} (-1) = -1$$

   Since the limit from the right and the limit from the left are unequal, the limit does not exist and the given statement is false.

47. Determine the value of c so that the function

$$f(x) = \begin{cases} x + 3, & x \le 2 \\ cs + 6, & x > 2 \end{cases}$$

is continuous for all x.

Solution

Since f is linear to the right and left of 2, it is continuous for all values of x other than possibly at $x = 2$. Furthermore, since $f(2) = 2 + 3 = 5$, we can make f continuous at $x = 2$ by finding c such that

$$c(2) + 6 = 5$$
$$2c = -1$$
$$c = -\frac{1}{2}$$

Now we have

$$\lim_{x \to 2^-} f(x) = \lim_{x \to 2^-} (x + 3) = 2 + 3 = 5$$

and

$$\lim_{x \to 2^+} f(x) = \lim_{x \to 2^+} \left(-\frac{x}{2} + 6 \right) = -1 + 6 = 5$$

Thus for $c = -\frac{1}{2}$, f is continuous at $x = 2$ and consequently f is continuous for all x.

3 Differentiation

3.1
The derivative and the tangent line problem

9. Find the derivative of $f(x) = 2x^2 + x - 1$ by the four-step process.

   Solution

   $$f(x) = 2x^2 + x - 1$$

   1. $$f(x + \Delta x) = 2(x + \Delta x)^2 + (x + \Delta x) - 1$$
   $$= 2x^2 + 4x\,\Delta x + 2(\Delta x)^2 + x$$
   $$+ \Delta x - 1$$

   2. $$f(x + \Delta x) - f(x) = 2x^2 + 4x\,\Delta x + 2(\Delta x)^2 + x$$
   $$+ \Delta x - 1 - (2x^2 + x - 1)$$
   $$= 4x\,\Delta x + 2(\Delta x)^2 + \Delta x$$

   3. $$\frac{f(x + \Delta x) - f(x)}{\Delta x} = \frac{\Delta x(4x + 2\,\Delta x + 1)}{\Delta x}$$
   $$= 4x + 2\,\Delta x + 1$$

   4. $$\lim_{\Delta x \to 0} \frac{f(x + \Delta x) - f(x)}{\Delta x} = 4x + 1$$

11. Find the derivative of $f(x) = 1/(x - 1)$ by the four-step process.

    Solution

    $$f(x) = \frac{1}{x - 1}$$

1. $$f(x + \Delta x) = \frac{1}{(x + \Delta x) - 1}$$

2. $$f(x + \Delta x) - f(x) = \frac{1}{(x + \Delta x - 1)} - \frac{1}{x - 1}$$

$$= \frac{x - 1 - (x + \Delta x - 1)}{(x + \Delta x - 1)(x - 1)}$$

$$= \frac{-\Delta x}{(x + \Delta x - 1)(x - 1)}$$

3. $$\frac{f(x + \Delta x) - f(x)}{\Delta x} = \frac{-\Delta x}{\Delta x(x + \Delta x - 1)(x - 1)}$$

$$= \frac{-1}{(x + \Delta x - 1)(x - 1)}$$

4. $$\lim_{\Delta x \to 0} \frac{f(x + \Delta x) - f(x)}{\Delta x} = \frac{-1}{(x - 1)(x - 1)} = \frac{-1}{(x - 1)^2}$$

17. Use the four-step process to find the derivative of $f(x) = x^3$. Sketch the graph of f and find the equation of the tangent line at $(2, 8)$.

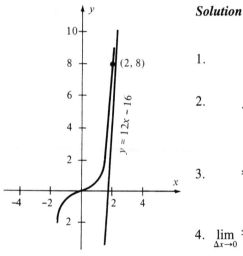

Solution

$$f(x) = x^3$$

1. $$f(x + \Delta x) = (x + \Delta x)^3$$

$$= x^3 + 3x^2 \, \Delta x + 3x \, (\Delta x)^2 + (\Delta x)^3$$

2. $$f(x + \Delta x) - f(x) = x^3 + 3x^2 \, \Delta x + 3x \, (\Delta x)^2$$
$$+ (\Delta x)^3 - x^3$$

$$= 3x^2 \, \Delta x + 3x \, (\Delta x)^2 + (\Delta x)^3$$

3. $$\frac{f(x + \Delta x) - f(x)}{\Delta x} = \frac{\Delta x[3x^2 + 3x \, \Delta x + (\Delta x)^2]}{\Delta x}$$

$$= 3x^2 + 3x \, \Delta x + (\Delta x)^2$$

4. $$\lim_{\Delta x \to 0} \frac{f(x + \Delta x) - f(x)}{\Delta x} = 3x^2$$

Thus $f'(x) = 3x^2$ and the slope of the tangent line at $(2, 8)$ is $f'(2) = 3(2)^2 = 12$. Finally, by the point-slope equation of a line, we have

$$y - y_1 = m(x - x_1)$$
$$y - 8 = 12(x - 2)$$

$$y - 8 = 12x - 24$$
$$y = 12x - 16$$

19. Use the four-step process to find the derivative of $f(x) = \sqrt{x + 1}$. Sketch the graph of f and find the equation of the tangent line at $(3, 2)$.

Solution

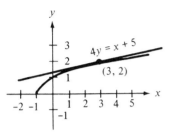

$$f(x) = \sqrt{x + 1}$$

1. $$f(x + \Delta x) = \sqrt{x + \Delta x + 1}$$

2. $$f(x + \Delta x) - f(x) = \sqrt{x + \Delta x + 1} - \sqrt{x + 1}$$
$$= (\sqrt{x + \Delta x + 1} - \sqrt{x + 1})$$
$$\times \left(\frac{\sqrt{x + \Delta x + 1} + \sqrt{x + 1}}{\sqrt{x + \Delta x + 1} + \sqrt{x + 1}} \right)$$
$$= \frac{x + \Delta x + 1 - (x + 1)}{\sqrt{x + \Delta x + 1} + \sqrt{x + 1}}$$
$$= \frac{\Delta x}{\sqrt{x + \Delta x + 1} + \sqrt{x + 1}}$$

3. $$\frac{f(x + \Delta x) - f(x)}{\Delta x} = \frac{\Delta x}{\Delta x(\sqrt{x + \Delta x + 1} + \sqrt{x + 1})}$$
$$= \frac{1}{\sqrt{x + \Delta x + 1} + \sqrt{x + 1}}$$

4. $$\lim_{\Delta x \to 0} \frac{f(x + \Delta x) - f(x)}{\Delta x} = \frac{1}{\sqrt{x + 0 + 1} + \sqrt{x + 1}}$$
$$= \frac{1}{2\sqrt{x + 1}}$$

Thus $f'(x) = 1/2\sqrt{x + 1}$ and the slope of the tangent line at $(3, 2)$ is $f'(3) = 1/2\sqrt{3 + 1} = \frac{1}{4}$. Finally, by the point-slope equation of a line, we have

$$y - y_1 = m(x - x_1)$$
$$y - 2 = \frac{1}{4}(x - 3)$$
$$4y - 8 = x - 3$$
$$4y = x + 5 \qquad \text{or} \qquad y = \frac{x}{4} + \frac{5}{4}$$

23. Use the alternative limit form to find the derivative (if it exists) of $f(x) = x^3 + 2x^2 + 1$ at $x = -2$.

Solution

$$f'(-2) = \lim_{x \to -2} \frac{f(x) - f(-2)}{x - (-2)}$$

$$= \lim_{x \to -2} \frac{(x^3 + 2x^2 + 1) - 1}{x + 2}$$

$$= \lim_{x \to -2} \frac{x^2(x + 2)}{x + 2}$$

$$= \lim_{x \to -2} x^2 = 4$$

The graph of f has a **cusp** at $x = c$ if:

(a)   there exists $a < c < b$ such that f is continuous for $a < x < b$,

(b) f is differentiable for all points in the interval (a, b) *except* $x = c$,

(c)   the one-sided limits

$$\lim_{x \to c^-} \frac{f(x) - f(c)}{x - c} \quad \text{and} \quad \lim_{x \to c^+} \frac{f(x) - f(c)}{x - c}$$

are infinite and of opposite signs.

The graph of f has a **node** at $x = c$ if:

(a)   there exists $a < c < b$ such that f is continuous for $a < x < b$,

(b) f is differentiable for all points in the interval (a, b) *except* $x = c$,

(c)   at least one of the one-sided limits

$$\lim_{x \to c^-} \frac{f(x) - f(c)}{x - c} \quad \text{and} \quad \lim_{x \to c^+} \frac{f(x) - f(c)}{x - c}$$

is finite.

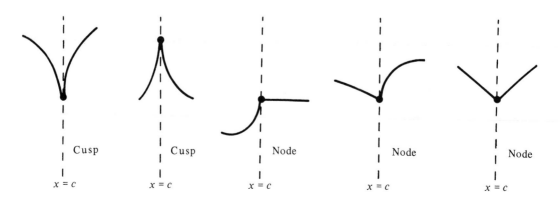

Cusp	Cusp	Node	Node	Node
$x = c$	$x = c$	$x = c$	$x = c$	$x = c$

31. Find every point at which the function $f(x) = (x - 3)^{2/3}$ is not differentiable.

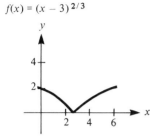

$f(x) = (x - 3)^{2/3}$

Solution

The graph of $f(x) = (x - 3)^{2/3}$ is continuous at $x = 3$. However, the one-sided limits

$$\lim_{x \to 3^+} \frac{f(x) - f(3)}{x - 3} = \lim_{x \to 3^+} \frac{(x - 3)^{2/3}}{x - 3} = \lim_{x \to 3^+} \frac{1}{(x - 3)^{1/3}} = \infty$$

$$\lim_{x \to 3^-} \frac{f(x) - f(3)}{x - 3} = \lim_{x \to 3^-} \frac{(x - 3)^{2/3}}{x - 3} = \lim_{x \to 3^-} \frac{1}{(x - 3)^{1/3}} = -\infty$$

are not equal. Therefore, f is not differentiable at $x = 3$ and the point $(3, 0)$ is a cusp.

33. Find every point at which the function $f(x) = \sqrt{x - 1}$ is not differentiable.

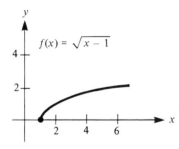

$f(x) = \sqrt{x - 1}$

Solution

We first observe that f is continuous in its domain $[1, \infty)$. From the one-sided limit

$$\lim_{x \to 1^+} \frac{f(x) - f(1)}{x - 1} = \lim_{x \to 1^+} \frac{\sqrt{x - 1}}{x - 1} = \lim_{x \to 1^+} \frac{1}{\sqrt{x - 1}} = \infty$$

we observe that f has a vertical tangent at $(1, 0)$ and therefore is not differentiable at $x = 1$.

41. Find the equation of a line that is tangent to the curve $y = x^3$ and is parallel to the line $3x - y + 1 = 0$.

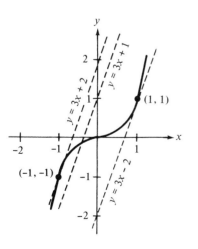

Solution

The slope of the line given by $3x - y + 1 = 0$ is 3 since

$$3x - y + 1 = 0$$
$$y = 3x + 1 = mx + b$$

Thus we wish to find a point (or points) on the graph of $y = x^3$ such that the tangent line at that point has a slope of 3. From Exercise 17, we know that $dy/dx = 3x^2$ is the slope at any point on the graph of $y = x^3$. Therefore, we have

$$\frac{dy}{dx} = 3x^2 = 3$$

$$x^2 = 1 \qquad \text{or} \qquad x = \pm 1$$

Finally, we conclude that the slope of the graph of $y = x^3$ is 3 at the points $(1, 1)$ and $(-1, -1)$. The tangent lines at these two points are

$$y - 1 = 3(x - 1) \qquad y - (-1) = 3[x - (-1)]$$
$$y = 3x - 2 \qquad\qquad y = 3x + 2$$

43. There are two tangent lines to the curve $y = 4x - x^2$ that pass through the point $(2, 5)$. Find the equations of those two lines and make a sketch to verify your results.

Solution

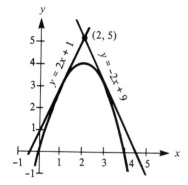

To begin, we find dy/dx as follows:

$$y = f(x) = 4x - x^2$$

1. $$f(x + \Delta x) = 4(x + \Delta x) - (x + \Delta x)^2$$
$$= 4x + 4\Delta x - x^2 - 2x\,\Delta x$$
$$- (\Delta x)^2$$

2. $$f(x + \Delta x) - f(x) = 4x + 4\Delta x - x^2 - 2x\,\Delta x$$
$$- (\Delta x)^2 - 4x + x^2$$
$$= (4 - 2x)\Delta x - (\Delta x)^2$$

3. $$\frac{f(x + \Delta x) - f(x)}{\Delta x} = \frac{\Delta x(x - 2x - \Delta x)}{\Delta x}$$
$$= 4 - 2x - \Delta x$$

4. $$\lim_{\Delta x \to 0} \frac{f(x + \Delta x) - f(x)}{\Delta x} = 4 - 2x$$

Now let (x, y) be a point on the graph of $y = 4x - x^2$. Since $dy/dx = 4 - 2x$, the slope of the tangent line at (x, y) is $m = 4 - 2x$. On the other hand, if the tangent line at (x, y) passes through the point $(2, 5)$, then its slope must be

$$m = \frac{y - 5}{x - 2} = \frac{4x - x^2 - 5}{x - 2}$$

Equating these two values of m, we have

$$\frac{4x - x^2 - 5}{x - 2} = 4 - 2x$$
$$4x - x^2 - 5 = (4 - 2x)(x - 2)$$
$$4x - x^2 - 5 = 4x - 2x^2 - 8 + 4x$$
$$x^2 - 4x + 3 = 0$$
$$(x - 3)(x - 1) = 0$$
$$x = 3 \quad \text{or} \quad x = 1$$

If $x = 3$, $y = 4(3) - 3^2 = 3$ and $m = 4 - 2(3) = -2$.

$$y - 3 = -2(x - 3) \quad \text{or} \quad y = -2x + 9$$

If $x = 1$, $y = 4(1) - 1^2 = 3$ and $m = 4 - 2(1) = 2$.

$$y - 3 = 2(x - 1) \quad \text{or} \quad y = 2x + 1$$

3.2
Velocity, acceleration, and other rates of change

3. Find the average rate of change of $f(x) = 1/(x + 1)$ between $(0, 1)$ and $(3, \frac{1}{4})$. Compare this average rate of change to the instantaneous rate of change at each point.

Solution

The average rate of change is given by

$$\frac{\Delta y}{\Delta x} = \frac{\frac{1}{4} - 1}{3 - 0} = \frac{-\frac{3}{4}}{3} = -\frac{1}{4}$$

To find the instantaneous rate of change, $f'(x)$, we use the four-step process as follows:

$$f(x) = \frac{1}{x + 1}$$

1. $$f(x + \Delta x) = \frac{1}{x + \Delta x + 1}$$

2. $$f(x + \Delta x) - f(x) = \frac{1}{x + \Delta x + 1} - \frac{1}{x + 1}$$

$$= \frac{x + 1 - (x + \Delta x + 1)}{(x + \Delta x + 1)(x + 1)}$$

$$= \frac{-\Delta x}{(x + \Delta x + 1)(x + 1)}$$

3. $$\frac{f(x + \Delta x) - f(x)}{\Delta x} = \frac{-\Delta x}{\Delta x(x + \Delta x + 1)(x + 1)}$$

$$= \frac{-1}{(x + \Delta x + 1)(x + 1)}$$

4. $$\lim_{\Delta x \to 0} \frac{f(x + \Delta x) - f(x)}{\Delta x} = \frac{-1}{(x + 0 + 1)(x + 1)} = \frac{-1}{(x + 1)^2}$$

Thus the instantaneous rate of change at $(0, 1)$ is given by

$$f'(0) = \frac{-1}{(0 + 1)^2} = -1$$

and the instantaneous rate of change at $(3, \frac{1}{4})$ is given by

$$f'(3) = \frac{-1}{(3 + 1)^2} = -\frac{1}{16}$$

7. The height s at time t of a silver dollar dropped from the World Trade Center is given by $s(t) = -16t^2 + 1350$, where s is measured in feet and t is measured in seconds. $[s'(t) = -32t]$

(a) Find the average velocity on the interval $[1, 2]$.

(b) Find the instantaneous velocity when $t = 1$ and $t = 2$.

(c) How long will it take the dollar to hit the ground?

(d) Find the velocity of the dollar when it hits the ground.

Solution

(a) The average velocity is given by

$$\frac{s(2) - s(1)}{2 - 1} = 1286 - 1334 = -48 \text{ ft/sec}$$

(b) Since $v(t) = s'(t) = -32t$, the instantaneous velocity at time $t = 1$ is

$$s'(1) = -32(1) = -32 \text{ ft/sec}$$

and at time $t = 2$ is

$$s'(2) = -32(2) = -64 \text{ ft/sec}$$

(c) The dollar will be at ground level when

$$s(t) = -16t^2 + 1350 = 0$$

$$t^2 = \frac{1350}{16}$$

$$t = \frac{1}{4}\sqrt{1350} \approx 9.2 \text{ sec}$$

(d) The velocity when it hits the ground is

$$v\left(\frac{1}{4}\sqrt{1350}\right) = -32\left(\frac{1}{4}\sqrt{1350}\right) = -8\sqrt{1350}$$

$$\approx -293.9 \text{ ft/sec}.$$

21. Find $f'''(x)$ if $f''(x) = (2x - 2)/x$.

Solution

We rewrite the second derivative as

$$f''(x) = \frac{2x - 2}{x} = 2 - \frac{2}{x}.$$

Then $f'''(x) = \lim_{\Delta x \to 0} \dfrac{f''(x + \Delta x) - f''(x)}{\Delta x}$

$$= \lim_{\Delta x \to 0} \frac{2 - 2/(x + \Delta x) - (2 - 2/x)}{\Delta x}$$

$$= \lim_{\Delta x \to 0} \frac{2(x + \Delta x) - 2x}{x(\Delta x)(x + \Delta x)}$$

$$= \lim_{\Delta x \to 0} \frac{2}{x(x + \Delta x)} = \frac{2}{x^2}$$

3.3
Differentiation rules for sums, constant multiples, powers, sines, and cosines

11.   Differentiate $s(t) = t^3 - 2t + 4$.

Solution

$$s(t) = t^3 - 2t + 4$$
$$s'(t) = 3t^2 - 2$$

15.   Differentiate $y = \dfrac{1}{x} - 3 \sin x$.

Solution

$$y = \frac{1}{x} - 3 \sin x = x^{-1} - 3 \sin x$$

$$y' = -x^{-2} - 3 \cos x$$

$$= -\frac{1}{x^2} - 3 \cos x$$

19.   Differentiate $f(t) = 3 - (3/5t)$ and evaluate the derivative at $(\frac{3}{5}, 2)$.

Solution

$$f(t) = 3 - \left(\frac{3}{5}\right)t^{-1}$$

$$f'(t) = 0 - (-1)\left(\frac{3}{5}\right)t^{-2} = \frac{3}{5t^2}$$

At the point $(\frac{3}{5}, 2)$ the derivative is

$$f'\left(\frac{3}{5}\right) = \frac{3}{5(\frac{3}{5})^2} = \frac{3}{5}\left(\frac{5}{3}\right)^2 = \frac{5}{3}$$

21. Differentiate $y = (2x + 1)^2$ and evaluate the derivative at $(0, 1)$.

Solution

$$y = (2x + 1)^2 = 4x^2 + 4x + 1$$
$$y' = 8x + 4$$

At the point $(0, 1)$ the derivative is

$$y' = 8(0) + 4 = 4$$

25. Find $f'(x)$ if $f(x) = x^2 - (4/x)$.

Solution

$$f(x) = x^2 - \frac{4}{x} = x^2 - 4x^{-1}$$

$$f'(x) = 2x - (-1)4x^{-2} = 2x + \frac{4}{x^2}$$

29. Find $f'(x)$ if $f(x) = (x^3 - 3x^2 + 4)/x^2$.

Solution

$$f(x) = \frac{x^3 - 3x^2 + 4}{x^2} = \frac{x^3}{x^2} - \frac{3x^2}{x^2} + \frac{4}{x^2} = x - 3 + 4x^{-2}$$

$$f'(x) = 1 + (-2)4x^{-3} = 1 - \frac{8}{x^3} = \frac{x^3 - 8}{x^3}$$

35. Find $f'(x)$ if $f(x) = \sqrt[3]{x} + \sqrt[5]{x}$.

Solution

$$f(x) = \sqrt[3]{x} + \sqrt[5]{x} = x^{1/3} + x^{1/5}$$

$$f'(x) = \left(\frac{1}{3}\right)x^{-2/3} + \left(\frac{1}{5}\right)x^{-4/5} = \frac{1}{3x^{2/3}} + \frac{1}{5x^{4/5}}$$

37. Find $f'(x)$ if $f(x) = 4\sqrt{x} + 3\cos x$.

Solution

$$f(x) = 4\sqrt{x} + 3 \cos x = 4x^{1/2} + 3 \cos x$$

$$f'(x) = 4\left(\frac{1}{2}\right)x^{-1/2} - 3 \sin x = \frac{2}{\sqrt{x}} - 3 \sin x$$

42.   Find $f'(x)$ if $f(x) = \pi/(3x)^2$.

Solution

Function: $$f(x) = \frac{\pi}{(3x)^2}$$

Rewrite: $$f(x) = \left(\frac{\pi}{9}\right)x^{-2}$$

Derivative: $$f'(x) = (-2)\left(\frac{\pi}{9}\right)x^{-3}$$

Simplify: $$f'(x) = \frac{-2\pi}{9x^3}$$

45.   Find the equation of the line tangent to $y = x^4 - 3x^2 + 2$ at the point $(1, 0)$.

Solution

$$y = x^4 - 3x^2 + 2$$
$$y' = 4x^3 - 6x$$

Thus the slope of the tangent line at $(1, 0)$ is

$$y' = 4(1)^3 - 6(1) = 4 - 6 = -2$$

The equation of the tangent line at $(1, 0)$ is

$$y - y_1 = m(x - x_1)$$
$$y - 0 = -2(x - 1)$$
$$y = -2x + 2$$

47.   At what points, if any, does $y = x^4 - 3x^2 + 2$ have horizontal tangents?

Solution

A tangent line is horizontal if the derivative (the slope) at the point of tangency is zero. Since $y' = 4x^3 - 6x$, we must find all values of x that satisfy the equation

$$y' = 4x^3 - 6x = 0$$

$$2x(2x^2 - 3) = 0$$

$$x = 0, \pm\sqrt{\frac{3}{2}}$$

$$y = \left(\pm\sqrt{\frac{3}{2}}\right)^4 - 3\left(\pm\sqrt{\frac{3}{2}}\right)^2 + 2 = \frac{9}{4} - \frac{9}{2} + 2$$

$$= \frac{9 - 18 + 8}{4} = -\frac{1}{4}$$

Thus the points of horizontal tangency are $(0, 2)$, $(\sqrt{\frac{3}{2}}, -\frac{1}{4})$, $(-\sqrt{\frac{3}{2}}, -\frac{1}{4})$.

53. Sketch the graphs of the two equations $y = x^2$ and $y = -x^2 + 6x - 5$ and sketch the two lines that are tangent to both graphs. Find equations for these lines.

Solution

Let (x_1, y_1) and (x_2, y_2) be the points of tangency on the graphs of $y = x^2$ and $y = -x^2 + 6x - 5$, respectively. We know that $y_1 = x_1^2$ and $y_2 = -x_2^2 + 6x_2 - 5$. Let m be the slope of the tangent line. Since the line passes through (x_1, y_1) and (x_2, y_2), we have

$$m = \frac{y_2 - y_1}{x_2 - x_1} = \frac{-x_2^2 + 6x_2 - 5 - x_1^2}{x_2 - x_1}$$

Since the line is tangent to $y = x^2$ at (x_1, y_1) and the slope of this curve is $y' = 2x$, we have

$$m = 2x_1$$

Since the line is tangent to $y = -x^2 + 6x - 5$ at (x_2, y_2) and the slope of this curve is $y' = -2x + 6$, we have

$$m = -2x_2 + 6$$

Thus, from the preceding two equations, we have

$$m = 2x_1 = -2x_2 + 6$$

$$x_1 = -x_2 + 3$$

From the first and third equations,

$$m = -2x_2 + 6 = \frac{-x_2^2 + 6x_2 - 5 - x_1^2}{x_2 - x_1}$$

$$-2x_2 + 6 = \frac{-x_2^2 + 6x_2 - 5 - (-x_2 + 3)^2}{x_2 - (-x_2 + 3)}$$

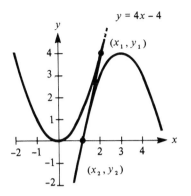

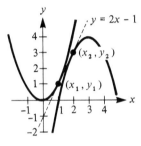

$$(2x_2 - 3)(-2x_2 + 6) = -x_2^2 + 6x_2 - 5 - (x_2^2 - 6x_2 + 9)$$
$$-4x_2^2 + 18x_2 - 18 = -2x_2^2 + 12x_2 - 14$$
$$-2x_2^2 + 6x_2 - 4 = 0$$
$$x_2^2 - 3x_2 + 2 = 0$$
$$(x_2 - 2)(x_2 - 1) = 0$$
$$x_2 = 1 \quad \text{or} \quad x_2 = 2$$

If $x_2 = 1$, $y_2 = -1^2 + 6(1) - 5 = 0$, $x_1 = -1 + 3 = 2$, and $y_1 = 2^2 = 4$. Thus the line containing $(1, 0)$ and $(2, 4)$ is tangent to both curves. The equation of this line is

$$y - 0 = \left(\frac{4 - 0}{2 - 1}\right)(x - 1) \quad \text{or} \quad y = 4x - 4$$

If $x_2 = 2$, $y_2 = -2^2 + 6(2) - 5 = -4 + 12 - 5 = 3$, $x_1 = -2 + 3 = 1$, and $y_1 = 1^2 = 1$. Thus the line containing $(2, 3)$ and $(1, 1)$ is tangent to both curves. The equation of this line is

$$y - 1 = \left(\frac{3 - 1}{2 - 1}\right)(x - 1)$$
$$= 2x - 2$$
$$y = 2x - 1$$

57.   Suppose a certain company finds that by charging p dollars per unit, its monthly revenue R will be

$$R = 12{,}000p - 1{,}000p^2, \quad 0 \le p \le 12$$

(Note that the revenue is zero when $p = 12$ since no one is willing to pay that much.) Find the rate of change of R with respect to p when:

(a) $p = 1$ (b) $p = 4$ (c) $p = 6$ (d) $p = 10$

Solution

$$R(p) = 12{,}000p - 1{,}000p^2$$
$$R'(p) = 12{,}000 - 2{,}000p$$

Therefore, the rate of change of R with respect to p is as follows:

(a) $R'(1) = 12{,}000 - 2{,}000(1) \quad = 10{,}000$

(b) $R'(4) = 12{,}000 - 2{,}000(4) \quad = 4{,}000$

(c) $R'(6) = 12{,}000 - 2{,}000(6) \quad = 0$

(d) $R'(10) = 12{,}000 - 2{,}000(10) = -8{,}000$

3.4

Differentiation rules for products, quotients, secants, and tangents

6.   Differentiate $f(x) = (x - 1)(x^2 - 3x + 2)$ and find $f'(0)$.

Solution

$$f(x) = (x - 1)(x^2 - 3x + 2)$$
$$f'(x) = (x - 1)(2x - 3) + (x^2 - 3x + 2)(1)$$
$$= 2x^2 - 3x - 2x + 3 + x^2 - 3x + 2$$
$$= 3x^2 - 8x + 5$$
$$f'(0) = 5$$

11.   Differentiate $f(x) = (3 - 2x - x^2)/(x^2 - 1)$.

Solution

$$f(x) = \frac{3 - 2x - x^2}{x^2 - 1} = -\frac{x^2 + 2x - 3}{x^2 - 1} = -\frac{(x + 3)(x - 1)}{(x + 1)(x - 1)}$$
$$= -\frac{x + 3}{x + 1}$$
$$f'(x) = -\frac{(x + 1)(1) - (x + 3)(1)}{(x + 1)^2} = \frac{2}{(x + 1)^2}$$

14.   Differentiate $f(x) = \sqrt[3]{x}(\sqrt{x} + 3)$.

Solution

$$f(x) = x^{1/3}(x^{1/2} + 3) = x^{5/6} + 3x^{1/3}$$
$$f'(x) = \left(\frac{5}{6}\right)x^{-1/6} + \left(\frac{1}{3}\right)(3)x^{-2/3} = \frac{5}{6x^{1/6}} + \frac{1}{x^{2/3}}$$

19.   Differentiate $g(x) = \left(\dfrac{x + 1}{x + 2}\right)(2x - 5)$.

Solution

$$g(x) = \frac{(x + 1)(2x - 5)}{x + 2} = \frac{2x^2 - 3x - 5}{x + 2}$$
$$g'(x) = \frac{(x + 2)(4x - 3) - (2x^2 - 3x - 5)(1)}{(x + 2)^2}$$
$$= \frac{4x^2 + 5x - 6 - 2x^2 + 3x + 5}{(x + 2)^2} = \frac{2x^2 + 8x - 1}{(x + 2)^2}$$

21. Differentiate $f(x) = (3x^3 + 4x)(x - 5)(x + 1)$.

Solution

$$f(x) = [(3x^3 + 4x)(x - 5)](x + 1)$$
$$f'(x) = [(3x^3 + 4x)(x - 5)](1)$$
$$\qquad + (x + 1)[(3x^3 + 4x)(1) + (x - 5)(9x^2 + 4)]$$
$$= (3x^3 + 4x)(x - 5) + (x + 1)(3x^3 + 4x)$$
$$\qquad + (x + 1)(x - 5)(9x^2 + 4)$$
$$= 15x^4 - 48x^3 - 33x^2 - 32x - 20$$

33. Differentiate $f(t) = \dfrac{\cos t}{t}$.

Solution

$$f(t) = \frac{\cos t}{t}$$

$$f'(t) = \frac{(t)(-\sin t) - (\cos t)(1)}{t^2} = \frac{-(t \sin t + \cos t)}{t^2}$$

39. Differentiate $y = -\csc x - \sin x$.

Solution

$$y = -\csc x - \sin x$$

$$y' = -(-\csc x \cot x) - \cos x = \frac{1}{\sin x}\left(\frac{\cos x}{\sin x}\right) - \cos x$$

$$= \cos x\left(\frac{1}{\sin^2 x} - 1\right)$$

$$= \frac{\cos x}{\sin^2 x}(1 - \sin^2 x) = \cot^2 x \cos x$$

45. Differentiate $f(x) = x^2 \tan x$.

Solution

$$f(x) = x^2 \tan x$$
$$f'(x) = x^2 \sec^2 x + (\tan x)(2x) = x(x \sec^2 x + 2 \tan x)$$

57. Find an equation of the line tangent to the graph of
$f(x) = x/(x - 1)$ at the point $(2, 2)$.

Solution

$$f(x) = \frac{x}{x-1}$$

$$f'(x) = \frac{(x-1)(1) - (x)(1)}{(x-1)^2} = \frac{x-1-x}{(x-1)^2} = \frac{-1}{(x-1)^2}$$

Therefore, the slope of the tangent line at $(2, 2)$ is $f'(2) = -1/(2-1)^2 = -1$ and the equation of the tangent line at $(2, 2)$ is

$$y - 2 = -1(x - 2)$$
$$= -x + 2$$
$$y = -x + 4$$

66. Suppose that the temperature T of food placed in a refrigerator drops according to the equation

$$T = 10\left(\frac{4t^2 + 16t + 75}{t^2 + 4t + 10}\right)$$

where t is the time in hours. What is the initial temperature of the food? What is the limit of T as t approaches infinity? Find the rate of change of T with respect to t when:

(a) $t = 1$ (b) $t = 3$ (c) $t = 5$ (d) $t = 10$

Solution

Initial temperature:

$$T(0) = 10\left(\frac{75}{10}\right) = 75$$

Temperature as $t \to \infty$:

$$\lim_{t \to \infty} 10\left(\frac{4t^2 + 16t + 75}{t^2 + 4t + 10}\right) = \lim_{t \to \infty} 10\left[\frac{4 + (16/t) + (75/t^2)}{1 + (4/t) + (10/t^2)}\right]$$

$$= 10\left(\frac{4}{1}\right) = 40$$

Rate of change of T with respect to t:

$$T'(t) = 10\left[\frac{(t^2 + 4t + 10)(8t + 16) - (4t^2 + 16t + 75)(2t + 4)}{(t^2 + 4t + 10)^2}\right]$$

$$= \frac{-700(t + 2)}{(t^2 + 4t + 10)^2}$$

(a) $T'(1) = \dfrac{-700(3)}{15^2} = -9.33$

(b) $T'(3) = \dfrac{-700(5)}{31^2} = -3.64$

(c) $T'(5) = \dfrac{-700(7)}{55^2} = -1.62$

(d) $T'(10) = \dfrac{-700(12)}{150^2} = -0.373$

67. A population of 500 bacteria is introduced into a culture and grows in number according to the equation

$$P(t) = 500\left(1 + \frac{4t}{50 + t^2}\right)$$

where t is measured in hours. Find the rate at which the population is growing when $t = 2$.

Solution

$$P'(t) = 500\left[\frac{(50 + t^2)(4) - (4t)(2t)}{(50 + t^2)^2}\right]$$

$$= \frac{500(200 + 4t^2 - 8t^2)}{(50 + t^2)^2} = \frac{2000(50 - t^2)}{(50 + t^2)^2}$$

Therefore, the rate of population growth when $t = 2$ is

$$P'(2) = \frac{2000(50 - 4)}{(50 + 4)^2} = 31.55$$

3.5
The chain rule

9. Find the first derivative of $f(x) = 2(x^2 - 1)^3$.

Solution

$$f(x) = 2(x^2 - 1)^3$$
$$f'(x) = 2(3)(x^2 - 1)^2(2x) = 12x(x^2 - 1)^2$$

13. Find the first derivative of $f(t) = 1/(t - 3)^2$.

Solution

$$f(t) = \frac{1}{(t - 3)^2} = (t - 3)^{-2}$$

$$f'(t) = (-2)(t - 3)^{-3}(1) = \frac{-2}{(t - 3)^3}$$

17. Find the first derivative of $f(x) = x^2(x - 2)^4$.

Solution

$$f(x) = x^2(x - 2)^4$$
$$f'(x) = x^2(4)(x - 2)^3(1) + (x - 2)^4(2x)$$
$$= 2x(x - 2)^3(2x + x - 2) = 2x(x - 2)^3(3x - 2)$$

23. Find the first derivative of $y = \sqrt[3]{9x^2 + 4}$.

Solution

$$y = \sqrt[3]{9x^2 + 4} = (9x^2 + 4)^{1/3}$$
$$\frac{dy}{dx} = \left(\frac{1}{3}\right)(9x^2 + 4)^{-2/3}(18x) = \frac{6x}{(9x^2 + 4)^{2/3}}$$

29. Find the first derivative of $y = \dfrac{1}{\sqrt{x + 2}}$.

Solution

$$y = \frac{1}{\sqrt{x + 2}} = (x + 2)^{-1/2}$$
$$y' = \left(-\frac{1}{2}\right)(x + 2)^{-3/2}(1) = \frac{-1}{2(x + 2)^{3/2}}$$

39. Find the first derivative of $g(t) = \dfrac{3t^2}{\sqrt{t^2 + 2t - 1}}$.

Solution

$$g(t) = \frac{3t^2}{\sqrt{t^2 + 2t - 1}} = \frac{3t^2}{(t^2 + 2t - 1)^{1/2}}$$
$$g'(t) = \frac{(t^2 + 2t - 1)^{1/2}(6t) - (3t^2)(\frac{1}{2})(t^2 + 2t - 1)^{-1/2}(2t + 2)}{t^2 + 2t - 1}$$
$$= \frac{(3t)(t^2 + 2t - 1)^{-1/2}[2(t^2 + 2t - 1) - (t)(t + 1)]}{t^2 + 2t - 1}$$
$$= \frac{3t(2t^2 + 4t - 2 - t^2 - t)}{(t^2 + 2t - 1)^{3/2}}$$
$$= \frac{3t(t^2 + 3t - 2)}{(t^2 + 2t - 1)^{3/2}}$$

47. Find an equation of the tangent line to the graph of $f(x) = \sin 2x$ at the point $(\pi, 0)$.

Solution

$$f(x) = \sin 2x$$
$$f'(x) = (\cos 2x)(2) = 2 \cos 2x$$

Therefore, the slope of the tangent line at $(\pi, 0)$ is
$f'(\pi) = 2 \cos 2\pi = 2$ and the equation of the tangent line at $(\pi, 0)$ is

$$y - 0 = 2(x - \pi)$$
$$y = 2(x - \pi)$$

51.   Find $f''(x)$ if $f(x) = \sin x^2$.

Solution

$$f'(x) = (\cos x^2)(2x) = 2x \cos x^2$$
$$f''(x) = 2[x(-\sin x^2)(2x) + (\cos x^2)(1)]$$
$$= 2(\cos x^2 - 2x^2 \sin x^2)$$

52.   Find $f''(t)$ if $f(t) = \sqrt{t^2 + 1}/t$.

Solution

$$f'(t) = \frac{t(\tfrac{1}{2})(t^2 + 1)^{-1/2}(2t) - \sqrt{t^2 + 1}\,(1)}{t^2}$$

$$= \frac{(t^2/\sqrt{t^2 + 1}) - \sqrt{t^2 + 1}}{t^2}$$

$$= \frac{[t^2 - (t^2 + 1)]/\sqrt{t^2 + 1}}{t^2}$$

$$= \frac{-1}{t^2\sqrt{t^2 + 1}} = (-t^{-2})(t^2 + 1)^{-1/2}$$

$$f''(t) = (-t^{-2})(-\tfrac{1}{2})(t^2 + 1)^{-3/2}(2t) + (t^2 + 1)^{-1/2}(2t^{-3})$$

$$= \frac{1}{t(t^2 + 1)^{3/2}} + \frac{2}{t^3(t^2 + 1)^{1/2}}$$

$$= \frac{t^2 + 2(t^2 + 1)}{t^3(t^2 + 1)^{3/2}} = \frac{3t^2 + 2}{t^3(t^2 + 1)^{3/2}}$$

58.   Find dy/dx given $y = 2 \cos (x/2)$.

Solution

$$y = 2 \cos \frac{x}{2}$$

$$\frac{dy}{dx} = 2\left(-\sin \frac{x}{2}\right)\left(\frac{1}{2}\right) = -\sin \frac{x}{2}$$

61. Find dy/dx given $y = \frac{1}{4} \sin^2 2x$.

   Solution

   $$y = \frac{1}{4}(\sin 2x)^2$$

   $$\frac{dy}{dx} = \frac{1}{4}(2)(\sin 2x)(\cos 2x)(2)$$

   $$= \frac{1}{2}(2 \sin 2x \cos 2x) = \frac{1}{2} \sin 4x$$

71. Find dy/dx given $y = \sec^3 (2x)$.

   Solution

   $$y = (\sec 2x)^3$$

   $$\frac{dy}{dx} = 3(\sec 2x)^2[(\sec 2x \tan 2x)(2)] = 6 \sec^3 2x \tan 2x$$

3.6
Implicit differentiation

3. Given $xy = 4$, find dy/dx by implicit differentiation and evaluate the derivative at $(-4, -1)$.

   Solution

   $$xy = 4$$
   $$xy' + y(1) = 0$$
   $$xy' = -y$$
   $$y' = -\frac{y}{x}$$

   At $(-4, -1)$,

   $$\frac{dy}{dx} = -\frac{(-1)}{-4} = -\frac{1}{4}$$

7. Given $x^3 - xy + y^2 = 4$, find dy/dx by implicit differentiation and evaluate the derivative at $(0, -2)$.

   Solution

   $$x^3 - xy + y^2 = 4$$
   $$3x^2 - [xy' + y(1)] + 2yy' = 0$$

$$3x^2 - xy' - y + 2yy' = 0$$

$$y'(2y - x) = y - 3x^2$$

$$y' = \frac{y - 3x^2}{2y - x}$$

At $(0, -2)$,

$$\frac{dy}{dx} = \frac{-2 - 3(0^2)}{2(-2) - 0} = \frac{-2}{-4} = \frac{1}{2}$$

21. Given $\tan (x + y) = x$, find dy/dx by implicit differentiation and evaluate the derivative at $(0, 0)$.

Solution

$$\tan (x + y) = x$$

$$\sec^2 (x + y)(1 + y') = 1$$

$$\sec^2 (x + y)y' = 1 - \sec^2 (x + y)$$

$$y' = \frac{1 - \sec^2 (x + y)}{\sec^2 (x + y)}$$

$$= \frac{-\tan^2 (x + y)}{\tan^2 (x + y) + 1} = \frac{-x^2}{x^2 + 1}$$

At $(0, 0)$,

$$\frac{dy}{dx} = \frac{0}{1} = 0$$

27. Sketch the graph of the equation $9x^2 + 16y^2 = 144$. Find dy/dx implicitly and explicitly and show that the two results are equivalent.

Solution

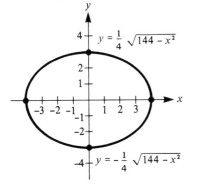

Differentiating implicitly we have

$$9x^2 + 16y^2 = 144$$

$$18x + 32yy' = 0$$

$$y' = \frac{-9x}{16y}$$

Solving the equation for y, we obtain

$$9x^2 + 16y^2 = 144$$

$$16y^2 = 144 - 9x^2$$

$$y^2 = \frac{1}{16}(144 - 9x^2)$$

$$y = \pm\frac{1}{4}\sqrt{144 - 9x^2}, \qquad -4 \le x \le 4$$

We now differentiate explicitly to obtain

$$\frac{dy}{dx} = \pm\left(\frac{1}{4}\right)\left(\frac{1}{2}\right)(144 - 9x^2)^{-1/2}(-18x) = \frac{\mp 9x}{4\sqrt{144 - 9x^2}}$$

$$= \frac{\mp 9x}{16(\frac{1}{4})\sqrt{144 - 9x^2}} = \frac{-9x}{16y}$$

31.   Given $x^2 - y^2 = 16$, find d^2y/dx^2 in terms of x and y.

Solution

$$x^2 - y^2 = 16$$

$$2x - 2yy' = 0$$

$$-2yy' = -2x$$

$$y' = \frac{x}{y}$$

By differentiating again, we have

$$y'' = \frac{y(1) - (x)y'}{y^2} = \frac{y - x(x/y)}{y^2} = \frac{y^2 - x^2}{y^3} = -\frac{16}{y^3}$$

39.   Show that the graphs of the equations $2x^2 + y^2 = 6$ and $y^2 = 4x$ are orthogonal (the curves intersect at right angles). Sketch the graph of each equation.

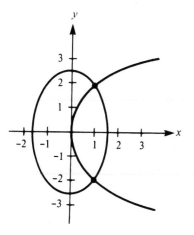

Solution

To find the points of intersection, we set $y^2 = 6 - 2x^2$ and $y^2 = 4x$ equal to each other.

$$4x = 6 - 2x^2$$

$$2x^2 + 4x - 6 = 0$$

$$x^2 + 2x - 3 = 0$$

$$(x + 3)(x - 1) = 0$$

$$x = -3 \qquad \text{and} \qquad x = 1$$

When $x = 1$, $y = \pm 2$, and when $x = -3$, y is undefined. Thus the two points of intersection are $(1, 2)$ and $(1, -2)$.

For $2x^2 + y^2 = 6$

$$4x + 2yy' = 0$$

$$y' = -\frac{2x}{y}$$

and for $y^2 = 4x$

$$2yy' = 4$$

$$y' = \frac{2}{y}$$

Thus at $(1, 2)$ the slopes of the two curves are

$$\frac{-2(1)}{2} = -1 \quad \text{and} \quad \frac{2}{2} = 1$$

which implies that the tangent lines at this point are perpendicular. Finally, at $(1, -2)$ the slopes of the two curves are

$$\frac{-2(1)}{-2} = 1 \quad \text{and} \quad \frac{2}{-2} = -1$$

and the tangent lines at this point are also perpendicular.

44. Show that the normal line (the line perpendicular to the tangent line to a curve) at any point on the circle $x^2 + y^2 = r^2$ passes through the origin.

Solution

By implicit differentiation,

$$x^2 + y^2 = r^2$$

$$2x + 2yy' = 0$$

$$y' = -\frac{x}{y}$$

Thus if (x, y) is a point on the circle $x^2 + y^2 = r^2$, the slope of the tangent line at (x, y) is $-x/y$. On the other hand, the slope of the line passing through (x, y) and $(0, 0)$ is

$$m = \frac{y - 0}{x - 0} = \frac{y}{x}$$

Since this slope is the negative reciprocal of y', the line passing through (x, y) and $(0, 0)$ must be perpendicular to the tangent line at (x, y).

3.7
Related rates

11. At a sand and gravel plant, sand is falling off a conveyer and onto a conical pile at the rate of 10 ft³/min. The diameter of the base of the cone is approximately three times the altitude. At what rate is the height of the pile changing when it is 15 ft high?

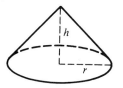

Solution

$$V = \text{volume of cone} = \frac{1}{3}\pi r^2 h$$

Since the diameter of the base is approximately three times the altitude, we have

$$2r = 3h \qquad \text{or} \qquad r = \frac{3}{2}h$$

Therefore,

$$V = \frac{1}{3}\pi\left(\frac{3}{2}h\right)^2 h = \frac{3\pi}{4}h^3$$

Differentiating with respect to t, we have

$$\frac{dV}{dt} = \frac{9\pi}{4}h^2\frac{dh}{dt} \qquad \text{or} \qquad \frac{4(dV/dt)}{9\pi h^2} = \frac{dh}{dt}$$

When $h = 15$,

$$\frac{dh}{dt} = \frac{4(10)}{9\pi(15)^2} = \frac{8}{405\pi}\,\text{ft/min}$$

19. A swimming pool is 40 ft long, 20 ft wide, 4 ft deep at the shallow end, and 9 ft deep at the deep end, the bottom being an inclined plane. Assume that water is being pumped into the pool at 10 ft³/min and there is 4 ft of water at the deep end.

(a) What percentage of the pool is filled?

(b) At what rate is the water level rising?

Solution

From the figure we see that x and y are related by the equation

$$m = \frac{y - 0}{x - 0} = \frac{5 - 0}{40 - 0}$$

$$\frac{y}{x} = \frac{1}{8}$$

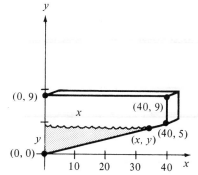

The volume of the inclined portion of the pool is given by

$$V_L = \frac{20xy}{2} = 10xy = 10(8y)y = 80y^2$$

When $y = 5$, the inclined portion of the pool has a volume of

$$V_L = 80(5^2) = 2000 \text{ ft}^3$$

Since the upper rectangular portion of the pool has a volume of

$$V_U = 4(40)(20) = 3200 \text{ ft}^3$$

the total volume of the pool is

$$V = V_L + V_U = 2000 + 3200 = 5200 \text{ ft}^3$$

(a)   When $y = 4$, the ratio of the filled portion of the pool to the total volume is

$$\frac{80y^2}{V} = \frac{80(4^2)}{5200} = 24.6\%$$

(b)   When $y = 4$, we can find dy/dt by differentiating $V_L = 80y^2$ and substituting $dV_L/dt = 10$.

$$V_L = 80y^2$$

$$\frac{dV_L}{dt} = 160y \frac{dy}{dt}$$

$$10 = 160(4) \frac{dy}{dt}$$

$$\frac{dy}{dt} = \frac{10}{640} = \frac{1}{64} \text{ ft/min}$$

21.   A ladder 25 ft long is leaning against a house. If the base of the ladder is pulled away from the house wall at a rate of 2 ft/s, how fast is the top moving down the wall when the base of the ladder is:

(a)   7 ft from the wall?

(b)   15 ft from the wall?

(c)   24 ft from the wall?

Solution

From the figure we see that x and y are related by the equation

$$x^2 + y^2 = (25)^2$$

Differentiating this equation with respect to t, we have

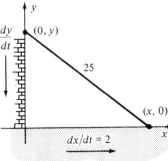

$$2x \frac{dx}{dt} + 2y \frac{dy}{dt} = 0$$

$$y \frac{dy}{dt} = -x \frac{dx}{dt}$$

$$\frac{dy}{dt} = -\frac{x}{y} \frac{dx}{dt}$$

(a) When $x = 7$, $y = \sqrt{(25)^2 - 7^2} = \sqrt{576} = 24$. Since $dx/dt = 2$, we have

$$\frac{dy}{dt} = -\frac{7}{24}(2) = -\frac{7}{12} \approx -0.583 \text{ ft/s}$$

(b) When $x = 15$, $y = \sqrt{(25)^2 - (15)^2} = \sqrt{400} = 20$. Thus

$$\frac{dy}{dt} = -\frac{15}{20}(2) = -\frac{3}{2} = -1.5 \text{ ft/s}$$

(c) When $x = 24$, $y = \sqrt{(25)^2 - (24)^2} = \sqrt{49} = 7$. Thus

$$\frac{dy}{dt} = -\frac{24}{7}(2) = -\frac{48}{7} \approx -6.857 \text{ ft/s}$$

27. A man 6 ft tall walks at a rate of 5 ft/s away from a light that is 15 ft above the ground. When the man is 10 ft from the base of the light:

(a) at what rate is the tip of his shadow moving?

(b) at what rate is the length of his shadow changing?

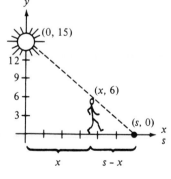

Solution

From the figure we see that x and s are related by similar triangles in such a way that

$$\frac{s - x}{6} = \frac{s}{15}$$

$$15s - 15x = 6s$$

$$9s = 15x$$

$$s = \frac{5}{3}x$$

(a) To find ds/dt, given that $dx/dt = 5$, we differentiate with respect to t as follows:

$$\frac{ds}{dt} = \frac{5}{3} \cdot \frac{dx}{dt} = \frac{5}{3}(5) = \frac{25}{3} \approx 8.3 \text{ ft/s}$$

(b)   The rate at which the shadow is increasing is

$$\frac{ds}{dt} - \frac{dx}{dt} = \frac{25}{3} - 5 = \frac{10}{3} \approx 3.3 \text{ ft/s}$$

(Note: The measurement 10 ft given in this problem is a "red herring" since the distance from the base of the light does not affect ds/dt.)

28.   An airplane is flying at 6 miles above ground level and the flight path passes directly over a radar antenna. The radar picks up the plane and determines that the distance s from the unit to the plane is changing at the rate of 4 mi/min when that distance is 10 mi. Compute the speed of the plane in miles per hour.

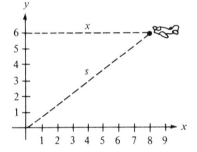

Solution

From the figure, we have

$$x^2 + y^2 = s^2 \qquad \text{or} \qquad x = \sqrt{s^2 - y^2}$$

We are given

$$\frac{ds}{dt} = 4 \text{ mi/min} = \frac{4 \text{ mi}}{\text{min}}\left(\frac{60 \text{ min}}{\text{h}}\right) = 240 \text{ mi/h}$$

Furthermore, y is always 6 mi. Therefore,

$$x = \sqrt{s^2 - 36}$$

$$\frac{dx}{dt} = \left(\frac{1}{2}\right)(s^2 - 36)^{-1/2}\left(2s\frac{ds}{dt}\right) = \frac{s}{\sqrt{s^2 - 36}} \cdot \frac{ds}{dt}$$

Finally, when $s = 10$,

$$\frac{dx}{dt} = \frac{10}{\sqrt{10^2 - 36}}(240) = \frac{10}{8}(240) = 300 \text{ mi/h}$$

35.   A wheel of radius 1 ft revolves at a rate of 10 rev/s. A dot is painted at a point P on the rim of the wheel as shown in the accompanying figure. Find the rate of horizontal movement of the dot for the following angles:

(a) $\theta = 0°$ (b) $\theta = 30°$ (c) $\theta = 60°$

Solution

From the figure, we have

$$x = \cos \theta$$

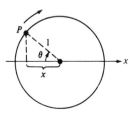

We are given

$$\frac{d\theta}{dt} = 10 \text{ rev/s} = 10(-2\pi) = -20\pi \text{ rad/s}$$

Therefore,

$$\frac{dx}{dt} = -\sin \theta \frac{d\theta}{dt}$$

$$= -\sin \theta(-20\pi) = 20\pi \sin \theta$$

(a) When $\theta = 0°$, $\dfrac{dx}{dt} = 0$.

(b) When $\theta = 30°$, $\dfrac{dx}{dt} = 20\pi\left(\dfrac{1}{2}\right) = 10\pi$ ft/s.

(c) When $\theta = 60°$, $\dfrac{dx}{dt} = 20\pi\left(\dfrac{\sqrt{3}}{2}\right) = 10\pi\sqrt{3}$ ft/s.

Review Exercises for Chapter 3

9. Find the derivative of $f(x) = (3x^2 + 7)(x^2 - 2x + 3)$.

Solution

$$f(x) = (3x^2 + 7)(x^2 - 2x + 3)$$
$$f'(x) = (3x^2 + 7)(2x - 2) + (x^2 - 2x + 3)(6x)$$
$$= 6x^3 - 6x^2 + 14x - 14 + 6x^3 - 12x^2 + 18x$$
$$= 2(6x^3 - 9x^2 + 16x - 7)$$

14. Find the derivative of $f(x) = (6x - 5)/(x^2 + 1)$.

Solution

$$f(x) = \frac{6x - 5}{x^2 + 1}$$

$$f'(x) = \frac{(x^2 + 1)(6) - (6x - 5)(2x)}{(x^2 + 1)^2}$$

$$= \frac{6x^2 + 6 - 12x^2 + 10x}{(x^2 + 1)^2}$$

$$= \frac{-6x^2 + 10x + 6}{(x^2 + 1)^2} = \frac{2(-3x^2 + 5x + 3)}{(x^2 + 1)^2}$$

27. Find the derivative of $y = \tan x - x$.

Solution

$$y = \tan x - x$$
$$y' = \sec^2 x - 1 = \tan^2 x$$

35. Find the derivative of $y = \dfrac{\sin x}{x^2}$.

Solution

$$y = \frac{\sin x}{x^2}$$

$$y' = \frac{x^2 \cos x - (\sin x)(2x)}{x^4}$$

$$= \frac{x(x \cos x - 2 \sin x)}{x^4} = \frac{x \cos x - 2 \sin x}{x^3}$$

41. Find the second derivative of $f(x) = \sqrt{x^2 + 9}$.

Solution

$$f(x) = \sqrt{x^2 + 9} = (x^2 + 9)^{1/2}$$

$$f'(x) = \left(\frac{1}{2}\right)(x^2 + 9)^{-1/2}(2x) = \frac{x}{(x^2 + 9)^{1/2}}$$

$$f''(x) = \frac{(x^2 + 9)^{1/2}(1) - (x)(\frac{1}{2})(x^2 + 9)^{-1/2}(2x)}{x^2 + 9}$$

$$= \left[\frac{(x^2 + 9)^{1/2} - x^2(x^2 + 9)^{-1/2}}{x^2 + 9}\right]\left[\frac{(x^2 + 9)^{1/2}}{(x^2 + 9)^{1/2}}\right]$$

$$= \frac{(x^2 + 9 - x^2)}{(x^2 + 9)^{3/2}} = \frac{9}{(x^2 + 9)^{3/2}}$$

45. Find the second derivative of $g(x) = (6x - 5)/(x^2 + 1)$.

Solution

$$g(x) = \frac{6x - 5}{x^2 + 1}$$

$$g'(x) = \frac{2(-3x^2 + 5x + 3)}{(x^2 + 1)^2} \qquad \text{See Exercise 14}$$

$$g''(x) = 2\left[\frac{(x^2 + 1)^2(-6x + 5) - (-3x^2 + 5x + 3)(2)(x^2 + 1)(2x)}{(x^2 + 1)^4}\right]$$

$$= 2\left[\frac{(x^2 + 1)[(x^2 + 1)(-6x + 5) - (4x)(-3x^2 + 5x + 3)]}{(x^2 + 1)^4}\right]$$

$$= 2\left[\frac{-6x^3 + 5x^2 - 6x + 5 + 12x^3 - 20x^2 - 12x}{(x^2 + 1)^3}\right]$$

$$= 2\left[\frac{6x^3 - 15x^2 - 18x + 5}{(x^2 + 1)^3}\right]$$

51. Use implicit differentiation to find dy/dx for
    $x^2 + 3xy + y^3 = 10$.

Solution

$$x^2 + 3xy + y^3 = 10$$

$$2x + 3xy' + 3y + 3y^2y' = 0$$

$$(3x + 3y^2)y' = -2x - 3y$$

$$y' = \frac{-(2x + 3y)}{3(x + y^2)}$$

63. Find the equations of the tangent line and the normal line to
    the graph of $y = \sqrt[3]{(x - 2)^2}$ at $(3, 1)$.

Solution

$$y = \sqrt[3]{(x - 2)^2} = (x - 2)^{2/3}$$

$$y' = \left(\frac{2}{3}\right)(x - 2)^{-1/3} = \frac{2}{3(x - 2)^{1/3}}$$

Thus the slope of the tangent line at $(3, 1)$ is

$$y' = \frac{2}{3(3 - 2)^{1/3}} = \frac{2}{3}$$

and the equation of the tangent line at $(3, 1)$ is

$$y - 1 = \frac{2}{3}(x - 3)$$

$$3y - 3 = 2x - 6$$

$$-2x + 3y + 3 = 0$$

Since the slope of the tangent line is $\frac{2}{3}$, the slope of the normal
line is $-\frac{3}{2}$ and the equation of the normal line is

$$y - 1 = -\frac{3}{2}(x - 3)$$

$$2y - 2 = -3x + 9$$

$$3x + 2y - 11 = 0$$

77. What is the smallest initial velocity that is required to throw a stone to the top of a 49-ft silo?

Solution

We assume that the stone is thrown from an initial height of $s_0 = 0$. Thus the position equation is

$$s = -16t^2 + v_0 t$$

The maximum value of s occurs when $ds/dt = 0$ and thus we have

$$\frac{ds}{dt} = -32t + v_0 = 0$$

$$-32t = -v_0$$

$$t = \frac{v_0}{32}$$

This means that the maximum height is

$$s = -16\left(\frac{v_0}{32}\right)^2 + v_0\left(\frac{v_0}{32}\right) = \frac{v_0^2}{64}$$

If s is to attain a value of 49, we must have

$$\frac{v_0^2}{64} = 49$$

$$v_0^2 = 3136$$

$$v_0 = 56 \text{ ft/s}$$

81. The path of a projectile, thrown at an angle of 45° with the ground, is given by $y = x - (32/v_0^2)(x^2)$, where the initial velocity is v_0 feet per second. Show that doubling the initial velocity of the projectile multiplies both the maximum height and the range by a factor of 4.

Solution

The maximum height occurs when the rate of change of y with respect to x is zero.

$$\frac{dy}{dx} = 1 - \left(\frac{64}{v_0^2}\right)x = 0$$

when $x = v_0^2/64$. Therefore, the maximum height of the projectile is

$$y = \frac{v_0{}^2}{64} - \left(\frac{32}{v_0{}^2}\right)\left(\frac{v_0{}^2}{64}\right)^2 = \frac{v_0{}^2}{64} - \frac{v_0{}^2}{128} = \frac{v_0{}^2}{128}$$

If the initial velocity is doubled, the maximum height is

$$\frac{(2v_0)^2}{128} = 4\left(\frac{v_0{}^2}{128}\right)$$

or four times the original maximum height.

The projectile is at ground level when $y = 0$ or

$$x - \left(\frac{32}{v_0{}^2}\right)x^2 = x\left[1 - \left(\frac{32}{v_0{}^2}\right)x\right] = 0$$

This occurs when $x = 0$ or $x = v_0{}^2/32$. Therefore, the range is $v_0{}^2/32$. When the initial velocity is doubled, the range is

$$\frac{(2v_0)^2}{32} = 4\left(\frac{v_0{}^2}{32}\right)$$

or four times the original range.

85. The cross section of a 5-ft trough is an isosceles trapezoid with lower base 2 ft, upper base 3 ft, and altitude 2 ft. Water is running into the trough at the rate of 1 ft³/min. How fast is the water level rising when the water is 1 ft deep?

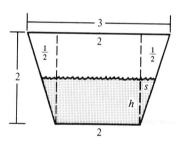

Solution

The figure is a cross section of the trough when the water is at a depth of h feet.

$$\frac{s}{h} = \frac{\frac{1}{2}}{2}$$

$$s = \frac{1}{4}h$$

A = area of cross section of water at depth h

$$= 2h + 2\left(\frac{1}{2}sh\right)$$

$$= 2h + \left(\frac{1}{4}h\right)h = 2h + \frac{1}{4}h^2$$

V = volume of water in trough at depth h

$$= 5A = 5\left(2h + \frac{1}{4}h^2\right)$$

Differentiating with respect to t, we have

$$\frac{dV}{dt} = 5\left(2 + \frac{1}{2}h\right)\frac{dh}{dt} = \frac{5}{2}(4 + h)\frac{dh}{dt}$$

$$\frac{2\,(dV/dt)}{5(4 + h)} = \frac{dh}{dt}$$

Therefore, when $dV/dt = 1$ and $h = 1$,

$$\frac{dh}{dt} = \frac{2(1)}{5(4 + 1)} = \frac{2}{25} \text{ ft/min}$$

4 Applications of differentiation

Extrema on an interval

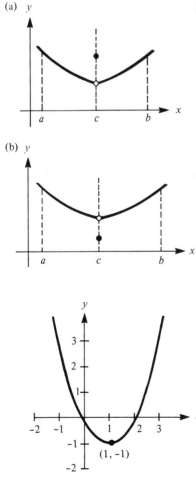

(a) y

(b) y

3. Determine from the graph of f if f possesses a relative minimum in the interval (a, b).

Solution

(a) Since f is decreasing in the interval (a, c) and increasing in the interval (c, b), the only possible minimum in the interval (a, b) would occur at $x = c$. However, $f(c)$ is greater than $f(x)$ for x near c. Thus, there is no minimum.

(b) Since $f(c) \leq f(x)$ for all x in (a, b), $f(c)$ is a minimum.

7. Locate the extrema of $f(x) = x^2 - 2x$ over the following intervals:

(a) $[-1, 2]$ (b) $(1, 3]$

(c) $(0, 2)$ (d) $[1, 3]$

Solution

To find the critical numbers, we differentiate to obtain

$$f'(x) = 2x - 2 = 2(x - 1) = 0$$

$$x = 1 \quad \text{Critical number}$$

We determine the extrema of f by evaluating f at the critical number and at the closed endpoints(s) of each interval.

(a) On the interval $[-1, 2]$ we have

$$f(-1) = (-1)^2 - 2(-1) = 3 \quad \text{Maximum}$$
$$f(1) = 1^2 - 2(1) = -1 \quad \text{Minimum}$$
$$f(2) = 2^2 - 2(2) = 0$$

(b) On the interval $(1, 3]$ we have

$$f(3) = 3^2 - 2(3) = 3 \qquad \text{Maximum}$$

(c) On the interval $(0, 2)$ we have

$$f(1) = -1 \text{ (from part a)} \qquad \text{Minimum}$$

(d) On the interval $[1, 3]$ we have

$$f(1) = -1 \text{ (from part a)} \qquad \text{Minimum}$$
$$f(3) = 3 \text{ (from part b)} \qquad \text{Maximum}$$

19. Locate the extrema of $f(x) = -x^2 + 3x$ on $[0, 3]$.

Solution

$$f(x) = -x^2 + 3x$$
$$f'(x) = -2x + 3 = 0$$

Therefore, $x = \frac{3}{2}$ is a critical number in $[0, 3]$.

We determine the extrema of f by evaluating f at the critical number and at the endpoints of $[0, 3]$.

$$f(0) = -0^2 + 3(0) = 0 \qquad \text{Minimum}$$
$$f\left(\frac{3}{2}\right) = -\left(\frac{3}{2}\right)^2 + 3\left(\frac{3}{2}\right) = \frac{9}{4} \qquad \text{Maximum}$$
$$f(3) = -3^2 + 3(3) = 0 \qquad \text{Minimum}$$

31. The error estimate for the Trapezoid Rule (see Section 8.9) involves the maximum of the absolute value of the second derivative in an interval. Find the maximum value of $|f''(x)|$ for $f(x) = \sqrt{1 + x^3}$ in $[0, 2]$.

Solution

To find the maximum value of $|f''(x)|$ in $[0, 2]$, we select the maximum of $|f''(0)|$, $|f''(2)|$, and $|f''(c)|$, where c is any critical number of $f''(x)$ [i.e., $f'''(c) = 0$ or $f'''(c)$ does not exist].

$$f(x) = \sqrt{1 + x^3} = (1 + x^3)^{1/2}$$

$$f'(x) = \frac{1}{2}(1 + x^3)^{-1/2}(3x^2) = \frac{3}{2}x^2(1 + x^3)^{-1/2}$$

$$f''(x) = \frac{3}{2}\left[x^2\left(-\frac{1}{2}\right)(1 + x^3)^{-3/2}(3x^2) + 2x(1 + x^3)^{-1/2}\right]$$

$$= \frac{3}{2}\left[\frac{3x^4}{2(1 + x^3)^{3/2}} + \frac{2x}{(1 + x^3)^{1/2}}\right] = \frac{3x(x^3 + 4)}{4(1 + x^3)^{3/2}}$$

$$f'''(x) = \frac{3}{4}\left[\frac{(1 + x^3)^{3/2}(4x^3 + 4) - (x^4 + 4x)(\frac{3}{2})(1 + x^3)^{1/2}(3x^2)}{(1 + x^3)^3}\right]$$

$$= \frac{3(1 + x^3)^{1/2}[(1 + x^3)(4x^3 + 4) - \frac{3}{2}(x^4 + 4x)(3x^2)]}{4(1 + x^3)^3}$$

$$= \frac{-3(x^6 + 20x^3 - 8)}{8(1 + x^3)^{5/2}}$$

Therefore, $f'''(x) = 0$ when

$$x^6 + 20x^3 - 8 = 0$$
$$(x^3)^2 + 20(x^3)^1 - 8 = 0$$
$$x^3 = \frac{-20 \pm \sqrt{(20)^2 - 4(1)(-8)}}{2(1)} \qquad \text{Quadratic Formula}$$
$$x = \sqrt[3]{\frac{-20 \pm \sqrt{432}}{2}}$$
$$= \sqrt[3]{\frac{-20 + \sqrt{432}}{2}}$$

which is the critical number in $[0, 2]$.

$$f''\left(\sqrt[3]{\frac{-20 + \sqrt{432}}{2}}\right) \approx 1.47$$

$$|f''(0)| = 0$$

$$|f''(2)| = \frac{2}{3}$$

The maximum value of $|f''(x)|$ in $[0, 2]$ is

$$\left|f''\left(\sqrt[3]{\frac{-20 + \sqrt{432}}{2}}\right)\right| \approx |f''(0.732)| \approx 1.47$$

35. The error estimate for Simpson's Rule (see Section 8.9) in-
     volves the maximum of the absolute value of the fourth deriva-
     tive in an interval. Find the maximum value of $|f^{(4)}(x)|$ for
     $f(x) = (x + 1)^{2/3}$ in $[0, 2]$.

Solution

To find the maximum value of $|f^{(4)}(x)|$ in $[0, 2]$, we select the
maximum of $|f^{(4)}(0)|$, $|f^{(4)}(2)|$, and $|f^{(4)}(c)|$, where c is any
critical number of $f^{(4)}(x)$ [i.e., $f^{(5)}(c) = 0$ or does not exist].

$$f(x) = (x + 1)^{2/3}$$

$$f'(x) = \frac{2}{3}(x + 1)^{-1/3}$$

$$f''(x) = -\frac{2}{9}(x + 1)^{-4/3}$$

$$f'''(x) = \frac{8}{27}(x + 1)^{-7/3}$$

$$f^{(4)}(x) = -\frac{56}{81}(x + 1)^{-10/3} = \frac{-56}{81(x + 1)^{10/3}}$$

$$f^{(5)}(x) = \frac{560}{243}(x + 1)^{-13/3} = \frac{560}{243(x + 1)^{13/3}}$$

We observe that $f^{(5)}(x)$ exists and is positive for all x in $[0, 2]$. Therefore, the maximum value of $|f^{(4)}(x)|$ occurs at one of the endpoints. The maximum value is

$$|f^{(4)}(0)| = \left|\frac{-56}{81}\right| = \frac{56}{81}$$

4.2
The Mean Value Theorem

3. For $f(x) = x^3 - 6x^2 + 11x - 6$, find all intervals $[a, b]$ on which $f(a) = f(b) = 0$ and Rolle's Theorem can be applied. For each such interval find all values of c such that $f'(c) = 0$.

Solution

Since f is a polynomial, it is continuous and differentiable for all x. To find the values of x such that $f(x) = x^3 - 6x^2 + 11x - 6 = 0$, we note that $f(1) = 0 - 6 + 11 - 6 = 0$. Dividing $(x - 1)$ into $x^3 - 6x^2 + 11x - 6$, we have

$$(x - 1)(x^2 - 5x + 6) = 0$$
$$(x - 1)(x - 2)(x - 3) = 0$$
$$x = 1, 2, 3$$

Thus the intervals on which Rolle's Theorem can be applied are $[1, 2]$ and $[2, 3]$. Setting $f'(x) = 0$ yields

$$f'(x) = 3x^2 - 12x + 11 = 0$$

$$x = \frac{12 \pm \sqrt{144 - 132}}{6} = \frac{12 \pm 2\sqrt{3}}{6} = \frac{6 \pm \sqrt{3}}{3}$$

Therefore, in the interval $[1, 2]$,

$$f'\left(\frac{6 - \sqrt{3}}{3}\right) = 0 \qquad \text{where} \qquad \frac{6 - \sqrt{3}}{3} \approx 1.423$$

and in the interval $[2, 3]$,

$$f'\left(\frac{6 + \sqrt{3}}{3}\right) = 0 \qquad \text{where} \qquad \frac{6 + \sqrt{3}}{3} \approx 2.577$$

7. For $f(x) = x^{2/3} - 1$, find all intervals $[a, b]$ on which $f(a) = f(b) = 0$ and Rolle's Theorem can be applied. For each such interval find all values of c such that $f'(c) = 0$.

Solution

$$f(x) = x^{2/3} - 1 = 0$$
$$x^{2/3} = 1$$
$$x^2 = 1$$
$$x = \pm 1$$

Since $f'(x) = 2/3x^{1/3}$ is not differentiable on the interval $[-1, 1]$ (in particular, $f'(0)$ is undefined), we cannot apply Rolle's Theorem to this function.

19. For $f(x) = \tan x$ find all intervals $[a, b]$ on which $f(a) = f(b) = 0$ and Rolle's Theorem can be applied. For each such interval find all values of c such that $f'(c) = 0$.

Solution

If $x = n\pi$ (n an integer), then $f(x) = \tan x = 0$. However, since f is discontinuous for $x = (2n + 1)\pi/2$, we cannot apply Rolle's Theorem to this function.

23. Apply the Mean Value Theorem to $f(x) = x^{2/3}$ on $[0, 1]$. Find all values of c in $[0, 1]$ such that

$$f'(c) = \frac{f(1) - f(0)}{1 - 0}$$

Solution

Since $f(x) = x^{2/3}$ is continuous for all x and differentiable for all x other than $x = 0$, we can apply the Mean Value Theorem on the interval $[0, 1]$.

$$f(x) = x^{2/3}$$
$$f'(x) = \frac{2}{3x^{1/3}} = \frac{f(1) - f(0)}{1 - 0}$$
$$\frac{2}{3x^{1/3}} = \frac{1 - 0}{1 - 0}$$

$$\frac{2}{3x^{1/3}} = 1$$

$$\frac{2}{3} = x^{1/3}$$

$$\frac{2^3}{3^3} = x$$

Therefore, $c = \frac{8}{27}$.

25. Apply the Mean Value Theorem to $f(x) = x/(x + 1)$ on $[-\frac{1}{2}, 2]$. Find all values of c in $[-\frac{1}{2}, 2]$ such that

$$f'(c) = \frac{f(2) - f(-\frac{1}{2})}{2 - (-\frac{1}{2})}$$

Solution

Since $f(x) = x/(x + 1)$ is continuous and differentiable for all x other than $x = -1$, we can apply the Mean Value Theorem on $[-\frac{1}{2}, 2]$.

$$f(x) = \frac{x}{x + 1}$$

$$f'(x) = \frac{(x + 1)(1) - x(1)}{(x + 1)^2} = \frac{f(2) - f(-\frac{1}{2})}{2 - (-\frac{1}{2})}$$

$$\frac{1}{(x + 1)^2} = \frac{(\frac{2}{3}) - (-1)}{\frac{5}{2}} = \frac{2}{3}$$

$$(x + 1)^2 = \frac{3}{2}$$

$$x + 1 = \pm\frac{\sqrt{3}}{\sqrt{2}}$$

$$x = -1 \pm \frac{\sqrt{6}}{2}$$

Therefore, on the interval $[-\frac{1}{2}, 2]$, $c = -1 + (\sqrt{6}/2)$.

37. Show that $x^{2n+1} + ax + b$ cannot have two real roots, where $a > 0$ and n is any positive integer.

Solution

The polynomial $f(x) = x^{2n+1} + ax + b$ is continuous and differentiable for all x. Therefore, by Rolle's Theorem, if $f(x) = 0$ for two distinct values of x, there would have to be at least one value of x such that $f'(x) = 0$. However,

$$f(x) = x^{2n+1} + ax + b$$
$$f'(x) = (2n + 1)x^{2n} + a = 0$$
$$(x^n)^2 = -\frac{a}{2n + 1}$$

(positive number) = (negative number)

Therefore, $f'(x) = 0$ has no solution and consequently $f(x) = 0$ cannot have two real zeros.

4.3
Increasing and decreasing functions and the First-Derivative Test

11. Find the critical numbers (if any) of $f(x) = 2x^3 + 3x^2 - 12x$, the intervals on which f is increasing or decreasing, and locate all relative extrema.

Solution

$$f(x) = 2x^3 + 3x^2 - 12x$$
$$f'(x) = 6x^2 + 6x - 12 = 0$$
$$6(x^2 + x - 2) = 0$$
$$6(x + 2)(x - 1) = 0$$
$$x = -2, 1$$

Since f is a polynomial, it is differentiable for all x and the only critical numbers are $x = -2$ and $x = 1$.

Interval	$(-\infty, -2)$	$(-2, 1)$	$(1, \infty)$
Test value	$x = -3$	$x = 0$	$x = 2$
Sign of $f'(x)$	$f'(-3) = 6(9 - 3 - 2)$ $= 24 > 0$	$f'(0) = -12$ < 0	$f'(2) = 6(4 + 2 - 2)$ $= 24 > 0$
Conclusion	f is increasing	f is decreasing	f is increasing

When $x = -2$, we have $f(-2) = 2(-2)^3 + 3(-2)^2 - 12(-2) = 20$.
When $x = 1$, we have $f(1) = 2 + 3 - 12 = -7$.
Therefore, we conclude that $(-2, 20)$ is a relative maximum and $(1, -7)$ is a relative minimum.

17. Find the critical numbers (if any) of $f(x) = x^{1/3} + 1$, the intervals on which f is increasing or decreasing, and locate all relative extrema.

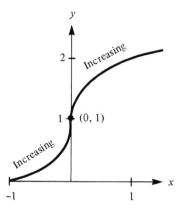

Solution

$$f(x) = x^{1/3} + 1$$

$$f'(x) = \left(\frac{1}{3}\right)x^{-2/3} = \frac{1}{3x^{2/3}} \neq 0$$

Since f is continuous for all x and differentiable for all x other than $x = 0$, the only critical number is $x = 0$.

Interval	$(-\infty, 0)$	$(0, \infty)$
Test value	$x = -1$	$x = 1$
Sign of $f'(x)$	$f'(-1) = \frac{1}{3} > 0$	$f'(1) = \frac{1}{3} > 0$
Conclusion	f is increasing	f is increasing

Therefore we conclude that f is increasing for all x and there are no relative extrema.

25. Locate all relative extrema for $f(x) = (x^2 - 2x + 1)/(x + 1)$.

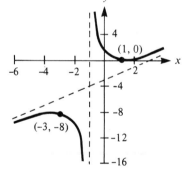

Solution

$$f(x) = \frac{x^2 - 2x + 1}{x + 1}$$

$$f'(x) = \frac{(x + 1)(2x - 2) - (x^2 - 2x + 1)(1)}{(x + 1)^2}$$

$$= \frac{(x + 1)(2)(x - 1) - (x - 1)^2}{(x + 1)^2}$$

$$= \frac{(x - 1)(2x + 2 - x + 1)}{(x + 1)^2} = \frac{(x - 1)(x + 3)}{(x + 1)^2} = 0$$

$$x = -3, 1$$

Since f is continuous and differentiable for all x other than $x = -1$, the only critical numbers are $x = 1$ and $x = -3$.

Interval	$(-\infty, -3)$	$(-3, -1)$	$(-1, 1)$	$(1, \infty)$
Test value	$x = -4$	$x = -2$	$x = 0$	$x = 2$
Sign of $f'(x)$	$f'(-4) = \frac{5}{9}$	$f'(-2) = -3$	$f'(0) = -3$	$f'(2) = \frac{5}{9}$
	> 0	< 0	< 0	> 0
Conclusion	f increasing	f decreasing	f decreasing	f increasing

When $x = -3$, we have $f(-3) = (9 + 6 + 1)/-2 = -8$, and when $x = 1$, we have $f(1) = (1 - 2 + 1)/2 = 0$. Therefore, we conclude that $(-3, -8)$ is a relative maximum and $(1, 0)$ is a relative minimum.

27. Find the critical numbers (if any) of $f(x) = (x/2) + \cos x$, $0 \le x < 2\pi$, the intervals where f is increasing or decreasing, and locate all relative extrema.

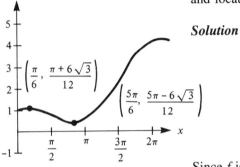

Solution

$$f(x) = \frac{x}{2} + \cos x$$

$$f'(x) = \frac{1}{2} - \sin x = 0$$

$$x = \frac{\pi}{6}, \frac{5\pi}{6}$$

Since f is differentiable for all x, the only critical numbers are $x = \pi/6$ and $5\pi/6$.

Interval	$(0, \pi/6)$	$(\pi/6, 5\pi/6)$	$(5\pi/6, 2\pi)$
Test value	$x = \pi/12$	$x = \pi/2$	$x = \pi$
Sign of $f'(x)$	$f'(\pi/12) > 0$	$f'(\pi/2) < 0$	$f'(\pi) > 0$
Conclusion	f increasing	f decreasing	f increasing

We conclude that a relative maximum occurs at
$$\left(\frac{\pi}{6}, \frac{\pi + 6\sqrt{3}}{12} \right) \text{ and a relative minimum at}$$
$$\left(\frac{5\pi}{6}, \frac{5\pi - 6\sqrt{3}}{12} \right).$$

39. The resistance R of a certain type of resistor is given by
$$R = \sqrt{0.001T^4 - 4T + 100}$$

where R is measured in ohms and the temperature T is measured in degress Celsius. What temperature produces a minimum resistance for this type of resistor?

Solution

$$R(T) = (0.001T^4 - 4T + 100)^{1/2}$$

$$R'(T) = \left(\frac{1}{2} \right)(0.001T^4 - 4T + 100)^{-1/2}(0.004T^3 - 4)$$

$$= \frac{0.004T^3 - 4}{2\sqrt{0.001T^4 - 4T + 100}}$$

We observe that $R'(T) = 0$ when

$$0.004T^3 - 4 = 0$$

$$T^3 = \frac{4}{0.004} = 1000$$

$$T = 10$$

Since R is continuous and differentiable for all T, the only critical number is $T = 10$.

Interval	$(-\infty, 10)$	$(10, \infty)$
Test value	$T = 0$	$T = 11$
Sign of $R'(T)$	$R'(0) < 0$	$R'(11) > 0$
Conclusion	R decreasing	R increasing

Therefore, at the temperature $T = 10°C$, the resistance is a minimum.

43. Find a, b, c, and d so that the function $f(x) = ax^3 + bx^2 + cx + d$ has a relative minimum at $(0, 0)$ and a relative maximum at $(2, 2)$.

Solution

In order for $(0, 0)$ and $(2, 2)$ to be solution points to f, we must have

$$f(0) = 0 = a(0^3) + b(0^2) + c(0) + d$$
$$0 = d$$
$$f(2) = 2 = a(2^3) + b(2^2) + c(2) + d$$
$$2 = 8a + 4b + 2c + 0$$
$$1 = 4a + 2b + c$$

Also, for f to have relative extrema at $(0, 0)$ and $(2, 2)$, $f'(x)$ must be zero at these points. Since $f'(x) = 3ax^2 + 2bx + c$, we have

$$f'(0) = 0 = 3a(0^2) + 2b(0) + c$$
$$0 = c$$
$$f'(2) = 0 = 3a(2^2) + 2b(2) + c$$

$$0 = 12a + 4b + 0$$

$$0 = 3a + b$$

Solving for a and b yields

$$4a + 2b + c = 1 \qquad 4a + 2b = 1$$

$$3a + b = 0 \qquad \underline{\quad 6a + 2b = 0\quad}$$

$$-2a \qquad = 1$$

$$a \qquad = -\frac{1}{2}$$

$$4\left(-\frac{1}{2}\right) + 2b = 1$$

$$2b = 3$$

$$b = \frac{3}{2}$$

Finally, we conclude that $a = -\frac{1}{2}$, $b = \frac{3}{2}$, $c = 0$, and $d = 0$. Thus, $f(x) = -\frac{1}{2}x^3 + \frac{3}{2}x^2$.

4.4
Concavity and the Second-Derivative Test

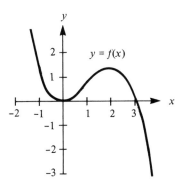

y

$y = f(x)$

3. Use the given graph of $f(x) = \frac{1}{3}x^3 + x^2$, to sketch the graph of f'. Find the intervals (if any) on which (a) f' is positive, (b) f' is negative, (c) f' is increasing, and (d) f' is decreasing. For each of these intervals describe the behavior of f.

Solution

$$f(x) = -\frac{1}{3}x^3 + x^2$$

$$f'(x) = -x^2 + 2x = x(2 - x) = 0 \qquad \text{when} \qquad x = 0, 2$$

$$f''(x) = -2x + 2 = 2(1 - x) = 0 \qquad \text{when} \qquad x = 1$$

Interval	$(-\infty, 0)$	$(0, 2)$	$(2, \infty)$
Test value	$x = -1$	$x = 1$	$x = 3$
Sign of f'	$f'(-1) < 0$	$f'(1) > 0$	$f'(3) < 0$
Conclusion	f decreasing	f increasing	f decreasing

To determine the intervals on which f' is increasing and the

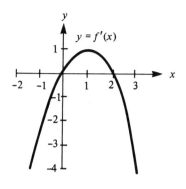

$y = f'(x)$

intervals on which f' is decreasing we determine the intervals on which f'' is positive and negative, respectively.

Interval	$(-\infty, 1)$	$(1, \infty)$
Test value	$x = 0$	$x = 2$
Sign of f''	$f''(0) > 0$	$f''(2) < 0$
Conclusion	f' increasing f concave upward	f' decreasing f concave downward

19.   Identify all relative extrema for $f(x) = x^3 - 3x^2 + 3$. Use the Second-Derivative Test when applicable.

Solution

$$f(x) = x^3 - 3x^2 + 3$$
$$f'(x) = 3x^2 - 6x = 0$$
$$3x(x - 2) = 0$$
$$x = 0, 2$$
$$f''(x) = 6x - 6$$

At $x = 0$, we have $f(0) = 3$, $f'(0) = 0$, and $f''(0) = -6$. Therefore, by the Second-Derivative Test, $(0, 3)$ is a relative maximum.

At $x = 2$, we have $f(2) = -2$, $f'(2) = 0$, and $f''(2) = 6$. Therefore, by the Second-Derivative Test, $(2, -1)$ is a relative minimum.

25.   Identify all relative extrema for $f(x) = x + (4/x)$. Use the Second-Derivative Test when applicable.

Solution

$$f(x) = x + \frac{4}{x}$$

$$f'(x) = 1 - \frac{4}{x^2} = 0$$

$$\frac{x^2 - 4}{x^2} = 0$$

$$x = \pm 2$$

$$f''(x) = (-4)(-2)x^{-3} = \frac{8}{x^3}$$

At $x = -2$, we have $f(-2) = -4$, $f'(-2) = 0$, and $f''(-2) = -1$. Therefore, by the Second-Derivative Test, $(-2, -4)$ is a relative maximum.

Since f is symmetrical with respect to the origin, $(2, 4)$ is a relative minimum.

31.   Sketch the graph of the function $f(x) = (x^4/4) - 2x^2$, and identify all relative extrema and points of inflection.

Solution

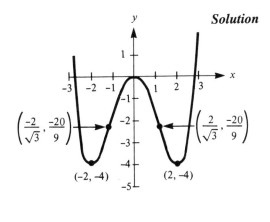

$$f(x) = \frac{x^4}{4} - 2x^2$$

$$f'(x) = x^3 - 4x = 0$$

$$x(x^2 - 4) = 0$$

$$x(x - 2)(x + 2) = 0$$

$$x = 0, \pm 2$$

$$f''(x) = 3x^2 - 4 = 0$$

$$x^2 = \frac{4}{3}$$

$$x = \pm \frac{2}{\sqrt{3}}$$

x	-2	$-2/\sqrt{3}$	0	$2/\sqrt{3}$	2
$f(x)$	-4	$-20/9$	0	$-20/9$	-4
$f'(x)$	0	$-16/3\sqrt{3}$	0	$16/3\sqrt{3}$	0
$f''(x)$	8	0	-4	0	8

Therefore:

$(-2, -4)$ is a relative minimum.
$(-2/\sqrt{3}, -20/9)$ is a point of inflection.
$(0, 0)$ is a relative maximum.
$(2/\sqrt{3}, -20/9)$ is a point of inflection.
$(2, -4)$ is a relative minimum.

35.   Sketch the graph of the function $f(x) = x\sqrt{x + 3}$, and identify all relative extrema and points of inflection.

Solution

$$f(x) = x\sqrt{x + 3} \qquad \text{domain: } [-3, \infty)$$

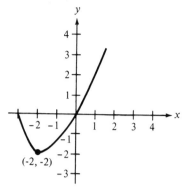

$$f'(x) = x\left(\frac{1}{2}\right)(x + 3)^{-1/2}(1) + (x + 3)^{1/2}(1) = 0$$

$$\frac{x}{2\sqrt{x + 3}} + \sqrt{x + 3} = 0$$

$$\frac{x + 2(x + 3)}{2\sqrt{x + 3}} = 0$$

$$\frac{3(x + 2)}{2\sqrt{x + 3}} = 0$$

$$x = -2$$

$$f''(x) = \frac{3}{2}\left[\frac{\sqrt{x + 3}\,(1) - (x + 2)(\frac{1}{2})(x + 3)^{-1/2}(1)}{x + 3}\right]$$

$$= \frac{3}{2}\left[\frac{\sqrt{x + 3} - (x + 2)/2\sqrt{x + 3}}{x + 3}\right]$$

$$= \frac{3(x + 4)}{4(x + 3)^{3/2}} > 0 \qquad \text{for all } x \text{ in } (-3, \infty)$$

We conclude that the graph is concave upward for each x in the domain of f and therefore that $(-2, -2)$ is a relative minimum.

48.   Sketch the graph of $f(x) = \cos x - x$ on the interval $[0, 4\pi]$, making use of relative extrema and points of inflection.

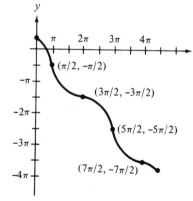

Solution

$$f(x) = \cos x - x$$

$$f'(x) = -\sin x - 1 = 0$$

$$\sin x = -1$$

$$x = \frac{3\pi}{2}, \frac{7\pi}{2} \qquad \text{in the interval } [0, 4\pi]$$

$$f''(x) = -\cos x = 0$$

$$x = \frac{\pi}{2}, \frac{3\pi}{2}, \frac{5\pi}{2}, \frac{7\pi}{2} \qquad \text{in the interval } [0, 4\pi]$$

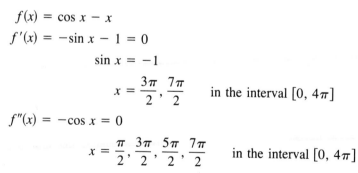

Interval	$f(x)$	$f'(x)$	$f''(x)$	Shape of graph
x in $[0, \pi/2)$		$-$	$-$	decreasing, concave down
$x = \pi/2$	$-\pi/2$	$-$	0	point of inflection
x in $(\pi/2, 3\pi/2)$		$-$	$+$	decreasing, concave up
$x = 3\pi/2$	$-3\pi/2$	0	0	point of inflection

Interval	$f(x)$	$f'(x)$	$f''(x)$	Shape of graph
x in $(3\pi/2, 5\pi/2)$		$-$	$-$	decreasing, concave down
$x = 5\pi/2$	$-5\pi/2$	$-$	0	point of inflection
x in $(5\pi/2, 7\pi/2)$		$-$	$+$	decreasing, concave up
$x = 7\pi/2$	$-7\pi/2$	0	0	point of inflection
x in $(7\pi/2, 4\pi]$		$-$	$-$	decreasing, concave down

51. Find a, b, c, and d so that the function $f(x) = ax^3 + bx^2 + cx + d$ has a relative maximum at $(3, 3)$, a relative minimum at $(5, 1)$, and a point of inflection at $(4, 2)$.

Solution

$$f(x) = ax^3 + bx^2 + cx + d$$
$$f'(x) = 3ax^2 + 2bx + c$$
$$f''(x) = 6ax + 2b = 2(3ax + b)$$

$f(3) = 3$ implies	$27a + 9b + 3c + d = 3$	(1)
$f(4) = 2$ implies	$64a + 16b + 4c + d = 2$	(2)
$f(5) = 1$ implies	$125a + 25b + 5c + d = 1$	(3)
$f'(3) = 0$ implies	$27a + 6b + c = 0$	(4)
$f'(5) = 0$ implies	$75a + 10b + c = 0$	(5)
$f''(4) = 0$ implies	$2(12a + b) = 0$	(6)

From (6) we get $b = -12a$. Substituting this into (5), we obtain

$$75a + 10(-12a) + c = 0$$
$$-45a + c = 0$$
$$c = 45a$$

Substitution into (1) and (2) yields

$$27a + 9(-12a) + 3(45a) + d = 3$$
$$54a + d = 3 \qquad (7)$$
$$64a + 16(-12a) + 4(45a) + d = 2$$
$$52a + d = 2 \qquad (8)$$

To solve simultaneously, we subtract (8) from (7) and obtain

$$2a = 1 \qquad a = \frac{1}{2}$$

From (7)

$$d = 3 - 54\left(\frac{1}{2}\right) = -24$$

$$c = 45\left(\frac{1}{2}\right) = \frac{45}{2}$$

$$b = -12\left(\frac{1}{2}\right) = -6$$

Therefore,

$$f(x) = \frac{1}{2}x^3 - 6x^2 + \frac{45}{2}x - 24 = \frac{1}{2}(x^3 - 12x^2 + 45x - 48)$$

54. Prove that a cubic function with three real zeros has a point of inflection whose x-coordinate is the average of the three zeros.

Solution

We assume the three zeros of the cubic are r_1, r_2, and r_3. Then

$$f(x) = a(x - r_1)(x - r_2)(x - r_3)$$
$$f'(x) = a[(x - r_1)(x - r_2) + (x - r_1)(x - r_3) + (x - r_2)(x - r_3)]$$
$$f''(x) = a[(x - r_1) + (x - r_2) + (x - r_1) + (x - r_3) + (x - r_2)$$
$$+ (x - r_3)]$$
$$= a[6x - 2(r_1 + r_2 + r_3)]$$

Consequently, $f''(x) = 0$ if

$$x = \frac{2(r_1 + r_2 + r_3)}{6} = \frac{r_1 + r_2 + r_3}{3} = \text{(average of } r_1, r_2 \text{ and } r_3)$$

58. Find the optimal order size if the total cost C for ordering and storing x units is $C = 2x + (300{,}000/x)$.

Solution

For this particular function the optimal order size would correspond to the value of x for which the cost, C, is a minimum.

$$C = 2x + \frac{300{,}000}{x}$$

$$\frac{dC}{dx} = 2 - \frac{300{,}000}{x^2} = 0$$

$$2 = \frac{300{,}000}{x^2}$$

$$x^2 = 150{,}000$$

$$x = 100\sqrt{15} \approx 387 \text{ units}$$

59. The graph of $y = x \sin (1/x)$ is shown in the accompanying figure. Show that the graph is concave downward to the right of $x = 1/\pi$.

Solution

$$f'(x) = -\frac{1}{x} \cos \left(\frac{1}{x}\right) + \sin \left(\frac{1}{x}\right)$$

$$f''(x) = -\frac{1}{x}\left[-\sin \left(\frac{1}{x}\right)\right]\left(-\frac{1}{x^2}\right) + \cos \left(\frac{1}{x}\right)\left(\frac{1}{x^2}\right) + \cos \left(\frac{1}{x}\right)\left(-\frac{1}{x^2}\right)$$

$$= -\frac{1}{x^3} \sin \left(\frac{1}{x}\right)$$

If $x > 1/\pi$, then $1/x < \pi$ and $\sin (1/x) > 0$. Since $x > 0$, we have

$$f''(x) = -\frac{1}{x^3} \sin \left(\frac{1}{x}\right) = (-)(+) < 0$$

Therefore, the graph is concave downward for $x > 1/\pi$.

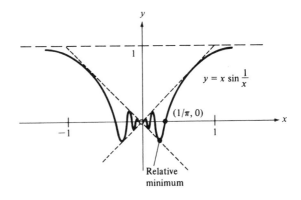

4.5
Limits at infinity

11. Evaluate $\lim\limits_{x \to \infty} \dfrac{x}{x^2 - 1}$.

Solution

$$\lim_{x \to \infty} \frac{x}{x^2 - 1} = \lim_{x \to \infty} \frac{x/x^2}{(x^2/x^2) - (1/x^2)} = \lim_{x \to \infty} \frac{1/x}{1 - (1/x^2)}$$

$$= \frac{0}{1 - 0} = 0$$

13.   Evaluate $\displaystyle\lim_{x \to -\infty} \frac{5x^2}{x + 3}$.

Solution

$$\lim_{x \to -\infty} \frac{5x^2}{x + 3} = \lim_{x \to -\infty} \frac{5x^2/x^2}{(x/x^2) + (3/x^2)}$$

$$= \lim_{x \to -\infty} \frac{5}{(1/x) + (3/x^2)} = -\infty$$

To determine that this is $-\infty$ rather than $+\infty$, we note that $5x^2/(x + 3) < 0$ whenever $x < -3$. (Note that the degree of the numerator is greater than the degree of the denominator. See Exercise 54 in this section.)

17.   Evaluate $\displaystyle\lim_{x \to -\infty} \left(\frac{2x}{x - 1} + \frac{3x}{x + 1} \right)$.

Solution

$$\lim_{x \to -\infty} \left(\frac{2x}{x - 1} + \frac{3x}{x + 1} \right) = \lim_{x \to -\infty} \left[\frac{2x/x}{(x/x) - (1/x)} + \frac{3x/x}{(x/x) + (1/x)} \right]$$

$$= \lim_{x \to -\infty} \left[\frac{2}{1 - (1/x)} + \frac{3}{1 + (1/x)} \right]$$

$$= \frac{2}{1 - 0} + \frac{3}{1 + 0} = 5$$

(Note that for each of the rational functions in this exercise, the denominator and numerator are of equal degree. See Exercise 54 in this section.)

21.   Evaluate $\displaystyle\lim_{x \to \infty} (x - \sqrt{x^2 + x})$.

Solution

$$\lim_{x \to \infty} (x - \sqrt{x^2 + x}) = \lim_{x \to \infty} (x - \sqrt{x^2 + x}) \frac{x + \sqrt{x^2 + x}}{x + \sqrt{x^2 + x}}$$

$$= \lim_{x \to \infty} \frac{x^2 - (x^2 + x)}{x + \sqrt{x^2 + x}}$$

$$= \lim_{x \to \infty} \frac{-x}{x + \sqrt{x^2 + x}}$$

$$= \lim_{x \to \infty} \frac{-x/x}{(x/x) + \sqrt{x^2 + x}/\sqrt{x^2}}$$

$$= \lim_{x \to \infty} \frac{-1}{1 + \sqrt{1 + (1/x)}}$$

$$= \frac{-1}{1 + \sqrt{1 + 0}} = -\frac{1}{2}$$

(Note: For $x > 0$, $x = \sqrt{x^2}$.)

23. Evaluate $\displaystyle\lim_{x \to -\infty} \frac{x}{\sqrt{x^2 - x}}$.

Solution

$$\lim_{x \to -\infty} \frac{x}{\sqrt{x^2 - x}} = \lim_{x \to -\infty} \frac{x/x}{\sqrt{x^2 - x}/(-\sqrt{x^2})} = \lim_{x \to -\infty} \frac{1}{-\sqrt{1 - (1/x)}}$$

$$= -\frac{1}{\sqrt{1 + 0}} = -1$$

(Note: For $x < 0$, $x = -\sqrt{x^2}$.)

34. Evaluate $\displaystyle\lim_{x \to \infty} x \tan \frac{1}{x}$.

Solution

If we let $x = 1/t$ and find the limit as $t \to 0^+$, we have

$$\lim_{x \to \infty} x \tan \frac{1}{x} = \lim_{t \to 0^+} \frac{\tan t}{t}$$

$$= \lim_{t \to 0^+} \left(\frac{\sin t}{t} \right) \left(\frac{1}{\cos t} \right) = 1$$

35. Sketch the graph of $y = (2 + x)/(1 - x)$. As a sketching aid examine the equation for intercepts, symmetry, and asymptotes.

Solution

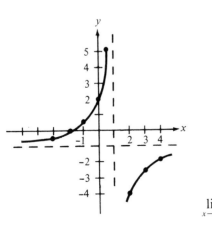

If $x = 0$, then $y = 2$ and the y-intercept occurs at $(0, 2)$. If $y = 0$, then $2 + x = 0$ and $x = -2$, so the x-intercept is $(-2, 0)$. There is no symmetry with respect to either axis or to the origin. Since the denominator of $(2 + x)/(1 - x)$ is zero when $x = 1$, there is a vertical asymptote at $x = 1$. Furthermore,

$$\lim_{x \to 1^-} \frac{2 + x}{1 - x} = \infty \quad \text{and} \quad \lim_{x \to 1^+} \frac{2 + x}{1 - x} = -\infty$$

$$\lim_{x \to \pm\infty} \frac{2 + x}{1 - x} = \lim_{x \to \pm\infty} \frac{(2/x) + (x/x)}{(1/x) - (x/x)} = \lim_{x \to \pm\infty} \frac{(2/x) + 1}{(1/x) - 1}$$

$$= \frac{0 + 1}{0 - 1} = -1$$

Therefore, there is a horizontal asymptote (to the right and left) at $y = -1$. To complete the graph, we add a few points as shown in the accompanying table.

x	-3	-1	0.5	2	3	4
y	-0.25	0.5	5	-4	-2.5	-2

By adding these points to the graph, we have the sketch shown.

45. Sketch the graph of $y = x^3/\sqrt{x^2 - 4}$. As a sketching aid examine the equation for intercepts, symmetry, and asymptotes.

Solution

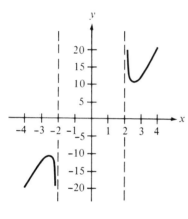

Since $x^2 - 4$ must be nonnegative, the domain is $(-\infty, -2)$ and $(2, \infty)$. There is no symmetry with respect to either axis. However, there is symmetry with respect to the origin since

$$(-y) = \frac{(-x)^3}{\sqrt{(-x)^2 - 4}}$$

$$-y = \frac{-x^3}{\sqrt{x^2 - 4}}$$

$$y = \frac{x^3}{\sqrt{x^2 - 4}}$$

which is equivalent to the original equation. Since the denominator of $x^3/\sqrt{x^2 - 4}$ is zero when $x = 2$ or $x = -2$, there are vertical asymptotes at $x = 2$ and $x = -2$.

x	2.25	2.50	2.75	3.00	4.00
y	11.05	10.45	11.02	12.07	18.48

4.6
A summary of curve sketching

9. Sketch the graph of $y = 3x^4 + 4x^3$, choosing a scale that allows all relative extrema and points of inflection to be identified on the sketch.

Solution

$$y = 3x^4 + 4x^3 = x^3(3x + 4) \qquad \text{Intercepts: } (0, 0), \left(-\tfrac{4}{3}, 0\right)$$

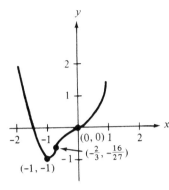

$$y' = 12x^3 + 12x^2 = 12x^2(x + 1)$$ Critical numbers: $x = 0$, $x = -1$

$$y'' = 36x^2 + 24x = 12x(3x + 2)$$ Possible points of inflection:
$$(-\tfrac{2}{3}, -\tfrac{16}{27})\ (0, 0)$$

x	y	y'	y''	*Shape of graph*
x in $(-\infty, -1)$		$-$	$+$	decreasing, concave up
$x = -1$	-1	0	$+$	relative minimum
x in $(-1, -\tfrac{2}{3})$		$+$	$+$	increasing, concave up
$x = -\tfrac{2}{3}$	$-\tfrac{16}{27}$	$+$	0	point of inflection
x in $(-\tfrac{2}{3}, 0)$		$+$	$-$	increasing, concave down
$x = 0$	0	0	0	point of inflection
x in $(0, \infty)$		$+$	$+$	increasing, concave up

18. Sketch the graph of $y = |x^2 - 6x + 5|$, choosing a scale that allows all relative extrema and points of inflection to be identified on the sketch.

Solution

Rather than find y' and y'', we begin by sketching $u = x^2 - 6x + 5$.

$$u = x^2 - 6x + 5 = (x - 5)(x - 1)$$

$$u' = 2x - 6 = 2(x - 3)$$ Critical number: $x = 3$

$$u'' = 2$$

Therefore $(3, -4)$ is a minimum.

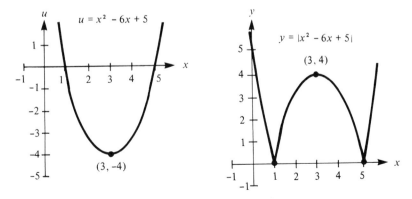

The graph of u is given above. Now by reflecting in the x-axis the portion of the graph that lies below the x-axis, we have the graph of $y = |u| = |x^2 - 6x + 5|$.

21. Sketch the graph of $y = 3x^{2/3} - 2x$, choosing a scale that allows all relative extrema and points of inflection to be identified on the sketch.

Solution

$$y = 3x^{2/3} - 2x = x^{2/3}(3 - 2x^{1/3})$$ Intercepts: $(0, 0)$, $(\frac{27}{8}, 0)$

$$y' = 2x^{-1/3} - 2 = \frac{2(1 - x^{1/3})}{x^{1/3}}$$ Critical numbers: $x = 0$, $x = 1$

$$y'' = \left(\frac{-1}{3}\right)(2)x^{-4/3} = \frac{-2}{3x^{4/3}}$$

x	y	y'	y''	*Shape of graph*
x in $(-\infty, 0)$		$-$	$-$	decreasing, concave down
$x = 0$	0	undefined	undefined	relative minimum
x in $(0, 1)$		$+$	$-$	increasing, concave down
$x = 1$	1	0	$-$	relative maximum
x in $(1, \infty)$		$-$	$-$	decreasing, concave down

25. Sketch the graph of $y = \sin x - \frac{1}{3} \sin 3x$ on the interval $[0, 2\pi]$, choosing a scale that allows all relative extrema and points of inflection to be identified on the sketch.

Solution

$$y = \sin x - \frac{1}{3} \sin 3x$$

$$y' = \cos x - \cos 3x$$

$$= \cos x - (4 \cos^3 x - 3 \cos x)$$

$$= 4 \cos x(1 - \cos^2 x)$$ Critical numbers: $x = \pi/2$, $x = \pi$, $x = 3\pi/2$

$$y'' = 4 \sin x(3 \cos^2 x - 1) = 0$$

when $\sin x = 0$ or $\cos x = \pm\frac{\sqrt{3}}{3}$

Possible points of inflection: $(0.955, 0.726)$, $(2.186, 0.726)$, $(\pi, 0)$, $(4.097, -0.726)$, $(5.328, -0.726)$

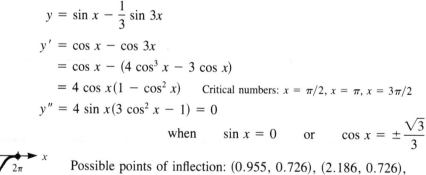

x	y	y'	y''	*Shape of graph*
x in $(0, 0.955)$		$+$	$+$	increasing, concave up
$x = 0.955$	0.726	$+$	0	point of inflection

x	y	y'	y''	Shape of graph
x in $(0.955, \pi/2)$		$+$	$-$	increasing, concave down
$x = \pi/2$	$4/3$	0	$-$	relative maximum
x in $(\pi/2, 2.186)$		$-$	$-$	decreasing, concave down
$x = 2.186$	0.726	$-$	0	point of inflection
x in $(2.186, \pi)$		$-$	$+$	decreasing, concave up
$x = \pi$	0	0	0	point of inflection
x in $(\pi, 4.097)$		$-$	$-$	decreasing, concave down
$x = 4.097$	-0.726	$-$	0	point of inflection
x in $(4.097, 3\pi/2)$		$-$	$+$	decreasing, concave up
$x = 3\pi/2$	$-4/3$	0	$+$	relative minimum
x in $(3\pi/2, 5.328)$		$+$	$+$	increasing, concave up
$x = 5.328$	-0.726	$+$	0	point of inflection
x in $(5.328, 2\pi)$		$+$	$-$	increasing, concave down

29.   Sketch the graph of $y = [1/(x - 2)] - 3$. Label the intercepts, relative extrema, points of inflection, and the domain.

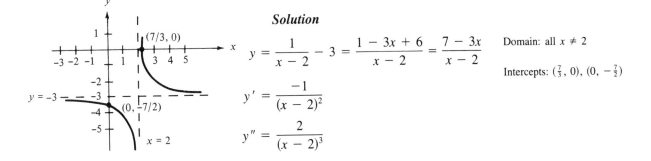

Solution

$$y = \frac{1}{x - 2} - 3 = \frac{1 - 3x + 6}{x - 2} = \frac{7 - 3x}{x - 2}$$

Domain: all $x \neq 2$

Intercepts: $(\frac{7}{3}, 0)$, $(0, -\frac{7}{2})$

$$y' = \frac{-1}{(x - 2)^2}$$

$$y'' = \frac{2}{(x - 2)^3}$$

x	y'	y''	Shape of graph
x in $(-\infty, 2)$	$-$	$-$	decreasing, concave down
x in $(2, \infty)$	$-$	$+$	decreasing, concave up

The graph has a vertical asymptote at $x = 2$ and a horizontal asymptote to the left and right at $y = -3$.

33.   Sketch the graph of $y = x\sqrt{4 - x}$. Label the intercepts, relative extrema, points of inflection, and the domain.

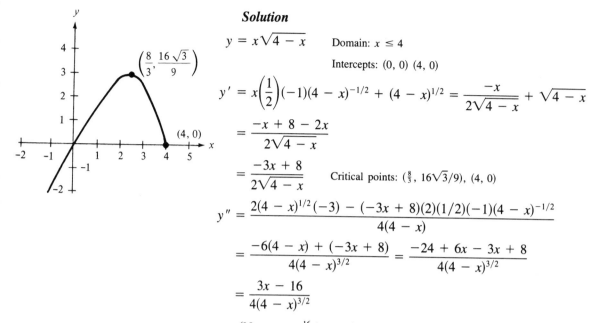

Solution

$$y = x\sqrt{4 - x} \qquad \text{Domain: } x \le 4$$

Intercepts: $(0, 0)$ $(4, 0)$

$$y' = x\left(\frac{1}{2}\right)(-1)(4 - x)^{-1/2} + (4 - x)^{1/2} = \frac{-x}{2\sqrt{4 - x}} + \sqrt{4 - x}$$

$$= \frac{-x + 8 - 2x}{2\sqrt{4 - x}}$$

$$= \frac{-3x + 8}{2\sqrt{4 - x}} \qquad \text{Critical points: } (\tfrac{8}{3}, 16\sqrt{3}/9), (4, 0)$$

$$y'' = \frac{2(4 - x)^{1/2}(-3) - (-3x + 8)(2)(1/2)(-1)(4 - x)^{-1/2}}{4(4 - x)}$$

$$= \frac{-6(4 - x) + (-3x + 8)}{4(4 - x)^{3/2}} = \frac{-24 + 6x - 3x + 8}{4(4 - x)^{3/2}}$$

$$= \frac{3x - 16}{4(4 - x)^{3/2}}$$

(Note: $x = \frac{16}{3}$ is not in the domain of the function.)

x	y	y'	y''	Shape of graph
x in $\left(-\infty, \dfrac{8}{3}\right)$		$+$	$-$	increasing, concave down
$x = \dfrac{8}{3}$	$\dfrac{16\sqrt{3}}{9}$	0	$-$	relative maximum
x in $\left(\dfrac{8}{3}, 4\right)$		$-$	$-$	decreasing, concave down
$x = 4$	0	undefined	undefined	

39. Sketch the graph of $y = x^3/(2x^2 - 8)$. Use any of the sketching aids that we have developed, including slant asymptotes.

Solution

Intercepts: $(0, 0)$. Symmetry: With respect to the origin. Asymptotes: Odd vertical asymptotes at $x = \pm 2$.

$$\lim_{x \to 2^-} \frac{x^3}{2x^2 - 8} = -\infty \qquad \text{and} \qquad \lim_{x \to 2^+} \frac{x^3}{2x^2 - 8} = \infty$$

Since the numerator of $x^3/(2x^2 - 8)$ is one degree higher than the denominator, we divide $2x^2 - 8$ into x^3 to find the slant asymptote.

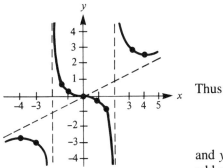

$$\begin{array}{r} x/2 \\ 2x^2 - 8\overline{)x^3 } \\ \underline{x^3 - 4x} \\ 4x \end{array}$$

Thus

$$\frac{x^3}{2x^2 - 8} = \frac{x}{2} + \frac{4x}{2x^2 - 8}$$

and $y = x/2$ is a slant asymptote. To complete the graph, we add a few points as shown in the accompanying table.

x	1	1.5	3	4
y	-0.16	-0.96	2.70	2.67

By adding these points to the graph and using symmetry, we have the sketch as shown.

41. Determine conditions on the coefficients a, b, and c such that the graph of $f(x) = ax^3 + bx^2 + cx + d$ will resemble the accompanying graph.

Solution

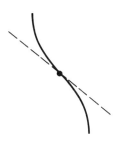

Since $\lim_{x \to \infty} f(x) = -\infty$, $a < 0$. Also, $f(x)$ is a decreasing function, and therefore $f'(x) = 3ax^2 + 2bx + c < 0$ for all x. Hence the discriminant must be negative and we have

$$(2b)^2 - 4(3a)(c) < 0$$
$$4(b^2 - 3ac) < 0$$
$$b^2 < 3ac$$

4.7
Optimization problems

3. The sum of one number and two times a second number is 24. What numbers should be selected so that the product is as large as possible?

Solution

1. Let $x = $ first number, $y = $ second number, and $P = $ product to be maximized.

2. To maximize P, we use the *primary* equation

$$P = xy$$

3. Since the sum of the first number and two times the second is 24, we have the *secondary* equation,

$$x + 2y = 24 \qquad \text{or} \qquad x = 24 - 2y$$

and therefore

$$
\begin{aligned}
P &= xy \\
&= (24 - 2y)y \\
&= 24y - 2y^2
\end{aligned}
$$

4. Differentiation yields

$$\frac{dP}{dy} = 24 - 4y = 0$$

$$y = 6$$

$$x = 24 - 2(6) = 12$$

Finally, since $(d^2P/dy^2) = -4$, the Second-Derivative Test can be used to determine that the point $(12, 6)$ is a maximum and we conclude that the two numbers are 12 and 6.

8. The product of two positive numbers is 192. What numbers should be chosen so that the sum of the first plus three times the second is a minimum?

Solution

1. Let x = first number, y = second number, and S = sum to be minimized.

2. To minimize S, we use the *primary* equation

$$S = x + 3y$$

3. Since the product of the two numbers is 192, we have the *secondary* equation

$$xy = 192 \qquad \text{or} \qquad x = \frac{192}{y}$$

and therefore

$$S = \frac{192}{y} + 3y$$

4. Differentiation yields

$$\frac{dS}{dy} = -\frac{192}{y^2} + 3 = 0$$

$$3y^2 = 192$$

$$y^2 = 64 \qquad \text{or} \qquad y = \pm 8$$

Choosing positive values for y and x, we have $y = 8$ and $x = \frac{192}{8} = 24$. Finally, for $y = 8$, we have

$$\frac{d^2S}{dy^2} = \frac{384}{y^3} = \frac{384}{8^3} > 0$$

which means that S is a minimum when $x = 24$ and $y = 8$.

12. An open box is to be made from a square piece of material, s inches on a side, by cutting equal squares from each corner and turning up the sides.

(a) Find the volume of the largest box that can be made in this manner.

(b) If the dimensions of the square piece of material are doubled, how does the volume change?

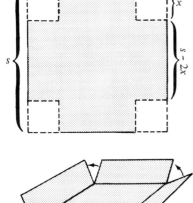

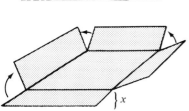

Solution

(a) The volume of the box is given by

$$V = x(s - 2x)^2 \qquad (0 < x < s/2)$$

Thus to find the maximum volume, we set $dV/dx = 0$ as follows:

$$\frac{dV}{dx} = x(2)(s - 2x)(-2) + (s - 2x)^2(1)$$

$$= (s - 2x)(-4x + s - 2x) = 0$$

$$x = \frac{s}{2} \qquad \text{or} \qquad \frac{s}{6}$$

Since $s/2$ is not in the domain of V ($V = 0$ if $x = s/2$), we test $x = s/6$ to determine that the volume is a maximum for this value of x. Therefore, we conclude that the maximum volume is

$$V = \left(\frac{s}{6}\right)\left[s - 2\left(\frac{s}{6}\right)\right]^2 = \left(\frac{s}{6}\right)\left(\frac{2s}{3}\right)^2 = \frac{4s^3}{54} = \frac{2s^3}{27}$$

(b) If s is replaced by $2s$, then the maximum volume is

$$V = \frac{2(2s)^3}{27} = \frac{16s^3}{27}$$

16. Find the dimensions of the largest isosceles triangle that can be inscribed in a circle of radius r.

Solution

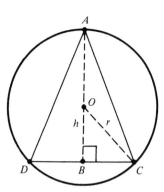

$$AO = OC = r = \text{radius of circle}$$
$$AB = h = \text{height of triangle}$$
$$OB = h - r$$

From triangle OBC we have

$$BC^2 + OB^2 = OC^2$$
$$BC^2 + (h - r)^2 = r^2$$
$$BC^2 = r^2 - r^2 + 2hr - h^2$$
$$BC^2 = 2hr - h^2$$
$$BC = \sqrt{2hr - h^2}$$

Therefore, the area of the triangle ACD is given by

$$a = h\sqrt{2hr - h^2} = \sqrt{2h^3r - h^4}$$

To maximize this area, we set $da/dh = 0$ as follows:

$$\frac{da}{dh} = \left(\frac{1}{2}\right)(2h^3r - h^4)^{-1/2}(6h^2r - 4h^3) = \frac{3h^2r - 2h^3}{\sqrt{2h^3r - h^4}}$$

$$= \frac{h^2(3r - 2h)}{\sqrt{2h^3r - h^4}} = 0$$

$$h = \frac{3r}{2} \qquad \text{We disregard } h = 0.$$

Therefore, $h = 3r/2$, and

$$CD = 2BC = 2\sqrt{2\left(\frac{3r}{2}\right)r - \left(\frac{3r}{2}\right)^2}$$

$$CD = 2\sqrt{3r^2 - \frac{9r^2}{4}} = 2\sqrt{\frac{3r^2}{4}} = \sqrt{3}r$$

Thus the triangle has a base of $\sqrt{3}r$ and a height of $3r/2$. Finally, since

$$AC = \sqrt{\left(\frac{3r}{2}\right)^2 + \left(\frac{\sqrt{3}r}{2}\right)^2} = \sqrt{\frac{9r^2 + 3r^2}{4}} = \sqrt{3r^2} = \sqrt{3}r$$

the triangle is equilateral.

19. A right triangle is formed in the first quadrant by the x- and y-axes and a line through the point $(2, 3)$. Find the vertices of the triangle so that its area is minimum.

Solution

The slope of the line is given by

$$m = \frac{y - 3}{0 - 2} = \frac{0 - 3}{x - 2}$$

Therefore,

$$y - 3 = \frac{6}{x - 2}$$

$$y = \frac{6}{x - 2} + 3 = \frac{6 + 3x - 6}{x - 2} = \frac{3x}{x - 2}$$

The area of the triangle is

$$a = \frac{1}{2}xy = \frac{1}{2}x\left(\frac{3x}{x - 2}\right) \qquad 2 < x$$

$$= \frac{3x^2}{2(x - 2)}$$

To minimize a, we set $da/dx = 0$ as follows:

$$\frac{da}{dx} = \frac{2(x - 2)(6x) - (3x^2)(2)}{4(x - 2)^2} = \frac{6x(2x - 4 - x)}{4(x - 2)^2}$$

$$= \frac{3x(x - 4)}{2(x - 2)^2} = 0$$

Disregarding $x = 0$, we have

$$x = 4 \qquad \text{and} \qquad y = \frac{3(4)}{4 - 2} = 6$$

Thus the vertices are $(0, 0)$, $(4, 0)$, and $(0, 6)$.

23.   Find the coordinates of the point on the curve $y = \sqrt{x}$ closest to the point $(4, 0)$.

Solution

1.   Consider (x, y) to be a point on the graph of $y = \sqrt{x}$.
2.   The distance between (x, y) and the point $(4, 0)$ is given by the *primary* equation

$$d(x) = \sqrt{(x - 4)^2 + (y - 0)^2} = [(x - 4)^2 + y^2]^{1/2}$$

3.   Since $y = \sqrt{x}$, we have

$$d(x) = [(x - 4)^2 + x]^{1/2}$$

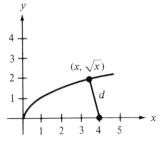

4. To minimize $d(x)$, we set $d'(x) = 0$ as follows:

$$d'(x) = \frac{1}{2}\left[(x-4)^2 + x\right]^{-1/2}\left[2(x-4) + 1\right]$$

$$= \frac{2(x-4) + 1}{2\sqrt{(x-4)^2 + x}}$$

$$= \frac{2x - 7}{2\sqrt{(x-4)^2 + x}} = 0$$

$$2x - 7 = 0$$

$$x = \frac{7}{2}$$

$$y = \sqrt{\frac{7}{2}} = \frac{\sqrt{14}}{2}$$

Therefore, the required point is $(\frac{7}{2}, \sqrt{14}/2)$.

25. A right circular cylinder is to be designed to hold 12 fluid ounces of a soft drink and to use the minimal amount of material in its construction. Find the required dimensions, assuming that 1 fluid ounce (oz) requires 1.80469 cubic inches.

Solution

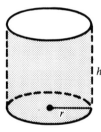

The volume of the cylinder is given by

$$V = \pi r^2 h = 12 \text{ oz}$$
$$= 12(1.80469) \text{ in.}^3 = 21.65628 \text{ in.}^3$$

The surface area of the cylinder is

$$S = 2(\text{area of base}) + (\text{lateral surface})$$
$$= 2\pi r^2 + 2\pi rh = 2\pi r(r + h)$$

Since $h = 21.65628/\pi r^2$, we have

$$S = 2\pi r\left(r - \frac{21.65628}{\pi r^2}\right) = 2\pi\left(r^2 + \frac{21.65628}{\pi r}\right)$$

$$\frac{dS}{dr} = 2\pi\left(2r - \frac{21.65628}{\pi r^2}\right) = 0$$

$$2r = \frac{21.65628}{\pi r^2}$$

$$r^3 = 3.44670$$

$$r = 1.51 \text{ in.}$$

$$h = \frac{21.65628}{\pi r^2} = 3.02 \text{ in.}$$

30. Find the volume of the largest right circular cylinder that can be inscribed in a sphere of radius r.

Solution

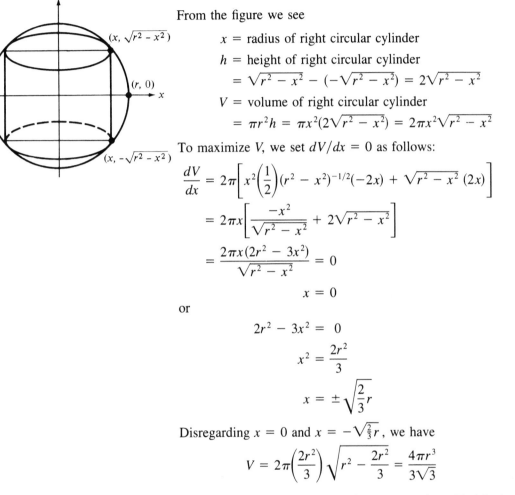

From the figure we see

x = radius of right circular cylinder

h = height of right circular cylinder

$\quad = \sqrt{r^2 - x^2} - (-\sqrt{r^2 - x^2}) = 2\sqrt{r^2 - x^2}$

V = volume of right circular cylinder

$\quad = \pi r^2 h = \pi x^2 (2\sqrt{r^2 - x^2}) = 2\pi x^2 \sqrt{r^2 - x^2}$

To maximize V, we set $dV/dx = 0$ as follows:

$$\frac{dV}{dx} = 2\pi \left[x^2 \left(\frac{1}{2}\right)(r^2 - x^2)^{-1/2}(-2x) + \sqrt{r^2 - x^2}\,(2x) \right]$$

$$= 2\pi x \left[\frac{-x^2}{\sqrt{r^2 - x^2}} + 2\sqrt{r^2 - x^2} \right]$$

$$= \frac{2\pi x (2r^2 - 3x^2)}{\sqrt{r^2 - x^2}} = 0$$

$$x = 0$$

or

$$2r^2 - 3x^2 = 0$$

$$x^2 = \frac{2r^2}{3}$$

$$x = \pm\sqrt{\frac{2}{3}}\,r$$

Disregarding $x = 0$ and $x = -\sqrt{\frac{2}{3}}r$, we have

$$V = 2\pi \left(\frac{2r^2}{3}\right)\sqrt{r^2 - \frac{2r^2}{3}} = \frac{4\pi r^3}{3\sqrt{3}}$$

37. A wooden beam has a rectangular cross section of height h and width w, as shown in the figure. The strength S of the beam is directly proportional to the width and the square of the height. What are the dimensions of the strongest beam that can be cut from a round log of diameter 24 in. (Hint: $S = kh^2w$, where k is the proportionality constant.)

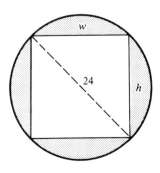

Solution

Letting S be the strength and k the constant of proportionality, we have $S = kwh^2$. Since $w^2 + h^2 = 24^2$, we have $h^2 = 24^2 - w^2$.

$$S = kw(24^2 - w^2) = k(576w - w^3)$$

Differentiating and letting $dS/dw = 0$, we obtain

$$\frac{dS}{dw} = k(576 - 3w^2) = 0$$

$$3w^2 = 576$$

$$w^2 = 192$$

$$w = \pm 8\sqrt{3}$$

Since w is positive, we conclude that $w = 8\sqrt{3}$ in. and
$h = \sqrt{24^2 - 192} = 8\sqrt{6}$ in. will produce the strongest beam.

39. A man is in a boat 2 mi from the nearest point on the coast.
     He is to go to a point Q, 3 mi down the coast and 1 mi inland.
     If he can row at 2 mi/h and walk at 4 mi/h, toward what point
     on the coast should he row in order to reach point Q in the
     least time?

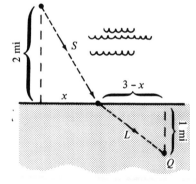

Solution

S = distance on water = $\sqrt{x^2 + 4}$

L = distance on land = $\sqrt{1 + (3 - x)^2} = \sqrt{x^2 - 6x + 10}$

Total time = (time on water) + (time on land)

$$T = \frac{S}{(\text{rate on water})} + \frac{L}{(\text{rate on land})}$$

$$= \frac{\sqrt{x^2 + 4}}{2} + \frac{\sqrt{x^2 - 6x + 10}}{4}$$

To minimize T, we set $dT/dx = 0$ as follows:

$$\frac{dT}{dx} = \left(\frac{1}{2}\right)\left(\frac{1}{2}\right)(x^2 + 4)^{-1/2}(2x)$$

$$+ \left(\frac{1}{4}\right)\left(\frac{1}{2}\right)(x^2 - 6x + 10)^{-1/2}(2x - 6)$$

$$= \frac{x}{2\sqrt{x^2 + 4}} + \frac{x - 3}{4\sqrt{x^2 - 6x + 10}} = 0$$

$$\frac{x}{2\sqrt{x^2 + 4}} = \frac{3 - x}{4\sqrt{x^2 - 6x + 10}}$$

$$\frac{x}{\sqrt{x^2 + 4}} = \frac{3 - x}{2\sqrt{x^2 - 6x + 10}}$$

$$\frac{x^2}{x^2 + 4} = \frac{9 - 6x + x^2}{4(x^2 - 6x + 10)}$$

$$x^2(4)(x^2 - 6x + 10) = (x^2 + 4)(9 - 6x + x^2)$$

$$4x^4 - 24x^3 + 40x^2 = 9x^2 + 36 - 6x^3 - 24x$$
$$+ x^4 + 4x^2$$

$$3x^4 - 18x^3 + 27x^2 + 24x - 36 = 0$$

$$x^4 - 6x^3 + 9x^2 + 8x - 12 = 0$$

This fourth-degree equation has only one positive real root, $x = 1$. Thus the man should row to a point 1 mi from the nearest point on the coast.

43. A component is designed to slide a block of steel of weight W across a table and into a chute as shown in the accompanying figure. The motion of the block is resisted by a frictional force proportional to its net weight. (Let k be the constant of proportionality.) Find the minimum force F needed to slide the block and find the corresponding value of θ.

Solution

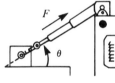

The force in the direction of motion is $F \cos \theta$ and the force tending to lift the block is $F \sin \theta$. Therefore, the net weight of the block is $W - F \sin \theta$, and

$$F \cos \theta = k(W - F \sin \theta)$$

$$F \cos \theta + kF \sin \theta = kW$$

$$F = \frac{kW}{\cos \theta + k \sin \theta}$$

$$= kW(\cos \theta + k \sin \theta)^{-1}$$

Differentiating and letting $dF/d\theta = 0$, we obtain

$$\frac{dF}{d\theta} = -kW(\cos \theta + k \sin \theta)^{-2}(-\sin \theta + k \cos \theta)$$

$$= \frac{-kW(k \cos \theta - \sin \theta)}{(\cos \theta + k \sin \theta)^2} = 0$$

$$\sin \theta = k \cos \theta$$

$$\tan \theta = k$$

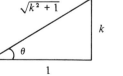

Therefore, F is minimum when $\theta = \arctan k$ and the minimum force is

$$F = \frac{kW}{\cos \theta + k \sin \theta}$$

$$= \frac{kW}{(1/\sqrt{k^2 + 1}) + k(k/\sqrt{k^2 + 1})}$$

$$= \frac{kW}{\sqrt{k^2 + 1}}$$

4.8
Business and economics applications

3. Find the number of units x that produce the maximum revenue R if $R = 1{,}000{,}000x/(0.02x^2 + 1800)$.

Solution

$$R = 1{,}000{,}000 \, \frac{x}{0.02x^2 + 1800}$$

To maximize R, we set $dR/dx = 0$ as follows:

$$\frac{dR}{dx} = 1{,}000{,}000 \, \frac{(0.02x^2 + 1{,}800)(1) - x(0.04x)}{(0.02x^2 + 1{,}800)^2}$$

$$= 1{,}000{,}000 \, \frac{1{,}800 - 0.02x^2}{(0.02x^2 + 1{,}800)^2} = 0$$

$$1{,}800 - 0.02x^2 = 0$$

$$0.02x^2 = 1800$$

$$x^2 = 90{,}000$$

$$x = 300$$

When 300 units are produced, the revenue will be maximum.

6. Find the number of units x that produce the minimum average cost per unit $\overline{C}(\overline{C} = C/x)$ if $C = 0.001x^3 - 5x + 250$.

Solution

$$C = \text{cost of producing } x \text{ units} = 0.001x^3 - 5x + 250$$

$$\overline{C} = \text{average cost per unit} = \frac{C}{x} = 0.001x^2 - 5 + \frac{250}{x}$$

To minimize $\overline{C}$, we set $d\overline{C}/dx = 0$ as follows:

$$\frac{d\overline{C}}{dx} = 0.002x - \frac{250}{x^2} = 0$$

$$0.002x^3 - 250 = 0$$

$$x^3 = 125{,}000$$
$$x = 50$$

Therefore, when 50 units are produced, the average cost will be minimum.

9. Find the price per unit p that produces the maximum profit $P (P = R - C)$ if $C = 100 + 30x$, $x = 90 - p$.

   Solution

   $$p = 90 - x$$
   $$P = R - C = (\text{number of units})(p) - (C)$$
   $$= x(90 - x) - (100 + 30x) = -x^2 + 60x - 100$$

   To maximize P, we set $dP/dx = 0$ as follows:

   $$\frac{dP}{dx} = -2x + 60 = 0$$
   $$2x = 60$$
   $$x = 30$$

   Therefore, profit will be maximum when the price is

   $$p = 90 - x = 90 - 30 = \$60$$

15. A manufacturer of radios charges $90 per unit when the average production cost per unit is $60. However, to encourage large orders from distributors, the manufacturer will reduce the charge by $0.10 per unit for each unit ordered in excess of 100 (for example, there would be a charge of $88 per radio for an order size of 120). Find the largest order size the manufacturer should allow so as to realize maximum profit.

    Solution

    Let $x =$ the number of units purchased, $p =$ the price per unit, and $P =$ the total profit. Then

    $$p = 90 - (0.10)(x - 100) = 90 - 0.1x + 10 = 100 - 0.1x$$

    Since each radio costs the manufacturer $60, the profit per radio is $p - 60$ and the total profit is

    $$P = x(p - 60) = x(100 - 0.1x - 60)$$
    $$= x(40 - 0.1x) = 40x - 0.1x^2$$

    To maximize P, we set $dP/dx = 0$ as follows:

$$\frac{dP}{dx} = 40 - 0.2x = 0$$

$$0.2x = 40$$

$$x = \frac{40}{0.2} = 200$$

Therefore, the manufacturer should not continue the discount for orders over 200 units.

19. Assume that the amount of money deposited in a bank is proportional to the square of the interest rate the bank pays on this money. Furthermore, the bank can reinvest this money at 12%. Find the interest rate the bank should pay to maximize profit. (Use the simple interest formula.)

Solution

Let

$$d = \text{amount in the bank}$$
$$i = \text{interest rate paid by the bank}$$
$$P = \text{profit}$$

The bank can take the deposited money d and reinvest to obtain 12% or $(0.12)d$. Since the bank pays out interest to its depositors, its profit is

$$P = (0.12)d - id$$

Finally, since d is proportional to the square of i, we have

$$d = ki^2$$

Thus

$$P = (0.12)(ki^2) - i(ki^2) = k[(0.12)i^2 - i^3]$$

To maximize P, we set $dP/di = 0$ as follows:

$$\frac{dP}{di} = k(0.24i - 3i^2) = 0$$

$$ki(0.24 - 3i) = 0$$

(We disregard the critical number $i = 0$.)

$$i = \frac{0.24}{3} = 0.08$$

Thus the bank can maximize its profit by setting $i = 8\%$.

29. The demand function for a certain product is given by
     $x = 20 - 2p^2$.

     (a) Consider the point $(2, 12)$. If the price decreases by 5%,
          determine the corresponding percentage increase in quan-
          tity demanded.

     (b) Find the exact elasticity at $(2, 12)$ by using the formula in
          this section.

     (c) Find an expression for total revenue $(R = xp)$, and find
          the values of x and p that maximize R.

     (d) For the value of x found in part (c), show that $|\eta| = 1$.

Solution

(a) If $p = 2$ and $x = 12$ and p decreases by 5%, then

$$p = 2 - (0.05)(2) = 1.9$$
$$x = 20 - 2(1.9)^2 = 20 - 7.22 = 12.78$$

Thus the percentage increase in x is

$$\frac{12.78 - 12}{12} = \frac{0.78}{12} = 0.065 = 6\tfrac{1}{2}\%$$

(b) The exact elasticity of demand at $(2, 12)$ is given by

$$\eta = \left(\frac{p}{x}\right)\left(\frac{dx}{dp}\right)$$

$$= \frac{p}{x}(-4p)$$

$$= \frac{2}{12}[-4(2)]$$

$$= \frac{-16}{12} = -\frac{4}{3} = -1.33$$

(c)

$$R = xp = (20 - 2p^2)p = 20p - 2p^3$$

To maximize R, we set $dR/dp = 0$ as follows:

$$\frac{dR}{dp} = 20 - 6p^2 = 0$$

$$6p^2 = 20$$

$$p^2 = \frac{20}{6} = \frac{10}{3}$$

$$p = \sqrt{\frac{10}{3}} \approx \$1.83$$

$$x = 20 - 2\left(\sqrt{\frac{10}{3}}\right)^2 = \frac{40}{3}$$

(d)   If $x = \frac{40}{3}$ and $p = \sqrt{\frac{10}{3}}$,

$$|\eta| = \left|\frac{p}{x}(-4p)\right| = \left|\frac{\sqrt{\frac{10}{3}}}{\frac{40}{3}}\left(-4\sqrt{\frac{10}{3}}\right)\right| = \left|\frac{-\frac{40}{3}}{\frac{40}{3}}\right| = 1$$

4.9
Newton's Method

5.   Approximate the zero of $f(x) = x^3 + x - 1$ in the interval $[0, 1]$. Use Newton's Method and continue the process until you are correct to three decimal places.

Solution

$$f(x) = x^3 + x - 1 \qquad f'(x) = 3x^2 + 1$$

n	x_n	$f(x_n)$	$f'(x_n)$	$f(x_n)/f'(x_n)$	$x_n - [f(x_n)/f'(x_n)]$
1	0.5000	−0.3750	1.7500	−0.2143	0.7143
2	0.7143	0.0787	2.5306	0.0311	0.6832
3	0.6832	0.0021	2.4002	0.0009	0.6823
4	0.6823	0.0000	2.3967	0.0000	0.6823

Therefore, we approximate the root to be $x = 0.682$.

7.   Approximate the zero of $f(x) = 3\sqrt{x - 1} - x$ in the interval $[1, 2]$. Use Newton's Method and continue the process until you are correct to three decimal places.

Solution

$$f(x) = 3\sqrt{x - 1} - x \qquad f'(x) = \frac{3}{2\sqrt{x - 1}} - 1$$

n	x_n	$f(x_n)$	$f'(x_n)$	$f(x_n)/f'(x_n)$	$x_n - [f(x_n)/f'(x_n)]$
1	1.2000	0.1416	2.3541	0.0602	1.1398
2	1.1398	−0.0180	3.0113	−0.0060	1.1458
3	1.1458	−0.0003	2.9283	−0.0001	1.1459

Therefore, we approximate the zero to be $x = 1.146$.

16. Apply Newton's Method to approximate the x-value of the point of intersection of the two graphs of the equations $f(x) = x^2$ and $g(x) = \cos x$. Continue the process until two successive approximations differ by less than 0.001.

Solution

To approximate the x-value of the point of intersection, we let $x^2 = \cos x$. Since this implies that $x^2 - \cos x = 0$, we need to find the zeros of the function given by

$$h(x) = x^2 - \cos x$$

Thus, the iterative formula for Newton's Method takes the form

$$x_{n+1} = x_n - \frac{x_n^2 - \cos x_n}{2x_n + \sin x_n}$$

The calculations are shown in the following table, beginning with an initial guess of $x_1 = 1$.

n	x_n	$f(x_n)$	$f'(x_n)$	$f(x_n)/f'(x_n)$	$x_n - [f(x_n)/f'(x_n)]$
1	1	0.4597	2.8415	0.1618	0.8382
2	0.8382	0.0338	2.4199	0.0140	0.8242
3	0.8242	0.0003	2.3825	0.0001	0.8241

Therefore, we approximate the zero to be $x = 0.824$ and by symmetry $x = -0.824$.

21. Use Newton's Method to obtain a general formula for approximating $\sqrt[n]{a}$. [Hint: Apply Newton's Method to the function $f(x) = x^n - a$.]

Solution

Let $f(x) = x^n - a$. Then $f'(x) = nx^{n-1}$. Since $\sqrt[n]{a}$ is a zero of $f(x) = 0$, we can use Newton's Method to approximate $\sqrt[n]{a}$ as follows:

$$x_{i+1} = x_i - \frac{x_i^n - a}{nx_i^{n-1}} = \frac{(n-1)x_i^n + a}{nx_i^{n-1}}$$

For example, if $a = 2$, $n = 2$, and $x_1 = 1$, then we can approximate $\sqrt{2}$ as follows:

$$x_1 = 1$$

$$x_2 = \frac{(2 - 1)(1^2) + 2}{2(1)} = \frac{1 + 2}{2} = 1.50000$$

$$x_3 = \frac{(2 - 1)(1.5)^2 + 2}{2(1.5)} = \frac{4.25}{3} = 1.41667$$

$$x_4 = \frac{(2 - 1)(1.41667)^2 + 2}{2(1.41667)} = \frac{4.00697}{2.8333} = 1.41421$$

(Note: To five decimal places, $\sqrt{2} = 1.41421$.)

27. Approximate the critical number of $f(x) = x \cos x$ on the interval $[0, 2\pi]$.

Solution

$$f(x) = x \cos x$$
$$f'(x) = x(-\sin x) + \cos x = 0$$
$$x - \cot x = 0$$

Thus the interative formula for Newton's Method takes the form

$$x_{n+1} = x_n - \frac{x_n - \cot x_n}{1 + \csc^2 x_n}$$

Since the critical number occurs at the x-value where $\cot x = x$, it must be in the interval $[0, \pi/2]$. Therefore, using the iterative formula and beginning with the initial guess $x_1 = 1$, we obtain the critical number $x \approx 0.860$.

29. Use the result of Exercise 27 to sketch the graph of $f(x) = x \cos x$ on the interval $[0, \pi]$.

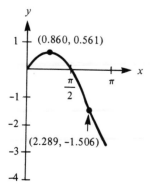

Solution

$$f(x) = x \cos x$$
$$f'(x) = -x \sin x + \cos x \qquad \text{Critical number: } x = 0.860$$
$$f''(x) = -(2 \sin x + x \cos x)$$

To find possible points of inflection we solve the equation $f''(x) = 0$. Newton's Method yields the iterative formula

$$x_{n+1} = x_n - \frac{-(2 \sin x + x \cos x)}{x \sin x - 3 \cos x}$$

Beginning with an initial guess $3\pi/4$, we obtain a possible point of inflection at $(2.289, -1.506)$.

Interval	$f(x)$	$f'(x)$	$f''(x)$	Shape of graph
x in $[0, 0.860)$		$+$	$-$	increasing, concave down
$x = 0.860$	0.561	0	$-$	relative maximum
x in $(0.860, 2.289)$		$-$	$-$	decreasing, concave down
$x = 2.289$	-1.506	$-$	0	point of inflection
x in $(2.289, \pi]$		$-$	$+$	decreasing, concave up

4.10
Differentials

5. Find the differential dy of $y = x\sqrt{1 - x^2}$.

Solution

$$y = x\sqrt{1 - x^2}$$
$$dy = \left[x(\tfrac{1}{2})(1 - x^2)^{-1/2}(-2x) + (1 - x^2)^{1/2}(1) \right] dx$$
$$= \left[\frac{-x^2}{\sqrt{1 - x^2}} + \frac{1 - x^2}{\sqrt{1 - x^2}} \right] dx = \frac{1 - 2x^2}{\sqrt{1 - x^2}} \, dx$$

21. The radius of a sphere is claimed to be 6 in., with a possible error of 0.02 in. Using differentials, approximate the maximum possible error in calculating the following:

(a) the volume of the sphere

(b) the surface area of the sphere

(c) What is the relative error in parts (a) and (b)?

Solution

The radius of the sphere is given by $r = 6 \pm 0.02$.

(a) The volume of the sphere is given by

$$V = \frac{4}{3} \pi r^3$$

To approximate ΔV by dV, we let $r = 6$ and $dr = \pm 0.02$

$$dV = \frac{4}{3} \pi (3r^2) \, dr = 4\pi(36)(\pm 0.02) = \pm 2.88\pi \text{ in.}^3$$

(b) The surface area of the sphere is given by

$$S = 4\pi r^2$$

To approximate ΔS by dS, we let $r = 6$ and $dr = \pm 0.02$.

$$dS = 8\pi r \, dr = 8\pi(6)(\pm 0.02) = \pm 0.96\pi \text{ in.}^2$$

(c) For part (a) the relative error is approximately

$$\frac{dV}{V} = \frac{2.88\pi}{\left(\frac{4}{3}\right)\pi(6^3)} = \frac{2.88}{288} = 0.01 = 1\%$$

For (b) the relative error is approximately

$$\frac{dS}{S} = \frac{0.96\pi}{4\pi(6^2)} = \frac{0.96}{144} = 0.0067 = \frac{2}{3}\%$$

23. The period of a pendulum is given by $T = 2\pi\sqrt{L/g}$, where L is the length of the pendulum in feet, g is the acceleration due to gravity, and T is time in seconds. Suppose that the pendulum has been subjected to an increase in temperature so that the length increases by $\frac{1}{2}\%$.

(a) What is the approximate percentage change in the period?

(b) Using the result of part (a), find the approximate error in this pendulum clock in one day.

Solution

(a) $T = 2\pi\sqrt{\dfrac{L}{g}}$

$$dT = 2\pi\left(\frac{1}{2}\right)\left(\frac{L}{g}\right)^{-1/2}\left(\frac{1}{g}\right)dL = \frac{\pi}{g\sqrt{L/g}}\,dL$$

$$\frac{dT}{T}(100) = \text{percentage error}$$

$$= \frac{(\pi/g\sqrt{L/g})\,dL}{2\pi\sqrt{L/g}}(100) = \frac{1}{2}\left(\frac{dL}{L}100\right)$$

$$= \frac{1}{2}\ (\text{percentage change in } L) = \frac{1}{2}\left(\frac{1}{2}\%\right) = \frac{1}{4}\%$$

(b) approximate error $= \left(\dfrac{1}{4}\%\right)(\text{number of seconds per day})$

$$= (0.0025)(60)(60)(24)$$

$$= 216 \text{ s} = 3.6 \text{ min}$$

Review Exercises for Chapter 4

5. Make use of domain, range, symmetry, asymptotes, intercepts, relative extrema, or points of inflection to obtain an accurate graph of $f(x) = (x + 1)/(x - 1)$.

Solution

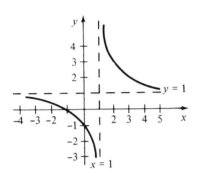

$$f(x) = \frac{x + 1}{x - 1} \qquad \text{Intercepts: } (0, -1), (-1, 0)$$

$$f'(x) = \frac{(x - 1)(1) - (x + 1)(1)}{(x - 1)^2} = \frac{-2}{(x - 1)^2} < 0$$

$$f''(x) = \frac{4}{(x - 1)^3} \neq 0$$

There are no critical values or possible points of inflection. Vertical asymptote at $x = 1$, horizontal asymptote at $y = 1$.

x	-4	-3	-2	-1	0	0.5	1.5	2	3	4
y	0.6	0.5	0.33	0	-1	-3	5	3	2	1.67

13. Make use of domain, range, symmetry, asymptotes, intercepts, relative extrema, or points of inflection to obtain an accurate graph of $f(x) = x^{1/3}(x + 3)^{2/3}$.

Solution

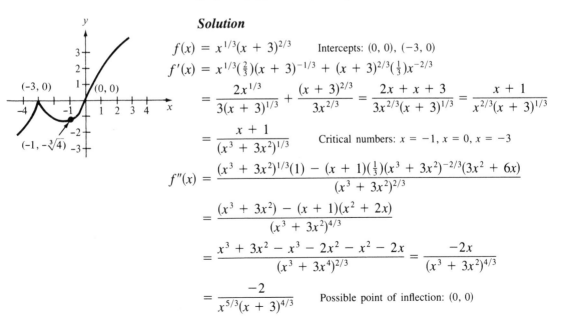

$$f(x) = x^{1/3}(x + 3)^{2/3} \qquad \text{Intercepts: } (0, 0), (-3, 0)$$

$$f'(x) = x^{1/3}(\tfrac{2}{3})(x + 3)^{-1/3} + (x + 3)^{2/3}(\tfrac{1}{3})x^{-2/3}$$

$$= \frac{2x^{1/3}}{3(x + 3)^{1/3}} + \frac{(x + 3)^{2/3}}{3x^{2/3}} = \frac{2x + x + 3}{3x^{2/3}(x + 3)^{1/3}} = \frac{x + 1}{x^{2/3}(x + 3)^{1/3}}$$

$$= \frac{x + 1}{(x^3 + 3x^2)^{1/3}} \qquad \text{Critical numbers: } x = -1, x = 0, x = -3$$

$$f''(x) = \frac{(x^3 + 3x^2)^{1/3}(1) - (x + 1)(\tfrac{1}{3})(x^3 + 3x^2)^{-2/3}(3x^2 + 6x)}{(x^3 + 3x^2)^{2/3}}$$

$$= \frac{(x^3 + 3x^2) - (x + 1)(x^2 + 2x)}{(x^3 + 3x^2)^{4/3}}$$

$$= \frac{x^3 + 3x^2 - x^3 - 2x^2 - x^2 - 2x}{(x^3 + 3x^4)^{2/3}} = \frac{-2x}{(x^3 + 3x^2)^{4/3}}$$

$$= \frac{-2}{x^{5/3}(x + 3)^{4/3}} \qquad \text{Possible point of inflection: } (0, 0)$$

x	$f(x)$	$f'(x)$	$f''(x)$	Shape of graph
x in $(-\infty, -3)$		$+$	$+$	increasing, concave up
$x = -3$	0	undefined	undefined	relative maximum
x in $(-3, -1)$		$-$	$+$	decreasing, concave up
$x = -1$	$-\sqrt[3]{4}$	0	$+$	relative minimum
x in $(-1, 0)$		$+$	$+$	increasing, concave up
$x = 0$	0	undefined	undefined	point of inflection
x in $(0, \infty)$		$+$	$-$	increasing, concave down

22. Make use of domain, range, symmetry, asymptotes, intercepts, relative extrema, or points of inflection to obtain an accurate graph of $f(x) = |x - 1| + |x - 3|$.

Solution

If $x < 1$, then $x - 1 < 0$ and $x - 3 < 0$. Therefore,

$$|x - 1| + |x - 3| = [-(x - 1)] + [-(x - 3)] = -2x + 4$$

If $1 \le x < 3$, then $x - 1 \ge 0$ and $x - 3 < 0$. Therefore,

$$|x - 1| + |x - 3| = (x - 1) + [-(x - 3)] = 2$$

If $x \ge 3$, then $x - 1 > 0$ and $x - 3 \ge 0$. Therefore,

$$|x - 1| + |x - 3| = (x - 1) + (x - 3) = 2x - 4$$

$$f(x) = \begin{cases} -2x + 4, & x < 1 \\ 2, & 1 \le x < 3 \\ 2x - 4, & x \ge 3 \end{cases}$$

Since the graph of f consists of line segments, it is not necessary to use additional sketching aids.

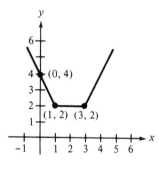

28. Consider the function $f(x) = x^n$ for positive integer values of n.

(a) For what values of n does the function have a relative minimum at the origin?

(b) For what values of n does the function have a point of inflection at the origin? Explain.

Solution

$$f(x) = x^n \qquad f'(x) = nx^{n-1} \qquad f''(x) = n(n - 1)x^{n-2}$$

(a) If n is even, then for x in $(-\infty, 0)$,

$$f'(x) = n(\text{negative number})^{(\text{odd power})} = (\text{negative number})$$

For x in $(0, \infty)$,

$$f'(x) = n(\text{positive number})^{(\text{odd power})} = (\text{positive number})$$

Thus by the First-Derivative Test, $(0, 0)$ is a relative minimum.

(b)   If n is odd, then for x in $(-\infty, 0)$,

$$f''(x) = n(n - 1)(\text{negative number})^{(\text{odd power})}$$
$$= (\text{negative number})$$

For x in $(0, \infty)$,

$$f''(x) = n(n - 1)(\text{positive number})^{(\text{odd power})}$$
$$= (\text{positive number})$$

Thus f is concave down in $(-\infty, 0)$ and concave up in $(0, \infty)$, which implies that $(0, 0)$ is a point of inflection.

37.   For the function $f(x) = Ax^2 + Bx + C$, determine the value of c guaranteed by the Mean Value Theorem on the interval $[x_1, x_2]$.

Solution

$$f(x) = Ax^2 + Bx + C$$
$$f'(x) = 2Ax + B$$
$$f'(c) = 2Ac + B = \frac{f(x_2) - f(x_1)}{x_2 - x_1}$$
$$= \frac{Ax_2^2 + Bx_2 + C - Ax_1^2 - Bx_1 - C}{x_2 - x_1}$$
$$= \frac{A(x_2^2 - x_1^2) + B(x_2 - x_1)}{x_2 - x_1}$$
$$= A(x_2 + x_1) + B$$
$$2c = x_2 + x_1$$
$$c = \frac{x_2 + x_1}{2}$$

41.   Find the maximum profit if the demand equation is $p = 36 - 4x$ and the total cost is $C = 2x^2 + 6$.

Solution

The profit is given by

$$P = (\text{price per unit})(\text{number of units}) - (\text{cost}) = px - C$$
$$= (36 - 4x)(x) - (2x^2 + 6) = 36x - 4x^2 - 2x^2 - 6$$
$$= -6x^2 + 36x - 6$$

To maximize P, we set $dP/dx = 0$ as follows:

$$\frac{dP}{dx} = -12x + 36 = 0$$

$$12x = 36$$

$$x = 3 \text{ units}$$

Thus the maximum profit is

$$P = -6(3)^2 + 36(3) - 6 = -54 + 108 - 6 = \$48$$

49. Find the length of the longest pipe that can be carried level around a right-angle corner if the two intersecting corridors are of widths 4 ft and 6 ft.

Solution

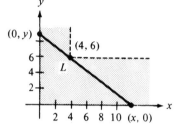

The longest pipe that will go around the corner will have a length equal to the minimum length of the hypotenuse [through the point $(4, 6)$] of the triangle whose vertices are $(0, 0)$, $(x, 0)$, and $(0, y)$. We begin by relating x and y as follows:

$$m = \frac{y - 6}{0 - 4} = \frac{6 - 0}{4 - x}$$

$$y - 6 = \frac{-24}{4 - x}$$

$$y = \frac{24}{x - 4} + 6 = \frac{6x}{x - 4}$$

[Note that $dy/dx = -24/(x - 4)^2$.] Now the length of the hypotenuse is given by

$$L = \sqrt{x^2 + y^2}$$

To minimize L, we set $dL/dx = 0$ as follows.

$$\frac{dL}{dx} = \frac{(\frac{1}{2})[2x + (2y)(dy/dx)]}{\sqrt{x^2 + y^2}} = 0$$

$$x = -y\frac{dy}{dx} = -\left(\frac{6x}{x - 4}\right)\left[\frac{-24}{(x - 4)^2}\right]$$

$$x(x - 4)^3 = 144x$$

$$(x - 4)^3 = 144$$

$$x - 4 = \sqrt[3]{144}$$

$$x = \sqrt[3]{144} + 4$$

Therefore, the minimum length of L and the maximum length of pipe are given by

$$L = \sqrt{x^2 + y^2} = \sqrt{x^2 + \frac{36x^2}{(x-4)^2}} = \frac{x}{x-4}\sqrt{(x-4)^2 + 36}$$

$$= \frac{\sqrt[3]{144} + 4}{\sqrt[3]{144}}\sqrt{144^{2/3} + 36} \approx 14.05 \text{ ft}$$

51. A hallway of width 6 ft meets a hallway of width 9 ft at right angles. Find the length of the longest pipe that can be carried horizontally around this corner. (Hint: If L is the length of the pipe, show that

$$L = 6 \csc \theta + 9 \csc \left(\frac{\pi}{2} - \theta\right)$$

where θ is the angle between the pipe and the wall of the narrower hallway.)

Solution

From the accompanying figure we observe that

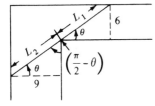

$$\csc \theta = \frac{L_1}{6} \quad \text{or} \quad L_1 = 6 \csc \theta$$

$$\csc \left(\frac{\pi}{2} - \theta\right) = \frac{L_2}{9} \quad \text{or} \quad L_2 = 9 \csc \left(\frac{\pi}{2} - \theta\right)$$

Therefore, the length of the pipe is given by

$$L = L_1 + L_2 = 6 \csc \theta + 9 \csc \left(\frac{\pi}{2} - \theta\right)$$

$$= 6 \csc \theta + 9 \sec \theta$$

{Note that $\csc[(\pi/2) - \theta] = \sec \theta$.} To maximize L, we set $dL/d\theta = 0$ as follows.

$$\frac{dL}{d\theta} = -6 \csc \theta \cot \theta + 9 \sec \theta \tan \theta = 0$$

$$9 \sec \theta \tan \theta = 6 \csc \theta \cot \theta$$

$$\frac{\sec \theta \tan \theta}{\csc \theta \cot \theta} = \frac{6}{9}$$

$$\tan^3 \theta = \frac{2}{3}$$

$$\tan \theta = \frac{2^{1/3}}{3^{1/3}}$$

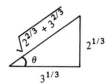

From the accompanying figure we observe that

$$\csc \theta = \frac{\sqrt{2^{2/3} + 3^{2/3}}}{2^{1/3}} \qquad \text{and} \qquad \sec \theta = \frac{\sqrt{2^{2/3} + 3^{2/3}}}{3^{1/3}}$$

$$L = (6)\left(\frac{\sqrt{2^{2/3} + 3^{2/3}}}{2^{1/3}}\right) + (9)\left(\frac{\sqrt{2^{2/3} + 3^{2/3}}}{3^{1/3}}\right)$$

$$= 3\sqrt{2^{2/3} + 3^{2/3}}\,(2^{2/3} + 3^{2/3}) = 3(2^{2/3} + 3^{2/3})^{3/2}$$

54. The general equation giving the height of an oscillating object attached to a spring is

$$y = A \sin \sqrt{\frac{k}{m}}\,t + B \cos \sqrt{\frac{k}{m}}\,t$$

where k is the spring constant and m is the mass of the object. Show that the maximum displacement of the object is $\sqrt{A^2 + B^2}$. Show that the frequency (number of oscillations per second) is $(1/2\pi)\sqrt{k/m}$. How is the frequency changed if the stiffness k of the spring is increased? How is the frequency changed if the mass m of the object is increased?

Solution

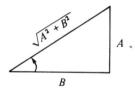

(a) $y = A \sin \sqrt{\dfrac{k}{m}}\,t + B \cos \sqrt{\dfrac{k}{m}}\,t$

$$y' = A\sqrt{\frac{k}{m}} \cos \sqrt{\frac{k}{m}}\,t - B\sqrt{\frac{k}{m}} \sin \sqrt{\frac{k}{m}}\,t = 0$$

$$A\sqrt{\frac{k}{m}} \cos \sqrt{\frac{k}{m}}\,t = B\sqrt{\frac{k}{m}} \sin \sqrt{\frac{k}{m}}\,t$$

$$\frac{A}{B} = \frac{\sin (\sqrt{k/m}\,)t}{\cos (\sqrt{k/m}\,)t}$$

$$\tan \sqrt{\frac{k}{m}}\,t = \frac{A}{B}$$

From the accompanying figure we observe that

$$\sin \sqrt{\frac{k}{m}}\,t = \frac{A}{\sqrt{A^2 + B^2}} \qquad \text{and} \qquad \cos \sqrt{\frac{k}{m}}\,t = \frac{B}{\sqrt{A^2 + B^2}}$$

Thus when $y' = 0$,

$$y = A\left[\frac{A}{\sqrt{A^2 + B^2}}\right] + B\left[\frac{B}{\sqrt{A^2 + B^2}}\right] = \sqrt{A^2 + B^2}$$

(b)   Since the frequency is the reciprocal of the period, we have

$$\text{period} = \frac{2\pi}{\sqrt{k/m}} \qquad \text{frequency} = \frac{\sqrt{k/m}}{2\pi} = \frac{1}{2\pi}\sqrt{\frac{k}{m}}$$

Therefore, if the stiffness k of the spring is increased, the frequency will increase. However, if the mass m of the object is increased, the frequency will decrease.

56.   The demand and cost equations for a certain product are

$$p = 600 - 3x \qquad \text{and} \qquad C = 0.3x^2 + 6x + 600$$

where p is the price per unit, x is the number of units, and C is the cost of producing x units. If t is the excise tax per unit, the profit for producing x units is

$$P = xp - C - xt$$

Find the maximum profit for the following:

(a) $t = 5$ (b) $t = 10$ (c) $t = 20$

Solution

$$P = xp - C - xt = x(600 - 3x) - (0.3x^2 + 6x + 600) - xt$$
$$= -3.3x^2 + (594 - t)x - 600$$

(a)   When $t = 5$,

$$P = -3.3x^2 + 589x - 600$$

To maximize P, set $dP/dx = 0$ as follows:

$$\frac{dP}{dx} = -6.6x + 589 = 0$$

$$6.6x = 589$$

$$x = 89.2$$

When $x = 89$, $P = -3.3(89)^2 + 589(89) - 600$
$= \$25,681.70$.

(b)   When $t = 10$,

$$P = -3.3x^2 + 584x - 600$$

To maximize P, set $dP/dx = 0$ as follows:

$$\frac{dP}{dx} = -6.6x + 584 = 0$$

$$6.6x = 584$$

$$x = 88.5$$

When $x = 88$, $P = -3.3(88)^2 + 584(88) - 600$
$= \$25,236.80$.

(c)   When $t = 20$,

$$P = -3.3x^2 + 574x - 600$$

To maximize P, set $dP/dx = 0$ as follows:

$$\frac{dP}{dx} = -6.6x + 574 = 0$$

$$6.6x = 574$$

$$x = 87$$

When $x = 87$, $P = -3.3(87)^2 + 574(87) - 600$
$= \$24,360.30$.

5 Integration

5.1
Antiderivatives and indefinite integration

3. Complete the following table for the indefinite integral

$$\int \frac{1}{x\sqrt{x}}\, dx.$$

Given	Rewrite	Integrate	Simplify
$\int \dfrac{1}{x\sqrt{x}}\, dx$			

Solution

Given	Rewrite	Integrate	Simplify
$\int \dfrac{1}{x\sqrt{x}}\, dx$	$\int x^{-3/2}\, dx$	$\dfrac{x^{-1/2}}{-\frac{1}{2}} + C$	$\dfrac{-2}{\sqrt{x}} + C$

9. Evaluate the indefinite integral $\int(x^{3/2} + 2x + 1)\, dx$ and check your results by differentiation.

Solution

$$\int(x^{3/2} + 2x + 1)\, dx = \frac{x^{5/2}}{\frac{5}{2}} + 2\left(\frac{x^2}{2}\right) + x + C$$

$$= \frac{2x^{5/2}}{5} + x^2 + x + C$$

Check

If $y = \dfrac{2x^{5/2}}{5} + x^2 + x + C$, then

$$\frac{dy}{dx} = \left(\frac{2}{5}\right)\left(\frac{5}{2}\right)x^{3/2} + 2x + 1 + 0 = x^{3/2} + 2x + 1$$

21. Evaluate the indefinite integral $\int \dfrac{t^2 + 2}{t^2}\, dt$ and check your results by differentiation.

Solution

$$\int \frac{t^2 + 2}{t^2}\, dt = \int \left(\frac{t^2}{t^2} + \frac{2}{t^2}\right) dt = \int (1 + 2t^{-2})\, dt$$

$$= t + \frac{2(t^{-1})}{-1} + C = t - \frac{2}{t} + C$$

Check

If $y = t - \dfrac{2}{t} + C = t - 2t^{-1} + C$, then

$$\frac{dy}{dt} = 1 - 2(-1)t^{-2} + 0 = 1 + \frac{2}{t^2} = \frac{t^2 + 2}{t^2}$$

33. Evaluate the indefinite integral $\int (\tan^2 y + 1)\, dy$ and check your results by differentiation.

Solution

$$\int (\tan^2 y + 1)\, dy = \int \sec^2 y\, dy = \tan y + C$$

Check

If $f(y) = \tan y + C$, then $f'(y) = \sec^2 y = \tan^2 y + 1$.

35. Find the equation of the curve such that $dy/dx = 2x - 1$ and the curve passes through the point $(1, 1)$.

Solution

$$\frac{dy}{dx} = 2x - 1$$

$$y = \int (2x - 1)\, dx = x^2 - x + C$$

Since the curve passes through the point $(1, 1)$, we have

$$y = x^2 - x + C$$
$$1 = (1)^2 - (1) + C \qquad \text{or} \qquad C = 1$$

Therefore the required equation is $y = x^2 - x + 1$.

41. Find $y = f(x)$ if $f''(x) = x^{-3/2}$, $f'(4) = 2$, and $f(0) = 0$.

Solution

$$f''(x) = x^{-3/2}$$

$$f'(x) = \int x^{-3/2} \, dx$$

$$= \frac{x^{-1/2}}{-\frac{1}{2}} + C_1 = \frac{-2}{\sqrt{x}} + C_1$$

$$f'(4) = \frac{-2}{\sqrt{4}} + C_1 = 2$$

$$\frac{-2}{2} + C_1 = 2$$

$$C_1 = 3$$

$$f'(x) = -2x^{-1/2} + 3$$

$$f(x) = \int(-2x^{-1/2} + 3) \, dx = \frac{(-2)x^{1/2}}{\frac{1}{2}} + 3x + C_2$$

$$= -4x^{1/2} + 3x + C_2$$

$$f(0) = -4(0)^{1/2} + 3(0) + C_2 = 0$$

$$C_2 = 0$$

Therefore,

$$f(x) = -4x^{1/2} + 3x$$

45. With what initial velocity must an object be thrown upward from the ground to reach a maximum height of 550 ft (approximate height of the Washington Monument)?

Solution

If $s = f(t)$, we know that

$$f''(t) = -32$$

(-32 ft/s^2 is the acceleration due to gravity.)

$$f'(t) = \int -32 \, dt = -32t + C_1 = -32t + v_0$$

where v_0 is the initial velocity. Furthermore,

$$f(t) = \int(-32t + v_0) \, dt = -16t^2 + v_0 t + C_2 = -16t^2 + v_0 t + s_0$$

where $s_0 = 0$ is the initial height. Thus,

$$s = f(t) = -16t^2 + v_0 t$$

Now since s is a maximum when $f'(t) = 0$, we have

$$-32t + v_0 = 0 \qquad \text{or} \qquad t = \frac{v_0}{32}$$

Finally, in order for s to attain a height of 550 ft, we must have

$$s = -16\left(\frac{v_0}{32}\right)^2 + v_0\left(\frac{v_0}{32}\right) = 550$$

$$\frac{v_0^2}{64} = 550$$

$$v_0^2 = 35{,}200$$

$$v_0 = \sqrt{35{,}200} = 40\sqrt{122}$$

$$\approx 187.62 \text{ ft/s}$$

51. At the instant the traffic light turns green, an automobile that has been waiting at an intersection starts ahead with a constant acceleration of 6 ft/s^2. At the same instant a truck traveling with a constant velocity of 30 ft/s overtakes and passes the car.

(a) How far beyond its starting point will the automobile overtake the truck?

(b) How fast will it be traveling?

Solution

Let $T(t)$ and $A(t)$ represent the position functions of the truck and auto. Then we know that

$$T'(t) = 30, \qquad T(0) = 0$$
$$A''(t) = 6, \qquad A'(0) = 0, \qquad A(0) = 0$$

For the truck, we have

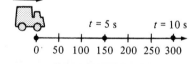

$$T(t) = \int 30 \, dt = 30t + C_1$$

$$T(0) = 30(0) + C_1 = 0$$

$$C_1 = 0$$

For the auto, we have

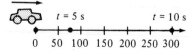

$$A'(t) = \int 6 \, dt = 6t + C_2$$

$$A'(0) = 6(0) + C_2 = 0$$

$$C_2 = 0$$

$$A'(t) = 6t$$

$$A(t) = \int 6t \, dt = 3t^2 + C_3$$

$$A(0) = 3(0)^2 + C_3 = 0$$

$$C_3 = 0$$

$$A(t) = 3t^2$$

Therefore, when the auto catches up with the truck, we have

$$A(t) = T(t)$$

$$3t^2 = 30t$$

$$3t^2 - 30t = 0$$

$$3t(t - 10) = 0$$

$$t = 10 \text{ s} \qquad \text{We disregard } t = 0.$$

(a)   When $t = 10$ s, the auto will have traveled

$$A(10) = 3(10^2) = 300 \text{ ft}$$

(b)   It will be traveling

$$A'(10) = 60 \text{ ft/s} = \left(60 \frac{\text{ft}}{\text{s}}\right)\left(\frac{1 \text{ mi}}{5280 \text{ ft}}\right)\left(\frac{3600 \text{ s}}{\text{h}}\right) \approx 40.9 \text{ mi/h}$$

55.   The marginal cost for product is $dC/dx = 2x - 12$. Find the total cost function and the average cost function if fixed costs are $50.

Solution

Since $dC/dx = 2x - 12$, we have

$$C(x) = \int (2x - 12) \, dx = x^2 - 12x + C_1$$

Since the fixed costs are $50, at $x = 0$ we have

$$C(0) = C_1 = 50$$

Therefore,

$$C(x) = x^2 - 12x + 50$$

$$\overline{C}(x) = \frac{C}{x} = x - 12 + \frac{50}{x}$$

5.2

Integration by substitution

9. Evaluate the indefinite integral $\int x^2(x^3 - 1)^4 \, dx$ and check your results by differentiation.

Solution

To evaluate $\int x^2(x^3 - 1)^4 \, dx$, we use the method of pattern recognition letting $u = x^3 - 1$, and $u' = 3x^2$.

$$\int x^2(x^3 - 1)^4 \, dx = \frac{1}{3} \int \overbrace{(x^3 - 1)^4}^{u^4}\overbrace{(3x^2)}^{u'} \, dx$$

$$= \left(\frac{1}{3}\right)\left[\frac{(x^3 - 1)^5}{5}\right] + C = \frac{1}{15}(x^3 - 1)^5 + C$$

Check

If $y = \frac{1}{15}(x^3 - 1)^5 + C$, then

$$\frac{dy}{dx} = \frac{1}{15}(5)(x^3 - 1)^4(3x^2) + 0 = x^2(x^3 - 1)^4$$

15. Evaluate the indefinite integral $\int 5x\sqrt[3]{1 + x^2} \, dx$ and check your results by differentiation.

Solution

To evaluate $\int 5x\sqrt[3]{1 + x^2} \, dx$, we use the method of pattern recognition letting $u = 1 + x^2$, and $u' = 2x$. Thus we have

$$\int 5x\sqrt[3]{1 + x^2} \, dx = \frac{5}{2} \int \overbrace{(1 + x^2)^{1/3}}^{u^{1/3}}\overbrace{(2x)}^{u'} \, dx$$

$$= \left(\frac{5}{2}\right)\left(\frac{(1 + x^2)^{4/3}}{4/3}\right) + C = \frac{15}{8}(1 + x^2)^{4/3} + C$$

Check

If $y = \frac{15}{8}(1 + x^2)^{4/3} + C$, then

$$\frac{dy}{dx} = \left(\frac{15}{8}\right)\left(\frac{4}{3}\right)(1 + x^2)^{1/3}(2x) + 0 = 5x(1 + x^2)^{1/3}$$

21. Evaluate the indefinite integral

$$\int \frac{1}{\sqrt{x}(1 + \sqrt{x})^2} \, dx$$

and check your results by differentiation.

Solution

To evaluate $\int \dfrac{1}{\sqrt{x}(1 + \sqrt{x})^2} \, dx$, we use the pattern

recognition method letting $u = 1 + \sqrt{x}$, and $u' = \dfrac{1}{2\sqrt{x}}$.

$$\int \frac{1}{\sqrt{x}(1 + \sqrt{x})^2} \, dx = 2 \int \overbrace{(1 + \sqrt{x})^{-2}}^{u^n} \overbrace{\left(\frac{1}{2\sqrt{x}}\right)}^{u'} \, dx$$

$$= 2\frac{(1 + \sqrt{x})^{-1}}{-1} + C = \frac{-2}{1 + \sqrt{x}} + C$$

Check

If $y = \dfrac{-2}{1 + \sqrt{x}} + C = -2(1 + \sqrt{x})^{-1} + C$, then

$$\frac{dy}{dx} = -2(-1)(1 + \sqrt{x})^{-2}\left(\frac{1}{2\sqrt{x}}\right) + 0 = \frac{1}{\sqrt{x}(1 + \sqrt{x})^2}$$

22. Evalute the indefinite integral

$$\int \left(1 + \frac{1}{t}\right)^3 \left(\frac{1}{t^2}\right) dt$$

and check your results by differentiation.

Solution

To evaluate this integral, we use the pattern recognition
method letting $u = 1 + (1/t)$, and $u' = (-1/t^2)$. Thus we
have

$$\int \left(1 + \frac{1}{t}\right)^3 \left(\frac{1}{t^2}\right) dt = -\int \overbrace{\left(1 + \frac{1}{t}\right)^3}^{u^n} \overbrace{\left(\frac{-1}{t^2}\right)}^{u'} dt$$

$$= -\frac{[1 + (1/t)]^4}{4} + C = -\frac{1}{4}\left(1 + \frac{1}{t}\right)^4 + C$$

Check

If $y = -\dfrac{1}{4}\left(1 + \dfrac{1}{t}\right)^4 + C$, then

$$\frac{dy}{dt} = \left(-\frac{1}{4}\right)(4)\left(1 + \frac{1}{t}\right)^3\left(-\frac{1}{t^2}\right) + 0 = \left(1 + \frac{1}{t}\right)^3\left(\frac{1}{t^2}\right)$$

29. Evaluate the indefinite integral $\int(1/\sqrt{2x})\,dx$ and check your results by differentiation.

Solution

$$\int \frac{1}{\sqrt{2x}}\,dx = \int \frac{1}{\sqrt{2}\sqrt{x}}\,dx = \frac{1}{\sqrt{2}}\int x^{-1/2}\,dx = \frac{1}{\sqrt{2}}\left(\frac{x^{1/2}}{\frac{1}{2}}\right) + C$$
$$= \sqrt{2x} + C$$

Check

If $y = \sqrt{2x} + C = (2x)^{1/2} + C$, then

$$\frac{dy}{dx} = \frac{1}{2}(2x)^{-1/2}(2) + 0 = \frac{1}{\sqrt{2x}}$$

33. Evaluate the indefinite integral

$$\int t^2\left(t - \frac{2}{t}\right)dt$$

and check your results by differentiation.

Solution

$$\int t^2\left(t - \frac{2}{t}\right)dt = \int(t^3 - 2t)\,dt = \frac{1}{4}t^4 - t^2 + C$$

Check

If $y = \frac{1}{4}t^4 - t^2 + C$, then

$$\frac{dy}{dt} = (4)\left(\frac{1}{4}\right)t^3 - 2t + 0$$
$$= t^3 - 2t = t^2\left(t - \frac{2}{t}\right)$$

41. Evaluate $\int x^2\sqrt{1 - x}\,dx$.

Solution

Let $u = \sqrt{1 - x}$. Then $u^2 = 1 - x$, $x = 1 - u^2$, and $dx = -2u\,du$. Thus

$$\int x^2\sqrt{1-x}\ dx = \int (1-u^2)^2 u(-2u)\ du$$

$$= -\int (2u^2 - 4u^4 + 2u^6)\ du$$

$$= -\left(\frac{2u^3}{3} - \frac{4u^5}{5} + \frac{2u^7}{7}\right) + C$$

$$= \frac{-2u^3}{105}(35 - 42u^2 + 15u^4) + C$$

$$= \frac{-2}{105}(1-x)^{3/2}[35 - 42(1-x)$$

$$+ \ 15(1-x)^2] + C$$

$$= \frac{-2}{105}(1-x)^{3/2}(15x^2 + 12x + 8) + C$$

43.   Evaluate $\displaystyle \int \frac{x^2 - 1}{\sqrt{2x - 1}}\ dx.$

Solution

Let $u = \sqrt{2x - 1}$. Then $u^2 = 2x - 1$, $x = \dfrac{u^2 + 1}{2}$, and $dx = u\ du$.

Thus

$$\int \frac{x^2 - 1}{\sqrt{2x - 1}}\ dx = \int \frac{[(u^2 + 1)/2]^2 - 1}{u}\, u\ du$$

$$= \frac{1}{4}\int (u^4 + 2u^2 - 3)\ du$$

$$= \frac{1}{4}\left(\frac{u^5}{5} + \frac{2u^3}{3} - 3u\right) + C$$

$$= \frac{u}{60}(3u^4 + 10u^2 - 45) + C$$

$$= \frac{1}{60}\sqrt{2x - 1}\ [3(2x - 1)^2$$

$$+ \ 10(2x - 1) - 45] + C$$

$$= \frac{1}{60}\sqrt{2x - 1}\ (12x^2 + 8x - 52) + C$$

$$= \frac{1}{15}\sqrt{2x - 1}\ (3x^2 + 2x - 13) + C$$

51.   Evaluate $\int x \cos x^2\ dx.$

Solution

If we let $u = x^2$, then $du = 2x\ dx$ and $\dfrac{du}{2} = x\ dx$. Therefore,

$$\int x \cos x^2\ dx = \int \cos u\ \frac{du}{2}$$

$$= \frac{1}{2} \int \cos u\ du = \frac{1}{2} \sin u + C = \frac{1}{2} \sin x^2 + C$$

56. Evaluate $\int \sin 2x \cos 2x\ dx$.

Solution

Using the double angle identity we obtain
$2 \sin 2x \cos 2x = \sin 4x$. If we let $u = 4x$, then $du = 4\ dx$
and $dx = \frac{1}{4}\ du$. Therefore,

$$\int \sin 2x \cos 2x\ dx = \frac{1}{2} \int \sin 4x\ dx$$

$$= \frac{1}{8} \int \sin u\ du$$

$$= -\frac{1}{8} \cos u + C = -\frac{1}{8} \cos 4x + C$$

59. Evaluate $\int \cot^2 x\ dx$.

Solution

$$\int \cot^2 x\ dx = \int (\csc^2 x - 1)\ dx = -\cot x - x + C$$

67. Find the equation of the curve $y = f(x)$, given $dy/dx = x\sqrt{1 - x^2}$ and $(0, \frac{4}{3})$ is one point on the curve.

Solution

$$\frac{dy}{dx} = f'(x) = x\sqrt{1 - x^2}$$

$$f(x) = \int x\sqrt{1 - x^2}\ dx \qquad \text{If we let } u = 1 - x^2, \text{ then}$$
$$\qquad\qquad\qquad\qquad\qquad u' = -2x.$$

$$= -\frac{1}{2} \int \overbrace{(1 - x^2)^{1/2}}^{u^n} \overbrace{(-2x)}^{u'}\ dx$$

$$= -\frac{1}{2} \left[\frac{(1 - x^2)^{3/2}}{3/2} \right] + C = -\frac{(1 - x^2)^{3/2}}{3} + C$$

Since $(0, \frac{4}{3})$ lies on the curve, we know that $f(0) = \frac{4}{3}$. Thus

$$f(0) = \frac{-(1 - 0^2)^{3/2}}{3} + C = \frac{4}{3} \quad \text{or} \quad C = \frac{5}{3}$$

Therefore, $f(x) = \dfrac{-(1 - x^2)^{3/2}}{3} + \dfrac{5}{3}$

5.3
Sigma notation and limit of a sequence

3. Find the sum

$$\sum_{k=0}^{4} \frac{1}{1 + k^2}$$

Solution

$$\sum_{k=0}^{4} \frac{1}{1 + k^2} = \frac{1}{1 + 0^2} + \frac{1}{1 + 1^2} + \frac{1}{1 + 2^2} + \frac{1}{1 + 3^2} + \frac{1}{1 + 4^2}$$

$$= \frac{1}{1} + \frac{1}{2} + \frac{1}{5} + \frac{1}{10} + \frac{1}{17}$$

$$= \frac{170 + 85 + 34 + 17 + 10}{170}$$

$$= \frac{316}{170} = \frac{158}{85}$$

15. Write the sum

$$\left[\left(\frac{2}{n}\right)^3 - \frac{2}{n} \right]\left(\frac{2}{n}\right) + \left[\left(\frac{4}{n}\right)^3 - \frac{4}{n} \right]\left(\frac{2}{n}\right) + \cdots + \left[\left(\frac{2n}{n}\right)^3 - \frac{2n}{n} \right]\left(\frac{2}{n}\right)$$

in sigma notation.

Solution

We begin by noting that the n terms in this sum are each of the form

$$f(i) = \left[\left(\frac{2i}{n}\right)^3 - \frac{2i}{n} \right]\left(\frac{2}{n}\right)$$

Furthermore, we observe that in the first term $i = 1$, in the second term $i = 2$, and so on until we reach the nth term. Thus our index i runs from 1 to n, and the sigma notation for the given sum is

$$\sum_{i=1}^{n} f(i) = \sum_{i=1}^{n} \left[\left(\frac{2i}{n}\right)^3 - \frac{2i}{n} \right]\left(\frac{2}{n}\right)$$

23. Use the properties of sigma notation and the expressions for the sums of powers of the first n positive integers to evaluate the sum

$$\sum_{i=1}^{15} \frac{1}{n^3}(i - 1)^2$$

Solution

$$\sum_{i=1}^{15} \frac{1}{n^3}(i - 1)^2 = \frac{1}{n^3} \sum_{i=1}^{15}(i^2 - 2i + 1)$$

$$= \frac{1}{n^3}\left[\sum_{i=1}^{15} i^2 - 2\sum_{i=1}^{15} i + \sum_{i=1}^{15} 1\right]$$

$$= \frac{1}{n^3}\left[\frac{15(16)(31)}{6} - 2\frac{15(16)}{2} + 15\right]$$

$$= \frac{1}{n^3}(1240 - 240 + 15) = \frac{1015}{n^3}$$

27. Find the limit of the sequence $s(n)$ as $n \to \infty$ where

$$s(n) = \frac{81}{n^4}\left[\frac{n^2(n + 1)^2}{4}\right]$$

Solution

$$\lim_{n\to\infty} s(n) = \lim_{n\to\infty} \frac{81}{n^4}\left[\frac{n^2(n + 1)^2}{4}\right]$$

$$= \lim_{n\to\infty} \frac{81}{4}\left(\frac{n^4 + 2n^3 + n^2}{n^4}\right)$$

$$= \lim_{n\to\infty} \frac{81}{4}\left(1 + \frac{2}{n} + \frac{1}{n^2}\right) = \frac{81}{4}$$

32. Find

$$\lim_{n\to\infty} \sum_{i=1}^{n}\left(\frac{1 + 2i}{n}\right)^2\left(\frac{2}{n}\right)$$

Solution

$$\sum_{i=1}^{n}\left(1 + \frac{2i}{n}\right)^2\left(\frac{2}{n}\right) = \sum_{i=1}^{n}\left(1 + \frac{4i}{n} + \frac{4i^2}{n^2}\right)\left(\frac{2}{n}\right)$$

$$= \frac{2}{n}\left[\sum_{i=1}^{n} 1 + \frac{4}{n}\sum_{i=1}^{n} i + \frac{4}{n^2}\sum_{i=1}^{n} i^2\right]$$

$$= \frac{2}{n}\left[n + \frac{4}{n}\left(\frac{n(n+1)}{2}\right) \right.$$

$$\left. + \frac{4}{n^2}\left(\frac{n(n+1)(2n+1)}{6}\right) \right]$$

$$= \frac{2n}{n} + \frac{4n^2 + 4n}{n^2} + \frac{8n^3 + 12n^2 + 4n}{3n^3}$$

$$= 2 + 4 + \frac{4}{n} + \frac{8}{3} + \frac{4}{n} + \frac{4}{3n^2}$$

$$= \frac{26}{3} + \frac{8}{n} + \frac{4}{3n^2}$$

Therefore

$$\lim_{n\to\infty} \sum_{i=1}^{n}\left(1 + \frac{2i}{n}\right)^2\left(\frac{2}{n}\right) = \lim_{n\to\infty}\left(\frac{26}{3} + \frac{8}{n} + \frac{4}{3n^2}\right) = \frac{26}{3}$$

5.4
Area

5. Use the upper and lower sums to approximate the area of the region between the graph of $y = \sqrt{1 - x^2}$ and the x-axis over the interval $[0, 1]$. Use five subdivisions.

Solution

Dividing the interval into five parts, we have

$$x_0 = 0, \qquad x_1 = 0.2, \qquad x_2 = 0.4, \qquad x_3 = 0.6,$$
$$x_4 = 0.8, \qquad x_5 = 1$$

Since y is decreasing from 0 to 1, the lower sum is obtained by using the *right* endpoints of the five subintervals. Thus

$$s = 0.2\sqrt{1 - (0.2)^2} + 0.2\sqrt{1 - (0.4)^2} + 0.2\sqrt{1 - (0.6)^2}$$
$$+ 0.2\sqrt{1 - (0.8)^2} + 0.2\sqrt{1 - (1)^2}$$
$$= 0.2(1 + \sqrt{0.96} + \sqrt{0.84} + \sqrt{0.64} + \sqrt{0.36})$$
$$\approx 0.2(0.9798 + 0.9165 + 0.8 + 0.6) \approx 0.659$$

Similarly, the upper sum is obtained by using the *left* endpoints of the five subdivisions. Thus

$$S = 0.2\sqrt{1 - 0^2} + 0.2\sqrt{1 - (0.2)^2} + 0.2\sqrt{1 - (0.4)^2}$$
$$+ 0.2\sqrt{1 - (0.6)^2} + 0.2\sqrt{1 - (0.8)^2}$$

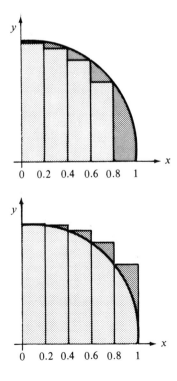

$$= 0.2(1 + \sqrt{0.96} + \sqrt{0.84} + \sqrt{0.64} + \sqrt{0.36})$$
$$\approx 0.2(1 + 0.9798 + 0.9165 + 0.8 + 0.6) \approx 0.859$$

9. Use the limit process to find the area of the region between the graph of $y = -2x + 3$ and the x-axis over the interval $[0, 1]$. Sketch the region.

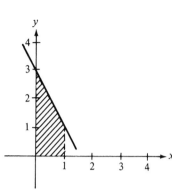

Solution

Let $\Delta x = (1 - 0)/n = 1/n$. Choosing right endpoints, we have

$$c_i = 0 + i\left(\frac{1}{n}\right) = \frac{i}{n}$$

Therefore,

$$s(n) = \sum_{i=1}^{n} f\left(\frac{i}{n}\right)\left(\frac{1}{n}\right) = \frac{1}{n} \cdot \sum_{i=1}^{n} \left[-2\left(\frac{i}{n}\right) + 3\right]$$

$$= \frac{1}{n}\left[\left(\frac{-2}{n}\right)\sum_{i=1}^{n} i + \sum_{i=1}^{n} 3\right]$$

$$= \frac{1}{n}\left[\left(\frac{-2}{n}\right)\left(\frac{n(n + 1)}{2}\right) + 3n\right] = -\frac{n + 1}{n} + 3 = 2 - \frac{1}{n}$$

Finally we have

$$\text{area} = \lim_{n \to \infty} s(n) = \lim_{n \to \infty} \left(2 - \frac{1}{n}\right) = 2$$

13. Use the limit process to find the area of the region between the graph of $y = 2x^2$ and the x-axis over the interval $[1, 3]$. Sketch the region.

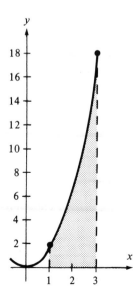

Solution

Let $\Delta x = (3 - 1)/n = 2/n$. Choosing right endpoints, we have $c_i = 1 + i(2/n)$. Therefore,

$$S(n) = \sum_{i=1}^{n} f\left(1 + \frac{2i}{n}\right)\left(\frac{2}{n}\right) = \sum_{i=1}^{n} 2\left(1 + \frac{2i}{n}\right)^2\left(\frac{2}{n}\right)$$

$$= \frac{4}{n}\left[\sum_{i=1}^{n} \left(1 + \frac{4i}{n} + \frac{4i^2}{n^2}\right)\right] = \frac{4}{n}\left[\sum_{i=1}^{n} 1 + \frac{4}{n}\sum_{i=1}^{n} i + \frac{4}{n^2}\sum_{i=1}^{n} i^2\right]$$

$$= \frac{4}{n}\left[n + \frac{4}{n} \cdot \frac{n(n + 1)}{2} + \frac{4}{n^2} \cdot \frac{n(n + 1)(2n + 1)}{6}\right]$$

$$= \frac{4n}{n} + \frac{8n^2 + 8n}{n^2} + \frac{16n^3 + 24n^2 + 8n}{3n^3}$$

$$= 4 + 8 + \frac{8}{n} + \frac{16}{3} + \frac{8}{n} + \frac{8}{3n^2} = \frac{52}{3} + \frac{16}{n} + \frac{8}{3n^2}$$

Finally, we have

$$\text{area} = \lim_{n \to \infty} S(n) = \lim_{n \to \infty} \left(\frac{52}{3} + \frac{16}{n} + \frac{8}{3n^2} \right) = \frac{52}{3}$$

17. Use the limit process to find the area of the region between the graph of $y = x^2 - x^3$ and the x-axis over the interval $[-1, 0]$. Sketch the region.

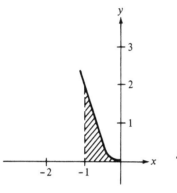

Solution

Let $\Delta x = [0 - (-1)]/n = 1/n$. Choosing right endpoints, we have

$$c_i = -1 + i\left(\frac{1}{n}\right) = -1 + \frac{i}{n}$$

Therefore,

$$s(n) = \sum_{i=1}^{n} f\left(-1 + \frac{i}{n}\right)\left(\frac{1}{n}\right) = \sum_{i=1}^{n} \left[\left(-1 + \frac{i}{n}\right)^2 - \left(-1 + \frac{i}{n}\right)^3\right]\frac{1}{n}$$

$$= \frac{1}{n}\sum_{i=1}^{n} \left[2 - \frac{5i}{n} + \frac{4i^2}{n^2} - \frac{i^3}{n^3}\right]$$

$$= \frac{1}{n}\sum_{i=1}^{n} 2 - \frac{5}{n^2}\sum_{i=1}^{n} i + \frac{4}{n^3}\sum_{i=1}^{n} i^2 - \frac{1}{n^4}\sum_{i=1}^{n} i^3$$

$$= \frac{1}{n}(2n) - \left(\frac{5}{n^2}\right)\left[\frac{n(n+1)}{2}\right] + \left(\frac{4}{n^3}\right)\left[\frac{n(n+1)(2n+1)}{6}\right]$$

$$\quad - \left(\frac{1}{n^4}\right)\left[\frac{n^2(n+1)^2}{4}\right]$$

$$= 2 - \frac{5}{2} - \frac{5}{2n} + \frac{4}{3} + \frac{2}{n} + \frac{2}{3n^3} - \frac{1}{4} - \frac{1}{2n} - \frac{1}{4n^2}$$

Finally we have

$$\text{area} = \lim_{n \to \infty} s(n) = 2 - \frac{5}{2} + \frac{4}{3} - \frac{1}{4} = \frac{7}{12}$$

5.5
Riemann sums and the definite integral

15. Sketch the region whose area is indicated by $\int_0^2 (2x + 5)\, dx$. Then use a geometric formula to evaluate the integral.

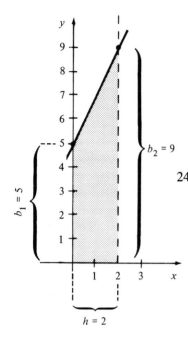

$b_1 = 5$

$h = 2$

$b_2 = 9$

Solution

The region whose area is given by $\int_0^2 (2x + 5)\, dx$ is shown by the accompanying figure to be a trapezoid. Since the height of the trapezoid is $h = 2$ and the lengths of the two bases are $b_1 = 5$ and $b_2 = 9$, the area of the trapezoid is

$$A = h\left[\frac{b_1 + b_2}{2}\right] = 2\left[\frac{5 + 9}{2}\right] = 14$$

24. If $\int_{-1}^1 f(x)\, dx = 0$ and $\int_0^1 f(x)\, dx = 5$, find:

(a) $\displaystyle\int_{-1}^0 f(x)\, dx$

(b) $\displaystyle\int_0^1 f(x)\, dx - \int_{-1}^0 f(x)\, dx$

(c) $\displaystyle\int_{-1}^1 3f(x)\, dx$

(d) $\displaystyle\int_0^1 3f(x)\, dx$

Solution

(a) $\displaystyle\int_{-1}^0 f(x)\, dx = \int_{-1}^1 f(x)\, dx - \int_0^1 f(x)\, dx$

$$= 0 - 5 = -5$$

(b) $\displaystyle\int_0^1 f(x)\, dx - \int_{-1}^0 f(x)\, dx = 5 - (-5)$ From part (a)

$$= 10$$

(c) $\displaystyle\int_{-1}^1 3f(x)\, dx = 3 \int_{-1}^1 f(x)\, dx = 3(0) = 0$

(d) $\displaystyle\int_0^1 3f(x)\, dx = 3 \int_0^1 f(x)\, dx = 3(5) = 15$

27. Evaluate the definite integral $\int_{-1}^1 x^3\, dx$.

Solution

Let $\Delta x = [1 - (-1)]/n = 2/n$. Using right-hand endpoints, we have $c_i = -1 + (2i/n)$, and the definite integral is given by the limit

$$\int_{-1}^1 x^3\, dx = \lim_{n \to \infty} \sum_{i=1}^n \left(-1 + \frac{2i}{n}\right)^3 \left(\frac{2}{n}\right)$$

$$= \lim_{n \to \infty} \sum_{i=1}^n \left(\frac{2}{n}\right)\left(-1 + \frac{6i}{n} - \frac{12i^2}{n^2} + \frac{8i^3}{n^3}\right)$$

$$= \lim_{n \to \infty} \left(\frac{2}{n}\right)\left[-\sum_{i=1}^{n} 1 + \frac{6}{n}\sum_{i=1}^{n} i - \frac{12}{n^2}\sum_{i=1}^{n} i^2 + \frac{8}{n^3}\sum_{i=1}^{n} i^3 \right]$$

$$= \lim_{n \to \infty} \left(\frac{2}{n}\right)\left[-n + \frac{6n(n+1)}{2n} - \frac{12n(n+1)(2n+1)}{6n^2} \right.$$
$$\left. + \frac{8n^2(n+1)^2}{4n^3} \right]$$

$$= \lim_{n \to \infty} \left[\frac{-2n}{n} + \frac{6n(n+1)}{n^2} - \frac{4n(n+1)(2n+1)}{n^3} \right.$$
$$\left. + \frac{4n^2(n+1)^2}{n^4} \right]$$

$$= -2 + 6 - 8 + 4 = 0$$

32. Use Example 1 as a model to evaluate the limit

$$\lim_{n \to \infty} \sum_{i=1}^{n} f(c_i)\,\Delta x_i$$

over the region bounded by the graphs of $f(x) = \sqrt[3]{x}$, $y = 0$, $x = 0$ and $x = 1$.

Solution

If we let $c_i = \dfrac{i^2}{n^3}$ then the width of the ith subinterval is given by

$$\Delta x_i = \frac{i^3}{n^3} - \frac{(i-1)^3}{n^3}$$

$$= \frac{i^3 - i^3 + 3i^2 - 3i + 1}{n^3} = \frac{3i^2 - 3i + 1}{n^3}$$

Thus, we have

$$\lim_{n \to \infty} \sum_{i=1}^{n} f(c_i)\Delta x_i = \lim_{n \to \infty} \sum_{i=1}^{n} \sqrt[3]{\frac{i^3}{n^3}}\left(\frac{3i^2 - 3i + 1}{n^3}\right)$$

$$= \lim_{n \to \infty} \sum_{i=1}^{n} \frac{3i^3 - 3i^2 + i}{n^4}$$

$$= \lim_{n \to \infty} \frac{1}{n^4}\left[\frac{3n^2(n+1)^2}{4} - \frac{3n(n+1)(2n+1)}{6} \right.$$
$$\left. + \frac{n(n+1)}{2} \right]$$

$$= \lim_{n \to \infty} \frac{3n^4 + 2n^3 - n^2}{4n^4} = \frac{3}{4}$$

5.6
The Fundamental Theorem of Calculus

7.   Evaluate the definite integral $\int_0^1 (2t - 1)^2 \, dt$.

Solution

$$\int_0^1 (2t - 1)^2 \, dt = \int_0^1 (4t^2 - 4t + 1) \, dt = \left[\frac{4t^3}{3} - \frac{4t^2}{2} + t \right]_0^1$$

$$= \left(\frac{4}{3} - \frac{4}{2} + 1 \right) - (0 - 0 + 0) = \frac{4}{3} - \frac{6}{3} + \frac{3}{3} = \frac{1}{3}$$

14.   Evaluate the indefinite integral $\int_{-2}^{-1} \sqrt{\dfrac{-2}{x}} \, dx$.

Solution

$$\int_{-2}^{-1} \sqrt{\frac{-2}{x}} \, dx = \int_{-2}^{-1} \sqrt{2}\,(-x)^{-1/2} \, dx$$

$$= -\sqrt{2} \int_{-2}^{-1} \overbrace{(-x)^{-1/2}}^{u^n}\,\overbrace{(-1)}^{u'} \, dx$$

$$= -\sqrt{2} \left[\frac{(-x)^{1/2}}{1/2} \right]_{-2}^{-1} = -2\sqrt{2}\,(1 - \sqrt{2})$$

$$= 4 - 2\sqrt{2}$$

15.   Evaluate the definite integral $\int_1^4 \dfrac{u - 2}{\sqrt{u}} \, du$.

Solution

$$\int_1^4 \frac{u - 2}{\sqrt{u}} \, du = \int_1^4 (u^{1/2} - 2u^{-1/2}) \, du = \left[\frac{2}{3}u^{3/2} - 4u^{1/2} \right]_1^4$$

$$= \left[\frac{2}{3}(\sqrt{4})^3 - 4\sqrt{4} \right] - \left(\frac{2}{3} - 4 \right) = \frac{2}{3}$$

23.   Evaluate the definite integral $\int_{-1}^1 x(x^2 + 1)^3 \, dx$.

Solution

$$\int_{-1}^1 x(x^2 + 1)^3 \, dx = \frac{1}{2} \int_{-1}^1 \overbrace{(x^2 + 1)^3}^{u^n}\,\overbrace{(2x)}^{u'} \, dx = \frac{1}{2} \left[\frac{(x^2 + 1)^4}{4} \right]_{-1}^1$$

$$= \frac{1}{8}(2^4 - 2^4) = 0$$

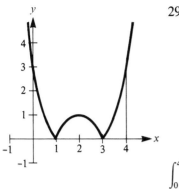

29. Evaluate the definite integral $\int_0^4 |x^2 - 4x + 3| \, dx$.

Solution

Since $x^2 - 4x + 3 = (x - 1)(x - 3)$, we know that $x^2 - 4x + 3$ is positive in the interval $(0, 1)$, negative in the interval $(1, 3)$, and positive in the interval $(3, 4)$. Therefore we can omit the absolute value signs by considering the three following integrals. (Note the change in signs.)

$$\int_0^4 |x^2 - 4x + 3| \, dx = \int_0^1 (x^2 - 4x + 3) \, dx + \int_1^3 (-x^2 + 4x - 3) \, dx$$

$$+ \int_3^4 (x^2 - 4x + 3) \, dx$$

$$= \left[\frac{x^3}{3} - \frac{4x^2}{2} + 3x \right]_0^1 + \left[-\frac{x^3}{3} + \frac{4x^2}{2} - 3x \right]_1^3$$

$$+ \left[\frac{x^3}{3} - \frac{4x^2}{2} + 3x \right]_3^4$$

$$= \left(\frac{1}{3} - 2 + 3 \right) - (0) + (-9 + 18 - 9)$$

$$- \left(-\frac{1}{3} + 2 - 3 \right) + \left(\frac{64}{3} - 32 + 12 \right)$$

$$- (9 - 18 + 9)$$

$$= \frac{4}{3} - 0 + 0 - \left(-\frac{4}{3} \right) + \frac{4}{3} - 0 = \frac{12}{3} = 4$$

35. Evaluate the definite integral $\int_0^7 x\sqrt[3]{x + 1} \, dx$.

Solution

Let $u = \sqrt[3]{x + 1}$. Then $u^3 = x + 1$, $x = u^3 - 1$, and $dx = 3u^2 \, du$. Furthermore, if $x = 0$, then $u = 1$, and if $x = 7$, then $u = 2$. Thus

$$\int_0^7 x\sqrt[3]{x + 1} \, dx = \int_1^2 (u^3 - 1)(u)(3u^2 \, du)$$

$$= 3 \int_1^2 (u^6 - u^3) \, du = 3 \left[\frac{u^7}{7} - \frac{u^4}{4} \right]_1^2$$

$$= 3 \left(\frac{128}{7} - \frac{16}{4} - \frac{1}{7} + \frac{1}{4} \right)$$

$$= 3 \left(\frac{127}{7} - \frac{15}{4} \right) = 3 \left(\frac{508 - 105}{28} \right) = \frac{1209}{28}$$

41. Evaluate the definite integral $\int_{\pi/2}^{2\pi/3} \sec^2\left(\dfrac{x}{2}\right) dx$.

Solution

$$\int_{\pi/2}^{2\pi/3} \sec^2\left(\frac{x}{2}\right) dx = 2 \int_{\pi/2}^{2\pi/3} \overset{u}{\sec^2\left(\frac{x}{2}\right)} \overset{u'}{\left(\frac{1}{2}\right)} dx$$

$$= 2\left[\tan\left(\frac{x}{2}\right) \right]_{\pi/2}^{2\pi/3}$$

$$= 2\left[\tan\frac{\pi}{3} - \tan\frac{\pi}{4} \right] = 2(\sqrt{3} - 1)$$

48. Determine the area of the region bounded by $y = -x^2 + 2x + 3$ and $y = 0$.

Solution

From the accompanying figure we see that the area of the region is given by the definite integral $\int_{-1}^{3}(-x^2 + 2x + 3)\, dx$.

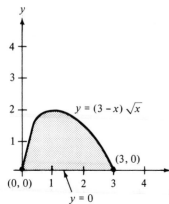

$$\text{area} = \int_{-1}^{3}(-x^2 + 2x + 3)\, dx = \left[-\frac{x^3}{3} + x^2 + 3x \right]_{-1}^{3}$$

$$= (-9 + 9 + 9) - \left(\frac{1}{3} + 1 - 3\right) = \frac{32}{3}$$

52. Determine the area of the region having the given boundaries in the accompanying figure.

Solution

From the accompanying figure we see that the area of the region is given by the definite integral $\int_{0}^{3}(3 - x)\sqrt{x}\, dx$. Thus

$$\text{area} = \int_{0}^{3}(3 - x)\sqrt{x}\, dx = \int_{0}^{3}(3x^{1/2} - x^{3/2})\, dx$$

$$= \left[\frac{3x^{3/2}}{3/2} - \frac{x^{5/2}}{5/2} \right]_{0}^{3} = \left[2(3^{3/2}) - \frac{2(3^{5/2})}{5} \right] - [0 - 0]$$

$$= 2\sqrt{3}\left(3 - \frac{9}{5}\right) = \frac{12\sqrt{3}}{5} \approx 4.157$$

55. Determine the area of the region having the given boundaries in the accompanying figure.

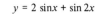

$y = 2 \sin x + \sin 2x$

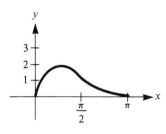

Solution

From the accompanying figure we see that the area of the region is given by the definite integral $\int_0^\pi (2 \sin x + \sin 2x)\, dx$.

$$\text{area} = \int_0^\pi (2 \sin x + \sin 2x)\, dx$$

$$= 2 \int_0^\pi \sin x \, dx + \frac{1}{2} \int_0^\pi \sin 2x (2)\, dx$$

$$= \left[-2 \cos x - \frac{1}{2} \cos 2x \right]_0^\pi = 4$$

63. Sketch the graph of $f(x) = x\sqrt{4 - x^2}$ over the interval $[0, 2]$. Find the average value of $f(x)$ over this interval and find all values of x where $f(x)$ equals its average.

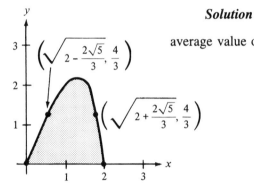

Solution

average value of $f(x)$ over $[0, 2]$ = $\dfrac{\int_0^2 x\sqrt{4 - x^2}\, dx}{2 - 0}$

$$= \left(\frac{1}{2} \right) \int_0^2 x\sqrt{4 - x^2}\, dx$$

$$= \left(\frac{1}{2} \right)\left(-\frac{1}{2} \right) \int_0^2 \overbrace{(4 - x^2)^{1/2}}^{u^n} \overbrace{(-2x)}^{u'}\, dx$$

$$= \frac{-1}{4} \left[\frac{(4 - x^2)^{3/2}}{\frac{3}{2}} \right]_0^2$$

$$= \frac{-1}{6}(0 - 8) = \frac{4}{3}$$

To find the values of x for which $f(x) = \frac{4}{3}$ in the interval $[0, 2]$, we solve the equation

$$f(x) = x\sqrt{4 - x^2} = \frac{4}{3}$$

$$x^2(4 - x^2) = \frac{16}{9}$$

$$36x^2 - 9x^4 = 16$$

$$9x^4 - 36x^2 + 16 = 0$$

$$x^2 = \frac{36 \pm \sqrt{36^2 - 4(16)(9)}}{2(9)} = \frac{36 \pm 12\sqrt{5}}{18}$$

$$x^2 = 2 \pm \frac{2\sqrt{5}}{3}$$

$$x = \sqrt{2 \pm \frac{2\sqrt{5}}{3}} \approx 1.868 \qquad \text{or} \qquad 0.714$$

75. The sales of a seasonal product are given by the model

$$S = 74.50 + 43.75 \sin \frac{\pi t}{6}$$

where S is measured in thousands of units and t is the time in months, with $t = 1$ corresponding to January. Find the average sales during: (a) the first quarter, (b) the second quarter, and (c) the entire year.

Solution

(a) The first quarter average

$$= \frac{1}{3 - 0} \int_0^3 \left(74.50 + 43.75 \sin \frac{\pi t}{6} \right) dt$$

$$= \frac{1}{3} \left[74.50t - 43.75 \left(\frac{6}{\pi} \right) \cos \frac{\pi t}{6} \right]_0^3$$

$$= \frac{1}{3} \left[(223.5 - 0) - \left(0 - \frac{43.75[6]}{\pi} \right) \right]$$

$$\approx 102.352 \text{ thousand units}$$

(b) The second quarter average

$$= \frac{1}{3 - 0} \int_3^6 \left(74.50 + 43.75 \sin \frac{\pi t}{6} \right) dt$$

$$= \frac{1}{3} \left[74.50t - 43.75 \left(\frac{6}{\pi} \right) \cos \frac{\pi t}{6} \right]_3^6$$

$$= \frac{1}{3} \left[\left(447 - \frac{-43.75[6]}{\pi} \right) \right.$$

$$\left. - (223.5 - 0) \right]$$

$$\approx 102.352 \text{ thousand units}$$

(c) The yearly average

$$= \frac{1}{12 - 0} \int_0^{12} \left(74.50 + 43.75 \sin \frac{\pi t}{6} \right) dt$$

$$= \frac{1}{12} \left[74.50t - 43.75 \left(\frac{6}{\pi} \right) \cos \frac{\pi t}{6} \right]_0^{12}$$

$$= \frac{1}{12} \left[\left(894 - \frac{43.75[6]}{\pi} \right) \right.$$

$$- \left(0 - \frac{43.75[6]}{\pi} \right) \Big]$$

$$\approx 74.5 \text{ thousand units}$$

5.7
Variable bounds of integration and the natural logarithmic function

3. (a)   Integrate

$$F(x) = \int_7^x \sqrt[3]{t + 1}\ dt$$

to find F as a function of x and (b) demonstrate Theorem 5.18 by finding $F'(x)$.

Solution

(a) $F(x) = \displaystyle\int_7^x \sqrt[3]{t + 1}\ dt = \int_7^x (t + 1)^{1/3}\ dt$

$$= \left[\frac{3}{4}(t + 1)^{4/3} \right]_7^x$$

$$= \frac{3}{4}(x + 1)^{4/3} - 12$$

(b) $F'(x) = \dfrac{d}{dx}\left[\displaystyle\int_7^x \sqrt[3]{t + 1}\ dt \right] = \dfrac{d}{dx}\left[\dfrac{3}{4}(x + 1)^{4/3} - 12 \right]$

$$= \sqrt[3]{x + 1}$$

11.   Use Theorem 5.18 to find $F'(x)$ if $F(x) = \int_0^x t \cos t\ dt$.

Solution

By Theorem 5.18, we have

$$F'(x) = \frac{d}{dx}\left[\int_0^x t \cos t\ dt \right] = x \cos x$$

17.   Using properties of logarithms and the fact that $\ln 2 \approx 0.6931$ and $\ln 3 \approx 1.0986$, find the following:

(a) $\ln 6$ (b) $\ln \frac{2}{3}$

(c) $\ln 81$ (d) $\ln \sqrt{3}$

Solution

(a) $\ln 6 = \ln 2(3) = \ln 2 + \ln 3$

$$\approx 0.6931 + 1.0986$$

$$= 1.7917$$

(b) $\ln \frac{2}{3} = \ln 2 - \ln 3$

$\approx 0.6931 - 1.0986$

$= -0.4055$

(c) $\ln 81 = \ln (3^4) = 4 \ln 3$

$\approx 4(1.0986) = 4.3944$

(d) $\ln \sqrt{3} = \ln (3^{1/2}) = \frac{1}{2} \ln 3$

$\approx \frac{1}{2}(1.0986) = 0.5493$

25. Use the properties of logarithms to write

$$\ln \left(\frac{x^2 - 1}{x^3} \right)^3$$

as a sum, difference, or multiple of logarithms.

Solution

$$\ln \left(\frac{x^2 - 1}{x^3} \right)^3 = 3[\ln (x^2 - 1) - \ln x^3]$$

$$= 3[\ln [(x + 1)(x - 1)] - 3 \ln x]$$

$$= 3[\ln (x + 1) + \ln (x - 1) - 3 \ln x]$$

32. Express $2[\ln x - \ln (x + 1) - \ln (x - 1)]$ as a single logarithm.

Solution

$2[\ln x - \ln (x + 1) - \ln (x - 1)]$

$$= 2\{\ln x - [\ln (x + 1) + \ln (x - 1)]\}$$

$$= 2[\ln x - \ln (x + 1)(x - 1)]$$

$$= 2[\ln x - \ln (x^2 - 1)]$$

$$= 2 \ln \frac{x}{x^2 - 1} = \ln \frac{x^2}{(x^2 - 1)^2}$$

5.8
The natural logarithmic function and differentiation

11. Find dy/dx given $y = \ln (x\sqrt{x^2 - 1})$.

Solution

$$y = \ln (x\sqrt{x^2 - 1}) = \ln x + \ln \sqrt{x^2 - 1}$$

$$= \ln x + \frac{1}{2} \ln (x^2 - 1)$$

$$\frac{dy}{dx} = \frac{1}{x} + \frac{1}{2}\left(\frac{1}{x^2 - 1}\right)(2x) = \frac{1}{x} + \frac{x}{x^2 - 1} = \frac{2x^2 - 1}{x(x^2 - 1)}$$

15.   Find dy/dx given $y = (\ln x)/x^2$.

Solution

$$y = \frac{\ln x}{x^2}$$

$$\frac{dy}{dx} = \frac{(x^2)(1/x) - (\ln x)(2x)}{x^4} = \frac{x - 2x \ln x}{x^4} = \frac{1 - 2 \ln x}{x^3}$$

23.   Find dy/dx given $y = -\sqrt{x^2 + 1}/x + \ln (x + \sqrt{x^2 + 1})$.

Solution

$$y = \frac{-\sqrt{x^2 + 1}}{x} + \ln (x + \sqrt{x^2 + 1})$$

$$\frac{dy}{dx} = -\frac{x(\frac{1}{2})(x^2 + 1)^{-1/2}(2x) - (x^2 + 1)^{1/2}(1)}{x^2}$$

$$+ \frac{1}{x + \sqrt{x^2 + 1}}\left[1 + \left(\frac{1}{2}\right)(x^2 + 1)^{-1/2}(2x)\right]$$

$$= -\frac{x^2(x^2 + 1)^{-1/2} - (x^2 + 1)^{1/2}}{x^2}$$

$$+ \left(\frac{1}{x + \sqrt{x^2 + 1}}\right)\left(1 + \frac{x}{\sqrt{x^2 + 1}}\right)$$

$$= \frac{-x^2 + (x^2 + 1)}{x^2\sqrt{x^2 + 1}} + \left(\frac{1}{x + \sqrt{x^2 + 1}}\right)\frac{\sqrt{x^2 + 1} + x}{\sqrt{x^2 + 1}}$$

$$= \frac{1}{x^2\sqrt{x^2 + 1}} + \frac{1}{\sqrt{x^2 + 1}} = \frac{1 + x^2}{x^2\sqrt{x^2 + 1}} = \frac{\sqrt{x^2 + 1}}{x^2}$$

27.   Find dy/dx given $y = \ln \left| \dfrac{\cos x}{\cos x - 1} \right|$.

Solution

$$y = \ln \left| \frac{\cos x}{\cos x - 1} \right| = \ln |\cos x| - \ln |\cos x - 1|$$

$$\frac{dy}{dx} = \frac{1}{\cos x}(-\sin x) - \frac{1}{\cos x - 1}(-\sin x)$$

$$= -\sin x \left(\frac{1}{\cos x} - \frac{1}{\cos x - 1}\right) = \frac{\sin x}{\cos x (\cos x - 1)}$$

34. Find dy/dx by using logarithmic differentiation given $y = \sqrt[3]{\dfrac{x^2 + 1}{x^2 - 1}}$.

Solution

$$y = \sqrt[3]{\frac{x^2 + 1}{x^2 - 1}}$$

$$\ln y = \ln \sqrt[3]{\frac{x^2 + 1}{x^2 - 1}} = \frac{1}{3}[\ln (x^2 + 1) - \ln (x^2 - 1)]$$

$$\frac{y'}{y} = \frac{1}{3}\left(\frac{2x}{x^2 + 1} - \frac{2x}{x^2 - 1}\right)$$

$$y' = \left(\frac{2xy}{3}\right)\left(\frac{x^2 - 1 - x^2 - 1}{x^4 - 1}\right) = \left(\frac{2xy}{3}\right)\left(\frac{-2}{x^4 - 1}\right)$$

$$= \frac{-4xy}{3(x^4 - 1)}$$

(Note: If the solution is desired in terms of x alone, we can simply replace y by its original value.)

$$\frac{dy}{dx} = \frac{-4x}{3(x^4 - 1)} \sqrt[3]{\frac{x^2 + 1}{x^2 - 1}}$$

37. Show that $y = 2(\ln x) + 3$ is a solution to the differential equation $x(y'') + y' = 0$.

Solution

$$y = 2(\ln x) + 3$$

$$y' = 2\left(\frac{1}{x}\right)$$

$$y'' = 2\left(\frac{-1}{x^2}\right) = \frac{-2}{x^2}$$

$$x(y'') + y' = x\left(\frac{-2}{x^2}\right) + \left(\frac{2}{x}\right) = \left(\frac{-2}{x}\right) + \left(\frac{2}{x}\right) = 0$$

39. Find any relative extrema and inflection points for $y = (x^2/2) - \ln x$, and sketch the graph of the function.

Solution

$$y = \frac{x^2}{2} - \ln x \qquad \text{Domain:} \quad 0 < x$$

$$y' = x - \frac{1}{x} = 0$$

$$x = \frac{1}{x}$$

$$x^2 = 1$$

$$x = 1 \qquad\qquad x = -1 \text{ is not in the domain.}$$

$$y'' = 1 + \frac{1}{x^2} > 0$$

Since y'' is positive for all x in the domain, the graph is concave up and $(1, \frac{1}{2})$ is a relative minimum point. Furthermore, since y'' is never zero, there are no points of inflection. By plotting a few points, we have the graph shown in the accompanying figure.

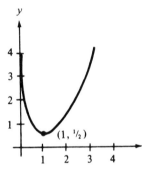

x	0.25	0.5	1	1.5	2	3
y	1.418	0.818	0.5	0.720	1.307	3.401

45. Use Newton's Method to approximate, to three decimal places, the value of x satisfying the equation $\ln x = -x$.

Solution

To find the value of x satisfying the equation $\ln x = -x$, we find the zeros of the function $f(x) = \ln x + x$.

$$f'(x) = \frac{1}{x} + 1$$

From the accompanying figure we choose $x = 0.5$ as the initial estimate.

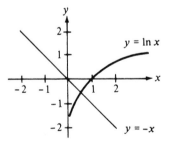

n	x_n	$f(x_n)$	$f'(x_n)$	$f(x_n)/f'(x_n)$	$x_n - \dfrac{f(x_n)}{f'(x_n)}$
1	0.5000	−0.1931	3	−0.0644	0.5644
2	0.5644	−0.0076	2.7718	−0.0027	0.5671
3	0.5671	0.0000	2.7632	0.0000	0.5671

Therefore, we approximate the root to be $x = 0.567$.

5.9
The natural logarithmic function and integration

5. Evaluate $\int \dfrac{x}{x^2 + 1}\, dx$.

Solution

Letting $u = x^2 + 1$, we have $du = 2x\, dx$. By multiplying and dividing by 2, we have

$$\int \frac{x}{x^2 + 1}\, dx = \frac{1}{2} \int \frac{2x}{x^2 + 1}\, dx = \frac{1}{2} \int \frac{1}{u}\, du$$

$$= \frac{1}{2} \ln u + C = \frac{1}{2} \ln (x^2 + 1) + C$$

$$= \ln \sqrt{x^2 + 1} + C$$

[Note that absolute value signs around $(x^2 + 1)$ are unnecessary since $x^2 + 1 > 0$ for all x.]

7. Evaluate $\int \dfrac{x^2 - 4}{x}\, dx$.

Solution

$$\int \frac{x^2 - 4}{x}\, dx = \int \left(\frac{x^2}{x} - \frac{4}{x} \right) dx = \int \left(x - \frac{4}{x} \right) dx$$

$$= \frac{x^2}{2} - 4 \ln |x| + C$$

11. Evaluate $\displaystyle\int_1^e \dfrac{(1 + \ln x)^2}{x}\, dx$.

Solution

Letting $u = 1 + \ln x$, we have $du = (1/x)\, dx$. When $x = 1$, $u = 1$ and when $x = e$, $u = 2$. Therefore,

$$\int_1^e \frac{(1 + \ln x)^2}{x}\, dx = \int_1^e (1 + \ln x)^2 \left(\frac{1}{x} \right) dx$$

$$= \int_1^2 u^2\, du = \left[\frac{u^3}{3} \right]_1^2 = \frac{7}{3}$$

15.   Evaluate $\displaystyle\int \frac{1}{\sqrt{x+1}}\, dx$.

Solution

$$\int \frac{1}{\sqrt{x+1}}\, dx = \int (x+1)^{-1/2}(1)\, dx$$

$$= \int u^{-1/2}\, du = 2(x+1)^{1/2} + C = 2\sqrt{x+1} + C$$

24.   Evaluate $\displaystyle\int_0^2 \frac{1}{1+\sqrt{2x}}\, dx$.

Solution

Let $u = \sqrt{2x}$. Then $u^2 = 2x$, $x = u^2/2$, and $dx = u\, du$.
Furthermore, when $x = 0$, then $u = 0$, and when $x = 2$, then
$u = 2$. Thus

$$\int_0^2 \frac{1}{1+\sqrt{2x}}\, dx = \int_0^2 \frac{u}{1+u}\, du$$

Now since the numerator and denominator are of equal degree,
we divide and obtain

$$\int_0^2 \frac{u}{1+u}\, du = \int_0^2 \left[1 - \frac{1}{1+u}\right] du$$

$$= \left[u - \ln(1+u)\right]_0^2 = 2 - \ln 3$$

25.   Evaluate $\displaystyle\int \frac{\sqrt{x}}{1-x\sqrt{x}}\, dx$.

Solution

Letting $u = 1 - x\sqrt{x} = 1 - x^{3/2}$, we have $du = -\frac{3}{2}\sqrt{x}\, dx$.
Therefore,

$$\int \frac{\sqrt{x}}{1-x\sqrt{x}}\, dx = \int \frac{\sqrt{x}}{1-x^{3/2}}\, dx$$

$$= \frac{-2}{3}\int \frac{1}{u}\, du = \frac{-2}{3}\ln|1-x\sqrt{x}| + C$$

27.   Evaluate $\displaystyle\int \frac{x(x-2)}{(x-1)^3}\, dx$.

Solution

$$\int \frac{x(x-2)}{(x-1)^3}\,dx = \int \frac{x^2 - 2x + 1 - 1}{(x-1)^3}\,dx = \int \frac{(x-1)^2 - 1}{(x-1)^3}\,dx$$

$$= \int \frac{1}{(x-1)}\,dx - \int \frac{1}{(x-1)^3}\,dx$$

$$= \int \frac{1}{(x-1)}\,dx - \int (x-1)^{-3}\,dx$$

$$= \int \frac{1}{u}\,du - \int u^{-3}\,du$$

$$= \ln|x-1| + \frac{1}{2(x-1)^2} + C$$

33.   Evaluate $\displaystyle\int \frac{\sec x \tan x}{\sec x - 1}\,dx$.

Solution

Letting $u = \sec x - 1$, we have $du = \sec x \tan x\,dx$.
Therefore,

$$\int \frac{\sec x \tan x}{\sec x - 1}\,dx = \int \frac{1}{u}\,du = \ln|u| + C = \ln|\sec x - 1| + C$$

40.   Evaluate $\int (\csc 2\theta - \cot 2\theta)^2\,d\theta$.

Solution

$$\int (\csc 2\theta - \cot 2\theta)^2\,d\theta$$

$$= \int (\csc^2 2\theta - 2\csc 2\theta \cot 2\theta + \cot^2 2\theta)\,d\theta$$

$$= \int (\csc^2 2\theta - 2\csc 2\theta \cot 2\theta + \csc^2 2\theta - 1)\,d\theta$$

$$= \int 2\csc^2 2\theta\,d\theta - \int 2\csc 2\theta \cot 2\theta\,d\theta - \int d\theta$$

$$= -\cot 2\theta + \csc 2\theta - \theta + C$$

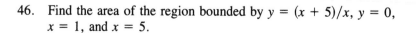

46.   Find the area of the region bounded by $y = (x+5)/x$, $y = 0$,
      $x = 1$, and $x = 5$.

Solution

$$A = \int_1^5 \frac{x + 5}{x} \, dx = \int_1^5 \left(1 + \frac{5}{x}\right) dx = [x + 5 \ln x]_1^5$$

$$= 4 + 5 \ln 5 \approx 12.047$$

Review Exercises for Chapter 5

5. Find the indefinite integral $\int \dfrac{(1 + x)^2}{\sqrt{x}} \, dx$.

Solution

$$\int \frac{(1 + x)^2}{\sqrt{x}} \, dx = \int \frac{1 + 2x + x^2}{\sqrt{x}} \, dx$$

$$= \int (x^{-1/2} + 2x^{1/2} + x^{3/2}) \, dx$$

$$= 2x^{1/2} + \frac{4}{3}x^{1/2} + \frac{2}{5}x^{5/2} + C$$

$$= \frac{2\sqrt{x}}{15}(15 + 10x + 3x^2) + C$$

7. Find the indefinite integral $\int \dfrac{x^2}{\sqrt{x^3 + 3}} \, dx$.

Solution

To solve $\int x^2/\sqrt{x^3 + 3} \, dx$, we let $u = x^3 + 1$, then $u' = 3x^2$.

$$\int \frac{x^2}{\sqrt{x^3 + 3}} \, dx = \frac{1}{3} \int \overbrace{(x^3 + 1)^{-1/2}}^{u^n} \overbrace{(3x^2)}^{u'} \, dx$$

$$= \frac{2}{3}(x^3 + 1)^{1/2} + C$$

$$= \frac{2}{3}\sqrt{x^3 + 1} + C$$

8. Find the indefinite integral $\int \dfrac{x^2 + 2x}{(x + 1)^2} \, dx$.

Solution

We begin by dividing the denominator into the numerator. Since $(x + 1)^2 = x^2 + 2x + 1$, we have

$$
\begin{array}{r}
1 \\
x^2 + 2x + 1 \overline{\smash{\big)}\, x^2 + 2x } \\
\underline{x^2 + 2x + 1} \\
-1
\end{array}
$$

Thus

$$
\int \frac{x^2 + 2x}{(x + 1)^2}\, dx = \int\left[1 - \frac{1}{(x + 1)^2}\right] dx = \int[1 - (x + 1)^{-2}]\, dx
$$

$$
= x - \frac{(x + 1)^{-1}}{-1} + C = x + \frac{1}{x + 1} + C
$$

21. Find the indefinite integral $\displaystyle\int \frac{1}{x \ln 3x}\, dx$.

Solution

Letting $u = \ln 3x$, we have $du = \dfrac{1}{x}\, dx$. Therefore,

$$
\int \frac{1}{x \ln 3x}\, dx = \int\left[\frac{1}{\ln 3x}\right]\left[\frac{1}{x}\right] dx = \int \frac{1}{u}\, du
$$

$$
= \ln |u| + C = \ln |\ln 3x| + C
$$

41. Use the Fundamental Theorem of Calculus to evaluate the definite integral

$$
\int_0^{\pi/3} \sec \theta\, d\theta
$$

Solution

$$
\int_0^{\pi/3} \sec \theta\, d\theta = \left[\ln |\sec \theta + \tan \theta|\right]_0^{\pi/3}
$$

$$
= \ln (2 + \sqrt{3}) - \ln (1 + 0) = \ln (2 + \sqrt{3})
$$

45. Use the Fundamental Theorem of Calculus to evaluate the definite integral

$$
\int_0^1 \frac{1 - \sqrt{y}}{1 + \sqrt{y}}\, dy
$$

Solution

Letting $u = \sqrt{y}$, we have $u^2 = y$ and $2u\, du = dy$. When

$y = 0$, $u = 0$, and when $y = 1$, $u = 1$. Therefore,

$$\int_0^1 \frac{1 - \sqrt{y}}{1 + \sqrt{y}}\,dy = \int_0^1 \frac{1 - u}{1 + u}(2u\,du)$$

$$= -2 \int_0^1 \frac{u^2 - u}{u + 1}\,du$$

$$= -2 \int_0^1 \left(u - 2 + \frac{2}{u + 1} \right) du$$

$$= -2 \left[\tfrac{1}{2}u^2 - 2u + 2 \ln |u + 1| \right]_0^1$$

$$= 3 - 4 \ln 2$$

53. Find dy/dx given $y \ln x + y^2 = 0$.

Solution

$$y \ln x + y^2 = 0$$

$$y\left(\frac{1}{x}\right) + (\ln x)y' + 2yy' = 0$$

$$y'(2y + \ln x) = -\frac{y}{x}$$

$$y' = \frac{-y}{x(2y + \ln x)}$$

57. Find dy/dx given $y = \dfrac{1}{b^2}\left[\ln(a + bx) + \dfrac{a}{a + bx} \right]$.

Solution

$$y = \frac{1}{b^2}\left[\ln(a + bx) + a(a + bx)^{-1} \right]$$

$$\frac{dy}{dx} = \frac{1}{b^2}\left[\left(\frac{1}{a + bx}\right)(b) + a(-1)(a + bx)^{-2}(b) \right]$$

$$= \frac{1}{b^2}\left[\frac{b}{a + bx} - \frac{ab}{(a + bx)^2} \right]$$

$$= \frac{1}{b^2}\left[\frac{b(a + bx) - ab}{(a + bx)^2} \right] = \frac{x}{(a + bx)^2}$$

62. A function $y = f(x)$ has a second derivative of $f''(x) = 6(x - 1)$. Find the function if its graph passes through the point $(2, 1)$ and at that point is tangent to the line $3x - y - 5 = 0$.

Solution

Since the graph of $y = f(x)$ is tangent to $3x - y - 5 = 0$ at $(2, 1)$, we know that $f(2) = 1$ and $f'(2) = 3$ (since the line $3x - y - 5 = 0$ has a slope of 3). Therefore,

$$f''(x) = 6(x - 1)$$

$$f'(x) = \int 6(x - 1)\, dx = \frac{6(x - 1)^2}{2} + C_1 = 3(x - 1)^2 + C_1$$

$$f'(2) = 3(2 - 1)^2 + C_1 = 3$$

$$C_1 = 0$$

$$f'(x) = 3(x - 1)^2$$

$$f(x) = \int 3(x - 1)^2\, dx = \frac{3(x - 1)^3}{3} + C_2 = (x - 1)^3 + C_2$$

$$f(2) = (2 - 1)^3 + C_2 = 1$$

$$C_2 = 0$$

$$f(x) = (x - 1)^3$$

64. The speed of a car, traveling in a straight line, is reduced from 45 to 30 mi/hr in a distance of 264 ft. Find the distance in which the car can be brought to rest from 30 mi/hr, assuming the same constant acceleration.

Solution

Let the position function of the car be given by $y = s(t)$. Let $s(0) = 0$ and $s'(0) = 45$. Then at time t_1,

$$s'(t_1) = 30 \qquad \text{and} \qquad s(t_1) = 264 \text{ ft} = 0.05 \text{ mi}$$

Finally, since we are given that the acceleration is constant, we have

$$s''(t) = k$$

$$s'(t) = \int k\, dt = kt + C_1$$

$$s'(0) = k(0) + C_1 = 45$$

$$C_1 = 45$$

$$s'(t) = kt + 45$$

$$s(t) = \int (kt + 45)\, dt = \frac{kt^2}{2} + 45t + C_2$$

$$s(0) = \frac{k(0^2)}{2} + 45(0) + C_2 = 0$$

$$C_2 = 0$$

$$s(t) = \frac{kt^2}{2} + 45t$$

Now since

$$s'(t_1) = kt_1 + 45 = 30$$

and

$$s(t_1) = \frac{kt_1^2}{2} + 45t_1 = 0.05$$

we have

$$kt_1 = -15$$

$$t_1 = \frac{-15}{k}$$

$$\frac{k(-15/k)^2}{2} + 45\left(\frac{-15}{k}\right) = 0.05$$

$$\frac{225}{2k} - \frac{675}{k} = 0.05$$

$$-\frac{1125}{0.1} = k$$

$$-11{,}250 = k$$

Therefore,

$$s'(t) = -11{,}250t + 45 = 0$$

$$t = \frac{45}{11{,}250} = \frac{1}{250} = 0.004 \text{ hr}$$

Thus the car is at rest when $t = 0.004$ hr and at this time its position is

$$s(0.004) = \frac{-11{,}250(0.004)^2}{2} + 45(0.004)$$

$$= -0.09 + 0.18 = 0.09 \text{ mi}$$

$$= (0.09)(5280) = 475.2 \text{ ft}$$

Consequently, the car traveled an additional

$$(475.2 - 264) = 211.2 \text{ ft}$$

to reduce its speed from 30 mi/hr to 0 mi/hr.

77. Find the average value of $f(x) = 1/\sqrt{x - 1}$ over the interval $[5, 10]$. Find the values of x where the function assumes its average value and sketch the graph of the function.

Solution

The average value is given by

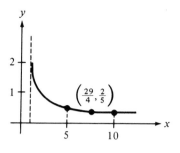

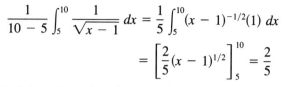

$$= \left[\frac{2}{5}(x - 1)^{1/2}\right]_5^{10} = \frac{2}{5}$$

To find the value of x where the function assumes its mean value on $[5, 10]$, solve

$$\frac{1}{\sqrt{x - 1}} = \frac{2}{5}$$

$$\sqrt{x - 1} = \frac{5}{2}$$

$$x - 1 = \frac{25}{4}$$

$$x = \frac{29}{4}$$

6 *Inverse functions*

6.1

Inverse functions

5. (a) Show that $f(x) = \sqrt{x-4}$ and $g(x) = x^2 + 4 \; (x \geq 0)$ are inverse functions by showing that $f(g(x)) = g(f(x)) = x$, and (b) graph f and g on the same set of coordinate axes.

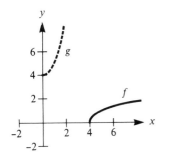

Solution

$$f(x) = \sqrt{x-4} \qquad \text{and} \qquad g(x) = x^2 + 4 \; (x \geq 0)$$

The composite of f with g is given by

$$f(g(x)) = f(x^2 + 4) = \sqrt{(x^2 + 4) - 4} = \sqrt{x^2} = x$$

The composite of g with f is given by

$$g(f(x)) = g(\sqrt{x-4}) = (\sqrt{x-4})^2 + 4 = x - 4 + 4 = x$$

Since $f(g(x)) = g(f(x)) = x$, we conclude that f and g are inverses of each other.

9. Find the inverse of $f(x) = 2x - 3$ and then graph both f and f^{-1}.

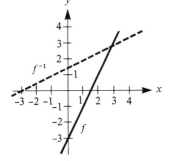

Solution

We begin by observing that f is increasing on its entire domain. To find an equation for the inverse, we let $y = f(x)$ and solve for x in terms of y.

$$2x - 3 = y$$

$$x = \frac{y + 3}{2}$$

$$f^{-1}(y) = \frac{y + 3}{2}$$

Finally, using x as the independent variable, we have

$$f^{-1}(x) = \frac{x + 3}{2}$$

19. Find the inverse of $f(x) = x^{2/3}$ $(x \geq 0)$ and then graph both f and f^{-1}.

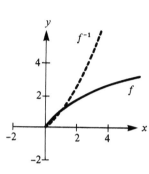

Solution

We begin by observing that f is increasing on $[0, \infty)$. To find an equation for the inverse, we let $y = f(x)$ and solve for x in terms of y.

$$
\begin{aligned}
x^{2/3} &= y & x &\geq 0 \\
x &= y^{3/2} & y &\geq 0 \\
f^{-1}(y) &= y^{3/2} & y &\geq 0
\end{aligned}
$$

Finally, using x as the independent variable, we have

$$f^{-1}(x) = x^{3/2} \qquad x \geq 0$$

39. Show that $f(x) = 4/x^2$ is strictly monotonic on the interval $(0, \infty)$ and therefore has an inverse on that interval.

Solution

$$f(x) = \frac{4}{x^2} \qquad \text{on} \qquad (0, \infty)$$

To verify that f is strictly monotonic we note that

$$f'(x) = \frac{-8}{x^3} < 0$$

on the interval $(0, \infty)$. Therefore, f is decreasing on $(0, \infty)$ and has an inverse.

45. Show that the slopes of the graphs of $f(x) = \sqrt{x - 4}$ and $f^{-1}(x) = x^2 + 4$ are reciprocals at the points $(5, 1)$ and $(1, 5)$, respectively.

Solution

For the function $f(x) = \sqrt{x - 4}$, we have

$$f'(x) = \frac{1}{2\sqrt{x - 4}} \qquad \text{and} \qquad f'(5) = \frac{1}{2\sqrt{5 - 4}} = \frac{1}{2}$$

For the function $f^{-1}(x) = x^2 + 4$, we have

$$(f^{-1})'(x) = 2x \qquad \text{and} \qquad (f^{-1})'(1) = 2$$

Therefore, $(f^{-1})'(1) = \dfrac{1}{f'(5)}$.

6.2
Exponential functions and differentiation

2. Write the following exponential equations as logarithmic equations:

   (a) $27^{2/3} = 9$ (b) $16^{3/4} = 8$

   Solution

   By definition,

   $$a^b = x \qquad \text{if and only if} \qquad \log_a x = b$$

   (a) By letting $a = 27$, $b = \frac{2}{3}$, and $x = 9$, we have

   $$27^{2/3} = 9 \qquad \text{if and only if} \qquad \log_{27} 9 = \tfrac{2}{3}$$

   (b) By letting $a = 16$, $b = \frac{3}{4}$, and $x = 8$, we have

   $$16^{3/4} = 8 \qquad \text{if and only if} \qquad \log_{16} 8 = \tfrac{3}{4}$$

5. Write the following logarithmic equations as exponential equations:

   (a) $\ln 2 = 0.6931\ldots$ (b) $\ln 8.4 = 2.128\ldots$

   Solution

   By definition,

   $$\log_a x = b \qquad \text{if and only if} \qquad a^b = x$$

   (a) By letting $a = e$, $x = 2$, and $b = 0.6931\ldots$, we have

   $$\ln 2 = \log_e 2 = 0.6931\ldots \qquad \text{if and only if} \qquad e^{0.6931\cdots} = 2$$

   (b) By letting $a = e$, $x = 8.4$, and $b = 2.128\ldots$, we have

   $$\ln 8.4 = \log_e 8.4 = 2.128\ldots \qquad \text{if and only if} \qquad e^{2.128\cdots} = 8.4$$

8. Find x if (a) $\log_4 (1/64) = x$, and (b) $\log_5 25 = x$.

   Solution

   (a) $x = \log_4 \dfrac{1}{64} = \log_4 \dfrac{1}{4^3} = \log_4 (4^{-3}) = -3$

   (b) $x = \log_5 25 = \log_5 (5^2) = 2$

13.   Find x if (a) $x^2 - x = \log_5 25$ and (b) $3x + 5 = \log_2 64$.

Solution

(a)
$$x^2 - x = \log_5 25$$
$$x^2 - x = \log_5 (5^2)$$
$$x^2 - x = 2$$
$$x^2 - x - 2 = 0$$
$$(x - 2)(x + 1) = 0$$
$$x = -1, 2$$

(b) $3x + 5 = \log_2 64$
$$3x + 5 = \log_2 (2^6)$$
$$3x + 5 = 6$$
$$3x = 1$$
$$x = \frac{1}{3}$$

19.   Sketch the graph of $y = e^{-x^2}$.

Solution

We begin by making the following observations:

(a)   The graph is symmetric with respect to the y-axis since
$$y = e^{-x^2} = e^{-(-x)^2}$$

(b)   The y-intercept is $(0, 1)$.

(c)   The x-axis is a horizontal asymptote since
$$\lim_{x \to \infty} e^{-x^2} = \lim_{x \to \infty} \frac{1}{e^{x^2}} = 0$$

(d)   The graph lies entirely above the x-axis, since for all x
$$0 < e^{-x^2}$$

(e)   For $0 < x$ the graph is decreasing, since $0 < x_1 < x_2$ implies that
$$\frac{1}{e^{x_1^2}} > \frac{1}{e^{x_2^2}}$$

Finally, by plotting a few points, we have the graph shown in the accompanying figure.

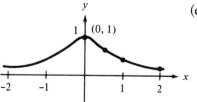

x	0	0.5	1	2
y	1	0.607	0.368	0.135

23. Show that $f(x) = e^{2x}$ and $g(x) = \ln\sqrt{x}$ are inverses of each other by sketching their graphs on the same coordinate system.

Solution

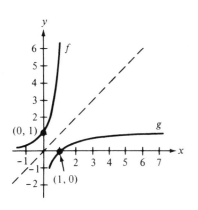

$$f(x) = e^{2x}$$

x	-2	-1	0	1	2
$f(x)$	0.01	0.14	1	7.39	54.60

$$g(x) = \ln\sqrt{x} = \frac{1}{2}\ln x \qquad \text{Domain: } x > 0$$

x	$\frac{1}{2}$	1	2	3	4	5
$g(x)$	-0.35	0	0.35	0.55	0.69	0.80

The graphs are reflections of each other through the line $y = x$. Therefore, the functions are inverses of each other.

32. Complete the accompanying table to find the time t necessary for P dollars to double if compounded continuously at the rate r.

r	2%	4%	6%	8%	10%	12%
t						

Solution

Since the interest is compounded continuously, use the formula $A = Pe^{rt}$. To double the deposit, we consider

$$Pe^{rt} = 2P$$
$$e^{rt} = 2$$
$$\ln e^{rt} = \ln 2$$
$$rt = \ln 2$$
$$t = \frac{\ln 2}{r}$$

Substituting the required values of r into this last equation, we complete the table (t in years).

r	2%	4%	6%	8%	10%	12%
t	34.66	17.33	11.55	8.66	6.93	5.78

41. Find dy/dx given $y = e^{\sqrt{x}}$.

Solution

$$y = e^{\sqrt{x}}$$

$$\frac{dy}{dx} = e^{\sqrt{x}}\frac{d}{dx}[\sqrt{x}]$$

$$= e^{\sqrt{x}}\left(\frac{1}{2\sqrt{x}}\right) = \frac{e^{\sqrt{x}}}{2\sqrt{x}}$$

45. Find dy/dx given $y = \ln(e^{x^2})$.

Solution

We observe that

$$y = \ln(e^{x^2}) = x^2 \ln e = x^2$$

Therefore, $dy/dx = 2x$.

55. Find dy/dx given $y = e^{-x} \ln x$.

Solution

$$y' = (e^{-x})\left(\frac{1}{x}\right) + (\ln x)(-e^{-x}) = e^{-x}\left(\frac{1}{x} - \ln x\right)$$

62. Find dy/dx given $y = \log_3(x\sqrt{x-1}/2)$

Solution

$$y = \log_3 x + \frac{1}{2}\log_3(x-1) - \log_3 2$$

$$\frac{dy}{dx} = (\log_3 e)\left[\frac{1}{x}\right] + \frac{1}{2}(\log_3 e)\left[\frac{1}{x-1}\right]$$

$$= (\log_3 e)\left[\frac{1}{x} + \frac{1}{2(x-1)}\right] = (\log_3 e)\left[\frac{3x-2}{2x(x-1)}\right]$$

65. Find dy/dx given $y = x^{2/x}$.

Solution

We use logarithmic differentiation. Taking the natural logarithm of both sides, we obtain

$$\ln y = \ln(x^{2/x}) = \frac{2}{x}\ln x$$

Differentiating both sides, we have

$$\frac{1}{y}\frac{dy}{dx} = \left[\frac{2}{x}\right]\left[\frac{1}{x}\right] + (\ln x)\left[\frac{-2}{x^2}\right]$$

$$\frac{dy}{dx} = y\left[\frac{2}{x^2} - \frac{2}{x^2}\ln x\right] = \frac{2y}{x^2}(1 - \ln x)$$

69. Show that $y = e^x(\cos \sqrt{2}x + \sin \sqrt{2}x)$ satisfies the differential equation $y'' - 2y' + 3y = 0$.

Solution

$$y = e^x(\cos \sqrt{2}x + \sin \sqrt{2}x)$$
$$y' = e^x(-\sqrt{2}\sin \sqrt{2}x + \sqrt{2}\cos \sqrt{2}x)$$
$$+ e^x(\cos \sqrt{2}x + \sin \sqrt{2}x)$$
$$y'' = e^x(-2\cos \sqrt{2}x - 2\sin \sqrt{2}x)$$
$$+ 2e^x(-\sqrt{2}\sin \sqrt{2}x + \sqrt{2}\cos \sqrt{2}x)$$
$$+ e^x(\cos \sqrt{2}x + \sin \sqrt{2}x)$$
$$= -e^x(\cos \sqrt{2}x + \sin \sqrt{2}x)$$
$$+ 2e^x(\sqrt{2}\cos \sqrt{2}x - \sqrt{2}\sin \sqrt{2}x)$$
$$y'' - 2y' + 3y = -e^x(\cos \sqrt{2}x + \sin \sqrt{2}x)$$
$$+ 2e^x(\sqrt{2}\cos \sqrt{2}x - \sqrt{2}\sin \sqrt{2}x)$$
$$- 2e^x(\cos \sqrt{2}x + \sin \sqrt{2}x)$$
$$- 2e^x(\sqrt{2}\cos \sqrt{2}x - \sqrt{2}\sin \sqrt{2}x)$$
$$+ 3e^x(\cos\sqrt{2}x + \sin\sqrt{2}x) = 0$$

75. Find (if any exist) the extrema and the points of inflection, and sketch the graph of $f(x) = x^2e^{-x}$.

Solution

$$f(x) = x^2e^{-x} \qquad \text{Intercepts: } (0, 0)$$
$$f'(x) = x^2(-e^{-x}) + e^{-x}(2x)$$
$$= xe^{-x}(-x + 2) \qquad \text{Critical points: } (0, 0), (2, 4/e^2)$$
$$f''(x) = (xe^{-x})(-1) + (-x + 2)[x(-e^{-x}) + e^{-x}(1)]$$
$$= e^{-x}[-x + (-x + 2)(-x + 1)] = e^{-x}(x^2 - 4x + 2)$$

Since $x^2 - 4x + 2 = 0$ if $x = 2 \pm \sqrt{2}$, the possible points of inflection are

$$\left(2 + \sqrt{2}, (2 + \sqrt{2})^2e^{-(2 + \sqrt{2})}\right) \qquad \text{and} \qquad \left(2 - \sqrt{2}, (2 - \sqrt{2})^2e^{-(2 - \sqrt{2})}\right)$$

$(2 - \sqrt{2}, (2 - \sqrt{2})^2e^{-(2 - \sqrt{2})})$

$(2, 4e^{-2})$

$(0, 0)$

$(2 + \sqrt{2}, (2 + \sqrt{2})^2e^{-(2 + \sqrt{2})})$

Interval	$f(x)$	$f'(x)$	$f''(x)$	Shape of graph
x in $(-\infty, 0)$		$-$	$+$	decreasing, concave up
$x = 0$	0	0	$+$	relative minimum
x in $(0, 2 - \sqrt{2})$		$+$	$+$	increasing, concave up
$x = 2 - \sqrt{2}$	0.191	$+$	0	point of inflection
x in $(2 - \sqrt{2}, 2)$		$+$	$-$	increasing, concave down
$x = 2$	0.541	0	$-$	relative minimum
x in $(2, 2 + \sqrt{2})$		$-$	$-$	decreasing, concave down
$x = 2 + \sqrt{2}$	0.384	$-$	0	point of inflection
x in $(2 + \sqrt{2}, \infty)$		$-$	$+$	decreasing, concave up

80. Find, to three decimal places, the value of x so that $e^{-x} = x$.
     [Use Newton's Method to find the zero of $f(x) = x - e^{-x}$.]

Solution

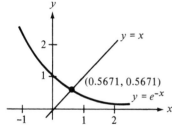

Letting $f(x) = x - e^{-x}$, we have $f'(x) = 1 + e^{-x}$. From the graphs of $y_1 = e^{-x}$ and $y_2 = x$, we approximate the point of intersection to occur when $x_1 = 1$. Thus by Newton's Method we have

$$x_2 = f(x_1) - \frac{f(x_1)}{f'(x_1)} = 0.5379$$

$$x_3 = f(x_2) - \frac{f(x_2)}{f'(x_2)} = 0.5670$$

$$x_4 = f(x_3) - \frac{f(x_3)}{f'(x_3)} = 0.5671$$

Therefore we approximate the point of intersection to occur when $x \approx 0.567$.

83. In a group project in learning theory, a mathematical model for the proportion P of correct responses after n trials was found to be

$$P = \frac{0.83}{1 + e^{-0.2n}}$$

(a) Find the limiting proportion of correct responses as $n \to \infty$.

(b) Find the rate at which P is increasing after $n = 3$ and $n = 10$ trials.

Solution

(a) $\lim\limits_{n \to \infty} \dfrac{0.83}{1 + e^{-0.2n}} = \dfrac{0.83}{1 + 0} = 0.83 = 83\%$

(b) $P(n) = \dfrac{0.83}{1 + e^{-0.2n}} = 0.83(1 + e^{-0.2n})^{-1}$

$P'(n) = 0.83(-1)(1 + e^{-0.2n})^{-2}(e^{-0.2n})(-0.2)$

$\qquad = \dfrac{0.166e^{-0.2n}}{(1 + e^{-0.2n})^2}$

$P'(3) = \dfrac{0.166e^{-0.6}}{(1 + e^{-0.6})^2} = 0.038$

$P'(10) = \dfrac{0.166e^{-2}}{(1 + e^{-2})^2} = 0.017$

6.3
Integration of exponential functions: Growth and decay

3.   Evaluate $\int_0^2 (x^2 - 1)\, e^{x^3 - 3x + 1}\, dx$.

Solution

Letting $u = x^3 - 3x + 1$, we have $du = (3x^2 - 3)\, dx$ or $(x^2 - 1)\, dx = du/3$. Therefore, when $x = 0$, $u = 1$ and when $x = 2$, $u = 3$.

$$\int_0^2 (x^2 - 1)\, e^{x^3 - 3x + 1}\, dx = \int_0^2 e^{x^3 - 3x + 1}(x^2 - 1)\, dx$$

$$= \int_1^3 e^u \left(\frac{du}{3}\right)$$

$$= \frac{1}{3}[e^u]_1^3 = \frac{e}{3}(e^2 - 1) \approx 5.789$$

9.   Evaluate $\displaystyle\int_1^3 \frac{e^{3/x}}{x^2}\, dx$.

Solution

Letting $u = 3/x$, we have $du = -3/x^2\, dx$.

$$\int_1^3 \frac{e^{3/x}}{x^2}\, dx = -\frac{1}{3} \int_1^3 e^{3/x}\left(\frac{-3}{x^2}\right) dx$$

$$= -\frac{1}{3}[e^{3/x}]_1^3 = -\frac{1}{3}(e - e^3) = \frac{e}{3}(e^2 - 1)$$

13.   Evaluate $\int e^x \sqrt{1 - e^x}\, dx.$

Solution

Letting $u = 1 - e^x$, we have $du = -e^x dx.$

$$\int e^x \sqrt{1 - e^x}\, dx = -\int \underbrace{(1 - e^x)^{1/2}}_{u^n}\underbrace{(-e^x)\, dx}_{du}$$

$$= -\left[\frac{(1 - e^x)^{3/2}}{3/2}\right] + C$$

$$= \frac{-2(1 - e^x)^{3/2}}{3} + C$$

15.   Evaluate

$$\int \frac{e^x + e^{-x}}{e^x - e^{-x}}\, dx.$$

Solution

Letting $u = e^x - e^{-x}$, we have $du = (e^x + e^{-x})\, dx$

$$\int \frac{e^x + e^{-x}}{e^x - e^{-x}}\, dx = \int \frac{1}{u}\, du = \ln |e^x - e^{-x}| + C$$

23.   Evaluate $\int e^{-x} \tan (e^{-x})\, dx.$

Solution

Letting $u = e^{-x}$, we have $du = -e^{-x} dx.$

$$\int e^{-x} \tan (e^{-x})\, dx = -\int \tan u\, du = \ln |\cos u| + C$$

$$= \ln |\cos (e^{-x})| + C$$

37.   The management at a certain factory has found that the maximum number of units a worker can produce in a day is 30. The learning curve for the number of units N produced per day after a new employee has worked t days is given by

$$N(t) = 30(1 - e^{kt})$$

After 20 days on the job, a particular worker produced 19 units.

(a)   Find the learning curve for this worker.

(b)   How many days should pass before this worker is producing 25 units per day?

Solution

(a)   Since $N(20) = 19$, we have

$$19 = 30(1 - e^{20k})$$

$$30e^{20k} = 11$$

$$e^{20k} = \frac{11}{30}$$

$$k = \frac{\ln (11/30)}{20} \approx -0.0502$$

Therefore,

$$N(t) = 30(1 - e^{-0.0502t})$$

(b)   To determine the time when the worker will be producing 25 units per day we solve the equation

$$25 = 30(1 - e^{-0.0502t})$$

$$e^{-0.0502t} = \frac{1}{6}$$

$$t = \frac{-\ln 6}{-0.0502} \approx 36 \text{ days}$$

43.   In 1960 the population of a town was 2500 and in 1970 it was 3350. Assuming the population increases continuously at a constant rate proportional to the existing population, estimate the population in 1990.

Solution

Let P be the population at time t. Then we have

$$\frac{dP}{dt} = kP$$

$$\left(\frac{1}{P}\right)\frac{dP}{dt} = k$$

$$\int \left(\frac{1}{P}\right)\frac{dP}{dt}\, dt = \int k\, dt$$

$$\ln P = kt + C$$

$$P = Ce^{kt}$$

Let $t = 0$ for the year 1960. Then

$$P(0) = C = 2500 \qquad \text{and} \qquad P(t) = 2500e^{kt}$$

In 1970 $t = 10$ and $P = 3350$. Therefore,

$$P(10) = 2500e^{10k} = 3350$$

$$e^{10k} = \frac{3350}{2500} = \frac{67}{50}$$

$$k = \tfrac{1}{10}(\ln 67 - \ln 50) \approx 0.029267$$

Therefore, in 1990 $t = 30$ and

$$P(30) = 2500 \exp\left[\left(\frac{1}{10}\right)\left(\ln 67 - \ln 50)(30)\right] \approx 6015$$

[Note: exp ( ) is another, and sometimes more convenient, way of writing $e^{()}$.]

46. If radioactive material decays continuously at a rate proportional to the amount present, find the half-life of the material if after 1 year 99.57% of an initial amount still remains.

Solution

$$\frac{dy}{dt} = ky$$

$$\left(\frac{1}{y}\right)\frac{dy}{dt} = k$$

$$\int \left(\frac{1}{y}\right)\frac{dy}{dt}\, dt = \int k\, dt$$

$$\ln y = kt + C$$

When $t = 0$ year, $y = 100\% = 1.00$. Thus

$$\ln (1) = k(0) + C \qquad \text{or} \qquad C = 0$$

When $t = 1$ year, $y = 99.57\% = 0.9957$. Thus

$$\ln (0.9957) = k(1)$$

Therefore,

$$\ln y = [\ln (0.9957)]\, t \qquad \text{or} \qquad t = \frac{\ln y}{\ln (0.9957)}$$

and when $y = 50\% = 0.5$, we have a half-life of

$$t = \frac{\ln (0.5)}{\ln (0.9957)} \approx 160.85 \text{ years}$$

49. Using Newton's Law of Cooling, determine the outdoor temperature if a thermometer is taken from a room where the temperaure is $68°$ to the outdoors, where after $\frac{1}{2}$ min and 1 min, the thermometer reads $53°$ and $42°$, respectively.

Solution

Let y be the temperature of the thermometer and T be the outdoor temperature. From Newton's Law of Cooling, we have

$$\frac{dy}{dt} = k(y - T)$$

$$\left(\frac{1}{y - T}\right) \frac{dy}{dt} = k$$

$$\int \left(\frac{1}{y - T}\right) \frac{dy}{dt}\, dt = \int k\, dt$$

$$\ln (y - T) = kt + C$$

When $t = 0$ min, $y = 68°$. Thus

$$\ln (68 - T) = k(0) + C \qquad \text{or} \qquad C = \ln (68 - T)$$

When $t = 0.5$ min, $y = 53°$. Thus

$$\ln (53 - T) = k(0.5) + \ln (68 - T)$$

$$k = 2[\ln (53 - T) - \ln (68 - T)]$$

$$= \ln \left[\left(\frac{53 - T}{68 - T}\right)^2 \right]$$

When $t = 1$ min, $y = 42°$. Thus

$$\ln (42 - T) = k(1) + \ln (68 - T)$$

$$= \ln \left[\left(\frac{53 - T}{68 - T}\right)^2 \right] + \ln (68 - T)$$

$$42 - T = \frac{(53 - T)^2}{68 - T}$$

$$(42 - T)(68 - T) = (53 - T)^2$$

$$2856 - 110T + T^2 = 2809 - 106T + T^2$$

$$-4T = -47$$

$$T = \frac{47}{4} = 11.75°$$

6.4

Inverse trigonometric functions and differentiation

3. Evaluate arccos $\left(\frac{1}{2}\right)$.

Solution

Since $y = \arccos \left(\frac{1}{2}\right)$ if and only if $\cos y = \frac{1}{2}$, we choose

$y = \pi/3$ in the interval $[0, \pi]$. Thus

$$\arccos\left(\frac{1}{2}\right) = \frac{\pi}{3}$$

6.   Evaluate arccot (-1).

Solution

Since $y = $ arccot (-1) if and only if cot $y = -1$, we choose $y = 3\pi/4$ in the interval $(0, \pi)$. Thus

$$\text{arccot } (-1) = \frac{3\pi}{4}$$

13.   Evaluate (a) sin $[\arctan\left(\frac{3}{4}\right)]$ and (b) sec $[\arcsin\left(\frac{4}{5}\right)]$ without the use of a calculator.

Solution

(a)   We begin by sketching a triangle to represent θ, "the angle whose tangent is $\frac{3}{4}$." Then

$$\theta = \arctan\left(\frac{3}{4}\right)$$

and

$$\sin\left[\arctan\left(\frac{3}{4}\right)\right] = \sin\theta = \frac{3}{5}$$

(b)   We begin by sketching a triangle to represent θ, "the angle whose sin is $\frac{4}{5}$." Then

$$\theta = \arcsin\left(\frac{4}{5}\right)$$

and

$$\sec\left[\arcsin\left(\frac{4}{5}\right)\right] = \sec\theta = \frac{5}{3}$$

18.   Write an algebraic equation that is equivalent to the equation

$$y = \sin(\arccos x)$$

Solution

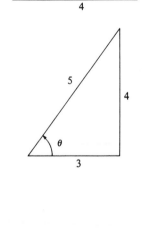

We begin by sketching a triangle to represent θ, "the angle whose cosine is x." Then

$$\theta = \arccos x$$

and
$$y = \sin (\arccos x) = \sin \theta = \sqrt{1 - x^2}$$
Therefore, the equation is $y = \sqrt{1 - x^2}$.

21. Write an algebraic equation that is equivalent to the equation $y = \sin (\text{arcsec } x)$.

Solution

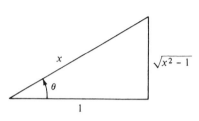

We begin by sketching a triangle to represent θ, "the angle whose secant is x." Then
$$\theta = \text{arcsec } x$$
and
$$y = \sin (\text{arcsec } x) = \sin \theta = \frac{\sqrt{x^2 - 1}}{x}$$
Therefore, the equation is
$$y = \frac{\sqrt{x^2 - 1}}{x}$$

37. Solve $\arcsin \sqrt{2x} = \arccos \sqrt{x}$ for x.

Solution

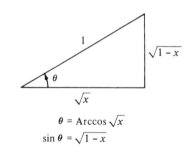

$\theta = \text{Arccos } \sqrt{x}$
$\sin \theta = \sqrt{1 - x}$

Taking the sine of each member of the equation, we have
$$\sin (\arcsin \sqrt{2x}) = \sin (\arccos \sqrt{x})$$
$$\sqrt{2x} = \sqrt{1 - x}$$
$$2x = 1 - x$$
$$3x = 1$$
$$x = \frac{1}{3}$$

41. Find the derivative of $f(x) = 2 \arcsin (x - 1)$.

Solution
$$f(x) = 2 \arcsin (x - 1)$$
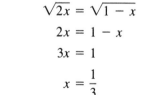

44. Find the derivative of $f(x) = \arctan \sqrt{x}$.

Solution

$$f(x) = \arctan \sqrt{x}$$

$$f'(x) = \frac{1}{1 + (\sqrt{x})^2} \left(\frac{1}{2\sqrt{x}}\right) = \frac{1}{2\sqrt{x}\,(1 + x)}$$

47.   Find the derivative of $f(x) = \text{arccos}\,(1/x)$.

Solution

$$f(x) = \text{arccos}\left(\frac{1}{x}\right)$$

$$f'(x) = \frac{-1}{\sqrt{1 - (1/x)^2}} \left(\frac{-1}{x^2}\right) = \frac{1}{x^2\sqrt{(x^2 - 1)/x^2}}$$

$$= \frac{1}{(x^2/|x|)/\sqrt{x^2 - 1}} = \frac{1}{|x|\sqrt{x^2 - 1}}$$

Note: We could have arrived at the same result more quickly had we recognized that $\text{arccos}\,(1/x) = \text{arcsec}\,x$. Then

$$f(x) = \text{arccos}\left(\frac{1}{x}\right) = \text{arcsec}\,x$$

$$f'(x) = \frac{1}{|x|\sqrt{x^2 - 1}}$$

51.   Find the derivative of $h(t) = \sin\,(\text{arccos}\,t)$.

Solution

From the accompanying figure we observe that

$$h(t) = \sin\,(\text{arccos}\,t) = \sqrt{1 - t^2}$$

Therefore,

$$h'(t) = \frac{1}{2}(1 - t^2)^{-1/2}(-2t) = \frac{-t}{\sqrt{1 - t^2}}$$

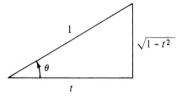

$\theta = \text{Arccos}\,t$

$\sin\theta = \sqrt{1 - t^2}$

55.   Find the derivative of $f(x) = \frac{1}{2}\left[\frac{1}{2}\ln\left(\frac{1 + x}{1 - x}\right) + \arctan x\right]$.

Solution

$$f(x) = \frac{1}{2}\left[\frac{1}{2}\ln\left(\frac{1 + x}{1 - x}\right) + \arctan x\right]$$

$$= \frac{1}{2}\left[\frac{1}{2}\ln\,(1 + x) - \frac{1}{2}\ln\,(1 - x) + \arctan x\right]$$

$$f'(x) = \frac{1}{2}\left[\frac{1}{2(1+x)} - \frac{-1}{2(1-x)} + \frac{1}{1+x^2}\right]$$

$$= \frac{1}{2}\left[\frac{1-x+1+x}{2(1-x^2)} + \frac{1}{1+x^2}\right] = \frac{1}{2}\left[\frac{1}{1-x^2} + \frac{1}{1+x^2}\right]$$

$$= \frac{1}{2}\left[\frac{1+x^2+1-x^2}{1-x^4}\right] = \frac{1}{1-x^4}$$

57. Find the derivative of $f(x) = x \arcsin x + \sqrt{1-x^2}$.

Solution

$$f(x) = x \arcsin x + \sqrt{1-x^2}$$

$$f'(x) = x\left(\frac{1}{\sqrt{1-x^2}}\right) + \arcsin x + \frac{1}{2}(1-x^2)^{-1/2}(-2x)$$

$$= \frac{x}{\sqrt{1-x^2}} + \arcsin x + \frac{-x}{\sqrt{1-x^2}} = \arcsin x$$

64. Find the point of intersection of the graphs of $y = \arcsin x$ and $y = \arccos x$.

Solution

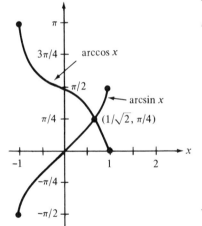

$$\arcsin x = \arccos x$$

$$\sin (\arcsin x) = \sin (\arccos x)$$

$$x = \sqrt{1-x^2}$$

$$x^2 = 1 - x^2$$

$$2x^2 = 1$$

$$x = \frac{1}{\sqrt{2}}$$

(We disregard the value $x = -1/\sqrt{2}$ since it is not a solution to $x = \sqrt{1-x^2}$.) Thus the point of intersection is $(1/\sqrt{2}, \pi/4)$, as shown in the accompanying figure.

66. An observer is standing 300 ft from the point at which a balloon is released. If the balloon rises at a rate of 5 ft/s, how fast is the angle of elevation of the observer's line of sight increasing when the balloon is 100 ft high?

Solution

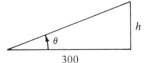

From the accompanying figure we observe that

$$\tan \theta = \frac{h}{300} \quad \text{or} \quad \theta = \arctan\left(\frac{h}{300}\right)$$

Differentiating with respect to t, we have

$$\frac{d\theta}{dt} = \left[\frac{1}{1 + (h/300)^2}\right]\left(\frac{1}{300}\right)\frac{dh}{dt} = \left(\frac{300}{h^2 + 300^2}\right)\frac{dh}{dt}$$

When $dh/dt = 5$ ft/s and $h = 100$ ft, we have

$$\frac{d\theta}{dt} = \left(\frac{300}{100^2 + 300^2}\right)(5) = \frac{3}{200} \text{ rad/s}$$

6.5

Inverse trigonometric functions: Integration and completing the square

1. Evaluate $\displaystyle\int_0^{1/6} \frac{1}{\sqrt{1 - 9x^2}}\, dx$.

    ### Solution

    If we let $a = 1$ and $u = 3x$, then $u' = 3$. Thus

    $$\int_0^{1/6} \frac{1}{\sqrt{1 - 9x^2}}\, dx = \frac{1}{3}\int_0^{1/6} \left(\frac{1}{\sqrt{1 - (3x)^2}}\right)(3)\, dx$$

    $$= \frac{1}{3}[\arcsin (3x)]_0^{1/6}$$

    $$= \frac{1}{3}\left[\arcsin \left(\frac{1}{2}\right) - \arcsin 0\right] = \frac{1}{3}\left(\frac{\pi}{6}\right) = \frac{\pi}{18}$$

7. Evaluate $\displaystyle\int \frac{x^3}{x^2 + 1}\, dx$.

    ### Solution

    Since the degree of the numerator is greater than the degree of the denominator, we divide to obtain

    $$\frac{x^3}{x^2 + 1} = x - \frac{x}{x^2 + 1}$$

    Therefore,

    $$\int \frac{x^3}{x^2 + 1}\, dx = \int\left(x - \frac{x}{x^2 + 1}\right)dx = \int x\, dx - \frac{1}{2}\int \frac{2x}{x^2 + 1}\, dx$$

    $$= \frac{x^2}{2} - \frac{1}{2}\ln (x^2 + 1) + C = \frac{1}{2}[x^2 - \ln (x^2 + 1)] + C$$

12. Evaluate $\displaystyle\int \frac{1}{x\sqrt{x^4 - 4}}\, dx$.

Solution

If we let $a = 2$ and $u = x^2$, then $du = 2x\ dx$. Thus

$$\int \frac{1}{x\sqrt{x^4 - 4}}\ dx = \int \frac{1}{x\sqrt{(x^2)^2 - 2^2}}\ dx$$

$$= \frac{1}{2}\int \frac{2x}{x^2\sqrt{(x^2)^2 - 2^2}}\ dx = \frac{1}{2}\int \frac{du}{u\sqrt{u^2 - a^2}}$$

$$= \frac{1}{2}\left[\frac{1}{2}\ \text{arcsec}\ \frac{x^2}{2}\right] + C = \frac{1}{4}\ \text{arcsec}\ \frac{x^2}{2} + C$$

14. Evaluate $\displaystyle\int \frac{1}{(x - 1)\sqrt{(x - 1)^2 - 4}}\ dx$.

Solution

If we let $a = 2$ and $u = x - 1$, then $du = dx$ and

$$\int \frac{1}{(x - 1)\sqrt{(x - 1)^2 - 4}}\ dx = \int \frac{1}{(x - 1)\sqrt{(x - 1)^2 - 2^2}}\ (1)\ dx$$

$$= \frac{1}{2}\ \text{arcsec}\ \left|\frac{x - 1}{2}\right| + C$$

15. Evaluate $\displaystyle\int_0^{1/\sqrt{2}} \frac{\arcsin x}{\sqrt{1 - x^2}}\ dx$.

Solution

If we let $u = \arcsin x$, then $du = \dfrac{1}{\sqrt{1 - x^2}}\ dx$. Thus

$$\int_0^{1/\sqrt{2}} \frac{\arcsin x}{\sqrt{1 - x^2}}\ dx = \int_0^{1/\sqrt{2}} (\arcsin x)^1\ \frac{1}{\sqrt{1 - x^2}}\ dx$$

$$= \left[\frac{(\arcsin x)^2}{2}\right]_0^{1/\sqrt{2}}$$

$$= \frac{1}{2}\left\{\left[\arcsin\left(\frac{1}{\sqrt{2}}\right)\right]^2 - [\arcsin (0)]^2\right\}$$

$$= \frac{1}{2}\left[\left(\frac{\pi}{4}\right)^2 - (0)^2\right] = \frac{\pi^2}{32}$$

23. Evaluate $\displaystyle\int \frac{1}{\sqrt{x}(1 + x)}\ dx$.

Solution

If we let $a = 1$ and $u = \sqrt{x}$, then $du = (1/2\sqrt{x})\ dx$. Thus

$$\int \frac{1}{\sqrt{x}(1 + x)}\, dx = 2 \int \frac{1}{2\sqrt{x}[1 + (\sqrt{x})^2]}\, dx = 2 \int \frac{1/2\sqrt{x}}{1 + (\sqrt{x})^2}\, dx$$

$$= 2 \int \frac{du}{1 + u^2} = 2 \arctan (\sqrt{x}) + C$$

29.   Evaluate $\int \frac{2x}{x^2 + 6x + 13}\, dx$.

Solution

$$\int \frac{2x}{x^2 + 6x + 13}\, dx = \int \frac{(2x + 6) - 6}{x^2 + 6x + 13}\, dx$$

$$= \int \frac{2x + 6}{x^2 + 6x + 13}\, dx - \int \frac{6}{x^2 + 6x + 13}\, dx$$

$$= \int \frac{2x + 6}{x^2 + 6x + 13}\, dx$$

$$- \int \frac{6}{(x^2 + 6x + 9) + 4}\, dx$$

$$= \int \frac{2x + 6}{x^2 + 6x + 13}\, dx - 6\int \frac{1}{(x + 3)^2 + 2^2}\, dx$$

$$= \ln (x^2 + 6x + 13) - 3 \arctan \left(\frac{x + 3}{2}\right) + C$$

31.   Evaluate $\int \frac{1}{\sqrt{-x^2 - 4x}}\, dx$.

Solution

$$\int \frac{1}{\sqrt{-x^2 - 4x}}\, dx = \int \frac{1}{\sqrt{-(x^2 + 4x)}}\, dx$$

$$= \int \frac{1}{\sqrt{4 - (x^2 + 4x + 4)}}\, dx$$

$$= \int \frac{1}{\sqrt{2^2 - (x + 2)^2}}\, dx = \arcsin \left(\frac{x + 2}{2}\right) + C$$

35.   Evaluate $\int_2^3 \frac{2x - 3}{\sqrt{4x - x^2}}\, dx$.

Solution

$$\int_2^3 \frac{2x - 3}{\sqrt{4x - x^2}}\, dx$$

$$= \int_2^3 \frac{(2x - 4) + 1}{\sqrt{4x - x^2}} \, dx$$

$$= \int_2^3 \frac{2x - 4}{\sqrt{4x - x^2}} \, dx + \int_2^3 \frac{1}{\sqrt{4x - x^2}} \, dx$$

$$= -\int_2^3 (4x - x^2)^{-1/2}(4 - 2x) \, dx + \int_2^3 \frac{1}{\sqrt{4 - (x^2 - 4x + 4)}} \, dx$$

$$= -\int_2^3 (4x - x^2)^{-1/2}(4 - 2x) \, dx + \int_2^3 \frac{1}{\sqrt{2^2 - (x - 2)^2}} \, dx$$

$$= \left[-\frac{(4x - x^2)^{1/2}}{\frac{1}{2}} + \arcsin\left(\frac{x - 2}{2}\right) \right]_2^3$$

$$= -2\sqrt{3} + \frac{\pi}{6} - (-4 + 0) = 4 - 2\sqrt{3} + \frac{\pi}{6}$$

43.   Evaluate $\int \sqrt{e^t - 3} \, dt$.

Solution

Let $u = \sqrt{e^t - 3}$. Then $e^t = u^2 + 3$, $t = \ln(u^2 + 3)$, and $dt = 2u/(u^2 + 3) \, du$. Therefore,

$$\int \sqrt{e^t - 3} \, dt = \int \frac{2u^2}{u^2 + 3} \, du$$

Now since the numerator and denominator are of equal degree, we divide to obtain

$$2 \int \frac{u^2}{u^2 + 3} \, du = 2 \int \left[1 - \frac{3}{u^2 + 3} \right] du$$

$$= 2 \left[\int du - 3 \int \frac{1}{(\sqrt{3})^2 + u^2} \, du \right]$$

$$= 2 \left[u - 3\left(\frac{1}{\sqrt{3}}\right) \arctan\left(\frac{u}{\sqrt{3}}\right) \right] + C$$

$$= 2 \left[\sqrt{e^t - 3} - \sqrt{3} \arctan\left(\frac{\sqrt{e^t - 3}}{\sqrt{3}}\right) \right] + C$$

48.   Find the area of the region bounded by

$$y = \frac{1}{\sqrt{3 + 2x - x^2}}, \qquad y = 0, \qquad x = 0, \qquad \text{and} \qquad x = 2$$

Solution

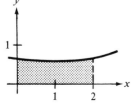

$$A = \int_0^2 \frac{1}{\sqrt{3 + 2x - x^2}} \, dx = \int_0^2 \frac{1}{\sqrt{3 - (x^2 - 2x)}} \, dx$$

$$= \int_0^2 \frac{1}{\sqrt{4 - (x^2 - 2x + 1)}}\, dx = \int_0^2 \frac{1}{\sqrt{2^2 - (x - 1)^2}}\, dx$$

$$= \left[\arcsin \left(\frac{x - 1}{2} \right) \right]_0^2 = \frac{\pi}{6} - \left(-\frac{\pi}{6} \right) = \frac{\pi}{3}$$

6.6
Hyperbolic functions

3. Evaluate (a) csch (ln 2) and (b) coth (ln 5). If the function value is not a rational number, give the answer to three decimal place accuracy.

   Solution

   (a) $\operatorname{csch}(\ln 2) = \dfrac{1}{\sinh(\ln 2)} = \dfrac{2}{e^{\ln 2} - e^{-\ln 2}} = \dfrac{2}{2 - (\frac{1}{2})} = \dfrac{4}{3}$

   (b) $\operatorname{coth}(\ln 5) = \dfrac{\cosh(\ln 5)}{\sinh(\ln 5)} = \dfrac{e^{\ln 5} + e^{-\ln 5}}{e^{\ln 5} - e^{-\ln 5}} = \dfrac{5 + (\frac{1}{5})}{5 - (\frac{1}{5})} = \dfrac{13}{12}$

17. If $y = \ln[\tanh(x/2)]$, find y' and simplify.

   Solution

   $$y = \ln\left[\tanh\left(\frac{x}{2} \right) \right]$$

   $$y' = \frac{1}{\tanh(x/2)} \left[\frac{1}{2} \operatorname{sech}^2\left(\frac{x}{2} \right) \right] = \frac{1}{2}\left[\frac{\cosh(x/2)}{\sinh(x/2)} \right]\left[\frac{1}{\cosh^2(x/2)} \right]$$

   $$= \frac{1}{2 \sinh(x/2) \cosh(x/2)} = \frac{1}{\sinh x} = \operatorname{csch} x$$

23. If $y = x^{\cosh x}$, find y' and simplify.

   Solution

   Using logarithmic differentiation, we have

   $$y = x^{\cosh x}$$

   $$\ln y = \ln(x^{\cosh x}) = (\cosh x)(\ln x)$$

   $$\frac{y'}{y} = (\cosh x)\left(\frac{1}{x} \right) + (\ln x)(\sinh x)$$

   $$y' = y\left[\frac{\cosh x}{x} + (\ln x)(\sinh x) \right]$$

30. If $y = \text{sech}^{-1} (\cosh 2x)$, find y' and simplify. (Assume that $0 < x < \pi/2$.)

Solution

$$y = \text{sech}^{-1} (\cos 2x)$$

$$y' = \left(\frac{-1}{\cos 2x \sqrt{1 - \cos^2 2x}} \right) (-2 \sin 2x)$$

$$= \frac{2 \sin 2x}{\cos 2x \sqrt{\sin^2 2x}} = \frac{2 \sin 2x}{\cos 2x \, |\sin 2x|}$$

(Since we are given $0 < x < \pi/2$, we can eliminate the absolute value signs.)

$$y' = \frac{2}{\cos 2x} = 2 \sec 2x$$

33. If $y = 2x \sinh^{-1} (2x) - \sqrt{1 + 4x^2}$, find y' and simplify.

Solution

$$y = 2x \sinh^{-1} (2x) - \sqrt{1 + 4x^2}$$

$$y' = 2x \left[\frac{1}{\sqrt{1 + (2x)^2}} \right] (2) + 2 \sinh^{-1} (2x) - \frac{1}{2}(1 + 4x^2)^{-1/2}(8x)$$

$$= \frac{4x}{\sqrt{1 + 4x^2}} + 2 \sinh^{-1} (2x) - \frac{4x}{\sqrt{1 + 4x^2}} = 2 \sinh^{-1} (2x)$$

38. Evaluate $\displaystyle\int \frac{\sinh x}{1 + \sinh^2 x} \, dx$.

Solution

$$\int \frac{\sinh x}{1 + \sinh^2 x} \, dx = \int \frac{\sinh x}{\cosh^2 x} \, dx$$

$$= \int (\cosh x)^{-2} (\sinh x) \, dx \qquad\qquad u = \cosh x$$

$$= \frac{(\cosh x)^{-1}}{-1} + C = -\text{sech} \, x + C$$

51. Evaluate $\displaystyle\int \frac{1}{\sqrt{1 + e^{2x}}} \, dx$.

Solution

$$\int \frac{1}{\sqrt{1 + e^{2x}}} \, dx = \int \frac{e^x}{e^x \sqrt{1 + (e^x)^2}} \, dx \qquad u = e^x$$

$$= \int \frac{du}{u\sqrt{1 + u^2}}$$

$$= -\ln \left(\frac{1 + \sqrt{1 + e^{2x}}}{e^x} \right) + C = -\operatorname{csch}^{-1}(e^x) + C$$

53. Evaluate $\int \frac{1}{\sqrt{x}\sqrt{1 + x}} \, dx$.

Solution

$$\int \frac{1}{\sqrt{x}\sqrt{1 + x}} \, dx = 2 \int \frac{1}{\sqrt{1 + (\sqrt{x})^2}} \left(\frac{1}{2\sqrt{x}} \right) dx \qquad u = \sqrt{x}$$

$$= 2 \int \frac{du}{\sqrt{1 + u^2}}$$

$$= 2 \sinh^{-1} \sqrt{x} + C$$

$$= 2 \ln (\sqrt{x} + \sqrt{1 + x}) + C$$

64. Find any relative extrema and points of inflection of the function $f(x) = x - \tanh x$. Sketch the graph of f.

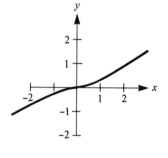

Solution

$$f(x) = x - \tanh x$$

$$f'(x) = 1 - \operatorname{sech}^2 x = 0 \qquad \text{when} \qquad x = 0$$

$$f''(x) = 2 \operatorname{sech}^2 x \tanh x = 0 \qquad \text{when} \qquad x = 0$$

Therefore, there is a point of inflection at $(0, 0)$. Since the range of hyperbolic secant is $(0, 1]$, we have $f'(x) > 0$ if $x \neq 0$ and f is a non-decreasing function.

67. Find dy/dx for the tractrix $y = a \operatorname{sech}^{-1}(x/a) - \sqrt{a^2 - x^2}$.

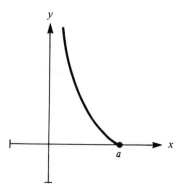

Solution

Note that the domain of this function restricts x to the interval $(0, a)$. (Assume $0 < a$.)

$$y = a \operatorname{sech}^{-1} \left(\frac{x}{a} \right) - \sqrt{a^2 - x^2}$$

$$\frac{dy}{dx} = (a) \left[\frac{-1}{|x/a|\sqrt{1 - (x/a)^2}} \right] \left(\frac{1}{a} \right) - \frac{1}{2}(a^2 - x^2)^{-1/2}(-2x)$$

$$= \frac{-1}{|x/a|\sqrt{(a^2 - x^2)/a^2}} + \frac{x}{\sqrt{a^2 - x^2}}$$

$$= \frac{-a^2}{x\sqrt{a^2 - x^2}} + \frac{x}{\sqrt{a^2 - x^2}}$$

(Note that we may drop the absolute value signs since $0 < x$.)

$$\frac{dy}{dx} = \frac{-a^2 + x^2}{x\sqrt{a^2 - x^2}} = \frac{-(a^2 - x^2)}{x\sqrt{a^2 - x^2}} = \frac{-\sqrt{a^2 - x^2}}{x}$$

74. Prove $\dfrac{d}{dx}[\tanh^{-1} x] = \dfrac{1}{1 - x^2}$.

Solution

$$y = \tahn^{-1} x \qquad \text{if and only if} \qquad \tanh y = x$$

Differentiating implicitly, we have

$$(\text{sech}^2 y)\, \frac{dy}{dx} = 1$$

$$\frac{dy}{dx} = \frac{1}{\text{sech}^2 y} = \frac{1}{1 - \tanh^2 y} = \frac{1}{1 - x^2}$$

Review Exercises for Chapter 6

11. Solve for x in the equation $\log_3 x + \log_3 (x - 1) - \log_3 (x - 2) = 2$.

Solution

$$\log_3 x + \log_3 (x - 1) - \log_3 (x - 2) = 2$$

$$\log_3 \frac{x(x - 1)}{x - 2} = 2$$

$$\frac{x(x - 1)}{x - 2} = 3^2$$

$$x^2 - x = 9x - 18$$

$$x^2 - 10x + 18 = 0$$

$$x = \frac{10 \pm \sqrt{100 - 72}}{2}$$

$$= 5 \pm \sqrt{7}$$

15. Find dy/dx given $y = x^2 e^x$.

Solution

$$y = x^2 e^x$$
$$y' = x^2(e^x) + e^x(2x) = xe^x(x + 2)$$

21. Find dy/dx given $ye^x + xe^y = xy$.

Solution

Using the Product Rule on each term and differentiating implicitly, we obtain

$$\left(ye^x + e^x \frac{dy}{dx}\right) + \left(xe^y \frac{dy}{dx} + e^y\right) = \left(x \frac{dy}{dx} + y\right)$$

Solving for dy/dx, we have

$$(e^x + xe^y - x) \frac{dy}{dx} = y - e^y - ye^x$$

$$\frac{dy}{dx} = \frac{y - e^y - ye^x}{e^x + xe^y - x}$$

25. Find dy/dx given $y = \tan(\arcsin x)$.

Solution

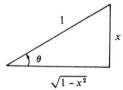

We begin by sketching a triangle to represent θ, "the angle whose sine is x." Then

$$\theta = \arcsin x$$

$$y = \tan(\arcsin x) = \tan \theta = \frac{x}{\sqrt{1 - x^2}}$$

Therefore,

$$\frac{dy}{dx} = \frac{\sqrt{1 - x^2}\,(1) - x(\tfrac{1}{2})(1/\sqrt{1 - x^2})(-2x)}{1 - x^2}$$

$$= \frac{(1 - x^2) + x^2}{(1 - x^2)\sqrt{1 - x^2}} = (1 - x^2)^{-3/2}$$

35. Evaluate $\displaystyle\int \frac{e^{4x} - e^{2x} + 1}{e^x}\, dx$.

Solution

$$\int \frac{e^{4x} - e^{2x} + 1}{e^x}\, dx = \int (e^{3x} - e^x + e^{-x})\, dx = \frac{e^{3x}}{3} - e^x - e^{-x} + C$$

42. Evaluate $\displaystyle\int \frac{\tan(1/x)}{x^2}\, dx$.

Solution

If we let $u = 1/x$, then $du = (-1/x^2)\, dx$. Thus

$$\int \frac{\tan(1/x)}{x^2} \, dx = -\int \tan(1/x)\left(-\frac{1}{x^2}\right) dx = -\int \tan u \, du$$

$$= -\left(-\ln\left[\cos\left(\frac{1}{x}\right)\right]\right) + C = \ln\left[\cos\left(\frac{1}{x}\right)\right] + C$$

43. Evaluate $\displaystyle\int \frac{1}{e^{2x} + e^{-2x}} \, dx$.

   ### Solution

   If we let $u = e^{2x}$, then $du = 2e^{2x} \, dx$. Thus

   $$\int \frac{1}{e^{2x} + e^{-2x}} \, dx = \int \left(\frac{1}{e^{2x} + e^{-2x}}\right)\left(\frac{e^{2x}}{e^{2x}}\right) dx = \int \frac{e^{2x}}{e^{4x} + 1} \, dx$$

   $$= \frac{1}{2}\int \left[\frac{1}{1 + (e^{2x})^2}\right](2e^{2x}) \, dx = \frac{1}{2}\int \frac{du}{1 + u^2}$$

   $$= \frac{1}{2}\arctan(e^{2x}) + C$$

50. Evaluate $\displaystyle\int \frac{4 - x}{\sqrt{4 - x^2}} \, dx$.

   ### Solution

   $$\int \frac{4 - x}{\sqrt{4 - x^2}} \, dx = \int \frac{4}{\sqrt{4 - x^2}} \, dx - \int \frac{x}{\sqrt{4 - x^2}} \, dx$$

   $$= 4\int \frac{1}{\sqrt{2^2 - x^2}} \, dx - \left(\frac{-1}{2}\right)\int (4 - x^2)^{-1/2}(-2x) \, dx$$

   $$= 4\arcsin\left(\frac{x}{2}\right) + \frac{1}{2}\left[\frac{(4 - x^2)^{1/2}}{\frac{1}{2}}\right] + C$$

   $$= 4\arcsin\left(\frac{x}{2}\right) + \sqrt{4 - x^2} + C$$

55. How large a deposit, at 7% interest compounded continuously, must be made to obtain a balance of $10,000 in 15 years?

   ### Solution

   Let A = amount of time t
   P = initial deposit
   t = time in years
   r = interest rate

   Then, using the formula for continuous interest, we have

$$A = Pe^{rt}$$

$$10,000 = Pe^{(0.07)(15)}$$

$$P = \frac{10,000}{e^{1.05}}$$

$$\approx \$3499.38$$

61. Trucks arrive at a terminal at an average rate of 3 per hour (thus $u = 20$ min is the average time between arrivals). If a truck has just arrived, find the probability that the next arrival will be:

(a) within the next 10 min

(b) within the next 30 min

(c) between 15 min and 30 min from now

(d) within the next hour

Solution

$$\int_{t_1}^{t_2} \left(\frac{1}{u}\right) e^{-t/u}\, dt = \int_{t_1}^{t_2} \left(\frac{1}{20}\right) e^{-t/20}\, dt = -\int_{t_1}^{t_2} e^{-t/20} \left(\frac{-1}{20}\right) dt$$

$$= [-e^{-t/20}]_{t_1}^{t_2} = e^{-t_1/20} - e^{-t_2/20}$$

(a) When $t_1 = 0$, $t_2 = 10$,

probability $= e^0 - e^{-1/2} = 1 - e^{-1/2} = 0.39$

(b) When $t_1 = 0$, $t_2 = 30$,

probability $= e^0 - e^{-3/2} = 1 - e^{-3/2} = 0.78$

(c) When $t_1 = 15$, $t_2 = 30$,

probability $= e^{-3/4} - e^{-3/2} = 0.25$

(d) When $t_1 = 0$, $t_2 = 60$,

probability $= e^0 - e^{-3} = 1 - e^{-3} = 0.95$

70. A certain automobile gets 28 mi/gal of gasoline for speeds up to 50 mi/h. Over 50 mi/h the miles per gallon drop at the rate of 12% for each 10 mi/h.

(a) If s is the speed and y is the miles per gallon, find y as a function of s by solving the differential equation

$$\frac{dy}{ds} = -0.012y \qquad s > 50$$

(b) Use the function in part (a) to complete the following table

Speed	50	55	60	65	70
Miles per gallon					

Solution

(a)
$$\frac{dy}{ds} = -0.012y$$

$$\left(\frac{1}{y}\right)\left(\frac{dy}{ds}\right) ds = -0.012 \, ds$$

$$\int \left(\frac{1}{y}\right)\left(\frac{dy}{ds}\right) ds = \int -0.012 \, ds$$

$$\ln y = -0.012s + C_1$$

$$y = Ce^{-0.012s}$$

When $s = 50$, $y = 28$. Therefore,

$$28 = Ce^{(-0.012)(50)}$$

$$28e^{0.6} = C$$

$$y = 28e^{0.6}e^{-0.012s} = 28e^{(0.6 - 0.012s)}$$

(b) We complete the table by substituting the required values of s into the function of part (a)

Speed (s)	50	55	60	65	70
Miles per gallon (y)	28.0	26.4	24.8	23.4	22.0

7 Applications of integration

7.1
Area of a region between two curves

6. Find the area of the region bounded by the graphs of $f(x) = (x - 1)^3$ and $g(x) = x - 1$.

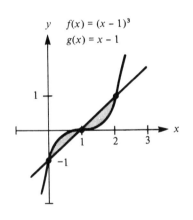

y $f(x) = (x - 1)^3$

$g(x) = x - 1$

Solution

From the accompanying figure we observe that $x - 1 \leq (x - 1)^3$ for $0 \leq x \leq 1$ and $(x - 1)^3 \leq x - 1$ for $1 \leq x \leq 2$. Therefore,

$$\text{area} = \int_0^1 [(x - 1)^3 - (x - 1)]\, dx$$

$$+ \int_1^2 [(x - 1) - (x - 1)^3]\, dx$$

$$= 2\int_0^1 [(x - 1)^3 - (x - 1)]\, dx \qquad \text{By symmetry}$$

$$= 2\left[\frac{1}{4}(x - 1)^4 - \frac{1}{2}(x - 1)^2 \right]_0^1 = \frac{1}{2}$$

9. Sketch the region bounded by the graphs of $f(x) = x^2 + 2x + 1$ and $g(x) = 3x + 3$ and find the area of this region by means of a definite integral.

Solution

The points of intersection of f and g are given by

$$f(x) = g(x)$$
$$x^2 + 2x + 1 = 3x + 3$$
$$x^2 - x - 2 = 0$$

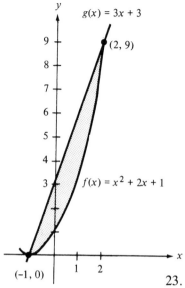

$g(x) = 3x + 3$

$(2, 9)$

$f(x) = x^2 + 2x + 1$

$(-1, 0)$

$$(x - 2)(x + 1) = 0$$

$$x = -1, 2$$

Since $x^2 + 2x + 1 \le 3x + 3$ for $-1 \le x \le 2$, we have

$$\text{area} = \int_{-1}^{2} [(3x + 3) - (x^2 + 2x + 1)]\, dx$$

$$= \int_{-1}^{2} (-x^2 + x + 2)\, dx$$

$$= \left[\frac{-x^3}{3} + \frac{x^2}{2} + 2x \right]_{-1}^{2}$$

$$= \left(\frac{-8}{3} + 2 + 4 \right) - \left(\frac{1}{3} + \frac{1}{2} - 2 \right)$$

$$= \frac{-16 + 12 + 24 - 2 - 3 + 12}{6} = \frac{27}{6} = \frac{9}{2}$$

23. Sketch the region bounded by the graphs of $f(y) = y^2 + 1$, $g(y) = 0$, $y = -1$, and $y = 2$ and find the area of this region by means of a definite integral.

Solution

$$\text{area} = \int_{-1}^{2} (y^2 + 1)\, dy = \left[\frac{y^3}{3} + y \right]_{-1}^{2}$$

$$= \left(\frac{8}{3} + 2 \right) - \left(\frac{-1}{3} - 1 \right)$$

$$= \frac{8 + 6 + 1 + 3}{3} = 6$$

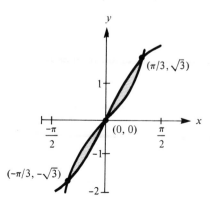

$(5, 2)$

$f(y)$

$(2, -1)$

29. Sketch the region bounded by the graphs of $f(x) = 2 \sin x$ and $g(x) = \tan x$ on the interval $-\pi/3 \le x \le \pi/3$ and find the area of the region.

Solution

Since f and g are symmetric to the origin, the area of the region bounded by their graphs for $-\pi/3 \le x \le \pi/3$ is twice the area of the region bounded by their graphs for $0 \le x \le \pi/3$. Since $\tan x \le 2 \sin x$ for $0 \le x \le \pi/3$, we have

$$\text{area} = 2 \int_{0}^{\pi/3} (2 \sin x - \tan x)\, dx$$

$$= 2[-2 \cos x + \ln |\cos x|]_{0}^{\pi/3}$$

$$= 2\left[-2\left(\frac{1}{2} \right) + \ln \left(\frac{1}{2} \right) - (-2) \right] = 2(1 - \ln 2)$$

$(\pi/3, \sqrt{3})$

$(0, 0)$

$(-\pi/3, -\sqrt{3})$

39. Use integration to find the area of the triangle with vertices $(0, 0)$, $(a, 0)$, and (b, c).

Solution

From the accompanying figure we observe that the triangular region is bounded by $y = (c/b)x$, $y = [c/(b - a)](x - a)$, and $y = 0$. Therefore, the area is given by

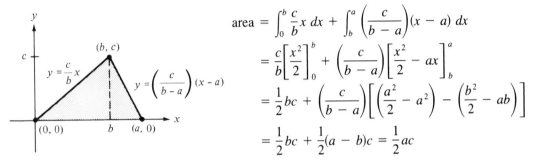

$$\text{area} = \int_0^b \frac{c}{b} x \, dx + \int_b^a \left(\frac{c}{b - a} \right)(x - a) \, dx$$

$$= \frac{c}{b} \left[\frac{x^2}{2} \right]_0^b + \left(\frac{c}{b - a} \right)\left[\frac{x^2}{2} - ax \right]_b^a$$

$$= \frac{1}{2}bc + \left(\frac{c}{b - a} \right)\left[\left(\frac{a^2}{2} - a^2 \right) - \left(\frac{b^2}{2} - ab \right) \right]$$

$$= \frac{1}{2}bc + \frac{1}{2}(a - b)c = \frac{1}{2}ac$$

47. Find the consumer surplus and producer surplus of the demand function $p_1 = 50 - 0.5x$ and supply function $p_2 = 0.125x$. The consumer surplus and producer surplus are represented by the areas shown in the accompanying figure.

Solution

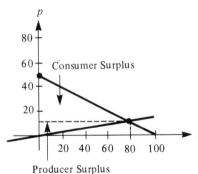

Solving the equations simultaneously we find the point of equilibrium to be $(80, 10)$. Therefore

$$\text{Consumer Surplus} = \int_0^{80} [(50 - 0.5x) - 10] \, dx$$

$$= [40x - 0.25x^2]_0^{80} = 1600$$

$$\text{Producer Surplus} = \int_0^{80} (10 - 0.125x) \, dx$$

$$= [10x - 0.0625x^2]_0^{80} = 400$$

7.2
Volume: The Disc Method

13. Find the volume of the solid generated by revolving the region bounded by $y = \sqrt{x}$, $y = 0$ and $x = 4$ about:

(a)   the x-axis   (b)   the y-axis

(c)   the line $x = 4$ (d)   the line $x = 6$

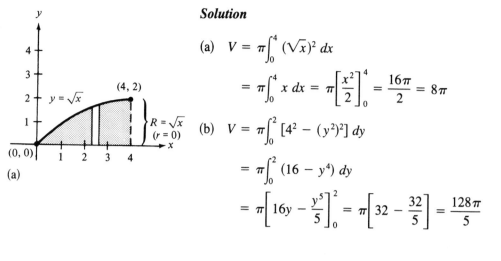

(a)

Solution

(a) $V = \pi \int_0^4 (\sqrt{x})^2 \, dx$

$\qquad = \pi \int_0^4 x \, dx = \pi \left[\dfrac{x^2}{2} \right]_0^4 = \dfrac{16\pi}{2} = 8\pi$

(b) $V = \pi \int_0^2 [4^2 - (y^2)^2] \, dy$

$\qquad = \pi \int_0^2 (16 - y^4) \, dy$

$\qquad = \pi \left[16y - \dfrac{y^5}{5} \right]_0^2 = \pi \left[32 - \dfrac{32}{5} \right] = \dfrac{128\pi}{5}$

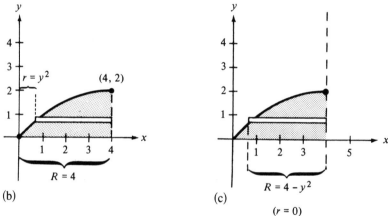

(b)

(c)

$(r = 0)$

(c) $V = \pi \int_0^2 (4 - y^2)^2 \, dy$

$\qquad = \pi \int_0^2 (16 - 8y^2 + y^4) \, dy$

$\qquad = \pi \left[16y - \dfrac{8y^3}{3} + \dfrac{y^5}{5} \right]_0^2 = \pi \left[32 - \dfrac{64}{3} + \dfrac{32}{5} \right] = \dfrac{256\pi}{15}$

(d) $V = \pi \int_0^2 [(6 - y^2)^2 - 2^2] \, dy$

$\qquad = \pi \int_0^2 (32 - 12y^2 + y^4) \, dy$

$\qquad = \pi \left[32y - 4y^3 + \dfrac{y^5}{5} \right]_0^2 = \pi \left[64 - 32 + \dfrac{32}{5} \right] = \dfrac{192\pi}{5}$

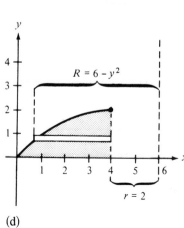

(d)

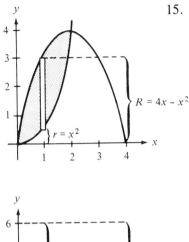

15. Find the volume of the solid formed by revolving the region bounded by $y = x^2$ and $y = 4x - x^2$ about:

(a)   the x-axis   (b)   the line $y = 6$

Solution

(a) $V = \pi \int_0^2 [(4x - x^2)^2 - (x^2)^2] \, dx$

$$= \pi \int_0^2 (16x^2 - 8x^3) \, dx$$

$$= \pi \left[\frac{16}{3}x^3 - 2x^4 \right]_0^2$$

$$= 8\pi \left[\frac{16}{3} - 4 \right] = \frac{32\pi}{3}$$

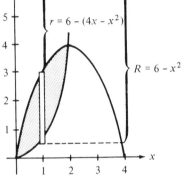

(b) $V = \pi \int_0^2 [(6 - x^2)^2 - (6 - 4x + x^2)^2] \, dx$

$$= 8\pi \int_0^2 (x^3 - 5x^2 + 6x) \, dx$$

$$= 8\pi \left[\frac{x^4}{4} - \frac{5}{3}x^3 + 3x^2 \right]_0^2$$

$$= 32\pi \left(1 - \frac{10}{3} + 3 \right) = \frac{64\pi}{3}$$

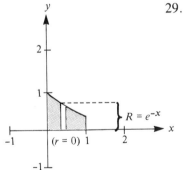

29. Find the volume of the solid generated by revolving the region bounded by the graphs of $y = e^{-x}$, $y = 0$, $x = 0$, and $x = 1$ about the x-axis.

Solution

$$V = \pi \int_0^1 (e^{-x})^2 \, dx$$

$$= \pi \int_0^1 e^{-2x} \, dx = \left[-\frac{\pi}{2}e^{-2x} \right]_0^1 = \frac{\pi}{2}(1 - e^{-2}) \approx 1.358$$

39. Use the Disc Method to verify that the volume of a sphere of radius r is $\frac{4}{3}\pi r^3$.

Solution

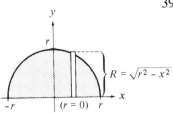

Let $y = \sqrt{r^2 - x^2}$ and let the region bounded by $y = \sqrt{r^2 - x^2}$ and $y = 0$ be revolved about the x-axis. The resulting solid of revolution is a sphere with volume

$$V = \pi \int_{-r}^{r} (\sqrt{r^2 - x^2})^2 \, dx = \pi \int_{-r}^{r} (r^2 - x^2) \, dx = \pi \left[r^2 x - \frac{x^3}{3} \right]_{-r}^{r}$$

$$= \pi \left[\left(r^3 - \frac{r^3}{3} \right) - \left(-r^3 + \frac{r^3}{3} \right) \right] = \pi \left[\frac{2r^3}{3} + \frac{2r^3}{3} \right] = \frac{4\pi r^3}{3}$$

44.   The tank on a water tower is a sphere of radius 50 ft. Determine the depth of the water when the tank is filled to 21.6% of its total capacity.

Solution

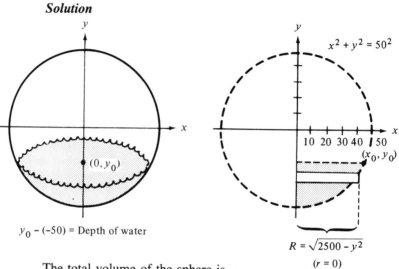

$y_0 - (-50) = $ Depth of water

The total volume of the sphere is

$$V = \frac{4\pi(50)^3}{3} = \frac{500{,}000\pi}{3} \text{ ft}^3$$

The volume of the portion filled with water is

$$(0.216)V = 36{,}000\pi = \pi \int_{-50}^{y_0} (\sqrt{2500 - y^2})^2 \, dy$$

$$= \pi \int_{-50}^{y_0} (2500 - y^2) \, dy = \pi \left[2500y - \frac{y^3}{3} \right]_{-50}^{y_0}$$

$$= \pi \left[\left(2{,}500y_0 - \frac{y_0^3}{3} \right) - \left(-125{,}000 + \frac{125{,}000}{3} \right) \right]$$

$$= \pi \left[2{,}500y_0 - \frac{y_0^3}{3} + \frac{250{,}000}{3} \right]$$

$$108{,}000 = 7{,}500y_0 - y_0^3 + 250{,}000$$

$$y_0^3 - 7{,}500y_0 - 142{,}000 = 0$$

$$(y_0 + 20)(y_0^2 - 20y_0 - 7100) = 0$$

$$y_0 = -20, \ 10 \pm 60\sqrt{2}$$

Since we are interested only in values of y_0 such that
$-50 < y_0 < 50$, we choose $y_0 = -20$ ft and conclude that the
depth of the water is $[-20 - (-50)] = 30$ ft.

49. The base of a solid is bounded by $y = x^3$, $y = 0$, and $x = 1$.
     Find the volume of the solid if the cross sections perpendicular
     to the y-axis are:

     (a) semicircles

     (b) squares

     (c) equilateral triangles

     (d) trapezoids for which $h = b_1 = \frac{1}{2}b_2$, where b_1 and b_2 are
          the lengths of the upper and lower bases, respectively

     (e) semiellipses whose heights are twice the length of their
          bases

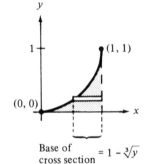

y

$(1, 1)$

1

$(0, 0)$

x

Base of
cross section $= 1 - \sqrt[3]{y}$

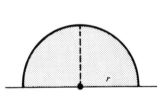

r

$1 - \sqrt[3]{y}$

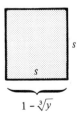

s

s

$1 - \sqrt[3]{y}$

Solution

The base of the solid is shown in the accompanying figure.
Since the cross sections are taken perpendicular to the y-axis,
the base of each cross section is given by $(1 - x) =$
$(1 - \sqrt[3]{y})$.

(a) The cross sections are semicircles whose radii are given
     by

$$r = \left(\frac{1}{2}\right)(1 - \sqrt[3]{y})$$

Thus

$$A(y) = \left(\frac{1}{2}\right)\pi\left[\left(\frac{1}{2}\right)(1 - \sqrt[3]{y})\right]^2 = \frac{\pi}{8}(1 - \sqrt[3]{y})^2$$

and

$$V = \frac{\pi}{8}\int_0^1 (1 - \sqrt[3]{y})^2\, dy = \frac{\pi}{8}\int_0^1 (1 - 2y^{1/3} + y^{2/3})\, dy$$

$$= \frac{\pi}{8}\left[y - \frac{2y^{4/3}}{\frac{4}{3}} + \frac{y^{5/3}}{\frac{5}{3}}\right]_0^1 = \frac{\pi}{8}\left(1 - \frac{3}{2} + \frac{3}{5}\right) = \frac{\pi}{8}\left(\frac{1}{10}\right) = \frac{\pi}{80}$$

(b) The cross sections are squares whose sides are given by

$$s = (1 - \sqrt[3]{y})$$

Thus

$$A(y) = s^2 = (1 - \sqrt[3]{y})^2$$

and

$$V = \int_0^1 (1 - \sqrt[3]{y})^2\, dy = \frac{1}{10} \qquad \text{From part (a)}$$

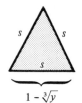

$1 - \sqrt[3]{y}$

(c) The cross sections are equilateral triangles whose sides are given by

$$s = (1 - \sqrt[3]{y})$$

Thus

$$A(y) = \frac{\sqrt{3}\,s^2}{4} = \frac{\sqrt{3}}{4}(1 - \sqrt[3]{y})^2$$

and

$$V = \frac{\sqrt{3}}{4} \int_0^1 (1 - \sqrt[3]{y})^2 \, dy = \frac{\sqrt{3}}{4}\left(\frac{1}{10}\right) = \frac{\sqrt{3}}{40}$$

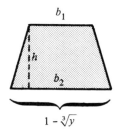

$1 - \sqrt[3]{y}$

(d) The cross sections are trapezoids for which $h = b_1 = b_2/2$, where

$$b_2 = (1 - \sqrt[3]{y})$$

Thus

$$A(y) = \left(\frac{b_1 + b_2}{2}\right)h = \left(\frac{3}{4}\right)(b_2)\left(\frac{1}{2}\right)(b_2) = \frac{3}{8}(1 - \sqrt[3]{y})^2$$

and

$$V = \frac{3}{8} \int_0^1 (1 - \sqrt[3]{y})^2 \, dy = \frac{3}{8}\left(\frac{1}{10}\right) = \frac{3}{8}$$

(e) The cross sections are semiellipses whose heights are twice the lengths of their bases. Thus $h = 2b$, where

$$b = (1 - \sqrt[3]{y})$$

Thus

$$A(y) = \left(\frac{1}{2}\right)\pi(h)\left(\frac{b}{2}\right) = \frac{\pi}{2}b^2 = \frac{\pi}{2}(1 - \sqrt[3]{y})^2$$

and

$$V = \frac{\pi}{2} \int_0^1 (1 - \sqrt[3]{y})^2 \, dy = \frac{\pi}{2}\left(\frac{1}{10}\right) = \frac{\pi}{20}$$

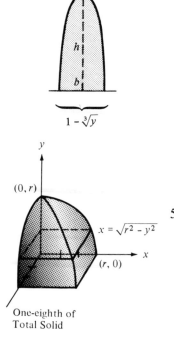

One-eighth of
Total Solid

50. Find the volume of the solid intersection (the solid common to both) of two right circular cylinders of radius r whose axes meet at right angles.

Solution

The cross sections are squares whose sides are given by

$$s = \sqrt{r^2 - y^2}$$

Thus

$$A(y) = s^2 = r^2 - y^2$$

and

$$\frac{1}{8}V = \int_0^r (r^2 - y^2)\, dy$$

$$V = 8\left[r^2 y - \frac{1}{3}y^3\right]_0^r = 8\left(\frac{2}{3}r^3\right) = \frac{16r^3}{3}$$

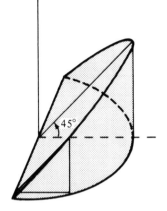

51.   A wedge is cut from a right circular cylinder of radius r inches by a plane through the diameter of the base, which makes a $45°$ angle with the plane of the base. Find the volume of the wedge cut out.

Solution

The cross sections are isosceles right triangles for which the base and height are equal. Thus

$$A(x) = \left(\frac{1}{2}\right)bh = \left(\frac{1}{2}\right)(\sqrt{r^2 - x^2})(\sqrt{r^2 - x^2}) = \left(\frac{1}{2}\right)(r^2 - x^2)$$

and

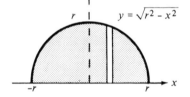

$$V = \frac{1}{2}\int_{-r}^r (r^2 - x^2)\, dx = \frac{1}{2}\left[r^2 x - \frac{x^3}{3}\right]_{-r}^r$$

$$= \frac{1}{2}\left[\left(r^3 - \frac{r^3}{3}\right) - \left(-r^3 + \frac{r^3}{3}\right)\right] = \frac{2r^3}{3}\ \text{in}^3$$

7.3
Volume: The Shell Method

3.   Use the Shell Method to find the volume generated by revolving the region bounded by $y = x$, $y = 0$, and $x = 2$ about the x-axis.

Solution

$$V = 2\pi \int_0^2 y(2 - y)\, dy = 2\pi \int_0^2 (2y - y^2)\, dy = 2\pi\left[y^2 - \frac{y^3}{3}\right]_0^2$$

$$= 2\pi\left(4 - \frac{8}{3}\right) = \frac{8\pi}{3}$$

9.   Use the Shell Method to find the volume generated by revolving

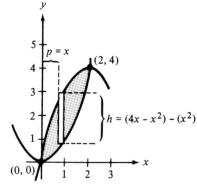

the region bounded by $y = x^2$ and $y = 4x - x^2$ about the
y-axis.

Solution

$$V = 2\pi \int_0^2 x[(4x - x^2) - (x^2)] \, dx = 2\pi \int_0^2 x(4x - 2x^2) \, dx$$

$$= 4\pi \int_0^2 (2x^2 - x^3) \, dx = 4\pi \left[\frac{2x^3}{3} - \frac{x^4}{4} \right]_0^2$$

$$= 4\pi \left[\frac{16}{3} - 4 \right] = \frac{16\pi}{3}$$

11. Use the Shell Method to find the volume generated by revolving the region bounded by $y = x^2$ and $y = 4x - x^2$ about
the line $x = 4$.

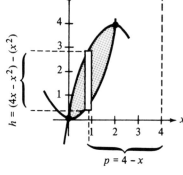

Solution

$$V = 2\pi \int_0^2 (4 - x)[(4x - x^2) - (x^2)] \, dx$$

$$= 2\pi \int_0^2 (4 - x)(4x - 2x^2) \, dx = 4\pi \int_0^2 (8x - 6x^2 + x^3) \, dx$$

$$= 4\pi \left[4x^2 - 2x^3 + \frac{x^4}{4} \right]_0^2 = 4\pi[16 - 16 + 4] = 16\pi$$

14. Use the Shell Method to find the volume generated by revolving the region bounded by $x + y^2 = 9$ and $x = 0$ about
the x-axis.

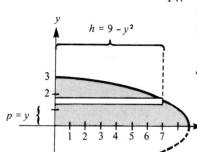

Solution

$$V = 2\pi \int_0^3 y(9 - y^2) \, dy = 2\pi \int_0^3 (9y - y^3) \, dy$$

$$= 2\pi \left[\frac{9y^2}{2} - \frac{y^4}{4} \right]_0^3 = 2\pi \left[\frac{81}{2} - \frac{8}{4} \right] = \frac{81\pi}{2}$$

21. Use the Disc or Shell Method to find the volume of the solid
generated by revolving the region bounded by $y = x^3$, $y = 0$,
and $x = 2$ about:

(a) the x-axis   (b) the y-axis

(c) the line $x = 4$ (d) the line $y = 8$

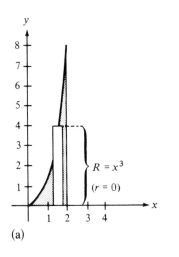

(a)

Solution

(a)   Disc Method

$$V = \pi \int_0^2 [(x^3)^2 - (0)^2] \, dx$$

$$= \pi \int_0^2 x^6 \, dx$$

$$= \pi \left[\frac{x^7}{7} \right]_0^2 = \frac{128\pi}{7}$$

(b)   Shell Method

$$V = 2\pi \int_0^2 x(x^3) \, dx$$

$$= 2\pi \int_0^2 x^4 \, dx$$

$$= 2\pi \left[\frac{x^5}{5} \right]_0^2 = \frac{64\pi}{5}$$

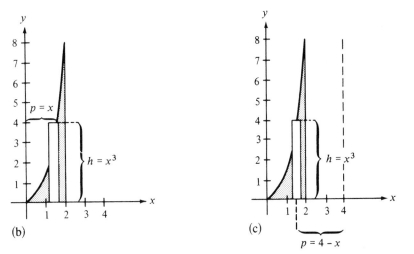

(b)   (c)

(c)   Shell Method

$$V = 2\pi \int_0^2 (4 - x)x^3 \, dx$$

$$= 2\pi \int_0^2 (4x^3 - x^4) \, dx$$

$$= 2\pi \left[x^4 - \frac{x^5}{5} \right]_0^2 = \frac{96\pi}{5}$$

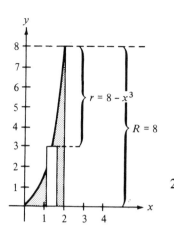

(d)   Disc Method

$$V = \pi \int_0^2 [8^2 - (8 - x^3)^2]\, dx$$

$$= \pi \int_0^2 (16x^3 - x^6)\, dx$$

$$= \pi \left[4x^4 - \frac{x^7}{7} \right]_0^2 = \frac{320\pi}{7}$$

25.   A solid is generated by revolving the region bounded by $y = \frac{1}{2}x^2$ and $y = 2$ about the x-axis. A hole, centered along the axis of revolution, is drilled through this solid so that $\frac{1}{4}$ of the volume is removed. Find the diameter of the hole.

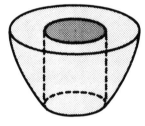

Solution

The total volume of the solid is given by

$$V = 2\pi \int_0^2 x\left(2 - \frac{x^2}{2}\right) dx = 2\pi \int_0^2 \left(2x - \frac{x^3}{2}\right) dx$$

$$= 2\pi \left[x^2 - \frac{x^4}{8} \right]_0^2 = 2\pi(4 - 2) = 4\pi$$

If a hole drilled in the center with a radius of x_0 removes $\frac{1}{4}$ of this volume, we would have

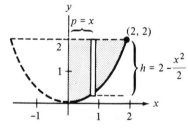

$$\left(\frac{3}{4}\right) V = 2\pi \int_{x_0}^2 x\left(2 - \frac{x^2}{2}\right) dx$$

$$3\pi = 2\pi \left[x^2 - \frac{x^4}{8} \right]_{x_0}^2$$

$$3\pi = 2\pi \left[(4 - 2) - \left(x_0^2 - \frac{x_0^4}{8} \right) \right]$$

$$3 = 4 - 2x_0^2 + \left(\frac{x_0^4}{4} \right)$$

$$x_0^4 - 8x_0^2 + 4 = 0$$

$$x_0^2 = \frac{8 \pm \sqrt{64 - 16}}{2}$$

$$x_0 = \sqrt{4 - 2\sqrt{3}}$$

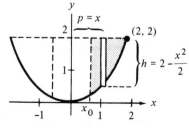

(Since $0 < x_0 < 2$, we are not interested in $x_0 = \sqrt{4 + 2\sqrt{3}} \approx 2.7$.) Finally, since

$$\text{radius} = x_0 = \sqrt{4 - 2\sqrt{3}} = 0.732$$

we have

$$\text{diameter} = 2x_0 = 2\sqrt{4 - 2\sqrt{3}} \approx 1.464$$

7.4
Work

7. A force of 60 lb stretches a spring 1 ft. How much work is done in stretching the spring from 9 in to 15 in?

Solution

Let $F(x)$ be the force required to stretch a spring x units. By Hooke's Law we know that $F(x) = kx$. Since 60 lb stretches the spring 1 ft (12 in), we have

$$60 = k(12) \qquad \text{or} \qquad k = 5$$

Therefore, the work done by stretching the spring from 9 to 15 in is

$$W = \underbrace{\int_9^{15} F(x)}_{} \underbrace{dx}_{} = \int_9^{15} 5x \, dx = \left[\frac{5x^2}{2}\right]_9^{15}$$

$$\text{(force)(distance)}$$

$$= \frac{5}{2}(225 - 81) = 360 \text{ in} \cdot \text{lb}$$

15. A hemispherical tank of radius 6 ft is positioned so that its base is circular. How much work is required to fill the tank with water through a hole in the base if the water source is at the base?

Solution

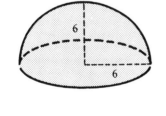

To fill the tank with water from a hole in the bottom, we do not have to move all the water a height of 6 ft. Some of the water must be moved 6 ft, some 5 ft, some 4 ft, and so on. In general, the "disc" of water that must be moved y feet has a volume of

$$\pi x^2 \Delta y = \pi(\sqrt{36 - y^2})^2 \, \Delta y = \pi(36 - y^2) \, \Delta y \text{ cubic feet}$$

The weight of this disc of water is

$$62.4(\pi)(36 - y^2) \, \Delta y \text{ pounds}$$

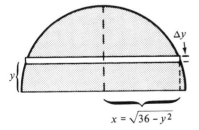

$x = \sqrt{36 - y^2}$

Thus the work done in filling the tank from the bottom is

$$W = \int_0^6 \underbrace{(y)}_{}\underbrace{[62.4\pi(36 - y^2) \, dy}_{}$$

$$\text{(distance)(force: weight of water)}$$

$$= 62.4\pi \int_0^6 (36y - y^3) \, dy = 62.4\pi\left[18y^2 - \frac{y^4}{4}\right]_0^6$$

$$= 62.4\pi(648 - 324) = 20{,}217.6\pi \text{ ft} \cdot \text{lb}$$

17. An open tank has the shape of a right circular cone. If the tank is 8 ft across the top and has a height of 6 ft, how much work is required to empty the tank of water by pumping the water over the top edge?

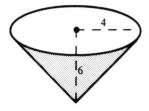

Solution

A disc of water at height y has to be moved $(6 - y)$ feet up and has a volume of $\pi x^2 \Delta y$. To find x in terms of y, we solve the equation

$$\frac{6 - 0}{4 - 0} = \frac{y - 0}{x - 0}$$

$$x = \frac{2y}{3}$$

Thus the work done in moving the water up over the top of the tank is

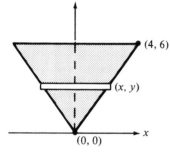

$$W = \int_0^6 \underbrace{(6 - y)}_{\text{(distance)}} \underbrace{\left[62.4\pi \left(\frac{2y}{3}\right)^2 dy \right]}_{\text{(force: weight of water)}}$$

$$= \frac{4}{9}(62.4)\pi \int_0^6 (6y^2 - y^3)\, dy = \frac{83.2}{3}\pi \left[2y^3 - \frac{y^4}{4} \right]_0^6$$

$$= \frac{83.2}{3}\pi (432 - 324) = 2995.2\pi \text{ ft} \cdot \text{lb}$$

29. A chain 15 ft long and weighing 3 lb/ft is suspended vertically from a height of 15 ft. How much work is required to take the bottom of the chain and raise it to the 15-ft level, leaving the chain doubled but still hanging vertically?

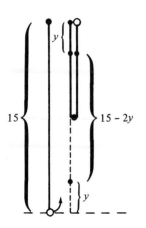

Solution

A small piece of chain of length Δy at height y must be moved so that it is y feet from the top. Therefore, the distance moved (as seen in the accompanying figure) is $15 - 2y$. (For example, the chain at an initial height of 7.5 is moved 0 ft.) The weight of a piece of chain of length Δy is

$$\frac{3 \text{ pounds}}{\text{foot}} \Delta y \text{ feet} = 3 \Delta y \text{ pounds}$$

Finally, since we are only moving chain that has an initial height between 0 and 7.5 ft, we have

$$W = \int_0^{7.5} \underbrace{(15 - 2y)}_{\text{(distance)}}\underbrace{(3\ dy)}_{\text{(force)}}$$

$$= 3\left[15y - y^2\right]_0^{7.5} = 3(112.5 - 56.25) = 168.75 \text{ ft} \cdot \text{lb}$$

7.5
Fluid pressure and fluid force

3. Find the force on a vertical side of a tank if the tank is full of water and the side has the shape of a trapezoid, as in the accompanying figure.

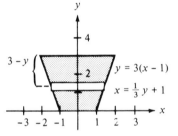

Solution

The force against a representative rectangle of length $2x$ is

$$\Delta F = (\text{density})(\text{depth})(\text{area}) = (62.4)(3 - y)(2x\ \Delta y)$$

$$= (62.4)(3 - y)(2)\left(\frac{1}{3}y + 1\right)\Delta y$$

Since y ranges from 0 to 3, the total force is

$$F = \int_0^3 (62.4)(3 - y)(2)\left(\frac{1}{3}y + 1\right)\ dy = 124.8\int_0^3\left(3 - \frac{1}{3}y^2\right)\ dy$$

$$= 124.8\left[3y - \frac{1}{9}y^3\right]_0^3 = 748.8 \text{ lb}$$

9. Find the total force on the given vertical plate submerged in water.

Solution

The force against a representative rectangle of length x is

$$\Delta F = (\text{density})(\text{depth})(\text{area}) = (62.4)(12 - y)\left(-\frac{2}{3}y + 6\right)\Delta y$$

Since y ranges from 0 to 9, the total force is

$$F = \int_0^9 62.4(12 - y)\left(-\frac{2}{3}y + 6\right)\ dy$$

$$= 62.4\int_0^9\left(\frac{2}{3}y^2 - 14y + 72\right)\ dy$$

$$= 62.4\left[\frac{2}{9}y^3 - 7y^2 + 72y\right]_0^9 = 15{,}163.2 \text{ lb}$$

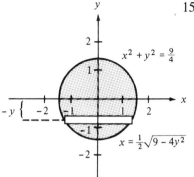

$x^2 + y^2 = \frac{9}{4}$

$x = \frac{1}{2}\sqrt{9 - 4y^2}$

15. A cylindrical gasoline tank is placed so that the axis of the cylinder is horizontal. If the tank is half full, find the force on a circular end of the tank, assuming that the diameter is 3 ft and gasoline has a density of 42 lb/ft³.

Solution

The force against a representative rectangle of length $2x$ is

$$\Delta F = (\text{density})(\text{depth})(\text{area}) = (42)(-y)(2x\,\Delta y)$$

$$= (42)(-y)(2)\left(\frac{1}{2}\right)(9 - 4y^2)^{1/2}\,\Delta y = -42y(9 - 4y^2)^{1/2}\,\Delta y$$

Since y varies from $-\frac{3}{2}$ to 0, the total force is

$$F = \int_{-3/2}^{0} (-42)y(9 - 4y^2)^{1/2}\,dy = \frac{42}{8}\int_{-3/2}^{0} (9 - 4y^2)^{1/2}(-8y)\,dy$$

$$= \frac{21}{4}\left(\frac{2}{3}\right)\left[(9 - 4y^2)^{3/2}\right]_{-3/2}^{0} = 94.5 \text{ lb}$$

21. A swimming pool is 20 ft wide, 40 ft long, 4 ft deep at one end, and 8 ft deep at the other. The bottom is an inclined plane. Find the total force on each of the vertical walls of the pool.

Solution

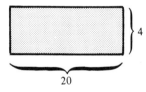

20

The force on the rectangular shallow end is given by

$$F = \int_{0}^{4} 62.4(4 - y)(20)\,dy = 1{,}248\int_{0}^{4} (4 - y)\,dy$$

$$= 1{,}248\left[4y - \frac{y^2}{2}\right]_{0}^{4} = 1248(8) = 9984 \text{ lb}$$

and for the deep end,

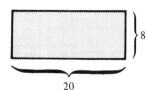

20

$$F = \int_{0}^{8} 62.4(8 - y)(20)\,dy = 1{,}248\int_{0}^{8} (8 - y)\,dy$$

$$= 1{,}248\left[8y - \frac{y^2}{2}\right]_{0}^{8} = 1{,}248(32) = 39{,}936 \text{ lb}$$

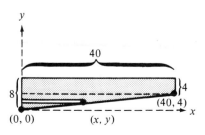

The force on each side is given by

$$F = \int_{0}^{4} 62.4(8 - y)(x)\,dy + \int_{4}^{8} 62.4(8 - y)40\,dy$$

where x and y are related by

$$\frac{4 - 0}{40 - 0} = \frac{y - 0}{x - 0} \qquad \text{or} \qquad x = 10y$$

$$\text{Thus } F = \int_0^4 62.4(8 - y)(10y)\, dy + \int_4^8 62.4(8 - y)(40)\, dy$$

$$= 624 \int_0^4 (8y - y^2)\, dy + 2{,}496 \int_4^8 (8 - y)\, dy$$

$$= 624 \left[4y^2 - \frac{y^3}{3} \right]_0^4 + 2{,}496 \left[8y - \frac{y^2}{2} \right]_4^8$$

$$= 624 \left(64 - \frac{64}{3} \right) + 2{,}496[(64 - 32) - (32 - 8)]$$

$$= 624 \left(\frac{128}{3} \right) + 2{,}496(8) = 46{,}592 \text{ lb}$$

7.6
Moments, centers of mass, and centroids

2. Find the center of mass for the masses located at the given points:

$$m_1 = 7, \qquad x_1 = -3, \qquad m_2 = 4, \qquad x_2 = -2,$$
$$m_3 = 3, \qquad x_3 = 5, \qquad m_4 = 8, \qquad x_4 = 6$$

Solution

$$\bar{x} = \frac{m_1 x_1 + m_2 x_2 + m_3 x_3 + m_4 x_4}{m_1 + m_2 + m_3 + m_4} = \frac{7(-3) + 4(-2) + 3(5) + 8(6)}{7 + 4 + 3 + 8}$$

$$= \frac{-21 - 8 + 15 + 48}{22} = \frac{34}{22} = \frac{17}{11}$$

10. Find the center of mass for the masses located at the given points:

$$m_1 = 4, \qquad P_1 = (2, 3), \qquad m_2 = 2, \qquad P_2 = (-1, 5),$$
$$m_3 = 2.5, \qquad P_3 = (6, 8), \qquad m_4 = 5, \qquad P_4 = (2, -2)$$

Solution

$$\bar{x} = \frac{m_1 x_1 + m_2 x_2 + m_3 x_3 + m_4 x_4}{m_1 + m_2 + m_3 + m_4} = \frac{4(2) + 2(-1) + (\frac{5}{2})(6) + 5(2)}{4 + 2 + (\frac{5}{2}) + 5}$$

$$= \frac{8 - 2 + 15 + 10}{\frac{27}{2}} = \frac{31}{\frac{27}{2}} = \frac{62}{27}$$

$$\bar{y} = \frac{m_1 y_1 + m_2 y_2 + m_3 y_3 + m_4 y_4}{m_1 + m_2 + m_3 + m_4} = \frac{4(3) + 2(5) + (\frac{5}{2})(8) + 5(-2)}{\frac{27}{2}}$$

$$= \frac{12 + 10 + 20 - 10}{\frac{27}{2}} = \frac{64}{27}$$

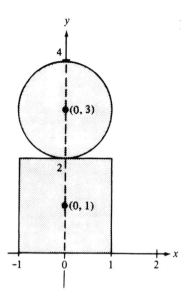

11. A plate of uniform density is formed by a circle and a square
     as shown in the accompanying figure. Introduce an appropriate
     rectangular coordinate system and find the coordinates of the
     center of mass.

Solution

Although a coordinate system may be introduced in many
different ways, the one shown in the accompanying figure is a
fairly natural choice. Since both the circle and square have a
uniform density, their masses are proportional to their areas, π
and 4, respectively. (For simplicity we assume the density to
be 1 unit of mass per 1 unit of area.) Again, because of the
uniform density, both the circle and the square have their
centers of mass at their geometrical centers, $(0, 3)$ and $(0, 1)$,
respectively. Therefore, we can find the center of mass of the
plate by considering a mass of π centered at $(0, 3)$ and a mass
of 4 centered at $(0, 1)$. Thus

$$\bar{x} = \frac{\pi(0) + 4(0)}{\pi + 4} = 0 \quad \text{and} \quad \bar{y} = \frac{\pi(3) + 4(1)}{\pi + 4} = \frac{3\pi + 4}{\pi + 4}$$

23. Find M_x, M_y, and $(\bar{x}, \bar{y})$ for the lamina of uniform density ρ
     bounded by $x = 4 - y^2$ and $x = 0$.

Solution

Since the region is symmetric with respect to the x-axis, we
know that

$$\bar{y} = 0$$

To find $\bar{x}$, we observe that x is a function of y, and we use the
formula

$$\bar{x} = \frac{\frac{1}{2} \int_a^b [f(y)^2 - g(y)^2]\, dy}{A} = \frac{M_y}{A}$$

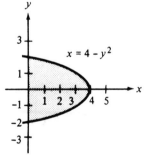

where $f(y) = 4 - y^2$, $g(y) = 0$, $a = -2$, $b = 2$, and

$$A = \int_{-2}^{2} (4 - y^2)\, dy = \left[4y - \frac{y^3}{3}\right]_{-2}^{2} = \frac{32}{3}$$

$$\bar{x} = \frac{\frac{1}{2} \int_{-2}^{2} (4 - y^2)^2\, dy}{\frac{32}{3}} = \frac{3}{64} \int_{-2}^{2} (16 - 8y^2 + y^4)\, dy$$

$$= \frac{3}{64}\left[16y - \frac{8y^3}{3} + \frac{y^5}{5}\right]_{-2}^{2}$$

$$= \frac{3}{64}\left[\left(32 - \frac{64}{3} + \frac{32}{5}\right) - \left(-32 + \frac{64}{3} - \frac{32}{5}\right)\right]$$

$$= \frac{3}{64}\left(\frac{512}{15}\right) = \frac{8}{5}$$

$$M_y = A\bar{x} = \frac{32}{3}\left(\frac{8}{5}\right) = \frac{256}{15} \quad \text{and} \quad M_x = A\bar{y} = 0$$

31. Find the centroid of the region bounded by $y = f(x) = x$ and $y = g(x) = x^2$.

Solution

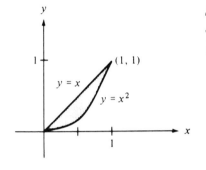

The two graphs intersect at the points $(0, 0)$ and $(1, 1)$. The area is given by

$$A = \int_0^1 [f(x) - g(x)] \, dx = \int_0^1 (x - x^2) \, dx$$

$$= \left[\frac{1}{2}x^2 - \frac{1}{3}x^3\right]_0^1 = \frac{1}{6}$$

$$\bar{x} = \frac{\int_0^1 x[f(x) - g(x)] \, dx}{A} = \frac{\int_0^1 x(x - x^2) \, dx}{\frac{1}{6}}$$

$$= 6\int_0^1 (x^2 - x^3) \, dx = 6\left[\frac{1}{3}x^3 - \frac{1}{4}x^4\right]_0^1 = \frac{1}{2}$$

$$\bar{y} = \frac{\frac{1}{2}\int_0^1 [f(x)^2 - g(x)^2] \, dx}{A} = \frac{\frac{1}{2}\int_0^1 [(x)^2 - (x^2)^2] \, dx}{\frac{1}{6}}$$

$$= 3\int_0^1 (x^2 - x^4) \, dx = 3\left[\frac{1}{3}x^3 - \frac{1}{5}x^5\right]_0^1 = \frac{2}{5}$$

$$(\bar{x}, \bar{y}) = \left(\frac{1}{2}, \frac{2}{5}\right)$$

35. Find the centroid of the triangle with vertices $(-a, 0)$, $(a, 0)$, (b, c). Show that it is at the point of intersection of the medians. (Assume that $-a < b < a$.)

Solution

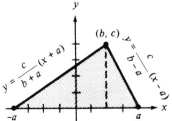

The equation of the line containing $(-a, 0)$ and (b, c) is

$$y = \left(\frac{c}{b + a}\right)(x + a)$$

The equation of the line containing $(a, 0)$ and (b, c) is

$$y = \left(\frac{c}{b - a}\right)(x - a)$$

Since the area of the triangle is

$$A = \left(\frac{1}{2}\right)(2a)(c) = ac$$

we have

$$\bar{x} = \frac{\displaystyle\int_{-a}^{b} x\left(\frac{c}{b + a}\right)(x + a)\, dx + \int_{b}^{a} x\left(\frac{c}{b - a}\right)(x - a)\, dx}{ac}$$

$$= \frac{1}{ac}\left[\frac{c}{b + a}\int_{-a}^{b}(x^2 + ax)\, dx + \frac{c}{b - a}\int_{b}^{a}(x^2 - ax)\, dx\right]$$

$$= \frac{1}{ac}\left(\frac{c}{b + a}\left[\frac{x^3}{3} + \frac{ax^2}{2}\right]_{-a}^{b} + \frac{c}{b - a}\left[\frac{x^3}{3} - \frac{ax^2}{2}\right]_{b}^{a}\right)$$

$$= \frac{1}{ac}\left[\frac{c}{b + a}\left(\frac{b^3}{3} + \frac{ab^2}{2} + \frac{a^3}{3} - \frac{a^3}{2}\right)\right.$$

$$\left. + \frac{c}{b - a}\left(\frac{a^3}{3} - \frac{a^3}{2} - \frac{b^3}{3} + \frac{ab^2}{2}\right)\right]$$

$$= \frac{2b^3 + 3ab^2 - a^3}{6a(b + a)} + \frac{-2b^3 + 3ab^2 - a^3}{6a(b - a)}$$

$$= \frac{(2b^2 + ab - a^2)(a + b)}{6a(b + a)} + \frac{(-2b^2 + ab + a^2)(b - a)}{6a(b - a)}$$

$$= \frac{2ab}{6a} = \frac{b}{3}$$

$$\bar{y} = \frac{\displaystyle\frac{1}{2}\int_{-a}^{b}\left[\frac{c}{b + a}(x + a)\right]^2 dx + \frac{1}{2}\int_{b}^{a}\left[\frac{c}{b - a}(x - a)\right]^2 dx}{ac}$$

$$= \frac{1}{2ac}\left\{\frac{c^2}{(b + a)^2}\left[\frac{(x + a)^3}{3}\right]_{-a}^{b} + \frac{c^2}{(b - a)^2}\left[\frac{(x - a)^3}{3}\right]_{b}^{a}\right\}$$

$$= \frac{1}{2ac}\left\{\frac{c^2}{(b + a)^2}\left[\frac{(b + a)^3}{3}\right] - \frac{c^2}{(b - a)^2}\left[\frac{(b - a)^3}{3}\right]\right\}$$

$$= \frac{1}{2ac}\left[\frac{c^2(b + a)}{3} - \frac{c^2(b - a)}{3}\right] = \frac{c}{3}$$

By Exercise 54, Section 1.4, we know that the point $(b/3, c/3)$ is the intersection of the medians of the triangle.

7.7
Arc length and surfaces of revolution

5. Find the arc length of $y = (x^4/8) + (1/4x^2)$ between $x = 1$ and $x = 2$.

   Solution

   $$y = \frac{x^4}{8} + \frac{1}{4x^2} \qquad y' = \frac{x^3}{2} - \frac{1}{2x^3}$$

   $$s = \int_1^2 \sqrt{1 + (y')^2}\, dx = \int_1^2 \sqrt{1 + \left(\frac{x^3}{2} - \frac{1}{2x^3}\right)^2}\, dx$$

   $$= \int_1^2 \sqrt{1 + \frac{x^6}{4} - \frac{1}{2} + \frac{1}{4x^6}}\, dx = \int_1^2 \sqrt{\frac{x^6}{4} + \frac{1}{2} + \frac{1}{4x^6}}\, dx$$

   $$= \int_1^2 \sqrt{\left(\frac{x^3}{2} + \frac{1}{2x^3}\right)^2}\, dx$$

   (Note that $0 < \dfrac{x^3}{2} + \dfrac{1}{2x^3}$ for $1 \le x \le 2$.)

   $$s = \int_1^2 \left(\frac{x^3}{2} + \frac{1}{2x^3}\right) dx = \left[\frac{x^4}{8} - \frac{1}{4x^2}\right]_1^2 = \left(\frac{16}{8} - \frac{1}{16}\right) - \left(\frac{1}{8} - \frac{1}{4}\right)$$

   $$= \frac{32 - 1 - 2 + 4}{16} = \frac{33}{16}$$

6. Find the arc length of $y = \frac{3}{2}x^{2/3}$ between $x = 1$ and $x = 8$.

   Solution

   $$y = \frac{3}{2}x^{2/3} \qquad y' = x^{-1/3}$$

   $$s = \int_1^8 \sqrt{1 + (y')^2}\, dx = \int_1^8 \sqrt{1 + (x^{-1/3})^2}\, dx$$

   $$= \int_1^8 \sqrt{1 + \frac{1}{x^{2/3}}}\, dx = \int_1^8 \sqrt{\frac{x^{2/3} + 1}{x^{2/3}}}\, dx$$

   (Note that $0 < x^{1/3}$ for $1 \le x \le 8$.)

   $$s = \int_1^8 \frac{1}{x^{1/3}}\sqrt{x^{2/3} + 1}\, dx = \frac{3}{2}\int_1^8 (x^{2/3} + 1)^{1/2}\left(\frac{2}{3}x^{-1/3}\right) dx$$

   $$= \frac{3}{2}\left[\frac{(x^{2/3} + 1)^{3/2}}{\frac{3}{2}}\right]_1^8 = (4 + 1)^{3/2} - (1 + 1)^{3/2}$$

   $$= 5\sqrt{5} - 2\sqrt{2} \approx 8.35$$

9. Find a definite integral that represents the arc length of $y = 1/x$ between $x = 1$ and $x = 3$. (Do not evaluate the integral.)

Solution

$$y = \frac{1}{x} \qquad y' = \frac{-1}{x^2}$$

$$s = \int_1^3 \sqrt{1 + (y')^2} \, dx = \int_1^3 \sqrt{1 + \left(\frac{1}{x^2}\right)^2} \, dx = \int_1^3 \sqrt{1 + \frac{1}{x^4}} \, dx$$

$$= \int_1^3 \sqrt{\frac{x^4 + 1}{x^4}} \, dx = \int_1^3 \frac{\sqrt{x^4 + 1}}{x^2} \, dx$$

19. A fleeing object leaves the origin and moves up the y-axis. At the same time a pursuer leaves the point $(1, 0)$ and moves always toward the fleeing object. If the pursuer's speed is twice that of the fleeing object, the equation of the path is

$$y = \frac{1}{3}(x^{3/2} - 3x^{1/2} + 2)$$

How far has the fleeing object traveled when it is caught? Show that the pursuer traveled twice as far.

Solution

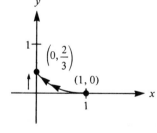

The y-intercept of $y = \frac{1}{3}(x^{3/2} - 3x^{1/2} + 2)$ is $(0, \frac{2}{3})$. Therefore, the fleeing object traveled from $(0, 0)$ to $(0, \frac{2}{3})$, a distance of $\frac{2}{3}$. The distance traveled by the pursuer is given by

$$s = \int_0^1 \sqrt{1 + (y')^2} \, dx$$

$$y = \frac{1}{3}(x^{3/2} - 3x^{1/2} + 2)$$

$$y' = \frac{1}{3}\left[\frac{3}{2}x^{1/2} - \frac{3}{2}x^{-1/2}\right] = \frac{1}{2}\left(\sqrt{x} - \frac{1}{\sqrt{x}}\right) = \frac{x - 1}{2\sqrt{x}}$$

$$1 + (y')^2 = 1 + \left(\frac{x - 1}{2\sqrt{x}}\right)^2 = \frac{4x + (x^2 - 2x + 1)}{4x}$$

$$= \frac{x^2 + 2x + 1}{4x} = \frac{(x + 1)^2}{4x}$$

Therefore,

$$s = \frac{1}{2}\int_0^1 \frac{x + 1}{\sqrt{x}} \, dx$$

$$= \frac{1}{2} \int_0^1 (x^{1/2} + x^{-1/2}) \, dx$$

$$= \frac{1}{2} \left[\frac{2}{3} x^{3/2} + 2x^{1/2} \right]_0^1 = \frac{4}{3}$$

Thus the pursuer traveled a distance of $\frac{4}{3}$, twice the distance of the fleeing object.

25. Find the area of the surface formed by revolving the graph of $y = (x^3/6) + (1/2x)$, $1 \le x \le 2$, about the x-axis.

Solution

$$y = \frac{x^3}{6} + \frac{1}{2x}$$

$$y' = \frac{1}{2} x^2 - \frac{1}{2x^2} = \frac{x^4 - 1}{2x^2}$$

$$1 + (y')^2 = 1 + \left(\frac{x^4 - 1}{2x^2} \right)^2 = \frac{4x^4 + (x^8 - 2x^4 + 1)}{4x^4}$$

$$= \frac{x^8 + 2x^4 + 1}{4x^4} = \left(\frac{x^4 + 1}{2x^2} \right)^2$$

$$S = 2\pi \int_1^2 y \sqrt{1 + (y')^2} \, dx$$

$$= 2\pi \int_1^2 \left(\frac{x^3}{6} + \frac{1}{2x} \right) \left(\frac{x^4 + 1}{2x^2} \right) \, dx$$

$$= 2\pi \int_1^2 \left(\frac{x^5}{12} + \frac{x}{3} + \frac{1}{4x^3} \right) \, dx$$

$$= 2\pi \left[\frac{x^6}{72} + \frac{x^2}{6} - \frac{1}{8x^2} \right]_1^2 = \frac{47\pi}{16}$$

27. Find the area of the surface formed by revolving the graph of $y = \sqrt[3]{x} + 2$, $1 \le x \le 8$, about the y-axis.

Solution

$$y = \sqrt[3]{x} + 2 \qquad y' = \frac{1}{3} x^{-2/3}$$

$$S = 2\pi \int_1^8 x \sqrt{1 + (y')^2} \, dx = 2\pi \int_1^8 x \sqrt{1 + \left(\frac{1}{3x^{2/3}} \right)^2} \, dx$$

$$= 2\pi \int_1^8 x \sqrt{\frac{9x^{4/3} + 1}{9x^{4/3}}} \, dx = 2\pi \int_1^8 \frac{x}{3x^{2/3}} \sqrt{9x^{4/3} + 1} \, dx$$

$$= \frac{2\pi}{3}\left(\frac{1}{12}\right)\int_1^8 (9x^{4/3} + 1)^{1/2}(12x^{1/3})\, dx$$

$$= \frac{\pi}{18}\left[\frac{(9x^{4/3} + 1)^{3/2}}{3/2}\right]_1^8 = \frac{\pi}{27}[145^{3/2} - 10^{3/2}] \approx 199.48$$

31. A right circular cone is generated by revolving the region bounded by $y = hx/r$, $y = h$, and $x = 0$ about the y-axis. Verify that the lateral surface area of the cone is $S = \pi r\sqrt{r^2 + h^2}$.

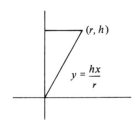

Solution

The distance between the y-axis and the graph of the line $y = hx/r$ is

$$r(y) = g(y) = \frac{ry}{h}$$

and since $g'(y) = \dfrac{r}{h}$, the surface area is given by

$$S = 2\pi \int_0^h r(y)\sqrt{1 + [g'(y)]^2}\, dy = 2\pi \int_0^h \left(\frac{ry}{h}\right)\sqrt{1 + \left(\frac{r}{h}\right)^2}\, dy$$

$$= \frac{2\pi r\sqrt{r^2 + h^2}}{h^2}\int_0^h y\, dy = \frac{2\pi r\sqrt{r^2 + h^2}}{h^2}\left[\frac{1}{2}y^2\right]_0^h = \pi r\sqrt{r^2 + h^2}$$

33. Find the area of the zone of a sphere formed by revolving the graph of $y = \sqrt{9 - x^2}$, $0 \le x \le 2$, about the y-axis.

Solution

$$y = \sqrt{9 - x^2} \qquad y' = \frac{-x}{\sqrt{9 - x^2}}$$

$$S = 2\pi \int_0^2 x\sqrt{1 + \left(\frac{-x}{\sqrt{9 - x^2}}\right)^2}\, dx = 2\pi \int_0^2 x\sqrt{\frac{9}{9 - x^2}}\, dx$$

$$= 2\pi \int_0^2 3x(9 - x^2)^{-1/2}\, dx = -3\pi \int_0^2 (9 - x^2)^{-1/2}(-2x)\, dx$$

$$= -3\pi\left[\frac{(9 - x^2)^{1/2}}{1/2}\right]_0^2 = -6\pi(\sqrt{5} - 3) = 6\pi(3 - \sqrt{5}) \approx 14.40$$

Review Exercises for Chapter 7

5. Sketch and find the area of the region bounded by the graphs of $x = y^2 - 2y$ and $x = 0$.

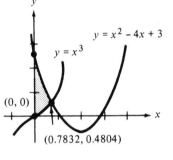

Solution

The points of intersection are given by

$$y^2 - 2y = 0$$
$$y(y - 2) = 0$$
$$y = 0, 2$$

Since $y^2 - 2y \leq 0$ for $0 \leq y \leq 2$, we have

$$A = \int_0^2 [0 - (y^2 - 2y)]\, dy = \int_0^2 (-y^2 + 2y)\, dy$$

$$= \left[\frac{-y^3}{3} + y^2\right]_0^2 = -\frac{8}{3} + 4 = \frac{4}{3}$$

10. Sketch and find the area of the region bounded by the graphs of $y = x^2 - 4x + 3$, $y = x^3$, and $x = 0$.

Solution

The points of intersection are given by

$$x^3 = x^2 - 4x + 3$$
$$x^3 - x^2 + 4x - 3 = 0$$

Since $x^3 - x^2 + 4x - 3$ does not factor easily, we apply Newton's Method to the function

$$f(x) = x^3 - x^2 + 4x - 3$$

By letting $x_1 = 1$, we have

$$x_2 = x_1 - \frac{f(x_1)}{f'(x_1)} = 0.8000$$

$$x_3 = x_2 - \frac{f(x_2)}{f'(x_2)} = 0.7833$$

$$x_4 = x_3 - \frac{f(x_3)}{f'(x_3)} = 0.7832$$

Since $x^3 \leq x^2 - 4x + 3$ for $0 \leq x \leq 0.7832$, we have

$$A \approx \int_0^{0.7832} (x^2 - 4x + 3 - x^3)\, dx = \left[\frac{x^3}{3} - 2x^2 + 3x - \frac{x^4}{4}\right]_0^{0.7832}$$

$$\approx 0.1601 - 1.2268 + 2.3496 - 0.0941 \approx 1.189$$

22. Find the volume of the solid generated by revolving the ellipse $(x^2/a^2) + (y^2/b^2) = 1$ about:
    (a) the y-axis (oblate spheroid)
    (b) the x-axis (prolate spheroid)

Solution

(a) Shell Method

$$V = 2\pi \int_0^a (x)\left(\frac{2b}{a}\sqrt{a^2 - x^2}\right) dx = \frac{-2\pi b}{a}\int_0^a (a^2 - x^2)^{1/2}(-2x)\, dx$$

$$= \frac{-4\pi b}{3a}[(a^2 - x^2)^{3/2}]_0^a = \frac{4}{3}\pi a^2 b$$

(b) Disc Method

$$V = \pi \int_{-a}^a \left(\frac{b}{a}\sqrt{a^2 - x^2}\right)^2 dx = \frac{2\pi b^2}{a^2}\int_0^a (a^2 - x^2)\, dx$$

$$= \frac{2\pi b^2}{a^2}\left[a^2 x - \frac{1}{3}x^3\right]_0^a = \frac{4}{3}\pi a b^2$$

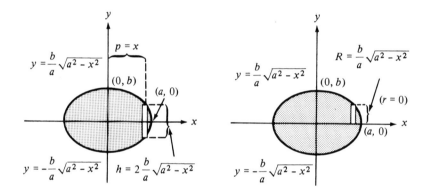

29. A water well has an 8-in casing (diameter) and is 175 ft deep. If the water is 25 ft from the top of the well, determine the amount of work done in pumping it dry, assuming that no water enters the well while it is being pumped.

Solution

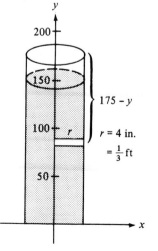

A disk of water at height y has to be moved $(175 - y)$ feet up and has a volume of $\pi(\frac{1}{3})^2 \Delta y$. Thus the work done in moving the water up over the top of the well is

$$W = \int_0^{150} \underbrace{(175 - y)}_{\text{(distance)}}\underbrace{\left[62.4\pi\left(\frac{1}{3}\right)^2 dy\right]}_{\text{(force: weight of water)}}$$

$$= \frac{62.4\pi}{9}\int_0^{150} (175 - y)\, dy = \frac{62.4\pi}{9}\left[175y - \frac{1}{2}y^2\right]_0^{150}$$

$$= 104{,}000\pi \text{ ft} \cdot \text{lb}$$

34. Show that the force against any vertical region in a liquid is the product of the density of the liquid, the area of the region, and the depth of the centroid of the region.

Solution

The force is given by

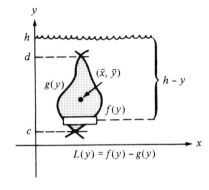

$$F = \int_c^d \rho(h - y)[f(y) - g(y)]\, dy$$

$$= \rho\left\{ h \int_c^d [f(y) - g(y)]\, dy - \int_c^d y[f(y) - g(y)]\, dy \right\}$$

$$= \rho\left\{ \int_c^d [f(y) - g(y)]\, dy \right\}\left\{ h - \frac{\int_c^d y[f(y) - g(y)]\, dy}{\int_c^d [f(y) - g(y)]\, dy} \right\}$$

$$= \rho(\text{area})(h - \bar{y}) = \rho(\text{area})(\text{depth of centroid})$$

37. Find the centroid of the region bounded by the graphs of $\sqrt{x} + \sqrt{y} = \sqrt{a}$, $x = 0$, and $y = 0$.

Solution

Solving the equation $\sqrt{x} + \sqrt{y} = \sqrt{a}$ for y we have $y = (\sqrt{a} - \sqrt{x})^2$. Therefore,

$$A = \int_0^a (\sqrt{a} - \sqrt{x})^2\, dx = \int_0^a (a - 2\sqrt{a}x^{1/2} + x)\, dx$$

$$= \left[ax - \frac{4}{3}\sqrt{a}x^{3/2} + \frac{1}{2}x^2 \right]_0^a = \frac{a^2}{6}$$

$$\bar{x} = \frac{1}{(a^2/6)} \int_0^a x(\sqrt{a} - \sqrt{x})^2\, dx = \frac{6}{a^2} \int_0^a (ax - 2\sqrt{a}x^{3/2} + x^2)\, dx$$

$$= \frac{6}{a^2}\left[\frac{ax^2}{2} - \frac{4}{5}\sqrt{a}x^{5/2} + \frac{1}{3}x^3 \right]_0^a = \frac{1}{5}a$$

By symmetry, we have $\bar{y} = \dfrac{a}{5}$. Therefore, $(\bar{x}, \bar{y}) = \left(\dfrac{a}{5}, \dfrac{a}{5} \right)$.

49. Find the arc length of the graph of $f(x) = \frac{4}{5}x^{5/4}$ from $x = 0$ to $x = 4$.

Solution

Since $f'(x) = x^{1/4}$, we have

$$s = \int_0^4 \sqrt{1 + [f'(x)]^2}\, dx = \int_0^4 \sqrt{1 + \sqrt{x}}\, dx$$

Let $u = \sqrt{1 + \sqrt{x}}$. Then $u^2 = 1 + \sqrt{x}$, $x = (u^2 - 1)^2$, and $dx = 2(u^2 - 1)(2u) \, du$. If $x = 0$, then $u = 1$, and if $x = 4$, then $u = \sqrt{3}$. Thus,

$$s = \int_0^4 \sqrt{1 + \sqrt{x}} \, dx = \int_1^{\sqrt{3}} u(2)(u^2 - 1)(2u) \, du$$

$$= 4 \int_1^{\sqrt{3}} (u^4 - u^2) \, du = 4 \left[\frac{u^5}{5} - \frac{u^3}{3} \right]_1^{\sqrt{3}} = \frac{8}{15}(6\sqrt{3} + 1)$$

8 Techniques of integration

Review of basic integration formulas

3. Evaluate $\int (-2x + 5)^{3/2}\, dx$.

    Solution

    If we let $u = -2x + 5$, then $du = -2\, dx$.

    $$\int (-2x + 5)^{3/2}\, dx = -\frac{1}{2} \int u^{3/2}\, du$$

    $$= -\frac{1}{2}\left[\frac{(-2x + 5)^{5/2}}{5/2}\right] + C$$

    $$= -\frac{1}{5}(-2x + 5)^{5/2} + C$$

7. Evaluate $\displaystyle\int \frac{t^2 - 3}{-t^3 + 9t + 1}\, dt$.

    Solution

    If we let $u = -t^3 + 9t + 1$, then $du = (-3t^2 + 9)\, dt = -3(t^2 - 3)\, dt$. Thus

    $$\int \frac{t^2 - 3}{-t^3 + 9t + 1}\, dt = -\frac{1}{3} \int \frac{-3(t^2 - 3)}{-t^3 + 9t + 1}\, dt$$

    $$= -\frac{1}{3} \int \frac{1}{u}\, du$$

    $$= -\frac{1}{3} \ln|-t^3 + 9t + 1| + C$$

13.   Evaluate $\int t \sin t^2 \, dt$.

Solution

If we let $u = t^2$, then $du = 2t \, dt$. Thus

$$\int t \sin t^2 \, dt = \frac{1}{2} \int \sin (t^2)(2t) \, dt$$

$$= \frac{1}{2} \int \sin u \, du$$

$$= \frac{1}{2}(-\cos t^2) + C$$

$$= -\frac{1}{2} \cos (t^2) + C$$

23.   Evaluate $\int \dfrac{1}{1 - \cos x} \, dx$.

Solution

$$\int \frac{1}{1 - \cos x} \, dx = \int \left(\frac{1}{1 - \cos x}\right)\left(\frac{1 + \cos x}{1 + \cos x}\right) dx$$

$$= \int \frac{1 + \cos x}{1 - \cos^2 x} \, dx$$

$$= \int \frac{1 + \cos x}{\sin^2 x} \, dx$$

$$= \int \frac{1}{\sin^2 x} \, dx + \int \frac{\cos x}{\sin^2 x} \, dx \qquad \begin{array}{l}\text{Let } u = \sin x \text{ and} \\ du = \cos x \, dx \\ \text{in the second} \\ \text{integral.}\end{array}$$

$$= \int \csc^2 x \, dx + \int (\sin x)^{-2}(\cos x) \, dx$$

$$= -\cot x + \frac{(\sin x)^{-1}}{-1} + C$$

$$= -\cot x - \csc x + C$$

25.   Evaluate $\int \dfrac{2t - 1}{t^2 + 4} \, dt$.

Solution

$$\int \frac{2t - 1}{t^2 + 4} \, dt = \int \frac{2t}{t^2 + 4} \, dt - \int \frac{1}{t^2 + 4} \, dt$$

Let $u = t^2 + 4$
and $du = 2t \, dt$
in the first
integral.

$$= \ln (t^2 + 4) - \frac{1}{2} \arctan \left(\frac{t}{2} \right) + C$$

31.   Evaluate $\displaystyle \int \frac{-1}{\sqrt{1 - (2t - 1)^2}} \, dt.$

Solution

$$\int \frac{-1}{\sqrt{1 - (2t - 1)^2}} \, dt = \frac{-1}{2} \int \frac{2}{\sqrt{1 - (2t - 1)^2}} \, dt$$

Let $u = 2t - 1$,
$du = 2 \, dt$

$$= \frac{-1}{2} \int \frac{1}{\sqrt{a^2 - u^2}} \, du$$

$$= -\frac{1}{2} \arcsin (2t - 1) + C$$

36.   Evaluate $\displaystyle \int \frac{\sin x}{\sqrt{\cos x}} \, dx.$

Solution

If we let $u = \cos x$, then $du = -\sin x \, dx$. Therefore,

$$\int \frac{\sin x}{\sqrt{\cos x}} \, dx = -\int u^{-1/2} \, du = -2\sqrt{\cos x} + C$$

47.   Evaluate $\displaystyle \int \frac{3}{\sqrt{6x - x^2}} \, dx.$

Solution

By completing the square we have

$$\int \frac{3}{\sqrt{6x - x^2}} \, dx = 3 \int \frac{1}{\sqrt{9 - (9 - 6x + x^2)}} \, dx$$

$$= 3 \int \frac{1}{\sqrt{9 - (x - 3)^2}} \, dx$$

$$= 3 \arcsin \left(\frac{x - 3}{3} \right) + C$$

53.   Evaluate $\displaystyle\int_1^e \frac{1 - \ln x}{x}\, dx$.

Solution

$$\int_1^e \frac{1 - \ln x}{x}\, dx = -\int_1^e (1 - \ln x)^1\left(\frac{-1}{x}\right) dx \qquad \text{Let } u = 1 - \ln x,\ du = -\frac{1}{x}\, dx$$

$$= -\left[\frac{(1 - \ln x)^2}{2}\right]_1^e$$

$$= -\frac{1}{2}[(1 - \ln e)^2 - (1 - \ln 1)^2]$$

$$= -\frac{1}{2}[(0)^2 - 1] = \frac{1}{2}$$

59.   Evaluate $\displaystyle\int_1^2 \frac{1}{2x\sqrt{4x^2 - 1}}\, dx$.

Solution

$$\int_1^2 \frac{1}{2x\sqrt{4x^2 - 1}}\, dx = \frac{1}{2}\int_1^2 \frac{2}{2x\sqrt{(2x)^2 - 1}}\, dx \qquad \text{Let } u = 2x,\ du = 2\, dx$$

$$= \frac{1}{2}[\operatorname{arcsec} 2x]_1^2$$

$$= \frac{1}{2}[\operatorname{arcsec} 4 - \operatorname{arcsec} 2]$$

$$= \frac{1}{2}\left[\operatorname{arcsec} 4 - \frac{\pi}{3}\right]$$

65.   The region bounded by $y = e^{-x^2}$, $y = 0$, and $x = b$ is revolved around the y-axis. Find b so that the volume of the generated solid is $\frac{4}{3}$ cubic units.

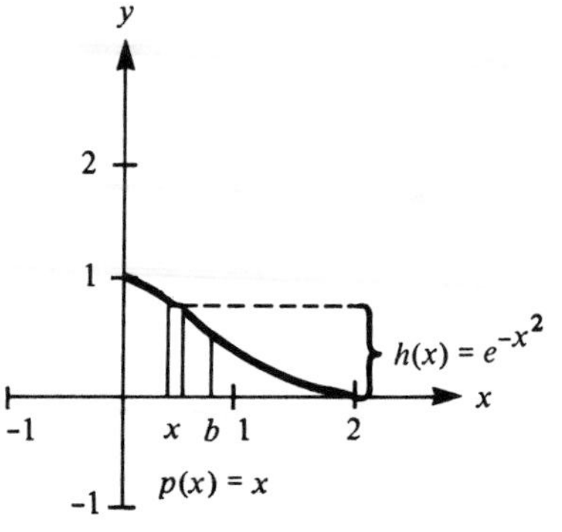

Solution

By using the Shell Method we have

$$V = 2\pi\int_0^b xe^{-x^2}\, dx$$

$$= -\pi\int_0^b e^{-x^2}(-2x\, dx) \qquad \text{Let } u = -x^2,\ du = -2x\, dx$$

$$= [-\pi e^{-x^2}]_0^b = \pi(1 - e^{-b^2}) = \frac{4}{3}$$

(figure on left showing curve $h(x) = e^{-x^2}$, line $p(x) = x$, with points x, b, 1, 2 on axes)

Solving for b we have

$$e^{-b^2} = \frac{3\pi - 4}{3\pi}$$

$$-b^2 = \ln\left(\frac{3\pi - 4}{3\pi}\right),$$

$$b = \sqrt{\ln\left(\frac{3\pi - 4}{3\pi}\right)^{-1}} = \sqrt{\ln\left(\frac{3\pi}{3\pi - 4}\right)} \approx 0.743$$

8.2
Integration by parts

5. Evaluate $\int xe^{-2x}\, dx$.

    Solution

    Let $dv = e^{-2x}\, dx \rightarrow\ v = -\frac{1}{2}e^{-2x}$

    $u = x \qquad \rightarrow du = dx$

    Therefore, we have

    $$\int xe^{-2x}\, dx = x\left(-\frac{1}{2}e^{-2x}\right) - \int -\frac{1}{2}e^{-2x}\, dx$$

    $$= -\frac{x}{2}e^{-2x} - \frac{1}{4}e^{-2x} + C$$

    $$= -\frac{e^{-2x}}{4}(2x + 1) + C$$

9. Evaluate $\int x^3 \ln x\, dx$.

    Solution

    Let $\quad dv = x^3 \quad \rightarrow\ v = \frac{x^4}{4}$

    $u = \ln x \rightarrow du = \frac{1}{x}\, dx$

    Therefore, we have

    $$\int x^3 \ln x\, dx = (\ln x)\left(\frac{x^4}{4}\right) - \int\left(\frac{x^4}{4}\right)\left(\frac{1}{x}\right) dx$$

    $$= \frac{1}{4}x^4 \ln x - \frac{1}{16}x^4 + C = \frac{x^4}{16}(4 \ln x - 1) + C$$

232

19.   Evaluate $\int x\sqrt{x-1}\ dx$.

Solution

Let $dv = \sqrt{x-1}\ dx \rightarrow\ v = \frac{2}{3}(x-1)^{3/2}$

$\qquad u = x \qquad\qquad \rightarrow du = dx$

Therefore

$$\int x\sqrt{x-1}\ dx = \frac{2}{3}x(x-1)^{3/2} - \int \frac{2}{3}(x-1)^{3/2}\ dx$$

$$= \frac{2}{3}x(x-1)^{3/2} - \frac{4}{15}(x-1)^{5/2} + C$$

$$= \frac{2}{15}(x-1)^{3/2}[5x - 2(x-1)] + C$$

$$= \frac{2}{15}(x-1)^{3/2}(3x+2) + C$$

31.   Evaluate $\int \arctan x\ dx$.

Solution

Let $dv = dx \qquad \rightarrow\ v = x$

$\qquad u = \arctan x \rightarrow du = \dfrac{1}{1+x^2}\ dx$

Therefore, we have

$$\int \arctan x\ dx = x \arctan x - \int \frac{x}{1+x^2}\ dx$$

$$= x \arctan x - \frac{1}{2} \ln(1+x^2) + C$$

33.   Evaluate $\int e^{2x} \sin x\ dx$.

Solution

Let $dv = \sin x\ dx \rightarrow\ v = -\cos x$

$\qquad u = e^{2x} \qquad\quad \rightarrow du = 2e^{2x}$

$$\int e^{2x} \sin x\ dx = -e^{2x} \cos x + \int \cos x (2e^{2x})\ dx$$

Using integration by parts again, we let

$$dv = \cos x\ dx \rightarrow\ v = \sin x$$

$$u = 2e^{2x} \qquad\quad \rightarrow du = 4e^{2x}$$

$$\int e^{2x} \sin x \, dx = -e^{2x} \cos x + 2e^{2x} \sin x - \int 4e^{2x} \sin x \, dx$$

By adding the right integral to both sides, we have

$$5 \int e^{2x} \sin x \, dx = -e^{2x} \cos x + 2e^{2x} \sin x + C_1$$

Finally, dividing both sides by 5, we conclude that

$$\int e^{2x} \sin x \, dx = \frac{e^{2x}}{5}(-\cos x + 2 \sin x) + C$$

36.   Evaluate $\int_0^1 x \arcsin x^2 \, dx$.

Solution

Let $dv = x \, dx \quad \rightarrow \quad v = \frac{x^2}{2}$

$$u = \arcsin x^2 \rightarrow du = \frac{2x}{\sqrt{1 - x^4}} \, dx$$

$$\int x \arcsin x^2 \, dx = \frac{x^2 \arcsin x^2}{2} - \int \frac{x^2}{2}\left(\frac{2x}{\sqrt{1 - x^4}}\right) dx$$

$$= \frac{x^2 \arcsin x^2}{2} - \int \frac{x^3}{\sqrt{1 - x^4}} \, dx$$

$$= \frac{x^2 \arcsin x^2}{2} + \frac{1}{4} \int (1 - x^4)^{-1/2}(-4x^3) \, dx$$

$$= \frac{x^2 \arcsin x^2}{2} + \frac{1}{4}\left[\frac{(1 - x^4)^{1/2}}{\frac{1}{2}}\right] + C$$

$$= \frac{1}{2}[x^2 \arcsin x^2 + \sqrt{1 - x^4}] + C$$

Finally,

$$\int_0^1 x \arcsin x^2 \, dx = \frac{1}{2}[x^2 \arcsin x^2 + \sqrt{1 - x^4}]_0^1$$

$$= \frac{1}{2}\left(\frac{\pi}{2} + 0 - 0 - 1\right) = \frac{1}{2}\left(\frac{\pi}{2} - 1\right)$$

37.   Evaluate $\int_0^1 e^x \sin x \, dx$.

Solution

Let $dv = e^x \, dx \rightarrow \quad v = e^x$

$$u = \sin x \rightarrow du = \cos x \, dx$$

$$\int e^x \sin x \, dx = e^x \sin x - \int e^x \cos x \, dx$$

Again letting $dv = e^x \, dx \rightarrow \quad v = e^x$

$$u = \cos x \rightarrow du = -\sin x \, dx$$

$$\int e^x \sin x \, dx = e^x \sin x - \left[e^x \cos x - \int e^x (-\sin x) \, dx \right]$$

$$= e^x \sin x - e^x \cos x - \int e^x \sin x \, dx$$

Thus $2 \int e^x \sin x \, dx = e^x \sin x - e^x \cos x + C$

$$\int e^x \sin x \, dx = \frac{1}{2}(e^x \sin x - e^x \cos x) + C_1$$

Finally, we have

$$\int_0^1 e^x \sin x \, dx = \frac{1}{2}[e^x \sin x - e^x \cos x]_0^1$$

$$= \frac{1}{2}\{e \sin (1) - e \cos (1) - [1(0) - 1(1)]\}$$

$$= \frac{1}{2}[e \sin (1) - e \cos (1) + 1] \approx 0.909$$

49. Use integration by parts to verify the formula

$$\int e^{ax} \sin bx \, dx = \frac{e^{ax}(a \sin bx - b \cos bx)}{a^2 + b^2} + C$$

Solution

Let $dv = \sin bx \, dx \rightarrow \quad v = -\frac{1}{b} \cos bx$

$$u = e^{ax} \qquad \rightarrow du = ae^{ax} \, dx$$

$$\int e^{ax} \sin bx \, dx = \frac{-e^{ax}}{b} \cos bx + \frac{a}{b} \int e^{ax} \cos bx \, dx$$

Using integration by parts again, we let

$$dv = \cos bx \, dx \rightarrow \quad v = \frac{1}{b} \sin bx$$

$$u = e^{ax} \qquad \rightarrow du = ae^{ax} \, dx$$

$$\int e^{ax} \sin bx \, dx = \frac{-e^{ax}}{b} \cos bx + \frac{a}{b}\left[\frac{e^{ax}}{b} \sin bx - \frac{a}{b} \int e^{ax} \sin bx \, dx \right]$$

$$= \frac{-e^{ax}}{b} \cos bx + \frac{ae^{ax}}{b^2} \sin bx - \frac{a^2}{b^2} \int e^{ax} \sin bx \, dx$$

$$b^2 \int e^{ax} \sin bx \, dx = -be^{ax} \cos bx + ae^{ax} \sin bx - a^2 \int e^{ax} \sin bx \, dx$$

Now by adding the right-hand integral to both sides of the equation, we have

$$(a^2 + b^2) \int e^{ax} \sin bx \, dx = e^{ax}(-b \cos bx + a \sin bx)$$

$$\int e^{ax} \sin bx \, dx = \frac{e^{ax}(a \sin bx - b \cos bx)}{a^2 + b^2} + C$$

55. Given the region bounded by the graphs of $y = \ln x$, $y = 0$, and $x = e$, find

(a) the area of the region.

(b) the volume of the solid generated by revolving the region about the x-axis.

(c) the volume of the solid generated by revolving the region about the y-axis.

(d) the centroid of the region.

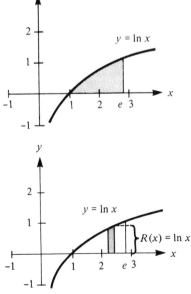

Solution

(a) $A = \displaystyle\int_1^e \ln x \, dx = [x(\ln x - 1)]_1^e = 1$ From Exercise 23

(b) Using the Disk Method we have

$$V = \pi \int_1^e (\ln x)^2 \, dx$$

Let $dv = dx \rightarrow v = x$

$$u = (\ln x)^2 \rightarrow du = \frac{2 \ln x}{x} \, dx$$

Therefore

$$\int (\ln x)^2 \, dx = x(\ln x)^2 - 2 \int \ln x \, dx$$

$$= x(\ln x)^2 - 2x(\ln x - 1)$$ From Exercise 23

Finally

$$V = \pi \int_1^e (\ln x)^2 \, dx = \pi[x(\ln x)^2 - 2x(\ln x - 1)]_1^e$$

$$= \pi(e - 2) \approx 2.26$$

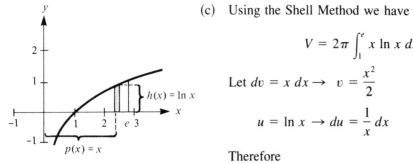

(c) Using the Shell Method we have

$$V = 2\pi \int_1^e x \ln x \, dx$$

Let $dv = x \, dx \rightarrow v = \dfrac{x^2}{2}$

$$u = \ln x \rightarrow du = \frac{1}{x} dx$$

Therefore

$$\int x \ln x \, dx = \frac{x^2}{2} \ln x - \frac{1}{2} \int x \, dx = \frac{x^2}{2} \ln x - \frac{x^2}{4}$$

Finally

$$V = 2\pi \int_1^e x \ln x \, dx$$

$$= 2\pi \left[\frac{x^2}{4}(2 \ln x - 1) \right]_1^e = \frac{\pi}{2}(e^2 + 1) \approx 13.18$$

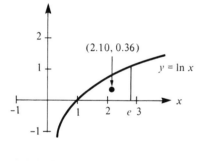

(d) $\bar{x} = \dfrac{1}{A} \displaystyle\int_1^e x \ln x \, dx = \left[\dfrac{x^2}{4}(2 \ln x - 1) \right]_1^e$ From part (c)

$$= \frac{1}{4}(e^2 + 1) \approx 2.10$$

$$\bar{y} = \frac{1}{2A} \int_1^e (\ln x)^2 \, dx$$

$$= \frac{1}{2}[x(\ln x)^2 - 2x(\ln x - 1)]_1^e$$ From part (b)

$$= \frac{e - 2}{2} \approx 0.36$$

8.3
Trigonometric integrals

5. Evaluate $\int \sin^5 x \cos^2 x \, dx$.

Solution

$$\int \sin^5 x \cos^2 x \, dx = \int \sin x \, (\sin^2 x)^2 \cos^2 x \, dx$$

$$= \int \sin x \, (1 - \cos^2 x)^2 \cos^2 x \, dx$$

$$= -\int (\cos^2 x - 2 \cos^4 x + \cos^6 x)(-\sin x) \, dx$$

$$= -\frac{1}{3} \cos^3 x + \frac{2}{5} \cos^5 x - \frac{1}{7} \cos^7 x + C$$

6.   Evaluate $\int \cos^3 \dfrac{x}{3} \, dx$.

Solution

$$\int \cos^3 \frac{x}{3} \, dx = \int \cos^2 \frac{x}{3} \cos \frac{x}{3} \, dx$$

$$= \int \left[1 - \sin^2 \frac{x}{3} \right] \cos \frac{x}{3} \, dx$$

$$= \int \cos \frac{x}{3} \, dx - \int \sin^2 \frac{x}{3} \cos \frac{x}{3} \, dx$$

$$= 3 \int \left(\cos \frac{x}{3} \right) \left(\frac{1}{3} \right) dx - 3 \int \sin^2 \frac{x}{3} \cos \frac{x}{3} \left(\frac{1}{3} \right) dx$$

$$= 3 \sin \frac{x}{3} - \sin^3 \frac{x}{3} + C$$

12.   Evaluate $\int \sin^2 \dfrac{x}{2} \cos^2 \dfrac{x}{2} \, dx$.

Solution

$$\int \sin^2 \frac{x}{2} \cos^2 \frac{x}{2} \, dx = \int \left(\frac{1 - \cos x}{2} \right) \left(\frac{1 + \cos x}{2} \right) dx$$

$$= \frac{1}{4} \int (1 - \cos^2 x) \, dx$$

$$= \frac{1}{4} \int \left(1 - \frac{1 + \cos 2x}{2} \right) dx$$

$$= \frac{1}{4} \int \frac{2 - 1 - \cos 2x}{2} \, dx$$

$$= \frac{1}{8} \int (1 - \cos 2x) \, dx = \frac{1}{8} x - \frac{1}{16} \sin 2x + C$$

31.   Evaluate $\int \sec^6 4x \tan 4x \, dx$.

Solution

$$\int \sec^6 4x \tan 4x \, dx = \int (\sec^2 4x)(\sec^2 4x)^2 \tan 4x \, dx$$

$$= \frac{1}{4} \int \tan 4x (\tan^2 4x + 1)^2 (4 \sec^2 4x)\, dx$$

$$= \frac{1}{4} \int (\tan^5 4x + 2 \tan^3 4x + \tan 4x)$$

$$\times (4 \sec^2 4x)\, dx$$

$$= \frac{1}{4} \left[\frac{\tan^6 4x}{6} + \frac{\tan^4 4x}{2} + \frac{\tan^2 4x}{2} \right] + C$$

$$= \frac{1}{24} (\tan^2 4x)(\tan^4 4x + 3 \tan^2 4x + 3) + C$$

or

$$\int \sec^6 4x \tan 4x\, dx = \frac{1}{4} \int \sec^5 4x (4 \sec 4x \tan 4x)\, dx$$

$$= \frac{1}{24} \sec^6 4x + C_1$$

(See Exercise 61 for a comparison of the two methods.)

35.   Evaluate $\int \tan^3 3x \sec 3x\, dx$.

Solution

$$\int \tan^3 3x \sec 3x\, dx = \int \tan^2 3x (\sec 3x \tan 3x)\, dx$$

$$= \int (\sec^2 3x - 1)(\sec 3x \tan 3x)\, dx$$

$$= \frac{1}{3} \int \sec^2 3x (3 \sec 3x \tan 3x)\, dx$$

$$- \frac{1}{3} \int \sec 3x \tan 3x\, (3)\, dx$$

$$= \frac{1}{3} \left(\frac{\sec^3 3x}{3} \right) - \frac{1}{3} \sec 3x + C$$

$$= \frac{1}{9} \sec^3 3x - \frac{1}{3} \sec 3x + C$$

43.   Evaluate $\int \sin 3x \cos 2x\, dx$.

Solution

Using the trigonometric identity $\sin u \cos v = \frac{1}{2}[\sin (u + v) + \sin (u - v)]$, we have

$$\int \sin 3x \cos 2x \, dx = \int \frac{1}{2}(\sin 5x + \sin x) \, dx$$

$$= \frac{1}{2}\left(\frac{1}{5}\right) \int \sin 5x \, (5) \, dx + \frac{1}{2} \int \sin x \, dx$$

$$= -\frac{1}{10} \cos 5x - \frac{1}{2} \cos x + C$$

53.   Evaluate $\int_0^{\pi/4} \tan^3 x \, dx$.

Solution

$$\int_0^{\pi/4} \tan^3 x \, dx = \int_0^{\pi/4} (\sec^2 x - 1)(\tan x) \, dx$$

$$= \int_0^{\pi/4} \tan x \sec^2 x \, dx - \int_0^{\pi/4} \tan x \, dx$$

$$= \left[\frac{1}{2} \tan^2 x + \ln|\cos x|\right]_0^{\pi/4} = \frac{1}{2}(1 - \ln 2)$$

61.   Evaluate $\int \sec^4 3x \tan^3 3x \, dx$ in two ways and show that the results differ only by a constant.

Solution

Method 1

Since there is an odd power of the tangent, we write

$$\int \sec^4 3x \tan^3 3x \, dx = \int \sec^3 3x \tan^2 3x (\sec 3x \tan 3x) \, dx$$

$$= \int \sec^3 3x (\sec^2 3x - 1) \sec 3x \tan 3x \, dx$$

$$= \frac{1}{3} \int (\sec^5 3x - \sec^3 3x)(3 \sec 3x \tan 3x \, dx)$$

$$= \frac{1}{3}\left(\frac{1}{6} \sec^6 3x - \frac{1}{4} \sec^4 3x\right) + C$$

Method 2

Since the power of the secant is even, we write

$$\int \sec^4 3x \tan^3 3x \, dx = \int \sec^2 3x \tan^3 3x (\sec^2 3x) \, dx$$

$$= \int (1 + \tan^2 3x)\tan^3 3x (\sec^2 3x) \, dx$$

$$= \frac{1}{3} \int (\tan^3 3x + \tan^5 3x)(3 \sec^2 3x) \, dx$$

$$= \frac{1}{3} \left(\frac{\tan^4 3x}{4} + \frac{\tan^6 3x}{6} \right) + C$$

Comparing the results of the two methods, we have

$$\frac{1}{3} \left[\frac{1}{4} \tan^4 3x + \frac{1}{6} \tan^6 3x \right] + C$$

$$= \frac{1}{3} \left[\frac{1}{4} (\sec^2 3x - 1)^2 + \frac{1}{6} (\sec^2 3x - 1)^3 \right] + C$$

$$= \frac{1}{3} \left[\frac{1}{4} (\sec^4 3x - 2 \sec^2 3x + 1) \right.$$

$$\left. + \frac{1}{6} (\sec^6 3x - 3 \sec^4 3x + 3 \sec^2 3x - 1) \right] + C$$

$$= \frac{1}{3} \left[\frac{1}{6} \sec^6 3x - \frac{1}{4} \sec^4 3x + \frac{1}{4} - \frac{1}{6} \right] + C$$

Therefore the results differ only by the constant $\frac{1}{3} (\frac{1}{4} - \frac{1}{6})$.

67. Find (a) the volume of the solid generated by revolving the region bounded by the graphs of $y = \sin x$, $y = 0$, $x = 0$, and $x = \pi$ about the x-axis and (b) find the centroid of the region.

Solution

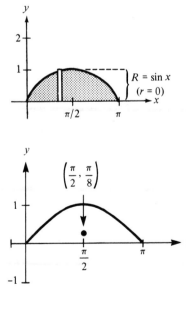

(a) $$V = \pi \int_0^\pi R^2 \, dx = \pi \int_0^\pi \sin^2 x \, dx = \pi \int_0^\pi \frac{1 - \cos 2x}{2} \, dx$$

$$= \frac{\pi}{2} \int_0^\pi (1 - \cos 2x) \, dx = \frac{\pi}{2} \left[x - \frac{1}{2} \sin 2x \right]_0^\pi$$

$$= \frac{\pi}{2} \left[\pi - \frac{1}{2} (0) - 0 \right] = \frac{\pi^2}{2}$$

(b) By symmetry $\bar{x} = \frac{\pi}{2}$.

The area of the region is given by

$$A = \int_0^\pi \sin x \, dx = [-\cos x]_0^\pi = 2$$

Therefore,

$$\bar{y} = \frac{1}{2A} \int_0^\pi \sin^2 x \, dx$$

$$= \frac{1}{8} \int_0^\pi (1 - \cos 2x) \, dx = \left[\frac{1}{8} \left(x - \frac{1}{2} \sin 2x \right) \right]_0^\pi = \frac{\pi}{8}$$

Therefore $(\bar{x}, \bar{y}) = (\pi/2, \pi/8)$.

77. Use integration by parts to verify the formula

$$\int \cos^m x \sin^n x \, dx$$

$$= -\frac{\cos^{m+1} x \sin^{n-1} x}{m + n} + \frac{n - 1}{m + n} \int \cos^m x \sin^{n-2} x \, dx$$

Solution

Let $dv = \cos^m x \sin x \, dx$ and $u = \sin^{n-1} x$. Then $v = (-\cos^{m+1} x)/(m + 1)$ and $du = (n - 1) \sin^{n-2} x \, (\cos x) \, dx$. Therefore

$$\int \cos^m x \sin^n x \, dx$$

$$= -\frac{\sin^{n-1} x \cos^{m+1} x}{m + 1} + \frac{n - 1}{m + 1} \int \sin^{n-2} x \cos^{m+2} x \, dx$$

$$= -\frac{\sin^{n-1} x \cos^{m+1} x}{m + 1} + \frac{n - 1}{m + 1} \int \sin^{n-2} x \cos^m x(1 - \sin^2 x) \, dx$$

$$= -\frac{\sin^{n-1} x \cos^{m+1} x}{m + 1} + \frac{n - 1}{m + 1} \int \sin^{n-2} x \cos^m x \, dx$$

$$- \frac{n - 1}{m + 1} \int \sin^n x \cos^m x \, dx$$

Now observe that the last integral is a multiple of the original. Adding yields

$$\frac{m + n}{m + 1} \int \cos^m x \sin^n x \, dx$$

$$= -\frac{\sin^{n-1} x \cos^{m+1} x}{m + 1} + \frac{n - 1}{m + 1} \int \cos^m x \sin^{n-2} x \, dx$$

$$\int \cos^m x \sin^n x \, dx$$

$$= -\frac{\cos^{m+1} x \sin^{n-1} x}{m + n} + \frac{n - 1}{m + n} \int \cos^m x \sin^{n-2} x \, dx$$

8.4
Trigonometric substitution

3. Evaluate $\int \dfrac{\sqrt{25 - x^2}}{x} \, dx$.

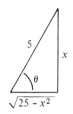

$\sqrt{25 - x^2}$

Solution

Let $x = 5 \sin \theta$. Then $\sqrt{25 - x^2} = 5 \cos \theta$ and $dx = 5 \cos \theta \, d\theta$. Thus

$$\int \frac{\sqrt{25 - x^2}}{x} \, dx = \int \frac{5 \cos \theta}{5 \sin \theta} 5 \cos \theta \, d\theta = 5 \int \frac{\cos^2 \theta}{\sin \theta} \, d\theta$$

$$= 5 \int \frac{1 - \sin^2 \theta}{\sin \theta} \, d\theta = 5 \int (\csc \theta - \sin \theta) \, d\theta$$

$$= 5(\ln |\csc \theta - \cot \theta| + \cos \theta) + C$$

$$= 5 \left(\ln \left| \frac{5}{x} - \frac{\sqrt{25 - x^2}}{x} \right| + \frac{\sqrt{25 - x^2}}{5} \right) + C$$

$$= 5 \ln \left| \frac{5 - \sqrt{25 - x^2}}{x} \right| + \sqrt{25 - x^2} + C$$

7. Evaluate $\int_0^2 \sqrt{16 - 4x^2} \, dx$.

Solution

Let $2x = 4 \sin \theta$. Then $\sqrt{16 - 4x^2} = 4 \cos \theta$ and $dx = 2 \cos \theta \, d\theta$. Furthermore, when $x = 0$, $\theta = 0$, and when $x = 2$, $\theta = \pi/2$. Thus

$$\int_0^2 \sqrt{16 - 4x^2} \, dx = \int_0^{\pi/2} 4 \cos \theta \, (2 \cos \theta) \, d\theta = 8 \int_0^{\pi/2} \cos^2 \theta \, d\theta$$

$$= 8 \int_0^2 \frac{1 + \cos 2\theta}{2} \, d\theta = 4 \left[\theta + \frac{1}{2} \sin 2\theta \right]_0^{\pi/2}$$

$$= 4 \left[\frac{\pi}{2} + \frac{1}{2}(0) - 0 \right] = 2\pi$$

14. Evaluate $\int \frac{\sqrt{4x^2 + 9}}{x^4} \, dx$.

Solution

Let $2x = 3 \tan \theta$. Then $\sqrt{4x^2 + 9} = 3 \sec \theta$ and $dx = \frac{3}{2} \sec^2 \theta \, d\theta$. Thus

$$\int \frac{\sqrt{4x^2 + 9}}{x^4} \, dx = \int \frac{(3 \sec \theta)(\frac{3}{2} \sec^2 \theta)}{(\frac{3}{2})^4 \tan^4 \theta} \, d\theta = \frac{8}{9} \int \frac{\sec^3 \theta}{\tan^4 \theta} \, d\theta$$

$$= \frac{8}{9} \int \left(\frac{1}{\cos^3 \theta} \right) \left(\frac{\cos^4 \theta}{\sin^4 \theta} \right) \, d\theta$$

$$= \frac{8}{9} \int (\sin \theta)^{-4}(\cos \theta)\, d\theta = \frac{8}{9}\left[\frac{(\sin \theta)^{-3}}{-3} \right] + C$$

$$= -\frac{8}{27} \csc^3 \theta + C$$

Now since $\tan \theta = 2x/3$, we have $\csc \theta = \sqrt{4x^2 + 9}/2x$.
Thus

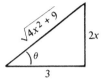

$$\int \frac{\sqrt{4x^2 + 9}}{x^4}\, dx = -\frac{8}{27}\left(\frac{\sqrt{x^2 + 9}}{2x} \right)^3 + C = \frac{-(4x^2 + 9)^{3/2}}{27x^3} + C$$

15.   Evaluate $\displaystyle\int \frac{1}{x\sqrt{4x^2 + 9}}\, dx.$

Solution

Let $2x = 3 \tan \theta$. Then $x = \frac{3}{2} \tan \theta$, $\sqrt{4x^2 + 9} = 3 \sec \theta$,
and $dx = \frac{3}{2} \sec^2 \theta\, d\theta$.

$$\int \frac{1}{x\sqrt{4x^2 + 9}}\, dx = \int \frac{\frac{3}{2} \sec^2 \theta\, d\theta}{\frac{3}{2} \tan \theta (3 \sec \theta)} = \frac{1}{3} \int \csc \theta\, d\theta$$

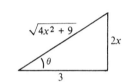

$$= \frac{1}{3} \ln |\csc \theta - \cot \theta| + C$$

$$= \frac{1}{3} \ln \left| \frac{\sqrt{4x^2 + 9} - 3}{2x} \right| + C$$

$$= -\frac{1}{3} \ln \left| \frac{2x}{\sqrt{4x^2 + 9} - 3} \right| + C$$

$$= -\frac{1}{3} \ln \left| \left(\frac{2x}{\sqrt{4x^2 + 9} - 3} \right) \right.$$

$$\left. \times \left(\frac{\sqrt{4x^2 + 9} + 3}{\sqrt{4x^2 + 9} + 3} \right) \right| + C$$

$$= -\frac{1}{3} \ln \left| \frac{3 + \sqrt{4x^2 + 9}}{2x} \right| + C$$

19.   Evaluate $\int x^3\sqrt{x^2 - 4}\, dx.$

Solution

Let $x = 2 \sec \theta$. Then $\sqrt{x^2 - 4} = 2 \tan \theta$ and
$dx = 2 \sec \theta \tan \theta\, d\theta$. Thus

$$\int x^3\sqrt{x^2 - 4}\, dx = \int (2 \sec \theta)^3 (2 \tan \theta)(2 \sec \theta \tan \theta)\, d\theta$$

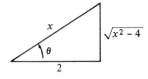

$$= 32 \int \sec^4 \theta \tan^2 \theta \, d\theta$$

$$= 32 \int \sec^2 \theta \tan^2 \theta \sec^2 \theta \, d\theta$$

$$= 32 \int (\tan^2 \theta + 1) \tan^2 \theta \sec^2 \theta \, d\theta$$

$$= 32 \int (\tan^4 \theta + \tan^2 \theta) \sec^2 \theta \, d\theta$$

$$= 32 \left[\frac{1}{5} \tan^5 \theta + \frac{1}{3} \tan^3 \theta \right] + C$$

$$= 32 \left[\frac{1}{5} \left(\frac{\sqrt{x^2 - 4}}{2} \right)^5 + \frac{1}{3} \left(\frac{\sqrt{x^2 - 4}}{2} \right)^3 \right] + C$$

$$= \frac{1}{15} (x^2 - 4)^{3/2} (3x^2 + 8) + C$$

26. Evaluate $\displaystyle \int \frac{x^2}{\sqrt{2x - x^2}} \, dx$.

Solution

We first complete the square within the radical.

$$\int \frac{x^2}{\sqrt{2x - x^2}} \, dx = \int \frac{x^2}{\sqrt{1 - (x^2 - 2x + 1)}} \, dx$$

$$= \int \frac{x^2}{\sqrt{1 - (x - 1)^2}} \, dx$$

Let $x - 1 = \sin \theta$. Then $\sqrt{1 - (x - 1)^2} = \cos \theta$, $dx = \cos \theta \, d\theta$, and $x^2 = (1 + \sin \theta)^2$. Thus

$$\int \frac{x^2}{\sqrt{2x - x^2}} \, dx = \int \frac{(1 + \sin \theta)^2 \cos \theta}{\cos \theta} \, d\theta$$

$$= \int (1 + 2 \sin \theta + \sin^2 \theta) \, d\theta$$

$$= \int \left(1 + 2 \sin \theta + \frac{1 - \cos 2\theta}{2} \right) d\theta$$

$$= \int \left(\frac{3}{2} + 2 \sin \theta - \frac{1}{2} \cos 2\theta \right) d\theta$$

$$= \frac{3}{2} \theta - 2 \cos \theta - \frac{1}{4} \sin 2\theta + C$$

$$= \frac{3}{2} \theta - 2 \cos \theta - \frac{1}{2} \sin \theta \cos \theta + C$$

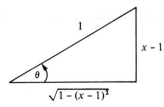

$$= \frac{3}{2}\,\theta - \frac{1}{2}\cos\theta(4 + \sin\theta) + C$$

$$= \frac{3}{2}\arcsin(x - 1) - \frac{1}{2}\sqrt{2x - x^2}(x + 3) + C$$

29. Evaluate $\displaystyle\int \frac{1}{4 + 4x^2 + x^4}\,dx$.

Solution

$$\int \frac{1}{4 + 4x^2 + x^4}\,dx = \int \frac{1}{(2 + x^2)^2}\,dx$$

Let $x = \sqrt{2}\tan\theta$. Then $2 + x^2 = 2\sec^2\theta$ and $dx = \sqrt{2}\sec^2\theta\,d\theta$. Thus

$$\int \frac{1}{(2 + x^2)^2}\,dx$$

$$= \int \frac{\sqrt{2}\sec^2\theta\,d\theta}{(2\sec^2\theta)^2} = \frac{\sqrt{2}}{4}\int \cos^2\theta\,d\theta$$

$$= \frac{\sqrt{2}}{4}\int \frac{1 + \cos 2\theta}{2}\,d\theta = \frac{\sqrt{2}}{8}\left(\theta + \frac{1}{2}\sin 2\theta\right) + C$$

$$= \frac{\sqrt{2}}{8}(\theta + \sin\theta\cos\theta) + C$$

$$= \frac{\sqrt{2}}{8}\left[\arctan\frac{x}{\sqrt{2}} + \left(\frac{x}{\sqrt{x^2 + 2}}\right)\left(\frac{\sqrt{2}}{\sqrt{x^2 + 2}}\right)\right] + C$$

$$= \frac{1}{4}\left[\frac{x}{x^2 + 2} + \frac{1}{\sqrt{2}}\arctan\frac{x}{\sqrt{2}}\right] + C$$

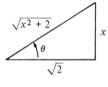

35. Evaluate $\int \operatorname{arcsec} 2x\,dx$.

Solution

Let $dv = dx \qquad \rightarrow \quad v = x$

$$u = \operatorname{arcsec} 2x \rightarrow du = \frac{1}{x\sqrt{4x^2 - 1}}\,dx$$

Therefore

$$\int \operatorname{arcsec} 2x\,dx = uv - \int v\,du$$

$$= x\operatorname{arcsec} 2x - \int \not{x}\left(\frac{1}{\not{x}\sqrt{4x^2 - 1}}\right)dx$$

$$= x \text{ arcsec } 2x - \frac{1}{2} \int \frac{2}{\sqrt{(2x)^2 - 1^2}} \, dx$$

Let $2x = \sec \theta$

$$= x \text{ arcsec } 2x - \frac{1}{2} \ln |2x + \sqrt{4x^2 - 1}| + C$$

$$= x \text{ arcsec } 2x + \frac{1}{2} \ln \left| \frac{1}{2x + \sqrt{4x^2 - 1}} \right| + C$$

$$= x \text{ arcsec } 2x + \frac{1}{2} \ln |2x - \sqrt{4x^2 - 1}| + C$$

43. The region bounded by the circle $(x - 3)^2 + y^2 = 1$ is revolved about the y-axis. The resulting doughnut-shaped solid is called a torus. Find the volume of this solid.

Solution

Shell Method

$$V = 2\pi \int_2^4 x \left[2\sqrt{1 - (x - 3)^2} \right] dx = 4\pi \int_2^4 x\sqrt{1 - (x - 3)^2} \, dx$$

Let $x - 3 = \sin \theta$. Then $\sqrt{1 - (x - 3)^2} = \cos \theta$ and $dx = \cos \theta \, d\theta$. Also when $x = 2$, $\sin \theta = -1$ and $\theta = -\pi/2$. When $x = 4$, $\sin \theta = 1$ and $\theta = \pi/2$. Therefore

$$V = 4\pi \int_{-\pi/2}^{\pi/2} (3 + \sin \theta)(\cos \theta)(\cos \theta) \, d\theta$$

$$= 4\pi \left[\int_{-\pi/2}^{\pi/2} 3 \cos^2 \theta \, d\theta + \int_{-\pi/2}^{\pi/2} \cos^2 \theta \sin \theta \, d\theta \right]$$

$$= 4\pi \left[\int_{-\pi/2}^{\pi/2} \frac{3}{2}(1 + \cos 2\theta) \, d\theta + \int_{-\pi/2}^{\pi/2} \cos^2 \theta \sin \theta \, d\theta \right]$$

$$= 4\pi \left[\frac{3}{2}\left(\theta + \frac{1}{2} \sin 2\theta\right) - \frac{1}{3} \cos^3 \theta \right]_{-\pi/2}^{\pi/2} = 6\pi^2$$

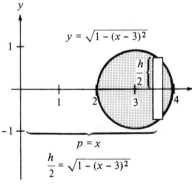

$y = \sqrt{1 - (x - 3)^2}$

$p = x$

$\dfrac{h}{2} = \sqrt{1 - (x - 3)^2}$

48. A projectile follows the path $y = -(x^2/72) + x$, where x and y are measured in meters. Find the distance traveled from $x = 0$ to $x = 72$.

Solution

$$y = -\frac{x^2}{72} + x \qquad y' = -\frac{x}{36} + 1$$

Therefore

$$s = \int_0^{72} \sqrt{1 + (y')^2} \, dx = \int_0^{72} \sqrt{1 + \left(1 - \frac{x}{36}\right)^2} \, dx$$

Let $1 - (x/36) = \tan \theta$. Then

$$\sqrt{1 + \left(1 - \frac{x}{36}\right)^2} = \sec \theta \quad \text{and} \quad dx = -36 \sec^2 \theta \, d\theta$$

Also, when $x = 0$, $\tan \theta = 1$ and $\theta = \pi/4$. When $x = 72$, $\tan \theta = -1$ and $\theta = -\pi/4$. Finally,

$$s = \int_{\pi/4}^{-\pi/4} \sec \theta \, (-36 \sec^2 \theta) \, d\theta = -36 \int_{\pi/4}^{-\pi/4} \sec^3 \theta \, d\theta$$

$$= \frac{-36}{2} [\sec \theta \tan \theta + \ln |\sec \theta + \tan \theta|]_{\pi/4}^{-\pi/4} \qquad \text{From Section 8.2}$$

$$= -18[-2\sqrt{2} + \ln (3 - 2\sqrt{2})] = 82.641 \text{ m}$$

49. Find the surface area of the solid generated by revolving the region bounded by $y = x^2$, $y = 0$, $x = 0$, and $x = \sqrt{2}$ about the x-axis.

Solution

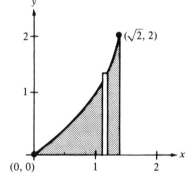

$$S = 2\pi \int_a^b y \sqrt{1 + (y')^2} \, dx = 2\pi \int_0^{\sqrt{2}} x^2 \sqrt{1 + 4x^2} \, dx$$

Using trigonometric substitution, we let $2x = \tan \theta$. Then $\sqrt{1 + 4x^2} \, dx = \sec \theta$, $x^2 = \frac{1}{4} \tan^2 \theta$, and $dx = \frac{1}{2} \sec^2 \theta \, d\theta$. Therefore

$$\int x^2 \sqrt{1 + 4x^2} \, dx = \int \frac{\tan^2 \theta}{4} (\sec \theta) \left(\frac{1}{2} \sec^2 \theta\right) d\theta$$

$$= \frac{1}{8} \int \sec^3 \theta \tan^2 \theta \, d\theta$$

$$= \frac{1}{8} \int \sec^3 \theta \, (\sec^2 \theta - 1) \, d\theta$$

$$= \frac{1}{8} \left[\int \sec^5 \theta \, d\theta - \int \sec^3 \theta \, d\theta \right]$$

Now using integration by parts on $\int \sec^5 \theta \, d\theta$, we let

$$dv = \sec^2 \theta \, d\theta \rightarrow \quad v = \tan \theta$$
$$u = \sec^3 \theta \quad \rightarrow du = 3 \sec^2 \theta \, (\sec \theta \tan \theta) \, d\theta$$
$$= 3 \sec^3 \theta \tan \theta \, d\theta$$

Thus,

$$\int \sec^5 \theta \, d\theta = \sec^3 \theta \tan \theta - 3 \int \sec^3 \theta \tan^2 \theta \, d\theta$$

$$= \sec^3 \theta \tan \theta - 3 \int \sec^3 \theta \, (\sec^2 \theta - 1) \, d\theta$$

$$= \sec^3 \theta \tan \theta - 3 \int \sec^5 \theta \, d\theta + 3 \int \sec^3 \theta \, d\theta$$

$$4 \int \sec^5 \theta \, d\theta = \sec^3 \theta \tan \theta + 3 \int \sec^3 \theta \, d\theta$$

$$\int \sec^5 \theta \, d\theta = \frac{1}{4} \left(\sec^3 \theta \tan \theta + 3 \int \sec^3 \theta \, d\theta \right)$$

Thus

$$\int x^2 \sqrt{1 + 4x^2} \, dx$$

$$= \frac{1}{8} \left[\frac{1}{4} \sec^3 \theta \tan \theta + \frac{3}{4} \int \sec^3 \theta \, d\theta - \int \sec^3 \theta \, d\theta \right]$$

$$= \frac{1}{32} \left(\sec^3 \theta \tan \theta - \int \sec^3 \theta \, d\theta \right)$$

Using the formula from Section 8.2 for $\int \sec^3 \theta \, d\theta$, we have

$$\int x^2 \sqrt{1 + 4x^2} \, dx$$

$$= \frac{1}{32} \left[\sec^3 \theta \tan \theta - \frac{1}{2} (\sec \theta \tan \theta + \ln |\sec \theta + \tan \theta|) \right] + C$$

Finally, when $x = 0$, $\theta = 0$, and when $x = \sqrt{2}$,
$\theta = \arctan 2\sqrt{2}$. Therefore,

$$S = 2\pi \int_0^{\sqrt{2}} x^2 \sqrt{1 + 4x^2} \, dx$$

$$= \frac{\pi}{16} \left[\sec^3 \theta \tan \theta - \frac{1}{2} (\sec \theta \tan \theta + \ln |\sec \theta + \tan \theta|) \right]_0^{\arctan 2\sqrt{2}}$$

$$= \frac{\pi}{16} \left[51\sqrt{2} - \frac{1}{2} \ln (2\sqrt{2} + 3) \right] \approx 13.989$$

8.5
Partial fractions

3. Evaluate $\int \dfrac{3}{x^2 + x - 2} \, dx$.

Solution

$$\frac{3}{x^2 + x - 2} = \frac{3}{(x - 1)(x + 2)} = \frac{A}{x - 1} + \frac{B}{x + 2}$$

Multiplying by $(x - 1)(x + 2)$, we have

$$3 = A(x + 2) + B(x - 1)$$

If $x = 1$, then

$$3 = A(3) \qquad \text{or} \qquad A = 1$$

If $x = -2$, then

$$3 = B(-3) \qquad \text{or} \qquad B = -1$$

Thus

$$\int \frac{3}{x^2 + x - 2}\, dx = \int \left(\frac{1}{x - 1} - \frac{1}{x + 2} \right) dx$$

$$= \ln |x - 1| - \ln |x + 2| + C$$

$$= \ln \left| \frac{x - 1}{x + 2} \right| + C$$

7. Evaluate $\int \dfrac{x^2 + 12x + 12}{x^3 - 4x}\, dx$.

Solution

$$\frac{x^2 + 12x + 12}{x^3 - 4x} = \frac{x^2 + 12x + 12}{x(x - 2)(x + 2)} = \frac{A}{x} + \frac{B}{x - 2} + \frac{C}{x + 2}$$

Multiplying by $(x)(x - 2)(x + 2)$, we have

$$x^2 + 12x + 12 = A(x - 2)(x + 2) + B(x)(x + 2) + C(x)(x - 2)$$

If $x = 0$, then

$$12 = A(-2)(2) \qquad \text{or} \qquad A = -3$$

If $x = 2$, then

$$4 + 24 + 12 = B(2)(4) \qquad \text{or} \qquad B = 5$$

If $x = -2$, then

$$4 - 24 + 12 = C(-2)(-4) \qquad \text{or} \qquad C = -1$$

Thus

$$\int \frac{x^2 + 12x + 12}{x^3 - 4x}\, dx = \int \left(\frac{-3}{x} + \frac{5}{x - 2} + \frac{-1}{x + 2} \right) dx$$

$$= -3 \ln |x| + 5 \ln |x - 2|$$

$$- \ln |x + 2| + C$$

$$= \ln \left| \frac{(x - 2)^5}{x^3(x + 2)} \right| + C$$

13. Evaluate $\int \dfrac{x^4}{(x-1)^3}\, dx.$

Solution

Division yields

$$\frac{x^4}{(x-1)^3} = x + 3 + \frac{6x^2 - 8x + 3}{(x-1)^3}$$

Furthermore, by partial fractions we have

$$\frac{6x^2 - 8x + 3}{(x-1)^3} = \frac{A}{x-1} + \frac{B}{(x-1)^2} + \frac{C}{(x-1)^3}$$

Multiplying by $(x-1)^3$, we have

$$6x^2 - 8x + 3 = A(x-1)^2 + B(x-1) + C$$
$$= A(x^2 - 2x + 1) + B(x-1) + C$$
$$= Ax^2 + (B - 2A)x + (A - B + C)$$

Now by equating coefficients of like terms, we have the three equations

$$6 = A, \qquad -8 = B - 2A, \qquad 3 = A - B + C$$

Solving these equations, we have

$$A = 6, \qquad B = 4, \qquad C = 1$$

Thus

$$\int \frac{x^4}{(x-1)^3}\, dx = \int \left[x + 3 + \frac{6}{x-1} + \frac{4}{(x-1)^2} + \frac{1}{(x-1)^3} \right] dx$$

$$= \frac{x^2}{2} + 3x + 6 \ln |x - 1|$$

$$- \frac{4}{x-1} - \frac{1}{2(x-1)^2} + C$$

17. Evaluate $\int \dfrac{x^2 - 1}{x^3 + x}\, dx.$

Solution

$$\frac{x^2 - 1}{x^3 + x} = \frac{x^2 - 1}{x(x^2 + 1)} = \frac{A}{x} + \frac{Bx + C}{x^2 + 1}$$

Multiplying by $x(x^2 + x)$, we have

$$x^2 - 1 = A(x^2 + 1) + (Bx + C)x = Ax^2 + A + Bx^2 + Cx$$
$$= (A + B)x^2 + Cx + A$$

Now equating coefficients of like terms, we have the three equations

$$A + B = 1, \qquad C = 0, \qquad A = -1$$

Therefore,

$$A = -1, \qquad B = 2, \qquad C = 0$$

Thus

$$\int \frac{x^2 - 1}{x^3 + x} \, dx = \int \left(\frac{-1}{x} + \frac{2x}{x^2 + 1} \right) dx$$

$$= -\ln |x| + \ln (x^2 + 1) + C = \ln \left| \frac{x^2 + 1}{x} \right| + C$$

25. Evaluate $\displaystyle\int \frac{x^2 + 5}{x^3 - x^2 + x + 3} \, dx$.

Solution

Since $x^3 - x^2 + x + 3 = (x + 1)(x^2 - 2x + 3)$, we have

$$\frac{x^2 + 5}{x^3 - x^2 + x + 3} = \frac{A}{x + 1} + \frac{Bx + C}{x^2 - 2x + 3}$$

Multiplying by $(x + 1)(x^2 - 2x + 3)$ yields

$$x^2 + 5 = A(x^2 - 2x + 3) + (Bx + C)(x + 1)$$

If $x = -1$,

$$1 + 5 = A(1 + 2 + 3) \qquad \text{or} \qquad A = 1$$

Therefore

$$x^2 + 5 = x^2 - 2x + 3 + Bx^2 + Bx + Cx + C$$
$$2x + 2 = Bx^2 + (B + C)x + C$$

Equating coefficients, we have $B = 0$, $B + C = 2$, and $C = 2$. Finally, we have

$$\int \frac{x^2 + 5}{x^3 - x^2 + x + 3} \, dx = \int \left(\frac{1}{x + 1} + \frac{2}{x^2 - 2x + 3} \right) dx$$

$$= \int \frac{1}{x + 1} \, dx + 2 \int \frac{1}{(x - 1)^2 + 2} \, dx$$

$$= \ln |x + 1| + 2 \left(\frac{1}{\sqrt{2}} \right) \arctan \left(\frac{x - 1}{\sqrt{2}} \right)$$

$$+ C$$

$$= \ln |x + 1| + \sqrt{2} \arctan \left(\frac{x - 1}{\sqrt{2}} \right) + C$$

33. Evaluate $\int_1^2 \dfrac{x + 1}{x(x^2 + 1)}\, dx.$

Solution

$$\frac{x + 1}{x(x^2 + 1)} = \frac{A}{x} + \frac{Bx + C}{x^2 + 1}$$

Multiplying by $(x)(x^2 + 1)$ yields

$$x + 1 = A(x^2 + 1) + (Bx + C)(x) = Ax^2 + A + Bx^2 + Cx$$
$$= (A + B)x^2 + Cx + A$$

By equating coefficients, we have $A + B = 0$, $C = 1$, and $A = 1$. Thus $B = -1$, and we get

$$\int_1^2 \frac{x + 1}{x(x^2 + 1)}\, dx = \int_1^2 \left(\frac{1}{x} + \frac{-x + 1}{x^2 + 1}\right) dx$$

$$= \int_1^2 \frac{1}{x}\, dx - \frac{1}{2}\int_1^2 \frac{2x}{x^2 + 1}\, dx + \int_1^2 \frac{1}{x^2 + 1}\, dx$$

$$= \left[\ln|x| - \frac{1}{2}\ln|x^2 + 1| + \arctan x\right]_1^2$$

$$= \ln 2 - \frac{1}{2}\ln 5 + \arctan 2 - \ln 1$$

$$+ \frac{1}{2}\ln 2 - \arctan 1$$

$$= \frac{3}{2}\ln 2 - \frac{1}{2}\ln 5 + \arctan 2 - \frac{\pi}{4}$$

$$= \frac{1}{2}\ln\frac{8}{5} + \arctan 2 - \frac{\pi}{4} \approx 0.557$$

39. Evaluate $\int \dfrac{e^x}{(e^x - 1)(e^x + 4)}\, dx$ by letting $u = e^x$.

Solution

If $u = e^x$, then $du = e^x\, dx$, and

$$\int \frac{e^x\, dx}{(e^x - 1)(e^x + 4)} = \int \frac{du}{(u - 1)(u + 4)}$$

By partial fractions,

$$\frac{1}{(u - 1)(u + 4)} = \frac{A}{u - 1} + \frac{B}{u + 4}$$

Multiplying by $(u - 1)(u + 4)$ yields

$$1 = A(u + 4) + B(u - 1)$$

If $u = 1$, then

$$1 = A(5) \quad \text{or} \quad A = \frac{1}{5}$$

If $u = -4$, then

$$1 = B(-5) \quad \text{or} \quad B = -\frac{1}{5}$$

Therefore

$$\int \frac{du}{(u - 1)(u + 4)} = \frac{1}{5} \int \left(\frac{1}{u - 1} - \frac{1}{u + 4} \right) du$$

$$= \frac{1}{5} (\ln |u - 1| - \ln |u + 4|) + C$$

$$= \frac{1}{5} \ln \left| \frac{u - 1}{u + 4} \right| + C = \frac{1}{5} \ln \left| \frac{e^x - 1}{e^x + 4} \right| + C$$

49. A single infected individual enters a community of n individuals susceptible to the disease. Let x be the number of newly infected individuals after time t. The common Epidemic Model assumes that the disease spreads at a rate proportional to the product of the total number infected and the number of susceptible not yet infected. Thus $dx/dt = k(x + 1)(n - x)$, and we obtain

$$\int \frac{1}{(x + 1)(n - x)} \, dx = \int k \, dt$$

Solve for x as a function of t.

Solution

$$\frac{1}{(x + 1)(n - x)} = \frac{A}{x + 1} + \frac{B}{n - x}$$

Multiplying by $(x + 1)(n - x)$, we have

$$1 = A(n - x) + B(x + 1)$$

If $x = -1$, then

$$1 = A(n + 1) \quad \text{or} \quad A = \frac{1}{n + 1}$$

If $x = n$, then

$$1 = B(n + 1) \qquad \text{or} \qquad B = \frac{1}{n + 1}$$

Therefore

$$\int \frac{1}{(x + 1)(n - x)} \, dx = \int k \, dt$$

$$\frac{1}{n + 1} \int \left(\frac{1}{x + 1} + \frac{1}{n - x} \right) dx = \int k \, dt$$

$$\frac{1}{n + 1} [\ln (x + 1) - \ln (n - x)] = kt + C$$

$$\left(\frac{1}{n + 1} \right) \ln \left(\frac{x + 1}{n - x} \right) = kt + C$$

When $t = 0$, $x = 0$. Thus

$$\left(\frac{1}{n + 1} \right) \ln \left(\frac{1}{n} \right) = C$$

Therefore

$$\ln \left(\frac{x + 1}{n - x} \right) = k(n + 1)t + \ln \left(\frac{1}{n} \right)$$

$$\frac{x + 1}{n - x} = \frac{1}{n} e^{k(n+1)t}$$

$$x = \frac{n[e^{k(n+1)t} - 1]}{e^{k(n+1)t} + n}$$

8.6
Summary and integration by tables

7. Use the integration table at the end of the text to evaluate

$$\int \frac{1}{x^2 \sqrt{x^2 - 4}} \, dx$$

Solution

Consider Formula 35, where $u = x$ and $a = 2$. Then

$$\int \frac{1}{x^2 \sqrt{x^2 - 4}} \, dx = \int \frac{du}{u^2 \sqrt{u^2 - a^2}} = \frac{\sqrt{u^2 - a^2}}{a^2 u} + C$$

$$= \frac{\sqrt{x^2 - 4}}{4x} + C$$

11.   Use the integration table at the end of the text to evaluate

$$\int \frac{2x}{(1 - 3x)^2} \, dx$$

Solution

Consider Formula 4, where $u = x$, $a = 1$, $b = -3$, and $a + bu = 1 - 3x$. Then

$$\int \frac{2x}{(1 - 3x)^2} \, dx = 2 \int \frac{x}{(1 - 3x)^2} \, dx = 2 \int \frac{u}{(a + bu)^2} \, du$$

$$= 2\left(\frac{1}{b^2}\right)\left(\frac{1}{a + bu} + \ln|a + bu|\right) + C$$

$$= \frac{2}{9}\left[\frac{1}{1 - 3x} + \ln|1 - 3x|\right] + C$$

21.   Use the integration table at the end of the text to evaluate

$$\int \frac{\cos x}{1 + \sin^2 x} \, dx$$

Solution

Consider Formula 23, with $u = \sin x$, $a = 1$, and $du = \cos x \, dx$. Then

$$\int \frac{\cos x}{1 + (\sin x)^2} \, dx = \arctan(\sin x) + C$$

23.   Use the integration table at the end of the text to evaluate

$$\int \frac{1}{1 + e^{2x}} \, dx$$

Solution

Consider Formula 84, where $u = 2x$ and $du = 2 \, dx$. Then

$$\int \frac{1}{1 + e^{2x}} \, dx = \frac{1}{2} \int \frac{2}{1 + e^{2x}} \, dx = \frac{1}{2}[2x - \ln(1 + e^{2x})] + C$$

$$= x - \frac{1}{2} \ln(1 + e^{2x}) + C$$

33.   Use the integration table at the end of the text to evaluate

$$\int \frac{1}{\sqrt{x}(1 - \cos \sqrt{x})} \, dx$$

Solution

Consider Formula 57, with $u = \sqrt{x}$ and $du = (1/2\sqrt{x})\,dx$.
Then

$$\int \frac{1}{\sqrt{x}(1 - \cos\sqrt{x})}\,dx = 2\int \frac{1/2\sqrt{x}}{1 - \cos\sqrt{x}}\,dx$$

$$= 2(-\cot\sqrt{x} - \csc\sqrt{x}) + C$$

39. Use the integration table at the end of the text to evaluate

$$\int \frac{\ln x}{x(3 + 2\ln x)}\,dx$$

Solution

Consider Formula 3, where $u = \ln x$, $a = 3$, $b = 2$, and
$du = (1/x)\,dx$. Then

$$\int \frac{\ln x}{x(3 + 2\ln x)}\,dx = \int \frac{(\ln x)(1/x)}{3 + 2\ln x}\,dx$$

$$= \frac{1}{4}[2\ln|x| - 3\ln(3 + 2\ln|x|)] + C$$

47. Use the integration table at the end of the text to evaluate

$$\int \frac{x^3}{\sqrt{4 - x^2}}\,dx$$

Solution

Consider Formula 21, with $u = x^2$, $a = 4$, $b = -1$,
$du = 2x\,dx$, and $\sqrt{a + bu} = \sqrt{4 - x^2}$. Then

$$\int \frac{x^2}{\sqrt{4 - x^2}}\,dx = \frac{1}{2}\int \frac{x^2(2x)}{\sqrt{4 - x^2}}\,dx = \frac{1}{2}\int \frac{u}{\sqrt{a + bu}}\,du$$

$$= \frac{1}{2}\left(\frac{(-2)(8 + x^2)}{3}\right)\sqrt{4 - x^2} + C$$

$$= -\left(\frac{x^2 + 8}{3}\right)\sqrt{4 - x^2} + C$$

52. Use the integration table at the end of the text to find
$\int \sec^5\theta\,d\theta$.

Solution

Using Formula 69 twice in succession, we obtain

$$\int \sec^5 \theta \, d\theta = \frac{\sec^3 \theta \tan \theta}{4} + \frac{3}{4} \int \sec^3 \theta \, d\theta$$

$$= \frac{\sec^3 \theta \tan \theta}{4} + \frac{3}{4} \left(\frac{\sec \theta \tan \theta}{2} + \frac{1}{2} \int \sec \theta \, d\theta \right)$$

Now using Formula 61, we finally obtain

$$\int \sec^5 \theta \, d\theta = \frac{\sec^3 \theta \tan \theta}{4} + \frac{3}{8} \sec \theta \tan \theta$$

$$+ \frac{3}{8} \ln |\sec \theta + \tan \theta| + C$$

53. Verify the formula

$$\int \frac{u^2}{(a + bu)^2} \, du = \frac{1}{b^3} \left(bu - \frac{a^2}{a + bu} - 2a \ln |a + bu| \right) + C$$

by the method of partial fractions.

Solution

Since the numerator and denominator are of the same degree, we begin by dividing

$$\frac{u^2}{(a + bu)^2} = \frac{1}{b^2} - \frac{(2a/b)u + (a^2/b^2)}{(a + bu)^2}$$

Now we use partial fractions to obtain

$$\frac{(2a/b)u + (a^2/b^2)}{(a + bu)^2} = \frac{A}{a + bu} + \frac{B}{(a + bu)^2}$$

Multiplying by $(a + bu)^2$, we have

$$\left(\frac{2a}{b} \right) u + \frac{a^2}{b^2} = A(a + bu) + B = bAu + (aA + B)$$

Equating the coefficients of like terms, we have

$$bA = \frac{2a}{b} \quad \text{and} \quad A = \frac{2a}{b^2}$$

$$aA + B = \frac{a^2}{b^2} \quad \text{and} \quad B = \frac{a^2}{b^2} - a \left(\frac{2a}{b^2} \right) = -\frac{a^2}{b^2}$$

Therefore

$$\int \frac{u^2}{(a + bu)^2} \, du$$

$$= \frac{1}{b^2} \int du - \frac{2a}{b^2} \left(\frac{1}{b} \right) \int \frac{b}{a + bu} \, du + \frac{a^2}{b^2} \left(\frac{1}{b} \right) \int \frac{b}{(a + bu)^2} \, du$$

$$= \left(\frac{1}{b^2}\right)u - \frac{2a}{b^3}(\ln |a + bu|) - \frac{a^2}{b^3}\left(\frac{1}{a + bu}\right) + C$$

$$= \frac{1}{b^3}\left[bu - \frac{a^2}{a + bu} - 2a \ln |a + bu|\right] + C$$

61. Evaluate $\int_0^{\pi/2} \dfrac{1}{1 + \sin \theta + \cos \theta} \, d\theta$.

Solution

Let $u = \dfrac{\sin \theta}{1 + \cos \theta}$. Then $\cos \theta = \dfrac{1 - u^2}{1 + u^2}$, $\sin \theta = \dfrac{2u}{1 + u^2}$,

and $d\theta = \dfrac{2 \, du}{1 + u^2}$. Furthermore, when $\theta = \pi/2$, $u = 1$, and

when $\theta = 0$, $u = 0$.

$$\int_0^{\pi/2} \frac{1}{1 + \sin \theta + \cos \theta} \, d\theta$$

$$= \int_0^1 \frac{2/(1 + u^2) \, du}{1 + [(2u)/(1 + u^2)] + [(1 - u^2)/(1 + u^2)]}$$

$$= \int_0^1 \frac{1}{u + 1} \, du = [\ln |u + 1|]_0^1 = \ln 2$$

8.7
Numerical integration

5. Use the Trapezoidal Rule and Simpson's Rule to approximate
    the value of $\int_0^2 x^3 \, dx$. Compare these results with the exact
    value of the definite integral. Round your answers to four deci-
    mal places.

Solution

(a) Trapezoidal Rule $(n = 8)$

$$\int_0^2 x^3 \, dx \approx \frac{2}{2(8)}\left[0 + 2\left(\frac{1}{4}\right)^3 + 2\left(\frac{2}{4}\right)^3 + 2\left(\frac{3}{4}\right)^3 + 2\left(\frac{4}{4}\right)^3 + 2\left(\frac{5}{4}\right)^3\right.$$

$$\left. + 2\left(\frac{6}{4}\right)^3 + 2\left(\frac{7}{4}\right)^3 + 2^3\right]$$

$$= \frac{1}{8}\left[\frac{2(1^3 + 2^3 + 3^3 + 4^3 + 5^3 + 6^3 + 7^3)}{4^3} + 8\right]$$

$$= \frac{1}{8}\left[\frac{2(784)}{164} + 8\right] = \frac{65}{16} = 4.0625$$

(b) Simpson's Rule ($n = 8$)

$$\int_0^2 x^3 \, dx \approx \frac{2}{3(8)}\left[0 + 4\left(\frac{1}{4}\right)^3 + 2\left(\frac{2}{4}\right)^3 + 4\left(\frac{3}{4}\right)^3 + 2\left(\frac{4}{4}\right)^3 + 4\left(\frac{5}{4}\right)^3 \right.$$

$$\left. + 2\left(\frac{6}{4}\right)^3 + 4\left(\frac{7}{4}\right)^3 + 2^3\right]$$

$$= \frac{1}{12}\left[\frac{4(1^3 + 3^3 + 5^3 + 7^3) + 2(2^3 + 4^3 + 6^3)}{4^3} + 8\right]$$

$$= \frac{1}{12}\left[\frac{4(496) + 2(288)}{4^3} + 8\right]$$

$$= \frac{1}{12}\left(\frac{2560}{64} + 8\right) = \frac{1}{12}(48) = 4$$

(c) In this particular case, we see that Simpson's Rule is exact since

$$\int_0^2 x^3 \, dx = \left[\frac{x^4}{4}\right]_0^2 = \frac{16}{4} = 4$$

11. Approximate $\int_0^{\sqrt{\pi/2}} \cos(x^2) \, dx$ using (a) the Trapezoidal Rule and (b) Simpson's Rule with $n = 4$.

Solution

(a) Trapezoidal Rule with $n = 4$

$$\int_0^{\sqrt{\pi/2}} \cos(x^2) \, dx \approx \frac{\sqrt{\pi/2}}{2(4)}\left[\cos(0) + 2\cos\left(\frac{\sqrt{\pi/2}}{4}\right)^2\right.$$

$$+ 2\cos\left(\frac{2\sqrt{\pi/2}}{4}\right)^2$$

$$\left. + 2\cos\left(\frac{3\sqrt{\pi/2}}{4}\right)^2 + \cos\left(\frac{4\sqrt{\pi/2}}{4}\right)^2\right]$$

$$\approx 0.9567$$

(b) Simpson's Rule with $n = 4$

$$\int_0^{\sqrt{\pi/2}} \cos(x^2) \, dx \approx \frac{\sqrt{\pi/2}}{3(4)}\left[\cos(0) + 4\cos\left(\frac{\sqrt{\pi/2}}{4}\right)^2\right.$$

$$+ 2\cos\left(\frac{2\sqrt{\pi/2}}{4}\right)^2$$

$$\left. + 4\cos\left(\frac{3\sqrt{\pi/2}}{4}\right)^2 + \cos\left(\frac{4\sqrt{\pi/2}}{4}\right)^2\right]$$

$$\approx 0.9782$$

15. Approximate $\int_0^1 \sqrt{x}\,\sqrt{1 - x} \, dx$ using (a) the Trapezoidal Rule and (b) Simpson's Rule with $n = 4$.

Solution

(a) Trapezoidal Rule ($n = 4$)

$$\int_0^1 \sqrt{x}\,\sqrt{1-x}\,dx \approx \frac{1}{2(4)}\left[0 + 2\sqrt{\frac{1}{4}}\sqrt{\frac{3}{4}} + 2\sqrt{\frac{2}{4}}\sqrt{\frac{2}{4}}\right.$$
$$\left. + 2\sqrt{\frac{3}{4}}\sqrt{\frac{1}{4}} + 0\right]$$
$$= \frac{1}{8}\left[\frac{2\sqrt{3}}{4} + \frac{2(2)}{4} + \frac{2\sqrt{3}}{4}\right]$$
$$= \frac{1}{8}(1 + \sqrt{3}) \approx 0.3415$$

(b) Simpson's Rule ($n = 4$)

$$\int_0^1 \sqrt{x}\,\sqrt{1-x}\,dx \approx \frac{1}{3(4)}\left[0 + 4\sqrt{\frac{1}{4}}\sqrt{\frac{3}{4}} + 2\sqrt{\frac{2}{4}}\sqrt{\frac{2}{4}}\right.$$
$$\left. + 4\sqrt{\frac{3}{4}}\sqrt{\frac{1}{4}} + 0\right]$$
$$= \frac{1}{12}\left[\frac{4\sqrt{3}}{4} + \frac{2(2)}{4} + \frac{4\sqrt{3}}{4}\right]$$
$$= \frac{1}{12}(2\sqrt{3} + 1) \approx 0.3720$$

23. Find the maximum possible error if the integral $\int_0^1 e^{x^3}\,dx$ is approximated by (a) the Trapezoidal Rule and (b) Simpson's Rule, using $n = 4$.

Solution

(a) Trapezoidal Rule ($n = 4$): Let $f(x) = e^{x^3}$. Then $f'(x) = 3x^2 e^{x^3}$ and $f''(x) = 3(3x^4 + 2x)e^{x^3}$. Since f'' is increasing on the interval $[0, 1]$, $f''(x)$ is at a maximum when $x = 1$. Therefore the maximum error for the Trapezoidal Rule is

$$|\text{error}| \le \frac{(1-0)^3}{12(4^2)}|f''(1)| = \frac{1}{192}(15e) = \frac{5e}{64} \approx 0.212$$

(b) Simpson's Rule ($n = 4$): Since $f''(x) = 3(3x^4 + 2x)e^{x^3}$, we have $f'''(x) = 3(9x^6 + 18x^3 + 2)e^{x^3}$ and $f^{(4)}(x) = 3(27x^8 + 108x^5 + 60x^2)e^{x^3}$. Since $f^{(4)}$ is increasing on the interval $[0, 1]$, it has a maximum when $x = 1$. Therefore the maximum error for Simpson's Rule is

$$|\text{error}| \le \frac{(1-0)^5}{180(4^4)}f^{(4)}(1) = \frac{1}{46,080}585e \approx 0.0345$$

27. Find n so that the error in the approximation of the definite integral

$$\int_0^1 e^{-x^2}\, dx$$

is less than 0.00001 using (a) the Trapezoidal Rule and (b) Simpson's Rule.

Solution

(a) Let $f(x) = e^{-x^2}$. Then $f'(x) = (-2x)e^{-x^2}$ and $f''(x) = (4x^2 - 2)e^{-x^2}$. By the methods used in Chapter 4, we determine that the maximum value of $|f''(x)|$ in the interval $[0, 1]$ occurs when $x = 0$. Thus for the Trapezoidal Rule, we have

$$|\text{error}| \le \frac{(1 - 0)^3}{12n^2}|f''(0)| \le 0.00001$$

$$\frac{1}{12n^2}(2) \le \frac{1}{100,000}$$

$$\frac{100,000}{6} \le n^2$$

$$16,667 \le n^2$$

$$129.1 \le n$$

Therefore we let $n = 130$.

(b) Since $f''(x) = (4x^2 - 2)e^{-x^2}$, we have $f'''(x) = -4(2x^3 - 3x)e^{-x^2}$ and $f^{(4)}(x) = 4(4x^4 - 12x^2 + 3)e^{-x^2}$. By the methods used in Chapter 4, we determine that the maximum value of $|f^{(4)}(x)|$ in $[0, 1]$ occurs when $x = 0$. Thus for Simpson's Rule, we have

$$|\text{error}| \le \frac{(1 - 0)^5}{180n^4}|f^{(4)}(0)| \le 0.00001$$

$$\frac{1}{180n^4}(12) \le \frac{1}{100,000}$$

$$\frac{20,000}{3} \le n^4$$

$$9.04 \le n$$

Therefore, we let $n = 10$.

32. The standard normal probability density function is

$$f(z) = \frac{1}{\sqrt{2\pi}}e^{-z^2/2}$$

The probability that z is in the interval $[a, b]$ is the area of the region defined by $y = f(z)$, $y = 0$, $z = a$, and $z = b$ and is denoted by $\Pr(a \leq z \leq b)$. Estimate the following probabilities. (Choose n so that the error is less than 0.0001.)

(a) $\Pr(0 \leq z \leq 1)$ (b) $\Pr(0 \leq z \leq 2)$

Solution

If $f(z) = (1/\sqrt{2\pi})e^{-z^2/2}$, then

$$f^{(4)}(z) = \frac{1}{\sqrt{2\pi}}(z^4 + 3)e^{-z^2/2}$$

By the methods of differential calculus, we can determine that $f^{(4)}(z)$ has a maximum value of $f^{(4)}(0) = 3/\sqrt{2\pi}$ on $[0, 2]$.

(a) By Theorem 8.5 (for Simpson's Rule), we have

$$E \leq \frac{(b-a)^5}{180n^4}f^{(4)}(z) \leq \frac{1}{180n^4}\left(\frac{3}{\sqrt{2\pi}}\right) < 0.0001$$

$$n > 2.86$$

Therefore we choose $n = 4$ and obtain

$$\Pr(0 \leq z \leq 1) = \frac{1}{\sqrt{2\pi}}\int_0^1 e^{-z^2/2}\,dz$$

$$\approx \frac{1}{\sqrt{2\pi}}\left[\frac{1}{3(4)}\right]\left[e^0 + 4e^{-(1/4)^2/2} + 2e^{-(2/4)^2/2}\right.$$
$$\left. + 4e^{-(3/4)^2/2} + e^{-1/2}\right]$$

$$\approx 0.3414$$

(b) By Theorem 8.5 (for Simpson's Rule), we have

$$E \leq \frac{(b-a)^5}{180n^4}f^{(4)}(z) \leq \frac{2^5}{180n^4}\left(\frac{3}{\sqrt{2\pi}}\right) < 0.00001$$

$$n > 6.79$$

Therefore we choose $n = 8$ and obtain

$$\Pr(0 \leq z \leq 2) = \frac{1}{\sqrt{2\pi}}\int_0^2 e^{-z^2/2}\,dz$$

$$\approx \frac{1}{\sqrt{2\pi}}\left[\frac{2}{3(8)}\right]\left[e^0 + 4e^{-(1/4)^2/2} + 2e^{-(2/4)^2/2}\right.$$
$$+ 4e^{-(3/4)^2/2} + 2e^{-(4/4)^2/2} + 4e^{-(5/4)^2/2}$$
$$\left. + 2e^{-(6/4)^2/2} + 4e^{-(7/4)^2/2} + e^{-(8/4)^2/2}\right]$$

$$\approx 0.4772$$

8.8
Indeterminate forms and L'Hôpital's Rule

3. Evaluate $\lim\limits_{x \to 0} \dfrac{\sqrt{4 - x^2} - 2}{x}$.

Solution

Since a direct substitution of $x = 0$ leaves us with the indeterminate form $0/0$, we apply L'Hôpital's Rule to obtain

$$\lim_{x \to 0} \frac{\sqrt{4 - x^2} - 2}{x} = \lim_{x \to 0} \frac{(\frac{1}{2})(4 - x^2)^{-1/2}(-2x)}{1}$$

$$= \lim_{x \to 0} \frac{-x}{\sqrt{4 - x^2}} = \frac{0}{2} = 0$$

7. Evaluate $\lim\limits_{x \to 0^+} \dfrac{e^x - (1 + x)}{x^n}$ where $n = 1, 2, 3, \ldots$.

Solution

Case 1: $n = 1$ (apply L'Hôpital's Rule once).

$$\lim_{x \to 0^+} \frac{e^x - (1 + x)}{x} = \lim_{x \to 0^+} \frac{e^x - 1}{1} = 0$$

Case 2: $n = 2$ (apply L'Hôpital's Rule twice).

$$\lim_{x \to 0^+} \frac{e^x - (1 + x)}{x^2} = \lim_{x \to 0^+} \frac{e^x - 1}{2x} = \lim_{x \to 0^+} \frac{e^x}{2} = \frac{1}{2}$$

Case 3: $n \geq 3$ (apply L'Hôpital's Rule twice).

$$\lim_{x \to 0^+} \frac{e^x - (1 + x)}{x^n} = \lim_{x \to 0^+} \frac{e^x - 1}{nx^{n-1}} = \lim_{x \to 0^+} \frac{e^x}{n(n - 1)x^{n-2}} = \infty$$

9. Evaluate $\lim\limits_{x \to \infty} \dfrac{\ln x}{x}$.

Solution

Since $\lim\limits_{x \to \infty} \dfrac{\ln x}{x} = \dfrac{\infty}{\infty}$ is an indeterminate form, we use L'Hôpital's Rule.

$$\lim_{x \to \infty} \frac{\ln x}{x} = \lim_{x \to \infty} \frac{1/x}{1} = \lim_{x \to \infty} \frac{1}{x} = 0$$

10. Evaluate $\lim_{x\to\infty} \dfrac{e^x}{x}$.

Solution

By L'Hôpital's Rule,

$$\lim_{x\to\infty} \frac{e^x}{x} = \lim_{x\to\infty} \frac{e^x}{1} = \infty$$

15. Evaluate $\lim_{x\to\infty} x^2 \ln x$.

Solution

Since $\lim_{x\to 0^+} x^2 \ln x = 0 \; (-\infty)$, we rewrite the limit and use L'Hôpital's Rule.

$$\lim_{x\to 0^+} x^2 \ln x = \lim_{x\to 0^+} \frac{\ln x}{1/x^2} = \lim_{x\to 0^+} \frac{1/x}{-2/x^3} = \lim_{x\to 0^+} \frac{-x^2}{2} = 0$$

17. Evaluate $\lim_{x\to 2} \left(\dfrac{8}{x^2 - 4} - \dfrac{x}{x - 2} \right)$.

Solution

$$\lim_{x\to 2}\left[\frac{8}{x^2 - 4} - \frac{x}{x - 2} \right] = \lim_{x\to 2}\left[\frac{8x - 16 - x^3 + 4x}{(x^2 - 4)(x - 2)} \right]$$

$$= \lim_{x\to 2}\left[\frac{(x - 2)(x - 2)(-x - 4)}{(x - 2)(x + 2)(x - 2)} \right] = -\frac{3}{2}$$

21. Evaluate $\lim_{x\to 0^+} x^{1/x}$.

Solution

$$\lim_{x\to 0^+} x^{1/x} = 0^\infty = 0$$

31. Evaluate $\lim_{x\to\infty} x \sin \dfrac{1}{x}$.

Solution

Since direct substitution yields an indeterminate form we use L'Hôpital's Rule.

$$\lim_{x\to\infty} x \sin \frac{1}{x} = \lim_{x\to\infty} \frac{\sin (1/x)}{1/x}$$

$$= \lim_{x\to\infty} \frac{(-1/x^2) \cos (1/x)}{-1/x^2}$$

$$= \lim_{x\to\infty} \cos \frac{1}{x} = 1$$

39.   Use L'Hôpital's Rule to evaluate $\lim\limits_{x\to\infty} \dfrac{(\ln x)^n}{x^m}$, where $0 < n, m$.

Solution

$$\lim_{x\to\infty} \frac{(\ln x)^n}{x^m} = \lim_{x\to\infty}\left[\frac{n(\ln x)^{n-1}(1/x)}{mx^{m-1}}\right] = \lim_{x\to\infty}\left[\frac{n(\ln x)^{n-1}}{mx^m}\right]$$

If $n - 1 \le 0$, this limit is zero. If $n - 1 > 0$, we repeat L'Hôpital's Rule to obtain

$$\lim_{x\to\infty}\left[\frac{n(n-1)(\ln x)^{n-2}}{m^2 x^m}\right]$$

Again, if $n - 2 \le 0$, this limit is zero. If $n - 2 > 0$, repeated applications of L'Hôpital's Rule will eventually yield a form where the numerator approaches a finite number and the denominator approaches infinity. Thus in every case the limit is 0.

51.   The velocity of an object falling in a resisting medium such as air or water (if the downward direction is positive) is given by

$$v = \frac{32}{k}\left(1 - e^{-kt} + \frac{v_0 k e^{-kt}}{32}\right)$$

where v_0 is the initial velocity, t is the time, and k is the resistance constant of the medium. Use L'Hôpital's Rule to find the formula for the velocity of a falling body in a vacuum by fixing v_0 and t and letting k approach zero.

Solution

$$\lim_{k\to 0}\frac{32}{k}\left(1 - e^{-kt} + \frac{v_0 k e^{-kt}}{32}\right) = \lim_{k\to 0}\frac{32(1 - e^{-kt})}{k} + \lim_{k\to 0}\frac{32}{k}\cdot\frac{v_0 k e^{-kt}}{32}$$

$$= \lim_{k\to 0}\frac{32te^{-kt}}{1} + v_0 = 32t + v_0$$

(Apply L'Hôpital's Rule to the first limit on the right-hand side.)

8.9
Improper integrals

4.   Determine the divergence or convergence of

$$\int_0^2 \frac{1}{(x-1)^2}\,dx$$

and evaluate it if it converges.

Solution

Since the integrand has an infinite discontinuity at $x = 1$, we write

$$\int_0^2 \frac{1}{(x-1)^2}\, dx = \lim_{b \to 1^-} \int_0^b \frac{1}{(x-1)^2}\, dx + \lim_{c \to 1^+} \int_c^2 \frac{1}{(x-1)^2}\, dx$$

$$= \lim_{b \to 1^-} \left[\frac{-1}{x-1}\right]_0^b + \lim_{c \to 1^+} \left[\frac{-1}{x-1}\right]_c^2$$

$$= \infty + \infty$$

Therefore this improper integral diverges.

13. Determine the divergence or convergence of

$$\int_0^\infty e^{-x} \cos x \, dx$$

and evaluate it if it converges.

Solution

Since

$$\int e^{-x} \cos x \, dx = \frac{1}{2} e^{-x}(-\cos x + \sin x) + C$$

we have

$$\int_0^\infty e^{-x} \cos x \, dx = \lim_{b \to \infty} \int_0^b e^{-x} \cos x \, dx$$

$$= \lim_{b \to \infty} \frac{1}{2}\left[\frac{(-\cos x + \sin x)}{e^x}\right]_0^b = \frac{1}{2}[0 - (-1)] = \frac{1}{2}$$

23. Determine the divergence or convergence of

$$\int_0^8 \frac{1}{\sqrt[3]{8-x}}\, dx$$

and evaluate it if it converges.

Solution

$$\int_0^8 \frac{1}{\sqrt[3]{8-x}}\, dx = \lim_{b \to 8^-} \int_0^b \frac{1}{\sqrt[3]{8-x}}\, dx$$

$$= \lim_{b \to 8^-} \left[-\frac{(8-x)^{2/3}}{\frac{2}{3}}\right]_0^b = -\frac{3}{2}(0) + \frac{3}{2}(4) = 6$$

29. Determine the divergence or convergence of

$$\int_2^4 \frac{1}{\sqrt{x^2 - 4}}\, dx$$

and evaluate it if it converges.

Solution

$$\int_2^4 \frac{1}{\sqrt{x^2 - 4}}\, dx = \lim_{a \to 2^+} \int_a^4 \frac{1}{\sqrt{x^2 - 4}}\, dx$$

$$= \lim_{a \to 2^+} \left[\ln |x + \sqrt{x^2 - 4}| \right]_a^4$$

$$= \ln (4 + \sqrt{12}) - \ln (2 + 0)$$

$$= \ln \left(\frac{4 + 2\sqrt{3}}{2} \right) = \ln (2 + \sqrt{3})$$

39. Use the result of Exercise 34 to determine if $\int_0^\infty e^{-x^2}\, dx$ converges.

Solution

On $[1, \infty)$ we have

$$x \leq x^2$$

$$e^x \leq e^{x^2}$$

$$\frac{1}{e^x} \geq \frac{1}{e^{x^2}}$$

$$e^{-x} \geq e^{-x^2}$$

Therefore by Exercise 34 we have

$$\int_0^\infty e^{-x^2}\, dx = \int_0^1 e^{-x^2}\, dx + \int_1^\infty e^{-x^2}\, dx < \int_0^1 e^{-x^2}\, dx + \int_1^\infty e^{-x}\, dx$$

$$= \int_0^1 e^{-x^2}\, dx + \left[-e^{-x} \right]_1^\infty = \int_0^1 e^{-x^2}\, dx + e^{-1}$$

Therefore the integral converges.

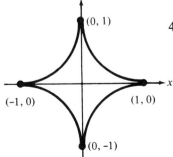

43. Sketch the graph of the hypocycloid of four cusps, $x^{2/3} + y^{2/3} = 1$, and find its perimeter.

Solution

By symmetry the perimeter of this figure is four times the length of the arc in the first quadrant. Thus

$$s = 4 \int_0^1 \sqrt{1 + (y')^2}\, dx$$

By implicit differentiation we have

$$x^{2/3} + y^{2/3} = 1$$

$$\frac{2}{3}x^{-1/3} + \frac{2}{3}y^{-1/3}y' = 0$$

$$y' = -\frac{y^{1/3}}{x^{1/3}}$$

Therefore

$$s = 4 \int_0^1 \sqrt{1 + \left(-\frac{y^{1/3}}{x^{1/3}}\right)^2}\, dx = 4 \int_0^1 \sqrt{\frac{x^{2/3} + y^{2/3}}{x^{2/3}}}\, dx$$

$$= 4 \lim_{b \to 0^+} \int_b^1 \sqrt{\frac{1}{x^{2/3}}}\, dx$$

$$= 4 \lim_{b \to 0^+} \int_b^1 x^{-1/3}\, dx = 4 \lim_{b \to 0^+} \left[\frac{x^{2/3}}{2/3}\right]_b^1 = 4\left(\frac{3}{2}\right) = 6$$

Review Exercises for Chapter 8

3. Evaluate $\displaystyle \int \frac{1}{1 - \sin \theta}\, d\theta$.

    Solution

    $$\int \frac{1}{1 - \sin \theta}\, d\theta = \int \frac{1}{1 - \sin \theta}\left(\frac{1 + \sin \theta}{1 + \sin \theta}\right) d\theta$$

    $$= \int \frac{1 + \sin \theta}{1 - \sin^2 \theta}\, d\theta = \int \frac{1 + \sin \theta}{\cos^2 \theta}\, d\theta$$

    $$= \int \left(\frac{1}{\cos^2 \theta} + \frac{\sin \theta}{\cos \theta \cos \theta}\right) d\theta$$

    $$= \int (\sec^2 \theta + \sec \theta \tan \theta)\, d\theta = \tan \theta + \sec \theta + C$$

7. Evaluate $\displaystyle \int \frac{\ln (2x)}{x^2}\, dx$.

    Solution

    Using integration by parts, let

    $$dv = \frac{1}{x^2}\, dx \;\rightarrow\; v = -\frac{1}{x}$$

    $$u = \ln (2x) \rightarrow du = \frac{1}{x}\, dx$$

$$\int \frac{\ln (2x)}{x^2}\, dx = -\frac{\ln (2x)}{x} - \int \frac{1}{x}\left(-\frac{1}{x}\right) dx = \frac{-\ln (2x)}{x} + \int x^{-2}\, dx$$

$$= \frac{-\ln (2x)}{x} - \frac{1}{x} + C$$

13. Evaluate $\int \sec^4 \frac{x}{2}\, dx$.

Solution

$$\int \sec^4 \frac{x}{2}\, dx = \int \sec^2 \frac{x}{2} \sec^2 \frac{x}{2}\, dx$$

$$= \int \left[\tan^2 \frac{x}{2} + 1\right] \sec^2 \frac{x}{2}\, dx$$

$$= 2 \int \tan^2 \frac{x}{2} \sec^2 \frac{x}{2}\left(\frac{1}{2}\right) dx + 2 \int \sec^2 \frac{x}{2}\left(\frac{1}{2}\right) dx$$

$$= \frac{2}{3} \tan^3 \frac{x}{2} + 2 \tan \frac{x}{2} + C$$

$$= \frac{2}{3}\left[\tan^3 \frac{x}{2} + 3 \tan \frac{x}{2}\right] + C$$

17. Evaluate $\int \dfrac{x^2 + 2x}{x^3 - x^2 + x - 1}\, dx$.

Solution

Since $x^3 - x^2 + x - 1 = (x - 1)(x^2 + 1)$, we have

$$\frac{x^2 + 2x}{(x - 1)(x^2 + 1)} = \frac{A}{x - 1} + \frac{Bx + C}{x^2 + 1}$$

Multiplying by $(x - 1)(x^2 + 1)$ yields

$$x^2 + 2x = A(x^2 + 1) + (Bx + C)(x - 1)$$

Let $x = 1$. Then

$$1 + 2 = A(2) + (B + C)(0) \qquad \text{or} \qquad A = \frac{3}{2}$$

Furthermore

$$x^2 + 2x = Ax^2 + A + Bx^2 - Bx + Cx - C$$

$$= (A + B)x^2 + (C - B)x + (A - C)$$

Now by equating coefficients, we have

$$1 = A + B = \frac{3}{2} + B \qquad \text{or} \qquad B = -\frac{1}{2}$$

$$2 = C - B = C + \frac{1}{2} \qquad \text{or} \qquad C = \frac{3}{2}$$

Therefore

$$\int \frac{x^2 + 2x}{(x - 1)(x^2 + 1)}\, dx = \int \frac{\frac{3}{2}}{x - 1} + \frac{(-x/2) + (\frac{3}{2})}{x^2 + 1}\, dx$$

$$= \frac{3}{2} \int \frac{1}{x - 1}\, dx - \frac{1}{2} \int \frac{x}{x^2 + 1}\, dx$$

$$+ \frac{3}{2} \int \frac{1}{x^2 + 1}\, dx$$

$$= \frac{3}{2} \ln |x - 1| - \frac{1}{4} \ln (x^2 + 1)$$

$$+ \frac{3}{2} \arctan x + C$$

31. Evaluate $\displaystyle \int \frac{x^{1/4}}{1 + x^{1/2}}\, dx$.

Solution

Let $u = x^{1/4}$. Then $u^4 = x$, $u^2 = x^{1/2}$, $dx = 4u^3\, du$, and

$$\int \frac{x^{1/4}}{1 + x^{1/2}}\, dx = \int \frac{u}{1 + u^2}(4u^3\, du) = 4 \int \frac{u^4}{1 + u^2}\, du$$

$$= 4 \int \left(u^2 - 1 + \frac{1}{1 + u^2} \right) du$$

$$= 4 \left(\frac{u^3}{3} - u + \arctan u \right) + C$$

$$= 4 \left(\frac{x^{3/4}}{3} - x^{1/4} + \arctan x^{1/4} \right) + C$$

37. Evaluate $\int \cos x \ln (\sin x)\, dx$.

Solution

Using integration by parts, let

$$dv = \cos x\, dx \;\rightarrow\; v = \sin x$$

$$u = \ln (\sin x) \rightarrow du = \frac{\cos x}{\sin x}\, dx$$

$$\int \cos x \ln (\sin x)\, dx = \sin x \ln (\sin x) - \int \sin x \left(\frac{\cos x}{\sin x}\right) dx$$

$$= \sin x \ln (\sin x) - \sin x + C$$

44. Integrate $\int x\sqrt{4 + x}\, dx$ by the following methods:

(a) Trigonometric substitution

(b) Substitution: $u^2 = 4 + x$

(c) Substitution: $u = 4 + x$

(d) By parts: $dv = \sqrt{4 + x}\, dx$

Solution

(a) Let $x = 4 \tan^2 \theta$. Then $\sqrt{4 + x} = 2 \sec \theta$ and
     $dx = 8 \tan \theta \sec^2 \theta\, d\theta$.

$$\int x\sqrt{4 + x}\, dx = \int 4 \tan^2 \theta (2 \sec \theta)(8 \tan \theta \sec^2 \theta)\, d\theta$$

$$= 64 \int \tan^2 \theta \sec^2 \theta (\sec \theta \tan \theta)\, d\theta$$

$$= 64 \int (\sec^2 \theta - 1)(\sec^2 \theta)(\sec \theta \tan \theta)\, d\theta$$

$$= 64 \int (\sec^4 \theta - \sec^2 \theta)(\sec \theta \tan \theta)\, d\theta$$

$$= 64\left[\left(\frac{1}{5}\right) \sec^5 \theta - \left(\frac{1}{3}\right) \sec^3 \theta\right] + C$$

$$= \frac{64 \sec^3 \theta}{15}(3 \sec^2 \theta - 5) + C$$

$$= \frac{2(4 + x)^{3/2}}{15}(3x - 8) + C$$

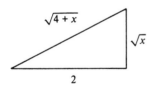

(b) Let $u^2 = 4 + x$. Then $x = u^2 - 4$ and $dx = 2u\, du$.

$$\int x\sqrt{4 + x}\, dx = \int (u^2 - 4)(u)(2u)\, du = 2 \int (u^4 - 4u^2)\, du$$

$$= 2\left[\frac{1}{5}u^5 - \frac{4}{3}u^3\right] + C = \frac{2u^3}{15}(3u^2 - 20) + C$$

$$= \frac{2(4 + x)^{3/2}}{15}(3x - 8) + C$$

(c) Let $u = 4 + x$. Then $du = dx$ and $x = u - 4$.

$$\int x\sqrt{4 + x}\, dx = \int (u - 4)(u^{1/2})\, du = \int (u^{3/2} - 4u^{1/2})\, du$$

$$= \frac{2}{5}u^{5/2} - \frac{8}{3}u^{3/2} + C = \frac{2u^{3/2}}{15}(3u - 20) + C$$

$$= \frac{2(4 + x)^{3/2}}{15}(3x - 8) + C$$

(d)   Let $dv = \sqrt{4 + x}\, dx \rightarrow v = \frac{2}{3}(4 + x)^{3/2}$

$$u = x \qquad\qquad \rightarrow du = dx$$

$$\int x\sqrt{4 + x}\, dx = x\left(\frac{2}{3}\right)(4 + x)^{3/2} - \frac{2}{3}\int (4 + x)^{3/2}\, dx$$

$$= \frac{2x}{3}(4 + x)^{3/2} - \frac{4}{15}(4 + x)^{5/2} + C$$

$$= \frac{2(4 + x)^{3/2}}{15}(3x - 8) + C$$

47.   Approximate $\displaystyle\int_0^2 \frac{1}{\sqrt{1 + x^3}}\, dx$ to two decimal places.

Solution

Since for this integral the error formulas of Theorem 8.7 are difficult to apply, we use Simpson's Rule for successively larger values of n until the desired accuracy is obtained. When $n = 4$, we have

$$\int_0^2 \frac{1}{\sqrt{1 + x^3}}\, dx \approx \frac{2}{3(4)}\left[\frac{1}{\sqrt{1 + 0^3}} + \frac{4}{\sqrt{1 + (\frac{1}{2})^3}} + \frac{2}{\sqrt{1 + (\frac{2}{2})^3}}\right.$$

$$\left. + \frac{4}{\sqrt{1 + (\frac{3}{2})^3}} + \frac{1}{\sqrt{1 + (\frac{4}{2})^3}}\right]$$

$$\approx 1.4052$$

When $n = 6$, we have

$$\int_0^2 \frac{1}{\sqrt{1 + x^3}}\, dx \approx \frac{2}{3(6)}\left[\frac{1}{\sqrt{1 + 0^3}} + \frac{4}{\sqrt{1 + (\frac{1}{3})^3}} + \frac{2}{\sqrt{1 + (\frac{2}{3})^3}}\right.$$

$$+ \frac{4}{\sqrt{1 + (\frac{3}{3})^3}} + \frac{2}{\sqrt{1 + (\frac{4}{3})^3}} + \frac{4}{\sqrt{1 + (\frac{5}{3})^3}}$$

$$\left. + \frac{1}{\sqrt{1 + (\frac{6}{3})^3}}\right] \approx 1.4024$$

When $n = 8$, we have

$$\int_0^2 \frac{1}{\sqrt{1 + x^3}} \, dx \approx \frac{2}{3(8)} \left[\frac{1}{\sqrt{1 + 0^3}} + \frac{4}{\sqrt{1 + (\frac{1}{4})^3}} + \frac{2}{\sqrt{1 + (\frac{2}{4})^3}} \right.$$

$$+ \frac{4}{\sqrt{1 + (\frac{3}{4})^3}} + \frac{2}{\sqrt{1 + (\frac{4}{4})^3}} + \frac{4}{\sqrt{1 + (\frac{5}{4})^3}}$$

$$\left. + \frac{2}{\sqrt{1 + (\frac{6}{4})^3}} + \frac{4}{\sqrt{1 + (\frac{7}{4})^3}} + \frac{1}{\sqrt{1 + (\frac{8}{4})^3}} \right]$$

$$\approx 1.4022$$

We therefore conclude that

$$\int_0^2 \frac{1}{\sqrt{1 + x^3}} \, dx \approx 1.40$$

57. Use L'Hôpital's Rule to evaluate $\lim_{x \to \infty} (\ln x)^{2/x}$.

Solution

We begin by taking the natural logarithm of both sides of the equation

$$y = \lim_{x \to \infty} (\ln x)^{2/x}$$

to obtain

$$\ln y = \ln \left[\lim_{x \to \infty} (\ln x)^{2/x} \right]$$

$$= \lim_{x \to \infty} \left[\frac{2}{x} \ln (\ln x) \right]$$

$$= 2 \lim_{x \to \infty} \frac{(1/\ln x)(1/x)}{1} = 0$$

Finally, as $\ln y \to 0$, we know that $y \to 1$ and conclude that

$$\lim_{x \to \infty} (\ln x)^{2/x} = 1$$

9 Infinite series

Introduction: Taylor polynomials and approximations

5. Find the Maclaurin polynomial of degree 5 for $f(x) = \sin x$.

    Solution

    We have

    $$
    \begin{aligned}
    f(x) &= \sin x & f(0) &= 0 \\
    f'(x) &= \cos x & f'(0) &= 1 \\
    f''(x) &= -\sin x & f''(0) &= 0 \\
    f'''(x) &= -\cos x & f'''(0) &= -1 \\
    f^{(4)}(x) &= \sin x & f^{(4)}(0) &= 0 \\
    f^{(5)}(x) &= \cos x & f^{(5)}(0) &= 1
    \end{aligned}
    $$

    Therefore, the expansion yields

    $$P_5(x) = f(0) + f'(0)x + \frac{f''(0)}{2!}x^2 + \frac{f'''(0)}{3!}x^3 + \frac{f^{(4)}(0)}{4!}x^4 + \frac{f^{(5)}(0)}{5!}x^5$$

    $$= x - \frac{x^3}{6} + \frac{x^5}{120}$$

9. Find the Maclaurin polynomial of degree 4 for $f(x) = \dfrac{1}{x + 1}$.

    Solution

    We have

    $$
    f(x) = \frac{1}{x + 1} \qquad f(0) = 1
    $$

$$f'(x) = \frac{-1}{(x + 1)^2} \qquad f'(0) = -1$$

$$f''(x) = \frac{2}{(x + 1)^3} \qquad f''(0) = 2$$

$$f'''(x) = \frac{-6}{(x + 1)^4} \qquad f'''(0) = -6$$

$$f^{(4)}(x) = \frac{24}{(x + 1)^5} \qquad f^{(4)}(0) = 24$$

Therefore, the expansion yields

$$P_4(x) = f(0) + f'(0)x + \frac{f''(0)}{2!}x^2 + \frac{f'''(0)}{3!}x^3 + \frac{f^{(4)}(0)}{4!}x^4$$

$$= 1 - x + x^2 - x^3 + x^4$$

11.   Find the Maclaurin polynomial of degree 2 for $f(x) = \sec x$.

Solution

We have

$$f(x) = \sec x \qquad\qquad f(0) = 1$$
$$f'(x) = \sec x \tan x \qquad\qquad f'(0) = 0$$
$$f''(x) = \sec^3 x + \sec x \tan^2 x \qquad f''(0) = 1$$

Therefore, the expansion yields

$$P_2(x) = f(0) + f'(0)x + \frac{f''(0)}{2!}x^2$$

$$= 1 + \frac{x^2}{2}$$

17.   Find the Taylor polynomial of degree 2, centered at $c = \pi$ for $f(x) = x^2 \cos x$.

Solution

We have

$$f(x) = x^2 \cos x \qquad\qquad f(\pi) = -\pi^2$$
$$f'(x) = -x^2 \sin x + 2x \cos x \qquad\qquad f'(\pi) = -2\pi$$
$$f''(x) = -x^2 \cos x - 4x \sin x + 2 \cos x \qquad f''(\pi) = \pi^2 - 2$$

Therefore, the expansion yields

$$P_2(x) = f(c) + f'(c)(x - c) + \frac{f''(c)}{2!}(x - c)^2$$

$$= -\pi^2 - 2\pi(x - \pi) + \frac{\pi^2 - 2}{2}(x - \pi)^2$$

23. Approximate $f(7\pi/8)$ for $f(x) = x^2 \cos x$ by using the polynomial found in Exercise 17.

Solution

$$f\left(\frac{7\pi}{8}\right) = \left(\frac{7\pi}{8}\right)^2 \cos\left(\frac{7\pi}{8}\right)$$

$$\approx -\pi^2 - 2\pi\left(\frac{7\pi}{8} - \pi\right) + \frac{\pi^2 - 2}{2}\left(\frac{7\pi}{8} - \pi\right)^2$$

$$\approx -6.7954$$

(The actual functional value accurate to four decimal places is -6.9812.)

27. For what values of $x < 0$ can e^x be replaced by the polynomial $1 + x + (x^2/2!) + (x^3/3!)$ if the error cannot exceed 0.001?

Solution

By Theorem 9.1

$$e^x = 1 + x + \frac{x^2}{2!} + \frac{x^3}{3!} + R_3$$

where

$$R_3 = \frac{f^{(4)}(z)}{4!}x^4 = \frac{e^z}{4!}x^4$$

For $z < 0$ we have

$$R_3 = \frac{e^z x^4}{4!} < \frac{x^4}{4!}$$

and we wish to find $x < 0$ such that

$$\frac{x^4}{4!} < 0.001$$

$$x^4 < 24(0.001) = 0.024$$

$$|x| < (0.024)^{1/4} \approx 0.39$$

Therefore for values of x such that $-0.39 < x < 0$, we have

$$e^x \approx 1 + x + \frac{x^2}{2!} + \frac{x^3}{3!}$$

31. (a) Find the Taylor polynomial $P(x)$ of degree 3 for
$f(x) = \arcsin x$.

(b) Complete the accompanying table for $f(x)$ and $P_3(x)$ of part (a).

x	-1	-0.75	-0.50	-0.25	0	0.25	0.50	0.75	1
$f(x)$									
$P_3(x)$									

(c) Use the table of part (b) to sketch the graphs of $f(x)$ and $P_3(x)$ on the same axis.

Solution

(a) $f(x) = \arcsin x$ $f(0) = 0$

$$f'(x) = \frac{1}{1 - x^2} \qquad f'(0) = 1$$

$$f''(x) = \frac{x}{(1 - x^2)^{3/2}} \qquad f''(0) = 0$$

$$f'''(x) = \frac{2x^2 + 1}{(1 - x^2)^{5/2}} \qquad f'''(0) = 1$$

$$P_3(x) = f(0) + f'(0)x + \frac{f''(0)}{2!}x^2 + \frac{f'''(0)}{3!}x^3 = x + \frac{x^3}{6}$$

(b)

x	-1.0	-0.75	-0.50	-0.25	0	0.25	0.50	0.75	1.0
$f(x)$	-1.571	-0.848	-0.524	-0.253	0	0.253	0.524	0.848	1.571
$P_3(x)$	-1.167	-0.820	-0.521	-0.253	0	0.253	0.521	0.820	1.167

(c)

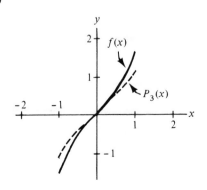

9.2

Sequences

5. Write out the first five terms of the sequence with $a_n = 3^n/n!$.

Solution

$$\left\{\frac{3^n}{n!}\right\} = \left\{\frac{3}{1!}, \frac{3^2}{2!}, \frac{3^3}{3!}, \frac{3^4}{4!}, \frac{3^5}{5!}, \cdots\right\} = \left\{3, \frac{9}{2}, \frac{27}{6}, \frac{81}{24}, \frac{243}{120}, \cdots\right\}$$

11. Write an expression for the nth term of the sequence
    $\{-1, 2, 7, 14, 23, \ldots\}$.

Solution

Compare the terms of the sequence $\{-1, 2, 7, 14, 23, \ldots\}$
with the sequence of squares

$$\{1^2, 2^2, 3^2, 4^2, 5^2, \ldots\} = \{1, 4, 9, 16, 25, \ldots\}$$

Observe that each term of the given sequence is two less than
the sequence of squares. Thus we write the nth term of the
given sequence as $a_n = n^2 - 2$.

15. Write an expression for the nth term of the sequence

$$\left\{2, -1, \frac{1}{2}, \frac{-1}{4}, \frac{1}{8}, \cdots\right\}$$

Solution

First observe that the denominators are powers of 2, where for
$n = 4$ the denominator is 2^2. This implies that the nth term has
a denominator 2^{n-2}. Note further that the signs alternate
starting with a positive sign, and thus the nth term is

$$a_n = (-1)^{n+1}\left(\frac{1}{2^{n-2}}\right) = \frac{(-1)^{n+1}}{2^{n-2}}$$

21. Write an expression for the nth term of the sequence

$$\left\{1, \frac{1}{1 \cdot 3}, \frac{1}{1 \cdot 3 \cdot 5}, \frac{1}{1 \cdot 3 \cdot 5 \cdot 7}, \cdots\right\}$$

Solution

The denominator of the nth term is the product of the first n
positive odd integers. Therefore

$$a_n = \frac{1}{1 \cdot 3 \cdot 5 \cdots (2n - 1)}$$

A second form of the nth term is obtained when we multiply the numerator and denominator by the n missing even integers. Then

$$a_n = \frac{1}{1 \cdot 3 \cdot 5 \cdots (2n-1)} = \frac{2 \cdot 4 \cdot 6 \cdot 8 \cdots (2n)}{1 \cdot 2 \cdot 3 \cdot 4 \cdot 5 \cdots (2n-1)(2n)}$$

$$= \frac{2^n(1 \cdot 2 \cdot 3 \cdot 4 \cdots n)}{1 \cdot 2 \cdot 3 \cdot 4 \cdot 5 \cdots (2n-1)(2n)} = \frac{2^n n!}{(2n)!}$$

27. Given $a_n = (-1)^n[n/(n+1)]$, determine if the sequence $\{a_n\}$ converges or diverges. If it converges, find its limit.

Solution

Since

$$\lim_{n \to \infty} a_n = \lim_{n \to \infty} (-1)^n\left(\frac{n}{n+1}\right) = \pm 1 \qquad \text{Limit does not exist}$$

the sequence $\{a_n\}$ diverges.

39. Given $a_n = f^{(n-1)}(2)$, $f(x) = \ln x$, determine if the sequence $\{a_n\}$ converges or diverges. If it converges, find its limit.

Solution

$$f(x) = \ln x \qquad\qquad a_1 = f(2) = \ln 2$$

$$f'(x) = \frac{1}{x} \qquad\qquad a_2 = f'(2) = \frac{1}{2}$$

$$f''(x) = \frac{-1}{x^2} \qquad\qquad a_3 = f''(2) = \frac{-1}{2^2}$$

$$f'''(x) = \frac{2}{x^3} \qquad\qquad a_4 = f'''(2) = \frac{2}{2^3}$$

$$f^{(4)}(x) = \frac{-6}{x^4} \qquad\qquad a_5 = f^{(4)}(2) = \frac{-6}{2^4}$$

$$f^{(5)}(x) = \frac{24}{x^5} \qquad\qquad a_6 = f^{(5)}(2) = \frac{24}{2^5}$$

$$\vdots \qquad\qquad\qquad\qquad \vdots$$

$$f^{(n-1)}(x) = \frac{(-1)^n(n-2)!}{x^{n-1}} \qquad a_n = f^{(n-1)}(2) = \frac{(-1)^n(n-2)!}{2^{n-1}}$$

Since

$$\lim_{n \to \infty} \frac{(-1)^n(n-2)!}{2^{n-1}}$$

does not exist, the sequence diverges.

41. Given $a_n = 3 - (1/2^n)$, determine if the sequence $\{a_n\}$ converges or diverges. If it converges, find its limit.

Solution

Since
$$\lim_{n \to \infty} a_n = \lim_{n \to \infty} \left(3 - \frac{1}{2^n}\right) = 3 - 0 = 3$$

the sequence converges to 3.

43. Given $a_n = [1 + (k/n)]^n$, determine if the sequence $\{a_n\}$ converges or diverges. If it converges, find its limit.

Solution

$$\lim_{n \to \infty} a_n = \lim_{n \to \infty} \left[1 + \left(\frac{k}{n}\right)^n\right] = \lim_{n \to \infty} \left[\left(1 + \frac{k}{n}\right)^{n/k}\right]^k$$

Now we let $u = k/n$. Then as n approaches infinity, u approaches zero, and

$$\lim_{n \to \infty} a_n = \lim_{u \to 0} [(1 + u)^{1/u}]^k = e^k$$

Therefore the sequence converges to e^k.

51. Determine if the sequence $a_n = (-1)^n\left(\dfrac{1}{n}\right)$ is monotonic.

Solution

Writing out the first few terms of the sequence we have

$$a_1 = -1, \; a_2 = \frac{1}{2}, \; a_3 = -\frac{1}{3}, \; a_4 = \frac{1}{4}, \; \cdots$$

Because of the alternating signs, we observe that the terms are neither nondecreasing nor nonincreasing. Therefore, the sequence is *not* monotonic.

63. A government program that currently costs taxpayers $2.5 billion per year is to be cut back by 20% per year.

(a) Write an expression for the amount budgeted for this program after n years.

(b) Compute the budgets for the first 4 years.

(c) Determine the convergence or divergence of the sequence of reduced budgets. If the sequence converges, find its limit.

Solution

(a)   If we let A_n be the amount budgeted after n years, we have

$$A_1 = 2.5 - 0.2(2.5) = 0.8(2.5) \text{ billion}$$
$$A_2 = A_1 - 0.2A_1 = 0.8A_1 = 0.8^2(2.5) \text{ billion}$$
$$A_3 = A_2 - 0.2A_2 = 0.8A_2 = 0.8^3(2.5) \text{ billion}$$
$$\vdots$$
$$A_n = 2.5(0.8)^n \text{ billion}$$

(b) $A_1 = 2.5(0.8) = \$2 \text{ billion}$
$$A_2 = 2.5(0.8)^2 = \$1.6 \text{ billion}$$
$$A_3 = 2.5(0.8)^3 = \$1.28 \text{ billion}$$
$$A_4 = 2.5(0.8)^4 = \$1.024 \text{ billion}$$

(c) $\lim\limits_{n \to \infty} 2.5(0.8)^n = 0$ and therefore the sequence converges.

9.3
Series and convergence

3.   Find the first five terms of the sequence of partial sums for the series

$$3 - \frac{9}{2} + \frac{27}{4} - \frac{81}{8} + \frac{243}{16} - \cdots$$

Solution

$$S_1 = 3$$

$$S_2 = 3 - \frac{9}{2} = -\frac{3}{2} = -1.5$$

$$S_3 = 3 - \frac{9}{2} + \frac{27}{4} = \frac{21}{4} = 5.25$$

$$S_4 = 3 - \frac{9}{2} + \frac{27}{4} - \frac{81}{8} = -\frac{39}{8} = -4.875$$

$$S_5 = 3 - \frac{9}{2} + \frac{27}{4} - \frac{81}{8} + \frac{243}{16} = \frac{165}{16} = 10.3125$$

13.   Verify that the series $\sum\limits_{n=0}^{\infty} 3\left(\frac{3}{2}\right)^n$ diverges.

Solution

Writing out a few terms of the series we obtain

$$\sum_{n=0}^{\infty} 3\left(\frac{3}{2}\right)^n = 3 + \frac{9}{2} + \frac{27}{4} + \frac{81}{8} + \cdots$$

The series is geometric with common ratio $r = \frac{3}{2}$. Since $|r| \geq 1$, the series diverges.

17. Verify that the infinite series $\frac{1}{4} + \frac{3}{8} + \frac{7}{16} + \frac{15}{32} + \frac{31}{64} + \cdots$ diverges.

Solution

$$\frac{1}{4} + \frac{3}{8} + \frac{7}{16} + \frac{15}{32} + \frac{31}{64} + \cdots = \sum_{n=1}^{\infty} \frac{2^n - 1}{2^{n+1}}$$

$$\lim_{n \to \infty} a_n = \lim_{n \to \infty} \frac{2^n - 1}{2^{n+1}} = \lim_{n \to \infty} \frac{1 - (1/2^n)}{2} = \frac{1}{2} \neq 0$$

Therefore by Theorem 9.8 the series diverges.

19. Verify that the infinite series $2 + \frac{3}{2} + \frac{9}{8} + \frac{27}{32} + \frac{81}{128} + \cdots$ converges.

Solution

$$2 + \frac{3}{2} + \frac{9}{8} + \frac{27}{32} + \frac{81}{128} + \cdots = \sum_{n=1}^{\infty} 2\left(\frac{3}{4}\right)^{n-1}$$

Therefore we have a geometric series with $a = 2$ and $r = \frac{3}{4}$. By Theorem 9.9 the series converges.

23. Verify that the infinite series

$$\frac{1}{1 \cdot 2} + \frac{1}{2 \cdot 3} + \frac{1}{3 \cdot 4} + \frac{1}{4 \cdot 5} + \frac{1}{5 \cdot 6} + \cdots$$

converges.

Solution

Since
$$a_n = \frac{1}{n(n+1)} = \frac{1}{n} - \frac{1}{n+1}$$

we can write the series as

$$\sum_{n=1}^{\infty} \frac{1}{n(n+1)} = \sum_{n=1}^{\infty} \left(\frac{1}{n} - \frac{1}{n+1}\right)$$

$$= \left(1 - \frac{1}{2}\right) + \left(\frac{1}{2} - \frac{1}{3}\right) + \left(\frac{1}{3} - \frac{1}{4}\right) + \cdots$$

where each term after the first term of 1 cancels. Therefore the series converges to 1.

31. Find the sum of the series $1 + 0.1 + 0.01 + 0.001 + \cdots$.

    Solution

    The series

    $$1 + 0.1 + 0.01 + 0.001 + \cdots = \sum_{n=0}^{\infty} (0.1)^n$$

    is geometric with $a = 1$ and $r = 0.1$. Therefore, the sum is

    $$S = \frac{a}{1-r} = \frac{1}{1-0.1} = \frac{10}{9}$$

37. Find the sum of the series $\displaystyle\sum_{n=1}^{\infty} \frac{4}{n(n+2)}$.

    Solution

    Using partial fractions we have

    $$\frac{4}{n(n+2)} = \frac{A}{n} + \frac{B}{n+2}$$

    Multiplying by $n(n+2)$, we obtain

    $$4 = A(n+2) + Bn$$

    If $n = 0$, then $4 = 2A$ or $A = 2$

    If $n = -2$, then $4 = -2B$ or $B = -2$

    Thus

    $$\sum_{n=1}^{\infty} \frac{4}{n(n+2)} = \sum_{n=1}^{\infty} \left[\frac{2}{n} - \frac{2}{n+2} \right]$$

    $$= \left[2 - \frac{2\!\!\!/}{3\!\!\!/} \right] + \left[\frac{2}{2} - \frac{2\!\!\!/}{4\!\!\!/} \right] + \left[\frac{2\!\!\!/}{3\!\!\!/} - \frac{2\!\!\!/}{5\!\!\!/} \right] + \left[\frac{2\!\!\!/}{4\!\!\!/} - \frac{2}{6} \right]$$

    $$+ \left[\frac{2\!\!\!/}{5\!\!\!/} - \frac{2}{7} \right] + \cdots = 2 + 1 = 3$$

49. Determine the convergence or divergence of the series

    $$\sum_{n=1}^{\infty} \frac{3n - 1}{2n + 1}$$

    Solution

    Since $\displaystyle\lim_{n \to \infty} a_n = \lim_{n \to \infty} \frac{3n - 1}{2n + 1} = \frac{3}{2} \neq 0$

    we conclude by Theorem 9.8 that the given series diverges.

54. Determine the convergence or divergence of the series

$$\sum_{n=1}^{\infty} \frac{2^n}{n^2}$$

Solution

$$\lim_{n\to\infty} a_n = \lim_{n\to\infty} \left(\frac{2^n}{n^2}\right)$$

$$= \lim_{n\to\infty} \left[\frac{(\ln 2)2^n}{2n}\right] \qquad \text{L'Hôpital's Rule}$$

$$= \lim_{n\to\infty} \left[\frac{(\ln 2)^2 2^n}{2}\right] \qquad \text{L'Hôpital's Rule}$$

$$= \infty$$

Therefore by Theorem 9.8 the series diverges.

63. A deposit of \$50 is made at the beginning of each month in an account that pays 12% compounded monthly. Find the balance after ten years by using the formula

$$\sum_{i=0}^{n-1} ar^i = \frac{a(1-r^n)}{1-r}$$

for the *n*th partial sum of a geometric series.

Solution

The balance *A* in the account at the end of the 10 years is given by

$$A = 50\left[1 + \frac{0.12}{12}\right] + 50\left[1 + \frac{0.12}{12}\right]^2$$

$$+ \cdots + 50\left[1 + \frac{0.12}{12}\right]^{120}$$

$$A + 50 = 50 + 50\left[1 + \frac{0.12}{12}\right] + 50\left[1 + \frac{0.12}{12}\right]^2$$

$$+ \cdots + 50\left[1 + \frac{0.12}{12}\right]^{120}$$

$$= \sum_{i=0}^{120} 50\left[1 + \frac{0.12}{12}\right]^i = \frac{50[1 - (1.01)^{121}]}{1 - 1.01} = 11{,}666.95$$

Therefore, $A = 11{,}666.95 - 50 = \$11{,}616.95.$

9.4
The Integral Test and p-series

7. Determine the convergence or divergence of the p-series

$$1 + \frac{1}{2\sqrt{2}} + \frac{1}{3\sqrt{3}} + \frac{1}{4\sqrt{4}} + \frac{1}{5\sqrt{5}} + \cdots$$

Solution

Since

$$1 + \frac{1}{2\sqrt{2}} + \frac{1}{3\sqrt{3}} + \frac{1}{4\sqrt{4}} + \frac{1}{5\sqrt{5}} + \cdots = \sum_{n=1}^{\infty} \frac{1}{n\sqrt{n}} = \sum_{n=1}^{\infty} \frac{1}{n^{3/2}}$$

we have a p-series with $p = \frac{3}{2} > 1$. Therefore by Theorem 9.11 the series converges.

17. Determine the convergence or divergence of the p-series

$$\frac{\ln 2}{2} + \frac{\ln 3}{3} + \frac{\ln 4}{4} + \frac{\ln 5}{5} + \cdots$$

by the Integral Test.

Solution

First observe that the nth term of the series is

$$a_n = \frac{\ln n}{n}$$

Now, since

$$f(x) = \frac{\ln x}{x}$$

satisfies the conditions for the Integral Test, we have

$$\int_1^\infty \frac{\ln x}{x}\, dx = \int_1^\infty (\ln x)\frac{1}{x}\, dx = \left[\frac{(\ln x)^2}{2}\right]_1^\infty = \infty$$

Therefore the given series diverges.

23. Determine the convergence or divergence of the series $\displaystyle\sum_{n=1}^{\infty} \frac{1}{n\sqrt[3]{n}}$.

Solution

The nth term of the series is $a_n = \dfrac{1}{n\sqrt[3]{n}} = \dfrac{1}{n^{4/3}}$. Thus, by Theorem 9.11, it is a convergent p-series.

29.   Determine the convergence or divergence of the series
$$\sum_{n=1}^{\infty} \left(1 + \frac{1}{n} \right)^n.$$

Solution

From the definition of the irrational number e we have

$$\lim_{n \to \infty} a_n = \lim_{n \to \infty} \left(1 + \frac{1}{n} \right)^n = e \neq 0$$

Thus, by Theorem 9.8, the series diverges.

35.   Approximate the sum of the series $\sum_{n=1}^{\infty} ne^{-n^2}$ using the first four terms. Include an estimate of the maximum error for your approximation.

Solution

Since $f(x) = xe^{-x^2}$ satisfies the conditions for the Integral Test, we have

$$\int_1^{\infty} xe^{-x^2} \, dx = \frac{1}{2} \int_1^{\infty} e^{-x^2}(-2x) \, dx = \frac{1}{2} \lim_{b \to \infty} [e^{-x^2}]_1^b = \frac{1}{2e}$$

Thus, the series converges and has a sum S. Furthermore, since

$$R_4 \leq \int_4^{\infty} xe^{-x^2} \, dx = \frac{1}{2e^{16}} \approx 0$$

and $S_4 = e^{-1} + 2e^{-2} + 3e^{-3} + 4e^{-4} \approx 0.40488$

we conclude that $S = 0.4049$.

37.   How many terms are required in approximating the series
$$\sum_{n=1}^{\infty} \frac{1}{n^4} \text{ so that } R_N < 0.001?$$

Solution

Since $p = 4$, we know the p-series converges, and from Theorem 9.11 we know

$$R_N < \frac{N^{1-p}}{p - 1} = \frac{N^{-3}}{3} < \frac{1}{1000}$$

$$N^3 > \frac{1000}{3} \text{ or } N \geq 7.$$

Therefore, let $N = 7$.

9.5
Comparisons of series

3. Use the Direct Comparison Test to determine the convergence or divergence of the series

$$\sum_{n=2}^{\infty} \frac{1}{n-1}$$

Solution

This series resembles

$$\sum_{n=2}^{\infty} \frac{1}{n} \qquad \text{Divergent } p\text{-series}$$

Term-by-term comparison yields

$$a_n = \frac{1}{n} < \frac{1}{n-1} = b_n$$

Thus, it follows by the Direct Comparison Test that the series diverges.

9. Use the Direct Comparison Test to determine the convergence or divergence of the series

$$\sum_{n=0}^{\infty} \frac{1}{n!}$$

Solution

We compare the given series to the convergent p-series $\sum_{n=1}^{\infty} \frac{1}{n^2}$.
For $n > 3$ we have $n^2 < n!$ and therefore $\frac{1}{n!} < \frac{1}{n^2}$. Therefore, by the Direct Comparison Test, the given series converges.

13. Use the Limit Comparison Test to determine the convergence or divergence of the series

$$\sum_{n=1}^{\infty} \frac{n}{n^2+1}$$

Solution

This series can be compared with the series

$$\sum_{n=1}^{\infty} \frac{1}{n}$$

since the degree of the denominator is only one greater than the degree of the numerator. Furthermore, since

$$\lim_{n \to \infty} \frac{n/(n^2 + 1)}{1/n} = \lim_{n \to \infty} \frac{n^2}{n^2 + 1} = 1$$

and since the series

$$\sum_{n=1}^{\infty} \frac{1}{n}$$

diverges, we conclude that the given series also diverges by the Limit Comparison Test.

21. Use the Limit Comparison Test to determine the convergence or divergence of the series

$$\sum_{n=1}^{\infty} \frac{1}{n\sqrt{n^2 + 1}}$$

Solution

Disregarding all but the highest power of n, we compare the series to

$$\sum_{n=1}^{\infty} \frac{1}{n\sqrt{n^2}} = \sum_{n=1}^{\infty} \frac{1}{n^2} \qquad \text{Convergent } p\text{-series}$$

Since

$$\lim_{n \to \infty} \frac{a_n}{b_n} = \lim_{n \to \infty} \left(\frac{1}{n\sqrt{n^2 + 1}} \right)\left(\frac{n^2}{1} \right) = \lim_{n \to \infty} \frac{1}{\sqrt{1 + (1/n)}} = 1$$

we conclude by the Limit Comparison Test that the given series converges.

31. Use one of the tests studied thus far to determine the convergence or divergence of the series

$$\sum_{n=1}^{\infty} \frac{n}{2n + 3}$$

Solution

Since

$$\lim_{n \to \infty} a_n = \lim_{n \to \infty} \frac{n}{2n + 3} = \frac{1}{2} \neq 0$$

we conclude that the series diverges by the nth-term Divergence Test.

35. Determine the values of p for which the series $\sum_{n=2}^{\infty} (\ln n)/n^p$ converges.

Solution

By the Integral Test $\sum_{n=2}^{\infty} (\ln n)/n^p$ converges if $\int_2^{\infty} [(\ln x)/x^p]\, dx$ converges. We integrate by parts, letting $dv = (1/x^p)\, dx$ and $u = \ln x$. Then $v = x^{1-p}/(1-p)$ and $du = (1/x)\, dx$. Thus

$$\int_2^{\infty} \frac{\ln x}{x^p}\, dx = \left[\frac{x^{1-p} \ln x}{1-p}\right]_2^{\infty} - \frac{1}{1-p} \int_2^{\infty} x^{-p}\, dx$$

$$= \left[\frac{x^{1-p} \ln x}{1-p} - \frac{x^{1-p}}{(1-p)^2}\right]_2^{\infty}$$

Therefore the integral, and thus the series, converges if $p > 1$.

40. Use the Polynomial Test of Exercise 38 to determine the convergence or divergence of the series

$$\frac{1}{3} + \frac{1}{8} + \frac{1}{15} + \frac{1}{24} + \frac{1}{35} + \cdots .$$

Solution

The nth term of the series is

$$a_n = \frac{1}{n^2 - 1} \qquad (n \geq 2)$$

and the series can be written

$$\sum_{n=2}^{\infty} \frac{1}{n^2 - 1} = \sum_{n=2}^{\infty} \frac{P(n)}{Q(n)}$$

In this case $P(n)$ has degree $j = 0$ and $q(n)$ has degree $k = 2$. Hence, we conclude by the Polynomial Test that the given series converges.

9.6

Alternating series

3. Use the Alternating Series Test to determine the convergence or divergence of

$$\sum_{n=1}^{\infty} \frac{(-1)^{n+1}}{2n - 1}$$

Solution

We observe that

$$\lim_{n \to \infty} \frac{1}{2n - 1} = 0$$

and

$$a_{n+1} = \frac{1}{2(n + 1) - 1} = \frac{1}{2n + 1} < \frac{1}{2n - 1} = a_n$$

for all n. Therefore, by the Alternating Series Test the given series converges.

8. Use the Alternating Series Test to determine the convergence or divergence of

$$\sum_{n=1}^{\infty} \frac{(-1)^{n+1} n^3}{n^3 + 6}$$

Solution

Since

$$\lim_{n \to \infty} a_n = \lim_{n \to \infty} \frac{n^3}{n^3 + 6} = 1 \neq 0$$

the series diverges.

21. Use the Alternating Series Test to determine the convergence or divergence of

$$\sum_{n=1}^{\infty} \frac{2(-1)^{n+1}}{e^n + e^{-n}}$$

Solution

We begin by using differentiation to establish that $a_{n+1} \leq a_n$. In this case we let $f(x) = \dfrac{2}{e^x + e^{-x}}$. Then

$$f'(x) = \frac{-2(e^x - e^{-x})}{(e^x + e^{-x})^2}$$

is negative for $x > 0$. Hence, f is a decreasing function and it follows that $a_{n+1} \leq a_n$ for $n \geq 1$. Furthermore, since

$$\lim_{n \to \infty} \frac{2}{e^n + e^{-n}} = 0$$

the series converges by the Alternating Series Test.

25. Examine

$$\sum_{n=1}^{\infty} \frac{(-1)^{n+1}}{\sqrt{n}}$$

for conditional convergence or absolute convergence.

Solution

In this case, the Alternating Series Test indicates that the given series converges. However, the series

$$\sum_{n=1}^{\infty} \left| \frac{(-1)^{n+1}}{\sqrt{n}} \right| = \sum_{n=1}^{\infty} \frac{1}{\sqrt{n}}$$

is a divergent p-series with $p = \frac{1}{2}$. Therefore, we conclude that the given series is conditionally convergent.

29. Examine

$$\sum_{n=2}^{\infty} \frac{(-1)^n n}{n^3 - 1}$$

for conditional convergence or absolute convergence.

Solution

In this case, the Alternating Series Test indicates that the given series converges. Also, the series

$$\sum_{n=2}^{\infty} \left| \frac{(-1)^{n+1} n}{n^3 - 1} \right| = \sum_{n=2}^{\infty} \frac{n}{n^3 - 1}$$

converges by the Limit Comparison Test with the series $\sum_{n=2}^{\infty} \frac{1}{n^2}$.

Therefore, we conclude that the given series is absolutely convergent.

41. Approximate the sum of the series

$$\sum_{n=0}^{\infty} \frac{(-1)^n}{n!}$$

with an error less than 0.001. (The actual sum is $1/e$.)

Solution

The series is alternating, and since

$$\lim_{n \to \infty} \frac{1}{n!} = 0 \quad \text{and} \quad \frac{1}{(n+1)!} < \frac{1}{n!}$$

the given series converges. By Theorem 9.15 the error, R_N, after N terms is such that $|R_N| \leq a_{N+1}$. Therefore, to ensure an error less than 0.001, we choose N sufficiently large so that

$$\frac{1}{(N+1)!} \leq 0.001$$

Since $1/6! = 1/720 = 0.0013888$ and $1/7! = 1/5040 < 0.001$, we choose $N = 6$ and obtain the approximation

$$\sum_{n=0}^{6} \frac{(-1)^n}{n!} = 1 - 1 + \frac{1}{2} - \frac{1}{6} + \frac{1}{24} - \frac{1}{120} + \frac{1}{720} \approx 0.368$$

which is within 0.001 of the actual sum.

9.7
The Ratio and Root Tests

2. Use the Ratio Test to test for convergence or divergence of $\sum_{n=1}^{\infty} n(\frac{2}{3})^n$.

Solution

Since

$$\lim_{n \to \infty} \left| \frac{a_{n+1}}{a_n} \right| = \lim_{n \to \infty} \frac{(n+1)(\frac{2}{3})^{n+1}}{n(\frac{2}{3})^n} = \lim_{n \to \infty} \left[\frac{n+1}{n} \left(\frac{2}{3} \right) \right] = \frac{2}{3} < 1$$

we conclude by the Ratio Test that the series $\sum_{n=1}^{\infty} n(\frac{2}{3})^n$ converges.

3. Use the Ratio Test to test for convergence or divergence of $\sum_{n=0}^{\infty} 3^n/n!$.

Solution

Since

$$\lim_{n \to \infty} \left| \frac{a_{n+1}}{a_n} \right|$$
$$= \lim_{n \to \infty} \left| \frac{3^{n+1}/(n+1)!}{3^n/n!} \right|$$
$$= \lim_{n \to \infty} \left| \frac{3^{n+1}}{1 \cdot 2 \cdot 3 \cdot 4 \cdots n \cdot (n+1)} \cdot \frac{1 \cdot 2 \cdot 3 \cdot 4 \cdots n}{3^n} \right|$$
$$= \lim_{n \to \infty} \left(\frac{3}{n+1} \right) = 0 < 1$$

we conclude by the Ratio Test that the series $\sum_{n=0}^{\infty} 3^n/n!$ converges.

7. Use the Ratio Test to test for convergence or divergence of

$$\sum_{n=0}^{\infty} \frac{(-1)^{n+1} n!}{1 \cdot 3 \cdot 5 \cdots (2n + 1)}$$

Solution

Since

$$\lim_{n \to \infty} \left| \frac{a_{n+1}}{a_n} \right|$$

$$= \lim_{n \to \infty} \left[\frac{(n + 1)!}{1 \cdot 3 \cdot 5 \cdots (2n + 1)(2n + 3)} \cdot \frac{1 \cdot 3 \cdot 5 \cdots (2n + 1)}{n!} \right]$$

$$= \lim_{n \to \infty} \frac{(n + 1)}{(2n + 3)} = \frac{1}{2}$$

we conclude by the Ratio Test that the given series converges.

8. Use the Ratio Test to test for convergence or divergence of

$$\sum_{n=1}^{\infty} \frac{(-1)^n 2 \cdot 4 \cdot 6 \cdots (2n)}{2 \cdot 5 \cdot 8 \cdots (3n - 1)}$$

Solution

We use the Ratio Test and first observe that

$$a_{n+1} = \frac{(-1)^{n+1} 2 \cdot 4 \cdot 6 \cdots (2n) \cdot [2(n + 1)]}{2 \cdot 5 \cdot 8 \cdots (3n - 1) \cdot [3(n + 1) - 1]}$$

$$= \frac{(-1)^{n+1} 2 \cdot 4 \cdot 6 \cdots (2n) \cdot (2n + 2)}{2 \cdot 5 \cdot 8 \cdots (3n - 1) \cdot (3n + 2)}$$

Now

$$\left| \frac{a_{n+1}}{a_n} \right|$$

$$= \left| \frac{(-1)^{n+1} 2 \cdot 4 \cdot 6 \cdots (2n) \cdot (2n + 2)}{2 \cdot 5 \cdot 8 \cdots (3n - 1) \cdot (3n + 2)} \cdot \frac{2 \cdot 5 \cdot 8 \cdots (3n - 1)}{(-1)^n 2 \cdot 4 \cdot 6 \cdots (2n)} \right|$$

$$= \frac{2n + 2}{3n + 2}$$

Therefore

$$\lim_{n \to \infty} \left| \frac{a_{n+1}}{a_n} \right| = \lim_{n \to \infty} \left(\frac{2n + 2}{3n + 2} \right) = \frac{2}{3} < 1$$

and we conclude by the Ratio Test that the series converges.

19. Use the Ratio Test to test for convergence or divergence of

$$\sum_{n=0}^{\infty} \frac{4^2}{3^n + 1}$$

Solution

Since $\lim\limits_{n \to \infty} \left| \dfrac{a_{n+1}}{a_n} \right| = \lim\limits_{n \to \infty} \left[\dfrac{4^2}{3^{n+1} + 1} \cdot \dfrac{3^n + 1}{4^2} \right]$

$$= \lim_{n \to \infty} \left[\frac{3^n + 1}{3^{n+1} + 1} \right]$$

$$= \lim_{n \to \infty} \left[\frac{1 + (1/3^n)}{3 + (1/3^n)} \right] = \frac{1}{3} < 1$$

we conclude by the Ratio Test that the given series converges.

21. Use the Root Test to test for convergence or divergence of

$$\sum_{n=0}^{\infty} \left(\frac{n}{2n + 1} \right)^n$$

Solution

$$\lim_{n \to \infty} \sqrt[n]{a_n} = \lim_{n \to \infty} \sqrt[n]{\left(\frac{n}{2n + 1} \right)^n} = \lim_{n \to \infty} \left(\frac{n}{2n + 1} \right) = \frac{1}{2} < 1$$

Therefore, we conclude by the Root Test that the series converges.

24. Use the Root Test to test for convergence or divergence of $\sum_{n=1}^{\infty} (\sqrt[n]{n} - 1)^n$.

Solution

Using the Root Test, we have

$$\lim_{n \to \infty} \sqrt[n]{a_n} = \lim_{n \to \infty} \sqrt[n]{(\sqrt[n]{n} - 1)^n} = \lim_{n \to \infty} (\sqrt[n]{n} - 1)$$

Now we let $y = \sqrt[n]{n} = n^{1/n}$. Then $\ln y = (1/n) \ln n$ and

$$\lim_{n \to \infty} \left(\frac{\ln n}{n} \right) = \lim_{n \to \infty} \left(\frac{1/n}{1} \right) \qquad \text{L'Hôpital's Rule}$$

$$= 0$$

Therefore as n approaches infinity, $\ln y$ approaches zero and $y = \sqrt[n]{n}$ approaches one. Thus

$$\lim_{n \to \infty} (\sqrt[n]{n} - 1) = 1 - 1 = 0$$

and the series converges.

29. Test for convergence or divergence using any appropriate test from this chapter for the series

$$\sum_{n=1}^{\infty} \frac{(-1)^n 3^{n-2}}{2^n}$$

Identify the test used.

Solution

For the given series the Ratio Test is convenient and we obtain

$$\lim_{n \to \infty} \left| \frac{a_{n+1}}{a_n} \right| = \lim_{n \to \infty} \left(\frac{3^{n-1}}{2^{n+1}} \right) \left(\frac{2^n}{3^{n-2}} \right) = \lim_{n \to \infty} \frac{3}{2} = \frac{3}{2} > 1$$

Therefore by the Ratio Test the given series diverges.

31. Test for convergence or divergence using any appropriate test from this chapter for the series

$$\sum_{n=1}^{\infty} \frac{10n + 3}{n \, 2^n}$$

Identify the test used.

Solution

We compare the given series to the convergent geometric series $\sum_{n=0}^{\infty} 1/2^n$. Since

$$\lim_{n \to \infty} \frac{(10n + 3)/n \, 2^n}{1/2^n} = \lim_{n \to \infty} \frac{10n + 3}{n} = 10$$

we conclude by the Limit Comparison Test that the given series also converges.

35. Test for convergence or divergence using any appropriate test from this chapter for the series $\sum_{n=1}^{\infty} \cos(n)/2^n$. Identify the test used.

Solution

This is not an alternating series; hence we use the Comparison Test and observe that

$$\left| \frac{\cos(n)}{2^n} \right| \le \frac{1}{2^n}$$

Now since the series $\sum_{n=0}^{\infty} 1/2^n$ is a convergent geometric series, we conclude by the Comparison Test that

$$\sum_{n=1}^{\infty} \left| \frac{\cos(n)}{2^n} \right|$$

also converges. Finally, by Theorem 9.16, since $\sum_{n=1}^{\infty} |\cos(n)/2^n|$ converges, then $\sum_{n=1}^{\infty} \cos(n)/2^n$ also converges.

9.8
Power series

3. Find the interval of convergence for the power series

$$\sum_{n=1}^{\infty} \frac{(-1)^n x^n}{n}$$

Include a check for convergence at the endpoints of the interval.

Solution

Since

$$\lim_{n \to \infty} \left| \frac{a_{n+1}}{a_n} \right| = \lim_{n \to \infty} \left[\frac{1/(n+1)}{1/n} \right] = \lim_{n \to \infty} \left[\frac{1}{n+1} \cdot \frac{n}{1} \right]$$

$$= \lim_{n \to \infty} \frac{n}{n+1} = 1$$

we conclude by the Ratio Test that the radius of convergence is 1 and that the interval of convergence includes $-1 < x < 1$. When $x = -1$, we have the harmonic series $\sum_{n=1}^{\infty} (1/n)$, which diverges. When $x = 1$, we have the alternating series $\sum_{n=1}^{\infty} (-1)^n/n$, which converges. Thus the interval of convergence is $-1 < x \leq 1$.

6. Find the interval of convergence for the power series

$$\sum_{n=0}^{\infty} \frac{(3x)^n}{(2n)!}$$

Include a check for convergence at the endpoints of the interval.

Solution

Since

$$\lim_{n \to \infty} \left| \frac{a_{n+1}}{a_n} \right| = \lim_{n \to \infty} \frac{(2n)!}{(2n+2)!} = \lim_{n \to \infty} \frac{1}{(2n+2)(2n+1)} = 0$$

we conclude by the Ratio Test that the radius of convergence is ∞ and that the interval of convergence is $-\infty < x < \infty$.

7. Find the interval of convergence for the power series

$$\sum_{n=0}^{\infty} (2n)! \left(\frac{x}{2}\right)^n$$

Include a check for convergence at the endpoints of the interval.

Solution

Since

$$\lim_{n \to \infty} \left| \frac{a_{n+1}}{a_n} \right| = \lim_{n \to \infty} \left| \frac{(2n+2)!/2^{n+1}}{(2n)!/2^n} \right| = \lim_{n \to \infty} \frac{(2n+2)(2n+1)}{2} = \infty$$

Therefore, by the Ratio Test we conclude that the series converges only when $x = 0$.

11. Find the interval of convergence for the power series

$$\sum_{n=1}^{\infty} \frac{(-1)^{n+1}(x-5)^n}{n\,5^n}$$

Include a check for convergence at the endpoints of the interval.

Solution

Since

$$\lim_{n \to \infty} \left| \frac{a_{n+1}}{a_n} \right| = \lim_{n \to \infty} \frac{n\,5^n}{(n+1)5^{n+1}} = \lim_{n \to \infty} \frac{1}{5}\left(\frac{n}{n+1}\right) = \frac{1}{5}$$

the radius of convergence is $R = 5$, and since the series is centered at $x = 5$, the series will converge in the interval $(0, 10)$. Furthermore, when $x = 0$ we have the series

$$\sum_{n=1}^{\infty} \frac{(-1)^{n+1}(-1)^n}{n} = \sum_{n=1}^{\infty} \frac{(-1)^{2n+1}}{n} = -\sum_{n=1}^{\infty} \frac{1}{n}$$

which diverges $(p = 1)$. When $x = 10$, we have the series

$$\sum_{n=1}^{\infty} \frac{(-1)^{n+1}}{n}$$

which converges by the Alternating Series Test. Hence the interval of convergence of the given series is $0 < x \le 10$.

17. Find the interval of convergence for the power series

$$\sum_{n=1}^{\infty} \frac{(-1)^{n+1} x^{2n-1}}{2n-1}$$

Include a check for convergence at the endpoints of the interval.

Solution

Since

$$\lim_{n \to \infty} \left| \frac{a_{n+1}}{a_n} \right| = \lim_{n \to \infty} \left| \frac{1/[2(n+1)-1]}{1/(2n-1)} \right| = \lim_{n \to \infty} \left(\frac{2n-1}{2n+1} \right) = 1$$

we conclude by the Ratio Test that the radius of convergence is 1 and that the interval of convergence includes $-1 < x < 1$. When $x = -1$, we have the series

$$\sum_{n=1}^{\infty} \frac{(-1)^{n+1}(-1)^{2n-1}}{2n-1} = \sum_{n=1}^{\infty} \frac{(-1)^n}{2n-1}$$

which converges by the Alternating Series Test. When $x = 1$, we have the series

$$\sum_{n=1}^{\infty} \frac{(-1)^{n+1}}{2n-1}$$

which converges by the Alternating Series Test. Therefore the interval of convergence is $-1 \le x \le 1$.

23. Find the interval of convergence for the power series

$$\sum_{n=1}^{\infty} \frac{k(k+1)(k+2) \cdots (k+n-1)x^n}{n!} \qquad (k \ge 1)$$

Include a check for convergence at the endpoints of the interval.

Solution

Since

$$\lim_{n \to \infty} \left| \frac{a_{n+1}}{a_n} \right| = \lim_{n \to \infty} \left[\frac{k(k+1)(k+2) \cdots (k+n-1)(k+n)}{(n+1)!} \right.$$
$$\left. \times \frac{n!}{k(k+1)(k+2) \cdots (k+n-1)} \right]$$
$$= \lim_{n \to \infty} \frac{k+n}{n+1} = 1$$

the radius of convergence is $R = 1$. Since the series is centered at $x = 0$, it will converge on the interval $(-1, 1)$. To test for convergence at the endpoints, we note that for $k \ge 1$, we have

$$\lim_{n \to \infty} a_n = \lim_{n \to \infty} \left[\left(\frac{k}{1} \right) \left(\frac{k+1}{2} \right) \left(\frac{k+2}{3} \right) \cdots \left(\frac{k+n-1}{n} \right) \right] \ne 0$$

Thus for $x = \pm 1$ the series diverges, and we conclude that the interval of convergence is $-1 < x < 1$.

25. Find the interval of convergence (check for convergence at the endpoints) of (a) $f(x)$, (b) $f'(x)$, (c) $f''(x)$, and (d) $\int f(x)\, dx$ if

$$f(x) = \sum_{n=0}^{\infty} \left(\frac{x}{2}\right)^2$$

Solution

(a) The given series is geometric with $r = x/2$ and converges if

$$\left|\frac{x}{2}\right| < 1$$

or $-2 < x < 2$

(b) $f'(x) = \sum_{n=1}^{\infty} n\left(\frac{x}{2}\right)^{n-1} \left(\frac{1}{2}\right) = \sum_{n=1}^{\infty} \left(\frac{n}{2}\right)\left(\frac{x}{2}\right)^{n-1}$

Therefore the series for $f'(x)$ diverges for $x = \pm 2$, and its interval of convergence is $-2 < x < 2$.

(c) $f''(x) = \sum_{n=2}^{\infty} \left(\frac{n}{2}\right)(n-1)\left(\frac{x}{2}\right)^{n-2}\left(\frac{1}{2}\right)$

$$= \sum_{n=2}^{\infty} \left[\frac{n(n-1)}{4}\right]\left(\frac{x}{2}\right)^{n-2}$$

Therefore the series for $f''(x)$ diverges for $x = \pm 2$, and its interval of convergence is $-2 < x < 2$.

(d) The series for $\int f(x)\, dx$ is

$$\sum_{n=0}^{\infty} \frac{2}{n+1}\left(\frac{x}{2}\right)^{n+1}$$

and converges (Alternating Series Test) for $x = -2$ and diverges (Limit Comparison Test with $\sum_{n=1}^{\infty} 1/n$) for $x = 2$. Therefore its interval of convergence is $-2 \leq x < 2$.

29. Let

$$f(x) = \sum_{n=0}^{\infty} \frac{(-1)^n x^{2n}}{(2n+1)!}$$

Approximate $\int_0^1 f(x)\, dx$, using $R_N \leq 0.001$.

Solution

Since

$$\lim_{n \to \infty} \left| \frac{a_{n+1}}{a_n} \right| = \lim_{n \to \infty} \left[\frac{1}{(2n+3)!} \cdot \frac{(2n+1)!}{1} \right]$$

$$= \lim_{n \to \infty} \frac{1}{(2n+3)(2n+2)} = 0$$

the interval of convergence is $(-\infty, \infty)$.

$$\int f(x) \, dx = \sum_{n=0}^{\infty} \left[\frac{(-1)^n}{(2n+1)!} \cdot \frac{x^{2n+1}}{2n+1} \right] = \sum_{n=0}^{\infty} \frac{(-1)^n x^{2n+1}}{(2n+1)(2n+1)!}$$

$$\int_0^1 f(x) \, dx \approx \left[\frac{x}{1} - \frac{x^3}{3(3!)} + \frac{x^5}{5(5!)} \right]_0^1 = 1 - \frac{1}{18} + \frac{1}{600} \approx 0.946$$

Since $R_3 \le a_4 = 1/35{,}280 < 0.001$, we let $N = 3$.

9.9
Power series for functions

3. Find a power series centered at $c = 5$ for $f(x) = 1/(2 - x)$, and find the interval of convergence.

Solution

If we write $\quad f(x) = \dfrac{1}{2 - x} = \dfrac{-1}{x - 2}$

then, by applying Theorem 9.21 we have $a = -1$, $b = 2$, and $c = 5$, and the series for $f(x)$ is

$$\frac{-1}{x - 2} = \frac{1}{-3} \sum_{n=0}^{\infty} \left(\frac{x - 5}{-3} \right)^n = \sum_{n=0}^{\infty} \frac{(x - 5)^n}{(-3)^{n+1}}$$

Since

$$\lim_{n \to \infty} \left| \frac{a_{n+1}}{a_n} \right| = \lim_{n \to \infty} \left| \frac{1}{(-3)^{n+2}} \cdot \frac{(-3)^{n+1}}{1} \right| = \frac{1}{3}$$

the radius of convergence is $R = 3$. Since the series is centered at $c = 5$, it converges in the interval $(2, 8)$. Finally, since the series diverges at both endpoints, the interval of convergence is $2 < x < 8$.

11. Find a power series centered at $c = 0$ for $f(x) = 3x/(x^2 + x - 2)$, and find the interval of convergence.

Solution

Using the method for finding partial fractions (Section 8.5), we have

$$\frac{3x}{x^2 + x - 2} = \frac{2}{x + 2} + \frac{1}{x - 1} = \frac{2}{x - (-2)} + \frac{1}{x - 1}$$

Therefore, by Theorem 9.21 we have

$$\frac{2}{x - (-2)} + \frac{1}{x - 1} = -\frac{-2}{2} \sum_{n=0}^{\infty} \left(\frac{x}{-2}\right)^n - \frac{1}{1} \sum_{n=0}^{\infty} \left(\frac{x}{1}\right)^n$$

$$= \sum_{n=0}^{\infty} \left[\frac{x^n}{(-2)^n} - x^n\right] = \sum_{n=0}^{\infty} \left[\frac{1}{(-2)^n} - 1\right] x^n$$

Now since

$$\lim_{n \to \infty} \left|\frac{a_{n+1}}{a_n}\right| = \lim_{n \to \infty} \left|\frac{[1/(-2)^{n+1}] - 1}{[1/(-2)^n] - 1}\right| = 1$$

we have $R = 1$. Finally, when $x = \pm 1$, we have

$$\lim_{n \to \infty} a_n = \lim_{n \to \infty} \left[\frac{1}{(-2)^n} - 1\right] \neq 0$$

Thus the series diverges when $x = \pm 1$ and the interval of convergence is $-1 < x < 1$.

13. Find a power series centered at $c = 0$ for $f(x) = 2/(1 - x^2)$, and find the interval of convergence.

Solution

Letting $u = x^2$, we have

$$\frac{2}{1 - x^2} = \frac{-2}{u - 1}$$

$$= 2 \sum_{n=0}^{\infty} u^n \qquad \text{Theorem 9.21}$$

$$= 2 \sum_{n=0}^{\infty} (x^2)^n = 2 \sum_{n=0}^{\infty} x^{2n}$$

The series will converge if $x^2 < 1$ or $-1 < x < 1$.

17. Use the power series

$$\frac{1}{x + 1} = \sum_{n=0}^{\infty} (-1)^n x^n$$

to determine a power series representation about $c = 0$ for $f(x) = \ln (x + 1)$. Specify the interval of convergence.

Solution

Since

$$\int \frac{1}{x + 1} \, dx = \ln (x + 1) + C$$

we can integrate the power series for $1/(x + 1)$ to obtain the
series for $\ln (x + 1)$ as follows:

$$\frac{1}{x + 1} = \sum_{n=0}^{\infty} (-1)^n x^n$$

$$\ln (x + 1) = \sum_{n=0}^{\infty} \frac{(-1)^n x^{n+1}}{n + 1} + C$$

By substituting $x = 0$ on both sides of this equation, we
determine that $C = 0$. Furthermore, since

$$\lim_{n \to \infty} \left| \frac{a_{n+1}}{a_n} \right| = \lim_{n \to \infty} \left[\frac{1}{n + 2} \cdot \frac{n + 1}{1} \right] = 1$$

we conclude that $R = 1$ and the series converges in the interval
$(-1, 1)$. At $x = -1$ we have the divergent series

$$\sum_{n=0}^{\infty} \frac{(-1)^{2n+1}}{n + 1} = -\sum_{n=0}^{\infty} \frac{1}{n + 1}$$

At $x = 1$ we have the convergent alternating series

$$\sum_{n=0}^{\infty} \frac{(-1)^n}{n + 1}$$

Therefore

$$\ln (x + 1) = \sum_{n=0}^{\infty} \frac{(-1)^n x^{n+1}}{n + 1} \qquad (-1 < x \leq 1)$$

27.   Use the series for the function $f(x) = \arctan x$ to approximate
(with $R_N \leq 0.001$)

$$\int_0^{1/2} \frac{\arctan x^2}{x} \, dx$$

Solution

From Example 5 we have

$$\arctan x = \sum_{n=0}^{\infty} \frac{(-1)^n x^{2n+1}}{2n + 1}$$

Therefore

$$\arctan x^2 = \sum_{n=0}^{\infty} \frac{(-1)^n x^{2(2n+1)}}{2n + 1}$$

and

$$\frac{\arctan x^2}{x} = \sum_{n=0}^{\infty} \frac{(-1)^n x^{4n+2}}{2n + 1}$$

Thus

$$\int_0^{1/2} \frac{\arctan x^2}{x} \, dx = \left[\sum_{n=0}^{\infty} \frac{(-1)^n x^{4n+2}}{(2n+1)(4n+2)} \right]_0^{1/2}$$

$$= \sum_{n=0}^{\infty} \frac{(-1)^n (1/2)^{4n+2}}{(2n+1)(4n+2)}$$

$$\approx \left(\frac{1}{2}\right)\left(\frac{1}{2}\right)^2 \approx \frac{1}{8}$$

Note that the second term of the series is $-(\frac{1}{18})(\frac{1}{2})^6 < 0.001$. Therefore, by Theorem 9.15 one term is sufficient to approximate the integral.

9.10
Taylor and Maclaurin Series

7. Find the Taylor Series centered at $c = 0$ for the function $f(x) = \sin 2x$.

Solution

Since

$$\begin{array}{ll} f(x) = \sin 2x & f(0) = 0 \\ f'(x) = 2 \cos 2x & f'(0) = 2 \\ f''(x) = -4 \sin 2x & f''(0) = 0 \\ f'''(x) = -8 \cos 2x & f'''(0) = -8 = -2^3 \\ f^{(4)}(x) = 16 \sin 2x & f^{(4)}(0) = 0 \\ f^{(5)}(x) = 32 \cos 2x & f^{(5)}(0) = 32 = 2^5 \end{array}$$

we can see that the signs alternate and that $|f^{(n)}(0)| = 2^n$ if n is odd. Therefore the Taylor Series is

$\sin 2x$

$$= f(0) + f'(0)x + \frac{f''(0)x^2}{2!} + \frac{f'''(0)x^3}{3!} + \frac{f^{(4)}(0)x^4}{4!} + \frac{f^{(5)}(0)x^5}{5!} + \cdots$$

$$= \frac{2x}{1!} - \frac{2^3 x^3}{3!} + \frac{2^5 x^5}{5!} - \cdots + \frac{(-1)^n (2x)^{2n+1}}{(2n+1)!} + \cdots$$

$$= \sum_{n=0}^{\infty} \frac{(-1)^n (2x)^{2n+1}}{(2n+1)!}$$

Note that we could have arrived at the same result by substituting $2x$ into the series for $\sin x$ as follows:

$$\sin x = x - \frac{x^3}{3!} + \frac{x^5}{5!} - \frac{x^7}{7!} + \cdots$$

$$\sin (2x) = (2x) - \frac{(2x)^3}{3!} + \frac{(2x)^5}{5!} - \frac{(2x)^7}{7!} + \cdots$$

$$= \sum_{n=0}^{\infty} \frac{(-1)^n (2x)^{2n+1}}{(2n + 1)!}$$

13. Use the binomial series to find the power series centered at $c = 0$ for the function $f(x) = 1/\sqrt{4 + x^2}$.

Solution

Consider f in the form

$$f(x) = \frac{1}{\sqrt{4 + x^2}} = \frac{1}{2\sqrt{1 + (x/2)^2}} = \frac{1}{2}\left[1 + \left(\frac{x}{2}\right)^2\right]^{-1/2}$$

which is similar to the binomial form $(1 + x)^{-k}$. Since

$$(1 + x)^{-k} = 1 - kx + \frac{k(k + 1)x^2}{2!} - \frac{k(k + 1)(k + 2)x^3}{3!} + \cdots$$

we have, for $k = \frac{1}{2}$,

$$(1 + x)^{-1/2} = 1 - \frac{1}{2}x + \frac{(\frac{1}{2})(\frac{3}{2})x^2}{2!} - \frac{(\frac{1}{2})(\frac{3}{2})(\frac{5}{2})x^3}{3!} + \cdots$$

$$= 1 - \frac{x}{2} + \frac{1 \cdot 3x^2}{2^2 2!} - \frac{1 \cdot 3 \cdot 5x^3}{2^3 3!} + \cdots$$

$$= 1 + \sum_{n=1}^{\infty} \frac{(-1)^{n+1} \, 1 \cdot 3 \cdot 5 \cdots (2n - 1)x^n}{2^n n!}$$

Now by substituting $(x/2)^2$ for x, we obtain

$$f(x) = \frac{1}{\sqrt{4 + x^2}} = \frac{1}{2}\left[1 + \left(\frac{x}{2}\right)^2\right]^{-1/2}$$

$$= \frac{1}{2}\left[1 + \sum_{n=1}^{\infty} \frac{(-1)^{n+1} \, 1 \cdot 3 \cdot 5 \cdots (2n - 1)(x/2)^{2n}}{2^n n!}\right]$$

$$= \frac{1}{2} + \sum_{n=1}^{\infty} \frac{(-1)^{n+1} \, 1 \cdot 3 \cdot 5 \cdots (2n - 1)x^{2n}}{2^{3n+1} n!}$$

19. Find the power series for $f(x) = e^{-x^2/2}$ by using the series for e^x.

Solution

Since $e^x = 1 + x + \frac{x^2}{2!} + \frac{x^3}{3!} + \frac{x^4}{4!} + \frac{x^5}{5!} + \cdots$

we can substitute $(-x^2/2)$ for x and obtain the series

$$e^{-x^2/2} = 1 - \frac{x^2}{2} + \frac{(-x^2/2)^2}{2!} + \frac{(-x^2/2)^3}{3!} + \frac{(-x^2/2)^4}{4!} + \cdots$$

$$= 1 - \frac{x^2}{2} + \frac{x^4}{2^2 2!} - \frac{x^6}{2^3 3!} + \frac{x^8}{2^4 4!} - \cdots$$

$$= \sum_{n=0}^{\infty} \frac{(-1)^n x^{2n}}{2^n n!}$$

25. Find the power series for $f(x) = (\sin x)/x$.

Solution

Since $\sin x = x - \dfrac{x^3}{3!} + \dfrac{x^5}{5!} - \dfrac{x^7}{7!} + \dfrac{x^9}{9!} - \cdots$

we can divide by x to obtain

$$\frac{\sin x}{x} = 1 - \frac{x^2}{3!} + \frac{x^4}{5!} - \frac{x^6}{7!} + \frac{x^8}{9!} - \cdots = \sum_{n=0}^{\infty} \frac{(-1)^n x^{2n}}{(2n + 1)!}$$

27. Use power series to approximate

$$\int_0^{\pi/2} \frac{\sin x}{x}\, dx$$

accurate to four decimal places.

Solution

From Exercise 25 we have

$$\frac{\sin x}{x} = \sum_{n=0}^{\infty} \frac{(-1)^n x^{2n}}{(2n + 1)!}$$

Therefore

$$\int_0^{\pi/2} \frac{\sin x}{x}\, dx = \left[\sum_{n=0}^{\infty} \frac{(-1)^n x^{2n+1}}{(2n + 1)(2n + 1)!} \right]_0^{\pi/2}$$

$$= \sum_{n=0}^{\infty} \frac{(-1)^n (\pi/2)^{2n+1}}{(2n + 1)(2n + 1)!}$$

$$\approx \frac{\pi}{2} - \frac{\pi^3}{3 \cdot 3! \cdot 2^3} + \frac{\pi^5}{5 \cdot 5! \cdot 2^5} + \frac{\pi^7}{7 \cdot 7! \cdot 2^7}$$

$$= 1.3708$$

Review Exercises for Chapter 9

6. Determine the convergence or divergence of $\{a_n\}$ when $a_n = n/\ln n$.

   Solution

   Since
   $$\lim_{n\to\infty} a_n = \lim_{n\to\infty} \frac{n}{\ln n} = \frac{\infty}{\infty}$$

   we can apply L'Hôpital's Rule and obtain
   $$\lim_{n\to\infty} \frac{n}{\ln n} = \lim_{n\to\infty} \frac{1}{(1/n)} = \infty$$

   Therefore the sequence $\{n/\ln n\}$ diverges.

7. Determine the convergence or divergence of $\{a_n\}$ when $a_n = \sqrt{n+1} - \sqrt{n}$.

   Solution
   $$\begin{aligned} \lim_{n\to\infty} a_n &= \lim_{n\to\infty} \left[\sqrt{n+1} - \sqrt{n}\right] \\ &= \lim_{n\to\infty} \left[(\sqrt{n+1} - \sqrt{n})\frac{\sqrt{n+1} + \sqrt{n}}{\sqrt{n+1} + \sqrt{n}}\right] \\ &= \lim_{n\to\infty} \frac{1}{\sqrt{n+1} + \sqrt{n}} = 0 \end{aligned}$$

   Therefore the sequence converges to 0.

13. Find the first five terms of the sequence of partial sums for the series
    $$\sum_{n=1}^{\infty} \frac{(-1)^{n+1}}{(2n)!}$$

    Solution

    Since
    $$\begin{aligned} \sum_{n=1}^{\infty} \frac{(-1)^{n+1}}{(2n)!} &= \frac{1}{2!} - \frac{1}{4!} + \frac{1}{6!} - \frac{1}{8!} + \frac{1}{10!} - \cdots \\ &= \frac{1}{2} - \frac{1}{24} + \frac{1}{720} - \frac{1}{40,320} + \frac{1}{3,628,800} + \cdots \end{aligned}$$

    we have
    $$S_1 = \frac{1}{2} = 0.5$$

$$S_2 = \frac{1}{2} - \frac{1}{24} \approx 0.45833$$

$$S_3 = \frac{1}{2} - \frac{1}{24} + \frac{1}{720} \approx 0.45972$$

$$S_4 = \frac{1}{2} - \frac{1}{24} + \frac{1}{720} - \frac{1}{40,320} \approx 0.45970$$

$$S_5 = \frac{1}{2} - \frac{1}{24} + \frac{1}{720} - \frac{1}{40,320} + \frac{1}{3,628,800} \approx 0.45970$$

17. Find the sum of the infinite series

$$\sum_{n=0}^{\infty} \left(\frac{1}{2^n} - \frac{1}{3^n} \right)$$

Solution

We begin by writing the given series as the difference of two geometric series

$$\sum_{n=0}^{\infty} \left(\frac{1}{2^n} - \frac{1}{3^n} \right) = \sum_{n=0}^{\infty} \left(\frac{1}{2} \right)^n - \sum_{n=0}^{\infty} \left(\frac{1}{3} \right)^n$$

Since these two geometric series have the values $a = 1$ and $r = \frac{1}{2}$ and $\frac{1}{3}$, respectively, we conclude that the sum of the given series is

$$\sum_{n=0}^{\infty} \left(\frac{1}{2^n} - \frac{1}{3^n} \right) = \frac{1}{1 - (\frac{1}{2})} - \frac{1}{1 - (\frac{1}{3})} = 2 - \frac{3}{2} = \frac{1}{2}$$

25. Determine the convergence or divergence of the series

$$\sum_{n=1}^{\infty} \frac{1}{(n^3 + 2n)^{1/3}}$$

Solution

Since the denominator is the cube root of a third-degree polynomial, we compare the given series to the divergent harmonic series $\sum_{n=1}^{\infty} 1/n$. Thus

$$\lim_{n \to \infty} \frac{[1/(n^3 + 2n)^{1/3}]}{1/n} = \lim_{n \to \infty} \frac{n}{(n^3 + 2n)^{1/3}}$$

$$= \lim_{n \to \infty} \frac{1}{[1 + (2/n^2)]^{1/3}} = 1$$

and we conclude by the Limit Comparison Test that the given series diverges.

35. Find the interval of convergence of the power series $\sum_{n=0}^{\infty} n!(x - 2)^n$.

Solution

Since
$$\lim_{n \to \infty} \left| \frac{a_{n+1}}{a_n} \right| = \lim_{n \to \infty} \frac{(n + 1)!}{n!} = \lim_{n \to \infty} (n + 1) = \infty$$
we conclude that the radius of convergence is $R = 0$. Therefore the given series will converge only at $x = 2$.

39. Find the power series for $f(x) = 3^x$ centered at $c = 0$.

Solution

Since
$$3^x = e^{\ln(3^x)} = e^{x(\ln 3)}$$
we can substitute $x(\ln 3)$ for the x in the series
$$e^x = 1 + x + \frac{x^2}{2!} + \frac{x^3}{3!} + \frac{x^4}{4!} + \cdots$$
to obtain
$$3^x = e^{x(\ln 3)} = 1 + (\ln 3)x + \frac{(\ln 3)^2 x^2}{2!} + \frac{(\ln 3)^3 x^3}{3!} + \frac{(\ln 3)^4 x^4}{4!}$$
$$+ \cdots$$
$$= \sum_{n=0}^{\infty} \frac{(\ln 3)^n x^n}{n!}$$

10 Conics

10.1
Parabolas

9. Find the vertex, focus, and directrix of the parabola given by $y^2 = -6x$, and sketch its graph.

Solution

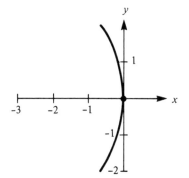

We begin by writing the equation of this parabola in standard form.

$$(y - k)^2 = 4p(x - h)$$
$$y^2 = -6x$$
$$(y - 0)^2 = 4\left(-\frac{3}{2}\right)(x - 0)$$

Thus $k = 0$, $h = 0$, and $p = -\frac{3}{2}$. We conclude that

$$\text{vertex, } (h, k): \quad (0, 0)$$
$$\text{focus, } (h + p, k): \quad \left(-\frac{3}{2}, 0\right)$$
$$\text{directrix, } x = h - p: \quad x = \frac{3}{2}$$

17. Find the vertex, focus, and directrix of the parabola given by $y = \frac{1}{4}(x^2 - 2x + 5)$, and sketch its graph.

Solution

We begin by writing the equation of this parabola in standard form.

$$(x - h)^2 = 4p(y - k)$$

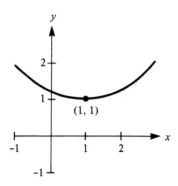

$$\frac{1}{4}(x^2 - 2x + 5) = y$$

$$x^2 - 2x + 5 = 4y$$

$$x^2 - 2x + 1 = 4y - 4$$

$$(x - 1)^2 = 4(1)(y - 1)$$

Thus $h = 1$, $k = 1$, and $p = 1$. We conclude that

$$\begin{aligned}
\text{vertex, } (h, k): &\quad (1, 1) \\
\text{focus, } (h, k + p): &\quad (1, 2) \\
\text{directrix, } y = k - p: &\quad y = 0
\end{aligned}$$

23. Find the vertex, focus, and directrix of the parabola given by $y^2 - 4y - 4x = 0$, and sketch its graph.

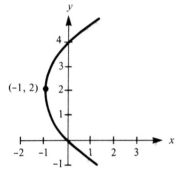

Solution

We begin by writing the equation of this parabola in standard form.

$$(y - k)^2 = 4p(x - h)$$

$$y^2 - 4y = 4x$$

$$y^2 - 4y + 4 = 4x + 4$$

$$(y - 2)^2 = 4(1)(x + 1)$$

Thus $h = -1$, $k = 2$, and $p = 1$. We conclude that

$$\begin{aligned}
\text{vertex, } (h, k): &\quad (-1, 2) \\
\text{focus, } (h + p, k): &\quad (0, 2) \\
\text{directrix, } x = h - p: &\quad x = -2
\end{aligned}$$

29. Find an equation of a parabola with vertex at $(3, 2)$ and focus at $(1, 2)$.

Solution

Since the vertex and focus lie on a horizontal line, the axis of this parabola must be horizontal and its standard form would be

$$(y - k)^2 = 4p(x - h)$$

Since the vertex lies at $(3, 2)$, we have $h = 3$ and $k = 2$. Furthermore, the *directed distance* from the focus to the vertex is

$$p = 1 - 3 = -2$$

Thus the standard equation is

$$(y - 2)^2 = 4(-2)(x - 3)$$
$$y^2 - 4y + 4 = -8x + 24$$
$$y^2 - 4y + 8x - 20 = 0$$

35. Find an equation for the parabola whose axis is parallel to the y-axis and passes through the points $(0, 3)$, $(3, 4)$, $(4, 11)$.

Solution

Since the axis of the parabola is vertical, the standard form is

$$(x - h)^2 = 4p(y - k)$$

A more convenient form for this equation is

$$y = ax^2 + bx + c$$

Now substituting the values of the given coordinates into this equation, we obtain three equations:

(i) $3 = a(0)^2 + b(0) + c$

(ii) $4 = a(3)^2 + b(3) + c$

(iii) $11 = a(4)^2 + b(4) + c$

Simplification yields

(i) $3 = c$

(ii) $4 = 9a + 3b + c \rightarrow 1 = 9a + 3b$

(iii) $11 = 16a + 4b + c \rightarrow 8 = 16a + 4b$

Solving (ii) and (iii) for a and b gives us

(ii) $9a + 3b = 1 \rightarrow \quad 9a + 3b = \quad 1$

(iii) $4a + \;\; b = 2 \rightarrow \underline{\;\;12a + 3b = \quad 6\;\;}$

$$-3a \qquad = \quad -5$$

$$a = \quad \frac{5}{3}$$

$$b = \quad -\frac{14}{3}$$

Finally, we have

$$y = ax^2 + bx + c$$
$$y = \frac{5}{3}x^2 - \frac{14}{3}x + 3$$
$$3y = 5x^2 - 14x + 9$$
$$5x^2 - 14x - 3y + 9 = 0$$

37. Find an equation of the parabola in the accompanying figure.

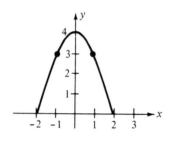

Solution

Since the axis of the parabola is vertical, the standard form is

$$(x - h)^2 = 4p(y - k)$$

Since the vertex is $(0, 4)$, $h = 0$ and $k = 4$. Therefore

$$x^2 = 4p(y - 4)$$

Since the parabola passes through the point $(2, 0)$, we have

$$4 = 4p(-4) \qquad \text{or} \qquad -\frac{1}{4} = p$$

Therefore $x^2 = 4\left(-\dfrac{1}{4}\right)(y - 4) \qquad$ or $\qquad y = 4 - x^2$

47. A cable of a parabolic suspension bridge is suspended between two towers that are 400 feet apart and 50 feet above the roadway as shown in the accompanying figure. The cable touches the roadway midway between the towers. Find the length of the cable.

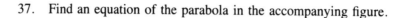

Solution

Since the axis of the parabola is vertical, the standard form is

$$(x - h)^2 = 4p(y - k)$$

Since the vertex is $(0, 0)$, $h = 0$ and $k = 0$. Therefore

$$x^2 = 4py$$

Since the parabola passes through the point $(200, 50)$ we have

$$200^2 = 4p(50) \qquad \text{or} \qquad 200 = p$$

Therefore

$$x^2 = 4(200)y \qquad \text{or} \qquad y = \frac{1}{800}x^2$$

$$y' = \frac{1}{400}x$$

$$s = \int_{-200}^{200} \sqrt{1 + (y')^2}\, dx = 2\int_0^{200} \sqrt{1 + \left(\frac{x}{400}\right)^2}\, dx$$

$$= \frac{1}{200}\int_0^{200} \sqrt{400^2 + x^2}\, dx$$

$$= \frac{1}{400}[x\sqrt{400^2 + x^2} + 400^2 \ln{(x + \sqrt{400^2 + x^2})}]_0^{200}$$

$$= 100\left[\sqrt{5} + 4\ln{\left(\frac{1 + \sqrt{5}}{2}\right)}\right] \approx 416.1 \text{ ft}$$

49. Water is flowing from a horizontal pipe 48 feet above the ground with a horizontal velocity of 10 feet per second. The falling stream of water has the shape of a parabola whose vertex, (0, 48), is the end of the pipe as shown in the accompanying figure. Where does the water hit the ground?

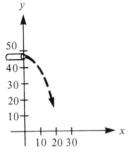

Solution

Since the water falls with an initial (vertical) velocity of $v_0 = 0$ from an initial height of $s_0 = 48$, we have the height of the water given by

$$s = 16t^2 + v_0 t + s_0 = -16t^2 + 48$$

When $s = 0$, we have

$$16t = 48$$
$$t^2 = 3$$
$$t = \sqrt{3}$$

Since the horizontal velocity of the water is constant during the fall, it would move $10\sqrt{3}$ ft horizontally before hitting the ground. Since the axis of the parabolic path is vertical and the vertex is (0, 48), the standard form of its equation is

$$(x - h)^2 = 4p(y - k)$$
$$x^2 = 4p(y - 48)$$

Finally, since the point $(10\sqrt{3}, 0)$ lies on the parabola, we have

$$(10\sqrt{3})^2 = 4p(0 - 48)$$
$$300 = -192p$$
$$p = \frac{-300}{192} = -\frac{25}{16}$$

We conclude that the equation of the parabola is

$$x^2 = 4\left(-\frac{25}{16}\right)(y - 48) \qquad \text{or} \qquad 4x^2 = -25y + 1200$$

10.2
Ellipses

13. Find the center, foci, vertices, eccentricity, and sketch the graph of the ellipse given by $x^2 + 4y^2 = 4$.

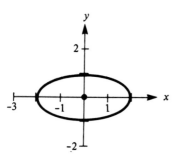

Solution

In standard form we have

$$x^2 + 4y^2 = 4$$

$$\frac{x^2}{4} + y^2 = 1$$

$$\frac{(x - 0)^2}{2^2} + \frac{(y - 0)^2}{1^2} = 1$$

$$\frac{(x - h)^2}{a^2} + \frac{(y - k)^2}{b^2} = 1$$

Thus $h = 0$, $k = 0$, $a = 2$, $b = 1$, and $c = \sqrt{2^2 - 1^2} = \sqrt{3}$. We conclude that

$$\begin{aligned}
\text{center, } (h, k)\text{:} \quad & (0, 0) \\
\text{foci, } (h \pm c, k)\text{:} \quad & (\pm\sqrt{3}, 0) \\
\text{vertices, } (h \pm a, k)\text{:} \quad & (\pm 2, 0)
\end{aligned}$$

$$e = \frac{c}{a} = \frac{\sqrt{3}}{2}$$

21. Find the center, foci, vertices, eccentricity, and sketch the graph of the ellipse given by $9x^2 + 4y^2 + 36x - 24y + 36 = 0$.

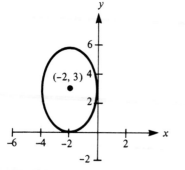

Solution

In standard form we have

$$9x^2 + 4y^2 + 36x - 24y + 36 = 0$$

$$9x^2 + 36x + 4y^2 - 24y = -36$$

$$9(x^2 + 4x + 4) + 4(y^2 - 6y + 9) = -36 + 36 + 36$$

$$9(x + 2)^2 + 4(y - 3)^2 = 36$$

$$\frac{(x + 2)^2}{4} + \frac{(y - 3)^2}{9} = 1$$

$$\frac{(x + 2)^2}{2^2} + \frac{(y - 3)^2}{3^2} = 1$$

$$\frac{(x - h)^2}{b^2} + \frac{(y - k)^2}{a^2} = 1$$

Thus $h = -2$, $k = 3$, $a = 3$, $b = 2$, and $c = \sqrt{3^2 - 2^2} = \sqrt{5}$. We conclude that

$$\begin{aligned}
\text{center, } (h, k)\text{:} &\qquad (-2, 3) \\
\text{foci, } (h, k \pm c)\text{:} &\qquad (-2, 3 \pm \sqrt{5}) \\
\text{vertices, } (h, k \pm a)\text{:} &\qquad (-2, 3 \pm 3)
\end{aligned}$$

$$e = \frac{c}{a} = \frac{\sqrt{5}}{3}$$

25. Find the center, foci, vertices, eccentricity, and sketch the graph of the ellipse given by $12x^2 + 20y^2 - 12x + 40y - 37 = 0$.

Solution

In standard form we have

$$12x^2 + 20y^2 - 12x + 40y - 37 = 0$$
$$12(x^2 - x + \tfrac{1}{4}) + 20(y^2 + 2y + 1) = 37 + 3 + 20$$
$$12(x - \tfrac{1}{2})^2 + 20(y + 1)^2 = 60$$
$$\frac{(x - \tfrac{1}{2})^2}{5} + \frac{(y + 1)^2}{3} = 1$$

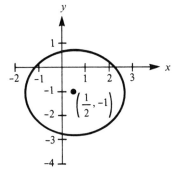

Thus $h = \tfrac{1}{2}$, $k = -1$, $a = \sqrt{5}$, $b = \sqrt{3}$, and $c = \sqrt{5 - 3} = \sqrt{2}$. We conclude that

$$\begin{aligned}
\text{center, } (h, k)\text{:} &\qquad (\tfrac{1}{2}, -1) \\
\text{foci, } (h \pm c, k)\text{:} &\qquad (\tfrac{1}{2} \pm \sqrt{2}, -1) \\
\text{vertices, } (h \pm a, k)\text{:} &\qquad (\tfrac{1}{2} \pm \sqrt{5}, -1)
\end{aligned}$$

$$e = \frac{c}{a} = \frac{\sqrt{2}}{\sqrt{5}} = \frac{\sqrt{10}}{5}$$

29. Find an equation for the ellipse with vertices $(5, 0)$ and $(-5, 0)$ and eccentricity $\tfrac{3}{5}$.

Solution

Since the vertices lie on the x-axis, the standard form for the ellipse is

$$\frac{(x - h)^2}{a^2} + \frac{(y - k)^2}{b^2} = 1$$

Since the center is the midpoint of the line segment connecting the vertices, we have $(h, k) = (0, 0)$. Furthermore, since a is the distance from the center to the vertices, we have $a = 5$.

Also,

$$e = \frac{c}{a} = \frac{c}{5} = \frac{3}{5}$$

Therefore $c = 3$ and $b^2 = a^2 - c^2 = 25 - 9 = 16$.
Finally, the equation is

$$\frac{x^2}{25} + \frac{y^2}{16} = 1$$

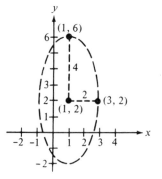

36. Find an equation for the ellipse whose center is $(1, 2)$, whose major axis is parallel to the y-axis, and which passes through the points $(1, 6)$ and $(3, 2)$.

Solution

By sketching these three points, we can see that $h = 1$, $k = 2$, $a = 4$, and $b = 2$. Thus we have

$$\frac{(x - h)^2}{b^2} + \frac{(y - k)^2}{a^2} = 1 \qquad \text{or} \qquad \frac{(x - 1)^2}{4} + \frac{(y - 2)^2}{16} = 1$$

41. A particle is traveling clockwise on the elliptical orbit given by

$$\frac{x^2}{10^2} + \frac{y^2}{5^2} = 1$$

The particle leaves the orbit at the point $(-8, 3)$ and travels in a straight line tangent to the ellipse. At which point will the particle cross the y-axis?

Solution

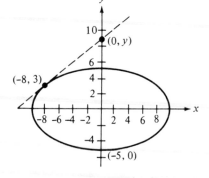

To find the slope of the tangent line at $(-8, 3)$, we differentiate implicitly as follows:

$$\frac{x^2}{100} + \frac{y^2}{25} = 1$$

$$\frac{x}{50} + \frac{2yy'}{25} = 0$$

$$x + 4yy' = 0$$

$$y' = \frac{-x}{4y}$$

Thus at $(-8, 3)$ the slope is

$$m = \frac{-(-8)}{4(3)} = \frac{8}{12} = \frac{2}{3}$$

The equation of the tangent line is

$$(y - 3) = \frac{2}{3}[x - (-8)]$$

$$3y - 9 = 2x + 16$$

$$3y = 2x + 25$$

Finally, when $x = 0$,

$$3y = 25 \qquad \text{or} \qquad y = \frac{25}{3}$$

43. Use Simpson's Rule with $n = 8$ to approximate the elliptic integral representing the circumference of the ellipse

$$\frac{x^2}{9} + \frac{y^2}{16} = 1$$

Solution

We begin by observing that the circumferences of the ellipses

$$\frac{x^2}{9} + \frac{y^2}{16} = 1 \qquad \text{and} \qquad \frac{x^2}{16} + \frac{y^2}{9} = 1$$

are the same, and therefore, we use the formula of Example 5. Since $a^2 = 16$ and $b^2 = 9$, we have $c^2 = 7$ and $e^2 = 7/16$. Hence the circumference is given by

$$C = 4(4) \int_0^{\pi/2} \sqrt{1 - \frac{7 \sin^2 \theta}{16}}\, d\theta$$

Applying Simpson's Rule with $n = 8$, the endpoints of the subintervals are $\dfrac{n\pi}{16}$, $n = 0, 1, 2, \ldots, 8$ and

$$C \approx 16\left(\frac{\pi}{6}\right)\left(\frac{1}{8}\right)[1 + 4(0.9916) + 2(0.9674) + 4(0.9300)$$

$$+ 2(0.8839) + 4(0.8359) + 2(0.7916) + 4(0.7610) + 0.7500]$$

$$\approx 22.10$$

54. A line segment, 9 in. long, moves so that one endpoint is always on the y-axis and the other always on the x-axis. Find the equation of the curve traced by a point (on the line segment) 6 in. from the endpoint on the y-axis.

Solution

From the accompanying figure we have

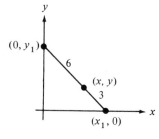

1. $(x - 0)^2 + (y - y_1)^2 = 36$

2. $(x - x_1)^2 + (y - 0)^2 = 9$

3. $\dfrac{y - y_1}{x} = \dfrac{y}{x - x_1}$ or $y - y_1 = \dfrac{xy}{x - x_1}$

Therefore

$$x^2 + \frac{x^2 y^2}{(x - x_1)^2} = 36 \qquad \text{From equations (1) and (3)}$$

$$x^2 + \frac{x^2 y^2}{9 - y^2} = 36 \qquad \text{From equation (2)}$$

Multiplying by $9 - y^2$, we obtain

$$x^2(9 - y^2) + x^2 y^2 = 36(9 - y^2)$$

$$9x^2 + 36y^2 = 36(9)$$

$$\frac{x^2}{36} + \frac{y^2}{9} = 1$$

10.3

Hyperbolas

9. Find the center, vertices, and foci of the hyperbola given by $y^2 - (x^2/4) = 1$ and sketch its graph, using asymptotes as an aid.

Solution

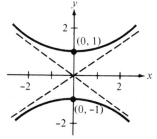

Writing the equation in standard form, we have

$$y^2 - \frac{x^2}{4} = 1$$

$$\frac{(y - 0)^2}{1^2} - \frac{(x - 0)^2}{2^2} = 1$$

$$\frac{(y - k)^2}{a^2} - \frac{(x - h)^2}{b^2} = 1$$

Thus $h = 0$, $k = 0$, $a = 1$, $b = 2$, and $c = \sqrt{1^2 + 2^2} = \sqrt{5}$. We conclude that

center, (h, k): $(0, 0)$

vertices, $(h, k \pm a)$: $(0, \pm 1)$

foci, $(h, k \pm c)$: $(0, \pm\sqrt{5})$

Finally, the asymptotes are given by

$$y = k \pm \frac{a}{b}(x - h) = \pm\frac{x}{2}$$

17.   Find the center, vertices, and foci of the hyperbola given by
$[(x - 1)^2/4] - [(y + 2)^2/1] = 1$ and sketch its graph, using
asymptotes as an aid.

Solution

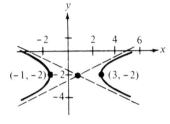

$$\frac{(x - 1)^2}{4} - \frac{(y + 2)^2}{1} = 1$$

$$\frac{(x - h)^2}{a^2} - \frac{(y + k)^2}{b^2} = 1$$

Thus $h = 1$, $k = -2$, $a = 2$, $b = 1$, and $c = \sqrt{4 + 1} = \sqrt{5}$. We conclude that

$$
\begin{array}{lll}
\text{center, } (h, k): & (1, -2) \\
\text{vertices, } (h \pm a, k): & (-1, -2), (3, -2) \\
\text{foci, } (h \pm c, k): & (1 \pm \sqrt{5}, -2)
\end{array}
$$

Finally, the asymptotes are given by

$$y = k \pm \frac{b}{a}(x - h) = -2 \pm \frac{1}{2}(x - 1)$$

$$y = \frac{1}{2}x - \frac{5}{2} \quad \text{and} \quad y = -\frac{1}{2}x - \frac{3}{2}$$

21.   Find the center, vertices, and foci of the hyperbola given by
$9x^2 - y^2 - 36x - 6y + 18 = 0$ and sketch its graph, using
asymptotes as an aid.

Solution

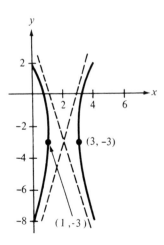

Writing the equation in standard form, we have

$$9x^2 - y^2 - 36x - 6y + 18 = 0$$

$$9x^2 - 36x - (y^2 + 6y) = -18$$

$$9(x^2 - 4x + 4) - (y^2 + 6y + 9) = -18 + 36 - 9$$

$$9(x - 2)^2 - (y + 3)^2 = 9$$

$$\frac{(x - 2)^2}{1^2} - \frac{(y + 3)^2}{3^2} = 1$$

Thus $h = 2$, $k = -3$, $a = 1$, $b = 3$, and $c = \sqrt{1^2 + 3^2} = \sqrt{10}$. We conclude that

$$
\begin{array}{lll}
\text{center, } (h, k): & (2, -3) \\
\text{vertices, } (h \pm a, k): & (3, -3) \quad \text{and} \quad (1, -3) \\
\text{foci, } (h \pm c, k): & (2 \pm \sqrt{10}, -3)
\end{array}
$$

Finally, the asymptotes are given by

$$y = k \pm \frac{b}{a}(x - h) = -3 \pm \frac{3}{1}(x - 2)$$

$$y = 3x - 9 \qquad \text{and} \qquad y = -3x + 3$$

23. Find the center, vertices, and foci of the hyperbola given by $9y^2 - x^2 + 2x + 54y + 62 = 0$ and sketch its graph, using asymptotes as an aid.

Solution

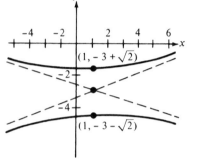

$$9y^2 - x^2 + 2x + 54y + 62 = 0$$

$$9(y^2 + 6y + 9) - (x^2 - 2x + 1) = -62 + 81 - 1$$

$$9(y + 3)^2 - (x - 1)^2 = 18$$

$$\frac{(y + 3)^2}{2} - \frac{(x - 1)^2}{18} = 1$$

Thus $h = 1$, $k = -3$, $a = \sqrt{2}$, $b = \sqrt{18} = 3\sqrt{2}$, and $c = \sqrt{2 + 18} = 2\sqrt{5}$. We conclude that

$$\begin{array}{ll} \text{center, } (h, k): & (1, -3) \\ \text{vertices, } (h, k \pm a): & (1, -3 \pm \sqrt{2}) \\ \text{foci, } (h, k \pm c): & (1, -3 \pm 2\sqrt{5}) \end{array}$$

Finally, the asymptotes are given by

$$y = k \pm \frac{a}{b}(x - h) = -3 \pm \frac{\sqrt{2}}{3\sqrt{2}}(x - 1)$$

$$y = \frac{1}{3}x - \frac{10}{3} \qquad \text{and} \qquad y = -\frac{1}{3}x - \frac{8}{3}$$

29. Find an equation for the hyperbola with vertices $(-1, 0)$ and $(1, 0)$ and whose asymptotes are given by $y = \pm 3x$.

Solution

Since the vertices lie on a horizontal line, the standard form is

$$\frac{(x - h)^2}{a^2} - \frac{(y - k)^2}{b^2} = 1$$

The center of the hyperbola lies at the midpoint of the line segment connecting the vertices. Thus

$$(h, k) = \left(\frac{1 + (-1)}{2}, \frac{0 + 0}{2} \right) = (0, 0)$$

and we have $h = 0$ and $k = 0$. Since the asymptotes are of the form

$$y = k \pm \frac{b}{a}(x - h) = \pm 3x$$

we have

$$\pm \frac{b}{a} = \pm 3 \qquad \text{or} \qquad b = 3a$$

Finally, since $a = 1$, we have $b = 3$ and the equation is

$$\frac{x^2}{1^2} - \frac{y^2}{3^2} = 1 \qquad \text{or} \qquad x^2 - \frac{y^2}{9} = 1$$

34. Find an equation for the hyperbola with asymptotes $y = \pm \frac{3}{4}x$ and one focus at $(10, 0)$.

Solution

We first observe that the asymptotes intersect at the origin. Therefore the center of the hyperbola is $(h, k) = (0, 0)$. Since one focus is $(10, 0)$, we know that the transverse axis is horizontal and that $c = 10$. From the slope of the asymptotes we have

$$\pm \frac{b}{a} = \pm \frac{3}{4} \qquad \text{or} \qquad b = \frac{3}{4}a$$

Also

$$c^2 = a^2 + b^2 = 100$$

$$a^2 + \left(\frac{3}{4}a\right)^2 = 100$$

$$\frac{25}{16}a^2 = 100$$

$$a^2 = 64$$

$$b^2 = 100 - a^2 = 36$$

Therefore

$$\frac{(x - h)^2}{a^2} - \frac{(y - k)^2}{b^2} = 1 \qquad \text{or} \qquad \frac{x^2}{64} - \frac{y^2}{36} = 1$$

38. Find the equation for (a) the tangent lines, and (b) the normal lines to the hyperbola $(y^2/4) - (x^2/2) = 1$ when $x = 4$.

Solution

(a) When $x = 4$,

$$y = \pm 2\sqrt{1 + \frac{4^2}{2}} = \pm 2(3) = \pm 6$$

By implicit differentiation, we have

$$\frac{y^2}{4} - \frac{x^2}{2} = 1$$

$$\left(\frac{1}{4}\right)(2y)y' - \left(\frac{1}{2}\right)(2x) = 0$$

$$y' = \frac{2x}{y}$$

Thus at $(4, 6)$ the slope of the tangent line is

$$m = y' = \frac{2(4)}{6} = \frac{4}{3}$$

and an equation of the tangent line is

$$y - 6 = \frac{4}{3}(x - 4) \qquad \text{or} \qquad 4x - 3y + 2 = 0$$

At $(4, -6)$ the slope of the tangent line is $\frac{-4}{3}$ and an equation of the tangent line is

$$y + 6 = \frac{-4}{3}(x - 4) \qquad \text{or} \qquad 4x + 3y + 2 = 0$$

(b)   At $(4, 6)$ the slope of the normal line is $-\frac{3}{4}$ (negative reciprocal of the slope of the tangent line) and an equation of the normal line is

$$y - 6 = -\frac{3}{4}(x - 4) \qquad \text{or} \qquad 3x + 4y - 36 = 0$$

At $(4, -6)$ the slope of the normal line is $\frac{3}{4}$ and an equation of the normal line is

$$y + 6 = \frac{3}{4}(x - 4) \qquad \text{or} \qquad 3x - 4y - 36 = 0$$

10.4

Rotation and the general second-degree equation

4.   Rotate the axes to eliminate the xy term in the equation $9x^2 + 24xy + 16y^2 + 80x - 60y = 0$. Sketch the graph of the resulting equation, showing both sets of axes.

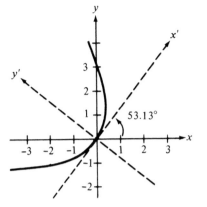

Solution

From the equations

$$9x^2 + 24xy + 16y^2 + 80x - 60y = 0$$

$$Ax^2 + Bxy + Cy^2 + Dx + Ey + F = 0$$

we have $A = 9$, $B = 24$, $C = 16$, $D = 80$, $E = -60$, and $F = 0$. Thus

$$\cot 2\theta = \frac{9 - 16}{24} = \frac{-7}{24} \qquad \theta \approx 53.13°$$

From the equation

$$\cot 2\theta = \frac{\cot^2 \theta - 1}{2 \cot \theta}$$

we have

$$\frac{\cot^2 \theta - 1}{2 \cot \theta} = \frac{-7}{24}$$

$$24 \cot^2 \theta - 24 = -14 \cot \theta$$

$$12 \cot^2 \theta + 7 \cot \theta - 12 = 0$$

$$(4 \cot \theta - 3)(3 \cot \theta + 4) = 0$$

$$\cot \theta = \frac{3}{4} \qquad \text{or} \qquad -\frac{4}{3}$$

Since $0 < \theta < 90°$, we choose $\cot \theta = \frac{3}{4}$ and conclude that $\sin \theta = \frac{4}{5}$ and $\cos \theta = \frac{3}{5}$. Therefore, using the equations

$$x = x' \cos \theta - y' \sin \theta = \frac{3}{5}x' - \frac{4}{5}y'$$

$$y = x' \sin \theta + y' \cos \theta = \frac{4}{5}x' + \frac{3}{5}y'$$

and

$$9x^2 + 24xy + 16y^2 + 80x - 60y = 0$$

we have

$$9\left(\frac{3}{5}x' - \frac{4}{5}y'\right)^2 + 24\left(\frac{3}{5}x' - \frac{4}{5}y'\right)\left(\frac{4}{5}x' + \frac{3}{5}y'\right) + 16\left(\frac{4}{5}x' + \frac{3}{5}y'\right)^2$$

$$+ 80\left(\frac{3}{5}x' - \frac{4}{5}y'\right) - 60\left(\frac{4}{5}x' + \frac{3}{5}y'\right) = 0$$

After expanding and combining like terms we have

$$25(x')^2 - 100y' = 0 \qquad \text{or} \qquad (x')^2 = 4y'$$

5. Rotate the axes to eliminate the xy term in the equation

$x^2 - 10xy + y^2 + 1 = 0$. Sketch the graph of the resulting equation, showing both sets of axes.

Solution

From the equations

$$x^2 - 10xy + y^2 + 1 = 0$$
$$Ax^2 + Bxy + Cy^2 + Dx + Ey + F = 0$$

we have $A = 1$, $B = -10$, $C = 1$, $D = 0$, $E = 0$, and $F = 1$. Thus

$$\cot 2\theta = \frac{1 - 1}{-10} = 0$$

and $\theta = 45°$. Now, since $\sin \theta = \sqrt{2}/2$ and $\cos \theta = \sqrt{2}/2$, we have

$$x = x' \cos \theta - y' \sin \theta = \frac{\sqrt{2}}{2}x' - \frac{\sqrt{2}}{2}y'$$

$$y = x' \sin \theta + y' \cos \theta = \frac{\sqrt{2}}{2}x' + \frac{\sqrt{2}}{2}y'$$

Substitution into $x^2 - 10xy + y^2 + 1 = 0$ yields

$$\left(\frac{\sqrt{2}}{2}x' - \frac{\sqrt{2}}{2}y'\right)^2 - 10\left(\frac{\sqrt{2}}{2}x' - \frac{\sqrt{2}}{2}y'\right)\left(\frac{\sqrt{2}}{2}x' + \frac{\sqrt{2}}{2}y'\right)$$
$$+ \left(\frac{\sqrt{2}}{2}x' + \frac{\sqrt{2}}{2}y'\right)^2 + 1 = 0$$

After expanding and combining like terms we have

$$-4(x')^2 + 6(y')^2 + 1 = 0 \quad \text{or} \quad \frac{(x')^2}{\frac{1}{4}} - \frac{(y')^2}{\frac{1}{6}} = 1$$

14. Rotate the axes to eliminate the xy term in the equation $40x^2 + 36xy + 25y^2 = 52$. Sketch the graph of the resulting equation, showing both sets of axes.

Solution

From the equations

$$40x^2 + 36xy + 25y^2 - 52 = 0$$
$$Ax^2 + Bxy + Cy^2 + Dx + Ey + F = 0$$

we have $A = 40$, $B = 36$, $C = 25$, $D = 0$, $E = 0$, and $F = -52$. Thus

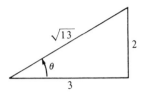

$$\cot 2\theta = \frac{40 - 25}{36} = \frac{5}{12} \qquad \theta \approx 33.69°$$

$$\frac{\cot^2 \theta - 1}{2 \cot \theta} = \frac{5}{12}$$

$$0 = 2(6 \cot^2 \theta - 5 \cot \theta - 6)$$

$$\cot \theta = \frac{5 \pm \sqrt{(-5)^2 - 4(6)(-6)}}{2(6)} = \frac{5 \pm 13}{12}$$

Since $0° < \theta < 90°$, we choose $\cot \theta = \frac{18}{12} = \frac{3}{2}$. From the accompanying figure we conclude that $\sin \theta = 2/\sqrt{13}$ and $\cos \theta = 3/\sqrt{13}$. Therefore, using the equations

$$x = x' \cos \theta - y' \sin \theta = \frac{3}{\sqrt{13}}x' - \frac{2}{\sqrt{13}}y'$$

$$y = x' \sin \theta + y' \cos \theta = \frac{2}{\sqrt{13}}x' + \frac{3}{\sqrt{13}}y'$$

and

$$40x^2 + 36xy + 25y^2 = 52$$

we have

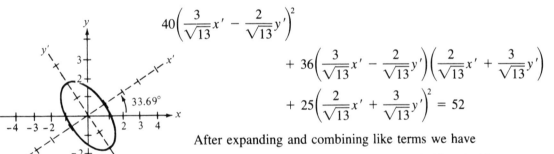

$$40\left(\frac{3}{\sqrt{13}}x' - \frac{2}{\sqrt{13}}y'\right)^2$$

$$+ 36\left(\frac{3}{\sqrt{13}}x' - \frac{2}{\sqrt{13}}y'\right)\left(\frac{2}{\sqrt{13}}x' + \frac{3}{\sqrt{13}}y'\right)$$

$$+ 25\left(\frac{2}{\sqrt{13}}x' + \frac{3}{\sqrt{13}}y'\right)^2 = 52$$

After expanding and combining like terms we have

$$52(x')^2 + 13(y')^2 - 52 = 0 \qquad \text{or} \qquad \frac{(x')^2}{1} + \frac{(y')^2}{4} = 1$$

Review Exercises for Chapter 10

15. Analyze the equation $3x^2 + 2y^2 - 12x + 12y + 29 = 0$ and sketch its graph.

Solution

In standard form the equation is

$$3x^2 + 2y^2 - 12x + 12y + 29 = 0$$

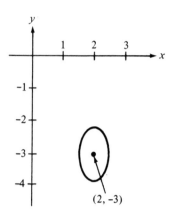

(2, -3)

$$3(x^2 - 4x) + 2(y^2 + 6y) = -29$$
$$3(x^2 - 4x + 4) + 2(y^2 + 6y + 9) = -29 + 12 + 18$$
$$3(x - 2)^2 + 2(y + 3)^2 = 1$$
$$\frac{(x - 2)^2}{(1/\sqrt{3})^2} + \frac{(y + 3)^2}{(1/\sqrt{2})^2} = 1$$

Therefore the graph is an ellipse where $h = 2$, $k = -3$, $a = 1/\sqrt{2}$, $b = 1/\sqrt{3}$, and $c = \sqrt{(1/2) - (1/3)} = 1/\sqrt{6}$. We conclude that

center, (h, k): $(2, -3)$

vertices, $(h, k \pm a)$: $\left(2, -3 \pm \dfrac{1}{\sqrt{2}}\right)$

foci, $(h, k \pm c)$: $\left(2, -3 \pm \dfrac{1}{\sqrt{6}}\right)$

27. Find an equation of the parabola whose vertex is at $(0, 0)$ and whose focus is at $(1, 1)$.

Solution

Since the vertex and focus lie on the line $y = x$, we let $\theta = 45°$. Then in the $x'y'$-coordinate system, the vertex lies at $(0, 0)$ and the focus is at $(\sqrt{2}, 0)$. Now p, the directed distance from the vertex to the focus, is $p = \sqrt{2}$, and we have

$$(y' - 0)^2 = 4\sqrt{2}(x' - 0) \qquad \text{or} \qquad (y')^2 = 4\sqrt{2}x'$$

Since $\theta = 45°$, we have

$$x' = x \cos \theta + y \sin \theta = \frac{1}{\sqrt{2}}(x + y)$$

$$y' = -x \sin \theta + y \cos \theta = \frac{1}{\sqrt{2}}(-x + y)$$

and

$$\left[\frac{1}{\sqrt{2}}(-x + y)\right]^2 = 4\sqrt{2}\left(\frac{1}{\sqrt{2}}\right)(x + y)$$

$$\frac{x^2 - 2xy + y^2}{2} = 4x + 4y$$

$$x^2 - 2xy + y^2 = 8x + 8y$$

$$x^2 - 2xy + y^2 - 8x - 8y = 0$$

31. Find an equation of the hyperbola with foci $(\pm 4, 0)$ where the

absolute value of the difference of the distances from a point on the hyperbola to the foci is 4.

Solution

We first observe that the center of the hyperbola is at the origin and therefore $(h, k) = (0, 0)$. Since the foci are $(\pm 4, 0)$ we know that the transverse axis is horizontal and $c = 4$. Also, from the definition of a hyperbola we know that the absolute value of the difference of the distances from a point on the hyperbola to the foci is $2a$. Thus $2a = 4$ or $a = 2$. Since $b^2 = c^2 - a^2$, we have $b^2 = 16 - 4 = 12$. Therefore, an equation of the hyperbola is

$$\frac{x^2}{4} - \frac{y^2}{12} = 1$$

34. Find the equations of the lines tangent to the ellipse $x^2 + 4y^2 - 4x - 8y - 24 = 0$ and parallel to the line $x - 2y - 12 = 0$.

Solution

The slope of the given line is $m = \frac{1}{2}$, since

$$x - 2y - 12 = 0$$
$$2y = x - 12$$
$$y = \frac{1}{2}x - 6 = mx + b$$

By implicit differentiation, we have

$$x^2 + 4y^2 - 4x - 8y - 24 = 0$$
$$2x + 8yy' - 4 - 8y' = 0$$
$$(8y - 8)y' = 4 - 2x$$
$$y' = \frac{4 - 2x}{8y - 8} = \frac{2 - x}{4y - 4}$$

Now, if $y' = \frac{1}{2}$, we have

$$\frac{1}{2} = \frac{2 - x}{4y - 4}$$
$$2y - 2 = 2 - x$$
$$x = 4 - 2y$$

Substituting for x in the equation of the ellipse, we obtain

$$(4 - 2y)^2 + 4y^2 - 4(4 - 2y) - 8y - 24 = 0$$
$$16 - 16y + 4y^2 + 4y^2 - 16 + 8y - 8y - 24 = 0$$

$$8y^2 - 16y - 24 = 0$$
$$y^2 - 2y - 3 = 0$$
$$(y - 3)(y + 1) = 0$$
$$y = -1, 3$$

If $y = -1$, $x = 4 - (2)(-1) = 6$ and we have

$$y + 1 = \frac{1}{2}(x - 6)$$
$$2y + 2 = x - 6$$
$$x - 2y - 8 = 0$$

If $y = 3$, $x = 4 - (2)(3) = -2$ and we have

$$y - 3 = \frac{1}{2}(x + 2)$$
$$2y - 6 = x + 2$$
$$x - 2y + 8 = 0$$

38. The ellipse $(x^2/a^2) + (y^2/b^2) = 1$ is revolved about its minor axis to form an oblate spheroid. Show that the volume of the spheroid is $\frac{4}{3}\pi a^2 b$ and its surface area is $2\pi a^2 + \pi(b^2/e) \ln [(1 + e)/(1 - e)]$.

Solution

Solving for x as a function of y, we have

$$x = \frac{a}{b}\sqrt{b^2 - y^2}$$

and

$$\frac{dx}{dy} = \left(\frac{a}{b}\right)\frac{-y}{\sqrt{b^2 - y^2}}$$

Revolving about the y-axis, we have

$$V = \pi \int_{-b}^{b} \left(\frac{a}{b}\sqrt{b^2 - y^2}\right)^2 dy = \frac{\pi a^2}{b^2} \int_{-b}^{b} (b^2 - y^2)\, dy$$

$$= \frac{\pi a^2}{b^2}\left[b^2 y - \frac{y^3}{3}\right]_{-b}^{b}$$

$$= \frac{\pi a^2}{b^2}\left(b^3 - \frac{b^3}{3} + b^3 - \frac{b^3}{3}\right) = \frac{4\pi a^2 b}{3}$$

$$S = 4\pi \int_{0}^{b} x\sqrt{1 + \left(\frac{dx}{dy}\right)^2}\, dy$$

$$= 4\pi \int_0^b \frac{a}{b} \sqrt{b^2 - y^2} \sqrt{1 + \frac{a^2 y^2}{b^2(b^2 - y^2)}} \, dy$$

$$= \frac{4\pi a}{b^2} \int_0^b \sqrt{b^4 + (a^2 - b^2)y^2} \, dy = \frac{4\pi a}{b^2} \int_0^b \sqrt{b^4 + c^2 y^2} \, dy$$

$$= \frac{4\pi a}{b^2} \left[\frac{1}{2c} \left(cy\sqrt{b^4 + c^2 y^2} + b^4 \ln \left| cy + \sqrt{b^4 + c^2 y^2} \right| \right) \right]_0^b$$

$$= \frac{2\pi a}{b^2 c} [bc\sqrt{b^4 + c^2 b^2} + b^4 \ln |bc + \sqrt{b^4 + b^2 c^2}| - b^4 \ln b^2]$$

$$= \frac{2\pi a}{b^2 c} \left(bc\sqrt{b^4 + (a^2 - b^2)b^2} + b^4 \ln \left(\frac{bc + \sqrt{b^4 + (a^2 - b^2)b^2}}{b^2} \right) \right)$$

$$= \frac{2\pi a}{b^2 c} \left[ab^2 c + b^4 \ln \left(\frac{cb + ab}{b^2} \right) \right] = 2\pi a^2 + \frac{2\pi a b^2}{c} \ln \left(\frac{a + c}{b} \right)$$

$$= 2\pi a^2 + \frac{\pi b^2}{c/a} \ln \frac{(a + c)^2}{b^2}$$

Note that

$$\frac{(a + c)^2}{a^2 - c^2} = \frac{(a + c)(a + c)}{(a + c)(a - c)} = \frac{a + c}{a - c} = \frac{1 + (c/a)}{1 - (c/a)} = \frac{1 + e}{1 - e}$$

Thus

$$S = 2\pi a^2 + \pi \frac{b^2}{e} \ln \left(\frac{1 + e}{1 - e} \right)$$

43. Consider a fire truck with a water tank 16 feet long whose vertical cross sections are ellipses as described by the equation

$$\frac{x^2}{16} + \frac{y^2}{9} = 1$$

Find the depth of water in the tank if it is $\frac{3}{4}$ full (by volume) and the truck is on level ground.

Solution

The truck will be carrying $\frac{3}{4}$ of its total capacity when the water covers $\frac{3}{4}$ of the area of a cross section of the tank. One-half of this area will be below the major axis and $\frac{1}{4}$ above the major axis of the ellipse

$$\frac{x^2}{16} + \frac{y^2}{9} = 1$$

The total area is given by

$$A = \pi ab = \pi(4)(3) = 12\pi$$

$$\frac{1}{4}A = 3\pi$$

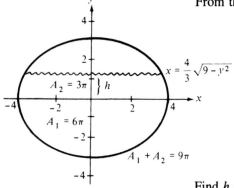

From the accompanying figure we have

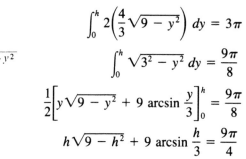

$$\int_0^h 2\left(\frac{4}{3}\sqrt{9 - y^2}\right) dy = 3\pi$$

$$\int_0^h \sqrt{3^2 - y^2}\, dy = \frac{9\pi}{8}$$

$$\frac{1}{2}\left[y\sqrt{9 - y^2} + 9 \arcsin \frac{y}{3}\right]_0^h = \frac{9\pi}{8}$$

$$h\sqrt{9 - h^2} + 9 \arcsin \frac{h}{3} = \frac{9\pi}{4}$$

Find h such that

$$f(h) = h\sqrt{9 - h^2} + 9 \arcsin \left(\frac{h}{3}\right) - \frac{9\pi}{4} = 0$$

By Newton's Method, with an initial estimate of $h = 1$, we have

$$h_{n+1} = h_n - \frac{f(h_n)}{f'(h_n)}$$

$$= h_n - \frac{h_n\sqrt{9 - h_n^2} + 9 \arcsin (h_n/3) - (9\pi/4)}{2\sqrt{9 - h_n^2}}$$

n	1	2	3	4
h_n	1.0000	1.2089	1.2119	1.2119

We conclude that $h \approx 1.212$ and therefore the total height of the water in the tank is $3 + 1.212 = 4.212$ ft.

11 Plane curves, parametric equations, and polar coordinates

11.1
Plane curves and parametric equations

7. Sketch the curve represented by the parametric equations $x = t^3$, $y = t^2/2$ and write the corresponding rectangular equation by eliminating the parameter.

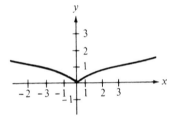

Solution

Since $x = t^3$ and $y = t^2/2$, we have

$$t = x^{1/3} \qquad \text{or} \qquad y = \frac{1}{2}(x^{1/3})^2 = \frac{1}{2}x^{2/3}$$

10. Sketch the curve represented by the parametric equations $x = t^2 + t$, $y = t^2 - t$ and write the corresponding rectangular equation by eliminating the parameter.

Solution

Since $x = t^2 + t$ and $y = t^2 - t$, we have

$$x - y = 2t \qquad \text{or} \qquad \frac{x - y}{2} = t$$

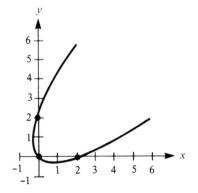

Thus

$$y = t^2 - t = \frac{(x - y)^2}{4} - \frac{x - y}{2} = \frac{x^2 - 2xy + y^2 - 2x + 2y}{4}$$

$$4y = x^2 - 2xy + y^2 - 2x + 2y$$

$$0 = x^2 - 2xy + y^2 - 2x - 2y$$

Since the discriminant of this equation is $B^2 - 4AC = 4 - 4 = 0$, the graph is a parabola. From the following table we make the accompanying sketch.

t	0	1	−1	2	−2
x	0	2	0	6	2
y	0	1	2	2	6

19. Sketch the curve represented by the parametric equations $x = \cos \theta$, $y = 2 \sin^2 \theta$ and write the corresponding rectangular equation by eliminating the parameter.

Solution

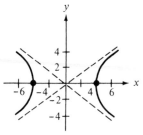

We first observe that $-1 \leq x \leq 1$ and $y \geq 0$.

$$x = \cos \theta \qquad y = 2 \sin^2 \theta$$

$$x^2 = \cos^2 \theta \qquad \frac{y}{2} = \sin^2 \theta$$

Therefore

$$x^2 + \frac{y}{2} = \cos^2 \theta + \sin^2 \theta = 1$$

$$y = 2 - 2x^2, \qquad -1 \leq x \leq 1$$

25. Sketch the curve represented by the parametric equations $x = 4 \sec \theta$, $y = 3 \tan \theta$ and write the corresponding rectangular equation by eliminating the parameter.

Solution

Since the parametric equations involve secants and tangents, we consider the identity $\sec^2 \theta - \tan^2 \theta = 1$. Therefore we write

$$\frac{x}{4} = \sec \theta \qquad \text{and} \qquad \frac{y}{3} = \tan \theta$$

or

$$\frac{x^2}{16} - \frac{y^2}{9} = \sec^2 \theta - \tan^2 \theta = 1$$

The graph of this equation is a hyperbola centered at the origin with vertices $(\pm 4, 0)$.

37. Sketch the curve (a prolate cycloid) represented by $x = \theta - \frac{3}{2} \sin \theta$, $y = 1 - \frac{3}{2} \cos \theta$.

Solution

Using the point-by-point plotting method, we have the following table:

θ	0	$\dfrac{\pi}{6}$	$\dfrac{\pi}{4}$	$\dfrac{\pi}{3}$	$\dfrac{\pi}{2}$	$\dfrac{2\pi}{3}$	π	$\dfrac{3\pi}{2}$	$\dfrac{5\pi}{3}$	$\dfrac{11\pi}{6}$
x	0	-0.23	-0.28	-0.25	0.07	0.795	3.14	6.21	6.54	6.51
y	$-\dfrac{1}{2}$	-0.30	-0.06	0.25	1.00	1.750	2.50	1.00	0.25	-0.30

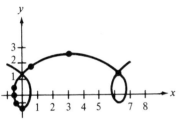

Using the values in the table, we make the accompanying sketch.

43.   Eliminate the parameter from the equations $x = h + a \sec \theta$ and $y = k + b \tan \theta$ to obtain the standard form of the equation of a hyperbola.

Solution

$$x = h + a \sec \theta \qquad\qquad y = k + b \tan \theta$$

$$\frac{x - h}{a} = \sec \theta \qquad\qquad \frac{y - k}{b} = \tan \theta$$

$$\frac{(x - h)^2}{a^2} = \sec^2 \theta \qquad\qquad \frac{(y - k)^2}{b^2} = \tan^2 \theta$$

$$\frac{(x - h)^2}{a^2} - \frac{(y - k)^2}{b^2} = \sec^2 \theta - \tan^2 \theta = 1$$

$$\frac{(x - h)^2}{a^2} - \frac{(y - k)^2}{b^2} = 1$$

45.   Find parametric equations for the line passing through the points $(0, 0)$ and $(5, -2)$.

Solution

The rectangular equation for this line has the form $y = mx$ where $m = -\frac{2}{5}$. Therefore if we let $x = 5t$, then

$y = -(\frac{2}{5})(5t) = -2t$ and the parametric equations are $x = 5t$, $y = -2t$.

11.2

Parametric equations and calculus

4. For the parametric equations $x = t^2 + 3t$, $y = t + 1$, find dy/dx and d^2y/dx^2 and evaluate these two derivatives when $t = 0$.

Solution

Since $x = t^2 + 3t$ and $y = t + 1$, we have

$$\frac{dy}{dx} = \frac{dy/dt}{dx/dt} = \frac{1}{2t + 3}$$

and

$$\frac{d^2y}{dx^2} = \frac{d[dy/dx]/dt}{dx/dt} = \frac{-2(2t + 3)^{-2}}{2t + 3} = \frac{-2}{(2t + 3)^3}$$

At $t = 0$,

$$\frac{dy}{dx} = \frac{1}{0 + 3} = \frac{1}{3} \quad \text{and} \quad \frac{d^2y}{dx^2} = \frac{-2}{(3)^3} = \frac{-2}{27}$$

7. For the parametric equations $x = 2 + \sec \theta$, $y = 1 + 2 \tan \theta$, find dy/dx and d^2y/dx^2 and evaluate these two derivatives when $\theta = \pi/6$.

Solution

Since $x = 2 + \sec \theta$ and $y = 1 + 2 \tan \theta$, we have

$$\frac{dy}{dx} = \frac{dy/d\theta}{dx/d\theta} = \frac{2 \sec^2 \theta}{\sec \theta \tan \theta} = \frac{2 \sec \theta}{\tan \theta} = 2 \csc \theta$$

$$\frac{d^2y}{dx^2} = \frac{d[dy/dx]/d\theta}{dx/d\theta} = \frac{-2 \csc \theta \cot \theta}{\sec \theta \tan \theta} = -2 \cot^3 \theta$$

At $\theta = \pi/6$,

$$\frac{dy}{dx} = 2 \csc \frac{\pi}{6} = 2(2) = 4$$

$$\frac{d^2y}{dx^2} = -2 \cot^3 \frac{\pi}{6} = -2(\sqrt{3})^3 = -6\sqrt{3}$$

14. Find an equation for the line tangent to the graph of
$x = 4 \cos \theta$, $y = 3 \sin \theta$ at $\theta = 3\pi/4$.

Solution

At $\theta = 3\pi/4$ the slope of the tangent line is

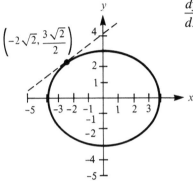

$$\frac{dy}{dx} = \frac{dy/d\theta}{dx/d\theta} = \frac{3 \cos \theta}{-4 \sin \theta} = \frac{3 \cos (3\pi/4)}{-4 \sin (3\pi/4)} = \frac{3(-1/\sqrt{2})}{-4(1/\sqrt{2})} = \frac{3}{4}$$

Furthermore, at $\theta = 3\pi/4$

$$x = 4 \cos \frac{3\pi}{4} = 4\left(-\frac{1}{\sqrt{2}}\right) = -2\sqrt{2}$$

$$y = 3 \sin \frac{3\pi}{4} = 3\left(\frac{1}{\sqrt{2}}\right) = \frac{3\sqrt{2}}{2}$$

Therefore the equation of the tangent line is

$$y - \frac{3\sqrt{2}}{2} = \frac{3}{4}(x + 2\sqrt{2})$$

$$4y - 6\sqrt{2} = 3x + 6\sqrt{2}$$

$$3x - 4y + 12\sqrt{2} = 0$$

19. Find all points of horizontal tangency on the graph of
$x = 1 - t$, $y = t^3 - 3t$.

Solution

Since
$$\frac{dy}{dx} = \frac{dy/dt}{dx/dt} = \frac{3t^2 - 3}{-1} = 3 - 3t^2$$

the horizontal tangents occur when

$$3 - 3t^2 = 0 \qquad \text{or} \qquad t = \pm 1$$

The corresponding points are $(0, -2)$ and $(2, 2)$.

25. Find the length of the arc given by $x = e^{-t} \cos t$, $y = e^{-t} \sin t$
for $0 \le t \le \pi/2$.

Solution

$$x = e^{-t} \cos t$$

$$\frac{dx}{dt} = -e^{-t}(\sin t + \cos t)$$

$$\left(\frac{dx}{dt}\right)^2 = e^{-2t}(\sin^2 t + 2 \sin t \cos t + \cos^2 t)$$

$$= e^{-2t}(1 + \sin 2t)$$

$$y = e^{-t} \sin t$$

$$\frac{dy}{dt} = e^{-t}(\cos t - \sin t)$$

$$\left(\frac{dy}{dt}\right)^2 = e^{-2t}(\cos^2 t - 2 \sin t \cos t + \sin^2 t)$$

$$= e^{-2t}(1 - \sin 2t)$$

$$\left(\frac{dx}{dt}\right)^2 + \left(\frac{dy}{dt}\right)^2 = 2e^{-2t}$$

Therefore

$$s = \int_0^{\pi/2} \sqrt{\left(\frac{dx}{dt}\right)^2 + \left(\frac{dy}{dt}\right)^2}\, dt = \int_0^{\pi/2} \sqrt{2e^{-2t}}\, dt$$

$$= -\sqrt{2} \int_0^{\pi/2} e^{-t}(-1)\, dt = -\sqrt{2}\,[e^{-t}]_0^{\pi/2} = \sqrt{2}\,(1 - e^{-\pi/2}) \approx 1.12$$

31. Find the perimeter of the hypocycloid $x = a \cos^3 \theta$, $y = a \sin^3 \theta$.

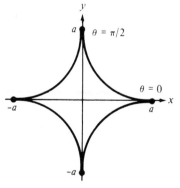

Solution

Since $dx/d\theta = -3a \cos^2 \theta \sin \theta$ and $dy/d\theta = 3a \sin^2 \theta \cos \theta$, then from the accompanying figure, we find the length of the hypocycloid to be

$$s = 4 \int_0^{\pi/2} \sqrt{\left(\frac{dx}{d\theta}\right)^2 + \left(\frac{dy}{d\theta}\right)^2}\, d\theta$$

$$= 4 \int_0^{\pi/2} \sqrt{9a^2 \cos^4 \theta \sin^2 \theta + 9a^2 \sin^4 \theta \cos^2 \theta}\, d\theta$$

$$= 4 \int_0^{\pi/2} \sqrt{9a^2 \sin^2 \theta \cos^2 \theta (\cos^2 \theta + \sin^2 \theta)}\, d\theta$$

$$= 4 \int_0^{\pi/2} 3a \sin \theta \cos \theta\, d\theta = 12a \left[\frac{\sin^2 \theta}{2}\right]_0^{\pi/2} = 6a$$

43. Sketch the graph of the cissoid $x = 2 \sin^2 \theta$, $y = 2 \sin^2 \theta \tan \theta$ and find the area of the region enclosed between the curve and its asymptote in the first quadrant.

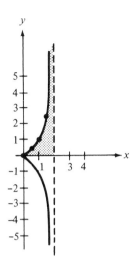

Solution

Using values from the following table, we make the accompanying sketch.

θ	0	$\dfrac{\pi}{6}$	$\dfrac{\pi}{4}$	$\dfrac{\pi}{3}$	$\dfrac{\pi}{2}$
x	0	$\dfrac{1}{2}$	1	$\dfrac{3}{2}$	2
y	0	$\dfrac{1}{2\sqrt{3}}$	1	$\dfrac{3\sqrt{3}}{2}$	undefined

From Theorem 11.4, we have the following convergent improper integral

$$A = \int_0^2 y \, dx = \int_0^{\pi/2} y \frac{dx}{d\theta} \, d\theta = \int_0^{\pi/2} 2 \sin^2 \theta \tan \theta (4 \sin \theta \cos \theta) \, d\theta$$

$$= 8 \int_0^{\pi/2} \sin^4 \theta \, d\theta = 8 \int_0^{\pi/2} \left(\frac{1 - \cos 2\theta}{2} \right)^2 d\theta$$

$$= 2 \int_0^{\pi/2} (1 - 2 \cos 2\theta + \cos^2 2\theta) \, d\theta$$

$$= 2 \int_0^{\pi/2} \left(1 - 2 \cos 2\theta + \frac{1 + \cos 4\theta}{2} \right) d\theta$$

$$= 2 \int_0^{\pi/2} \left(\frac{3}{2} - 2 \cos 2\theta + \frac{1}{2} \cos 4\theta \right) d\theta$$

$$= 2 \left[\frac{3}{2} \theta - \sin 2\theta + \frac{1}{8} \sin 4\theta \right]_0^{\pi/2} = 2 \left(\frac{3\pi}{4} \right) = \frac{3\pi}{2}$$

49. A portion of a sphere is removed by a circular cone with its vertex at the center of the sphere. Find the surface area removed from the sphere if the vertex of the cone forms an angle of 2θ.

Solution

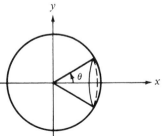

We represent the circle by

$$x = f(\phi) = a \cos \phi \qquad y = g(\phi) = a \sin \phi$$

$$\frac{dx}{d\phi} = -a \sin \phi \qquad \frac{dy}{d\phi} = a \cos \phi$$

From the integrals of Theorem 11.3 and the accompanying figure, we have

$$S = 2\pi \int_0^\theta g(\phi) \sqrt{\left(\frac{dx}{d\phi}\right)^2 + \left(\frac{dy}{d\phi}\right)^2} \, d\phi$$

$$= 2\pi \int_0^\theta a \sin \phi \sqrt{a^2 \sin^2 \phi + a^2 \cos^2 \phi} \, d\phi$$

$$= 2\pi a^2 \int_0^\theta \sin \phi \, d\phi = -2\pi a^2 [\cos \phi]_0^\theta = 2\pi a^2 (1 - \cos \theta)$$

11.3
Polar coordinates and polar graphs

7.   Plot the point $(\sqrt{2}, 2.36)$ in polar coordinates and find its corresponding rectangular coordinates.

Solution

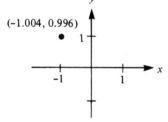

Using the conversion from polar to rectangular coordinates, we have

$$x = r \cos \theta = \sqrt{2} \cos (2.36) = -1.004$$
$$y = r \sin \theta = \sqrt{2} \sin (2.36) = 0.996$$

Therefore the rectangular coordinates are $(-1.004, 0.996)$.

11. Find two sets of polar coordinates for the point $(-3, 4)$ given in rectangular coordinates, using $0 \le \theta < 2\pi$.

Solution

First,

$$r = \pm\sqrt{x^2 + y^2} = \pm\sqrt{(-3)^2 + 4^2} = \pm 5$$

$$\tan \theta = -\frac{4}{3} \quad \text{and} \quad \arctan\left(-\frac{4}{3}\right) \approx 0.9273$$

Since $(-3, 4)$ lies in the second quadrant, we let $\theta = \pi - 0.9273 = 2.214$. Thus one polar representation is

$$(5, 2.214) \qquad (r > 0, 0 \le \theta < 2\pi)$$

To obtain a second representation, we rotate θ by π radians to obtain

$$(-5, 5.356) \qquad (r < 0, 0 \le \theta < 2\pi)$$

22. Find a polar equation of the horizontal line whose rectangular equation is $y = b$.

Solution

Since $y = r \sin \theta$, we have

$$y = b$$
$$r \sin \theta = b$$
$$r = \frac{b}{\sin \theta} = b \csc \theta$$

24. Find a polar equation of the vertical line whose rectangular equation is $x = a$.

Solution

Since $x = r \cos \theta$, we have

$$x = a$$
$$r \cos \theta = a$$
$$r = \frac{a}{\cos \theta} = a \sec \theta$$

25. Find a polar equation of the graph having the rectangular equation $3x - y + 2 = 0$.

Solution

Since $x = r \cos \theta$ and $y = r \sin \theta$, we have

$$3x - y + 2 = 0$$
$$3(r \cos \theta) - r \sin \theta + 2 = 0$$
$$r(3 \cos \theta - \sin \theta) = -2$$
$$r = \frac{-2}{3 \cos \theta - \sin \theta}$$

31. Find a polar equation of the graph having the rectangular equation $x^2 - 4ay - 4a^2 = 0$.

Solution

Since $x = r \cos \theta$ and $y = r \sin \theta$, we have

$$x^2 - 4ay - 4a^2 = 0$$
$$r^2 \cos^2 \theta - 4ar \sin \theta - 4a^2 = 0$$

$$r^2(1 - \sin^2 \theta) - 4ar \sin \theta - 4a^2 = 0$$

$$r^2 = r^2 \sin \theta + 4ar \sin \theta + 4a^2$$

$$r^2 = (r \sin \theta + 2a)^2$$

$$r = \pm(r \sin \theta + 2a)$$

$$r = \frac{2a}{1 - \sin \theta}$$

or

$$r = \frac{-2a}{1 + \sin \theta}$$

32. Find a polar equation of the graph having the rectangular equation $(x^2 + y^2)^2 - 9(x^2 - y^2) = 0$.

Solution

Since $r^2 = x^2 + y^2$, $x^2 = r^2 \cos^2 \theta$, and $y^2 = r^2 \sin^2 \theta$, we have

$$(x^2 + y^2)^2 - 9(x^2 - y^2) = 0$$

$$(r^2)^2 - 9(r^2 \cos^2 \theta - r^2 \sin^2 \theta) = 0$$

$$r^4 - 9r^2(\cos^2 \theta - \sin^2 \theta) = 0$$

$$r^2(r^2 - 9 \cos 2\theta) = 0$$

$$r^2 = 0 \quad \text{or} \quad r^2 = 9 \cos 2\theta$$

Since the pole is included in the equation $r^2 = 9 \cos 2\theta$, we can omit $r^2 = 0$ and conclude that a suitable polar equation is

$$r^2 = 9 \cos 2\theta$$

42. Find a rectangular equation of the graph having the polar equation $r = 2 \sin 3\theta$.

Solution

First we write $\sin 3\theta$ in terms of the single angle θ as follows:

$$\sin 3\theta = \sin (\theta + 2\theta) = \sin \theta \cos 2\theta + \cos \theta \sin 2\theta$$

$$= \sin \theta(2 \cos^2 \theta - 1) + \cos \theta(2 \sin \theta \cos \theta)$$

$$= 2 \sin \theta \cos^2 \theta - \sin \theta + 2 \sin \theta \cos^2 \theta$$

$$= 4 \sin \theta \cos^2 \theta - \sin \theta = \sin \theta(4 \cos^2 \theta - 1)$$

Now if we multiply the original equation by r, we have

$$r^2 = 2r \sin 3\theta = 2r \sin \theta(4 \cos^2 \theta - 1)$$

Since $r^2 = x^2 + y^2$, $r \sin \theta = y$, and $\cos^2 \theta = x^2/r^2 = x^2/(x^2 + y^2)$, we obtain

$$x^2 + y^2 = 2y\left(\frac{4x^2}{x^2 + y^2} - 1\right) = 2y\left(\frac{3x^2 - y^2}{x^2 + y^2}\right)$$

or $$(x^2 + y^2)^2 = 2y(3x^2 - y^2)$$

49. Sketch the graph of the polar equation $r = \sin \theta$, and indicate any symmetry possessed by the graph.

Solution

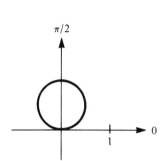

Since the sine is an odd function, we replace (r, θ) by $(-r, -\theta)$.

$$-r = \sin(-\theta) = -\sin \theta \rightarrow r = \sin \theta$$

Since the substitution produced an equivalent equation, the curve is symmetric with respect to the line $\theta = \pi/2$. This means that we need only use θ values from the first and fourth quadrants as shown in the following table.

θ	$-\dfrac{\pi}{2}$	$-\dfrac{\pi}{3}$	$-\dfrac{\pi}{4}$	$-\dfrac{\pi}{6}$	0	$\dfrac{\pi}{6}$	$\dfrac{\pi}{4}$	$\dfrac{\pi}{3}$	$\dfrac{\pi}{2}$
r	-1	$-\dfrac{\sqrt{3}}{2}$	$-\dfrac{\sqrt{2}}{2}$	$-\dfrac{1}{2}$	0	$\dfrac{1}{2}$	$\dfrac{\sqrt{2}}{2}$	$\dfrac{\sqrt{3}}{2}$	1

Plotting these points, we obtain the graph of the function.

11.4
Tangent lines and curve sketching in polar coordinates

1. Find the slope of the graph of $r = 3(1 - \cos \theta)$ at $\theta = \pi/2$.

Solution

Since $f(\theta) = 3(1 - \cos \theta)$, we have $f'(\theta) = 3 \sin \theta$ and

$$\frac{dy}{dx} = \frac{f'(\theta) \sin \theta + f(\theta) \cos \theta}{f'(\theta) \cos \theta - f(\theta) \sin \theta}$$

$$= \frac{(3 \sin \theta) \sin \theta + 3(1 - \cos \theta) \cos \theta}{(3 \sin \theta) \cos \theta - 3(1 - \cos \theta) \sin \theta}$$

$$= \frac{\sin^2 \theta + \cos \theta - \cos^2 \theta}{\sin \theta (2 \cos \theta - 1)}$$

When $\theta = \pi/2$, the slope is

$$\frac{dy}{dx} = \frac{1 + 0 - 0}{1(0 - 1)} = -1$$

5. Find the slope of the graph of $r = \theta$ at $\theta = \pi$.

Solution

Since $f(\theta) = \theta$, we have $f'(\theta) = 1$ and

$$\frac{dy}{dx} = \frac{f'(\theta) \sin \theta + f(\theta) \cos \theta}{f'(\theta) \cos \theta - f(\theta) \sin \theta} = \frac{\sin \theta + \theta \cos \theta}{\cos \theta - \theta \sin \theta}$$

When $\theta = \pi$, the slope is

$$\frac{dy}{dx} = \frac{0 + \pi(-1)}{(-1) - \pi(0)} = \frac{-\pi}{-1} = \pi$$

11. Find the horizontal and vertical tangents to the polar curve $r = 1 + \sin \theta$.

Solution

Since $f(\theta) = 1 + \sin \theta$ and $f'(\theta) = \cos \theta$, we have

$$\frac{dy}{dx} = \frac{f'(\theta) \sin \theta + f(\theta) \cos \theta}{f'(\theta) \cos \theta - f(\theta) \sin \theta}$$

$$= \frac{\cos \theta \sin \theta + (1 + \sin \theta) \cos \theta}{\cos^2 \theta - (1 + \sin \theta) \sin \theta}$$

$$= \frac{\cos \theta(1 + 2 \sin \theta)}{1 - \sin \theta - 2 \sin^2 \theta}$$

$$= \frac{\cos \theta(1 + 2 \sin \theta)}{(1 - 2 \sin \theta)(1 + \sin \theta)}$$

$dy/dx = 0$ when $\cos \theta(1 + 2 \sin \theta) = 0$ or $\theta = \pi/2, 7\pi/6, 11\pi/6$. Therefore the horizontal tangents are at the points

$$\left(2, \frac{\pi}{2}\right), \left(\frac{1}{2}, \frac{7\pi}{6}\right), \quad \text{and} \quad \left(\frac{1}{2}, \frac{11\pi}{6}\right)$$

dy/dx is undefined when $(1 - 2 \sin \theta)(1 + \sin \theta) = 0$ or $\theta = \pi/6, 5\pi/6, 3\pi/2$. Therefore the vertical tangents are at the points

$$\left(\frac{3}{2}, \frac{\pi}{6}\right), \left(\frac{3}{2}, \frac{5\pi}{6}\right), \quad \text{and} \quad \left(0, \frac{3\pi}{2}\right)$$

(Note that L'Hôpital's Rule must be used to show that dy/dx is undefined at $\theta = 3\pi/2$.)

23. Sketch and identify the graph of $r = 2 + 3 \sin \theta$ as a limacon, cardioid, rose curve, or conic. Find the tangents at the pole.

Solution

We first observe that the given equation has the form

$$r = a + b \sin \theta = 2 + 3 \sin \theta$$

with $b > a$, which means its graph is a limacon with two loops. Furthermore, since r is a function of $\sin \theta$, the graph has vertical axis symmetry. The relative extrema of r are $(-1, 3\pi/2)$ and $(5, \pi/2)$. Using this information and the points in the table, we make the accompanying sketch.

θ	0	$\dfrac{\pi}{6}$	$\dfrac{\pi}{2}$	$-\pi$	$\dfrac{3\pi}{2}$	$\dfrac{-\pi}{6}$
r	2	$\dfrac{7}{2}$	5	2	-1	$\dfrac{1}{2}$

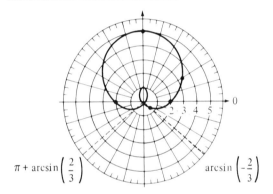

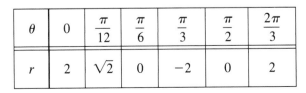

$$\pi + \arcsin\left(\frac{2}{3}\right) \qquad\qquad \arcsin\left(-\frac{2}{3}\right)$$

31.   Sketch and identify the graph of $r = 2 \cos 3\theta$ as a limacon, cardioid, rose curve, or conic. Find the tangents at the pole.

Solution

The equation has the form

$$r = a \cos (n\theta) = 2 \cos 3\theta$$

where n is odd. This means the graph is a rose curve with $n = 3$ petals. The curve has polar axis symmetry. The relative extrema of r are $(2, 0)$ $(-2, \pi/3)$, $(2, 2\pi/3)$. The tangents at the pole are $\theta = \pi/6$, $\theta = \pi/2$, and $\theta = 5\pi/6$, because $r = 0$ at these values. Using this information and the points in the table, we make the accompanying sketch.

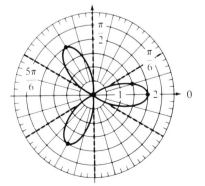

θ	0	$\dfrac{\pi}{12}$	$\dfrac{\pi}{6}$	$\dfrac{\pi}{3}$	$\dfrac{\pi}{2}$	$\dfrac{2\pi}{3}$
r	2	$\sqrt{2}$	0	-2	0	2

47. Sketch the graph of $r = 2 \cos (3\theta/2)$. Find the tangents at the pole.

Solution

Symmetry is with respect to the polar axis and the line $\theta = \pi/2$. Tangents at the pole: $r = 0$ when

$$\theta = \frac{\pi}{3}, \qquad \theta = \pi, \qquad \theta = \frac{5\pi}{3}$$

Relative extrema of r: $dr/d\theta = 0$ at

$$(2, 0), \qquad \left(-2, \frac{2\pi}{3}\right), \qquad \left(2, \frac{4\pi}{3}\right),$$

$$(-2, 2\pi), \qquad \left(2, \frac{8\pi}{3}\right), \qquad \left(-2, \frac{10\pi}{3}\right)$$

Using this information and the points in the table, we make the accompanying sketch.

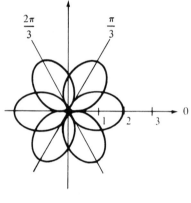

θ	0	$\dfrac{\pi}{9}$	$\dfrac{\pi}{6}$	$\dfrac{2\pi}{9}$	$\dfrac{\pi}{3}$
r	2	$\sqrt{3}$	$\sqrt{2}$	1	0

49. Sketch the graph of $r = 2 - \sec \theta$ and show that $x = -1$ is a vertical asymptote of the graph.

Solution

First, we write

$$r = 2 - \sec \theta = 2 - \frac{1}{\cos \theta}$$

and note that the graph will have polar axis symmetry. The tangents at the pole are $\theta = \pi/3$ and $\theta = -\pi/3$. Furthermore,

$$r \to -\infty \qquad \text{as} \qquad \theta \to \frac{\pi^-}{2}$$

and

$$r \to \infty \qquad \text{as} \qquad \theta \to \frac{-\pi^+}{2}$$

To see that the graph has a vertical asymptote at $x = -1$, we write

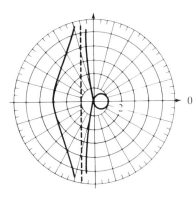

$$r = 2 - \frac{1}{\cos \theta} = 2 - \frac{r}{r \cos \theta} = 2 - \frac{r}{x}$$

$$rx = 2x - r$$

$$r(1 + x) = 2x$$

$$r = \frac{2x}{1 + x}$$

Thus $r \to \pm\infty$ as $x \to -1$, as shown in the accompanying figure.

11.5
Polar equations and conics

5. Sketch the graph of the equation $r = 2/(2 - \cos \theta)$ and identify the curve.

Solution

To determine the type of conic, we rewrite the equation as

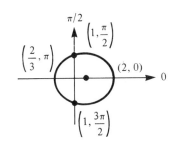

$$r = \frac{2}{2 - \cos \theta} = \frac{1}{1 - \frac{1}{2} \cos \theta} = \frac{ep}{1 - e \cos \theta}$$

From this form we conclude that the graph is an ellipse with $e = 1/2$. We sketch the upper half of the ellipse by plotting points from $\theta = 0$ to $\theta = \pi$ as shown in the accompanying figure. Then, using symmetry with respect to the polar axis we sketch the lower half.

9. Sketch the graph of the equation $r = -1/(1 - \sin \theta)$ and identify the curve.

Solution

From the form of the equation we have

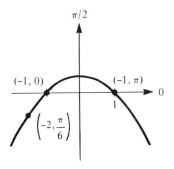

$$r = \frac{-1}{1 - \sin \theta} = \frac{ep}{1 - e \sin \theta}$$

We conclude that the graph of the equation is a parabola ($e = 1$). We sketch the left half of the parabola by plotting points for $-\pi/2 \le \theta < \pi/2$. Then using symmetry with respect to $\theta = \pi/2$, we sketch the right half.

13. Sketch the graph of the equation $r = 3/(2 - 6 \cos \theta)$ and identify the curve.

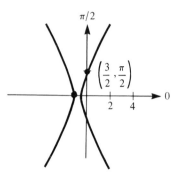

Solution

To determine the type of conic, we rewrite the equation as

$$r = \frac{3}{2 - 6 \cos \theta} = \frac{3/2}{1 - 3 \cos \theta} = \frac{ep}{1 - e \cos \theta}$$

From this form we conclude that the graph is a hyperbola with $e = 3$. We sketch the upper half of the right branch and the lower half of the left branch of the hyperbola by plotting points from $\theta = 0$ to $\theta = \pi$ as shown in the accompanying figure. Then, using symmetry with respect to the polar axis, we sketch the other half of the graph. Note that r is undefined when $\theta = \arccos(1/3) \approx 70.5°$.

17. Find a polar equation for the ellipse with focus at the pole, eccentricity $e = \frac{1}{2}$ and directrix $y = 1$.

Solution

Since the directrix is horizontal and above the pole (see the accompanying figure), we choose an equation of the form

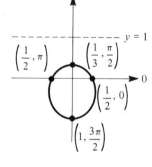

$$r = \frac{ep}{1 + e \sin \theta}$$

Moreover, since the eccentricity of the ellipse is $\frac{1}{2}$ and the directed distance from the focus to the directrix is $p = 1$, we have the equation

$$r = \frac{\frac{1}{2}}{1 + \frac{1}{2} \sin \theta} = \frac{1}{2 + \sin \theta}$$

23. Find a polar equation of the parabola with focus at the pole and vertex at $(1, -\pi/2)$.

Solution

Since the directrix is horizontal and below the pole (see the accompanying figure), we choose an equation of the form

$$r = \frac{ep}{1 - e \sin \theta}$$

Moreover, since the eccentricity of a parabola is $e = 1$ and the distance from the focus to the directrix is $p = 2$, we have the equation

$$r = \frac{2}{1 - \sin \theta}$$

27. Find an equation of the hyperbola with focus at the pole, vertex at $(2, 3\pi/2)$ and directrix $y = -3$.

Solution

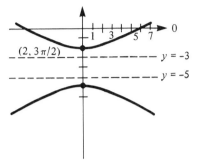

Since the directrix is horizontal and below the pole we choose an equation of the form

$$r = \frac{ep}{1 - e \sin \theta}$$

A hyperbola is the locus of a point that moves so that its distance from the focus is in constant ratio to its distance from the directrix. Therefore, when $\theta = 3\pi/2$ in the accompanying figure we have

$$e = \frac{2}{1} = 2$$

Moreover, since the distance between the pole and the directirx is $p = 3$, we have the equation

$$r = \frac{2(3)}{1 - 2 \sin \theta} = \frac{6}{1 - 2 \sin \theta}$$

29. Show that the corresponding polar equation of the ellipse $(x^2/a^2) + (y^3/b^2) = 1$ is

$$r^2 = \frac{b^2}{1 - e^2 \cos^2 \theta}$$

Solution

Since $x = r \cos \theta$ and $y = r \sin \theta$, we have

$$\frac{x^2}{a^2} + \frac{y^2}{b^2} = 1$$
$$b^2(r^2 \cos^2 \theta) + a^2(r^2 \sin^2 \theta) = a^2 b^2$$
$$r^2[b^2 \cos^2 \theta + a^2(1 - \cos^2 \theta)] = a^2 b^2$$
$$r^2\left[1 - \left(\frac{a^2 - b^2}{a^2}\right) \cos^2 \theta\right] = b^2$$
$$r^2(1 - e^2 \cos^2 \theta) = b^2$$
$$r^2 = \frac{b^2}{1 - e^2 \cos^2 \theta}$$

43. Find the angle ψ for the graph of $r = 6/(1 - \cos \theta)$ at $\theta = 2\pi/3$.

Solution

Since

$$f(\theta) = r = \frac{6}{1 - \cos \theta}$$

we get

$$f'(\theta) = \frac{-6 \sin \theta}{(1 - \cos \theta)^2}$$

Now from the definition of the angle between the radial line and the tangent line (see Figure 11.42), we have

$$\tan \psi = \left| \frac{f(\theta)}{f'(\theta)} \right| = \left| \frac{6/(1 - \cos \theta)}{-6 \sin \theta/(1 - \cos \theta)^2} \right| = \left| \frac{1 - \cos \theta}{-\sin \theta} \right|$$

At $\theta = 2\pi/3$,

$$\tan \psi = \left| \frac{1 - (-\frac{1}{2})}{-\sqrt{3}/2} \right| = \left| \frac{\frac{3}{2}}{-\sqrt{3}/2} \right| = \sqrt{3}$$

Therefore we conclude that $\psi = \pi/3$.

11.6
Area and arc length in polar coordinates

5. Find the points of intersection of the graphs of $r = 4 - 5 \sin \theta$ and $r = 3 \sin \theta$.

Solution

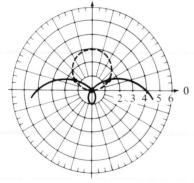

Solving the two equations simultaneously, we have

$$4 - 5 \sin \theta = 3 \sin \theta$$
$$8 \sin \theta = 4$$
$$\sin \theta = \frac{1}{2}$$
$$\theta = \frac{\pi}{6}, \frac{5\pi}{6} \qquad (0 \leq \theta < 2\pi)$$

From these values we obtain the points $(3/2, \pi/6)$ and $(3/2, 5\pi/6)$. To test for additional points of intersection, we replace r by $-r$ and θ by $\pi + \theta$ in $r = 4 - 5 \sin \theta$ to obtain

$$-r = 4 - 5 \sin (\pi + \theta) = 4 + 5 \sin \theta$$

Solving this equation simultaneously with $r = 3 \sin \theta$, we have

$$-4 - 5 \sin \theta = 3 \sin \theta$$

$$8 \sin \theta = -4$$

$$\sin \theta = -\frac{1}{2}$$

$$\theta = \frac{7\pi}{6}, \frac{11\pi}{6}$$

which yields the points $(-3/2, 7\pi/6)$ and $(-3/2, 11\pi/6)$. However, these two points coincide with the previous two points. Finally, we observe that both curves pass through the pole. Hence there are three points of intersection, $(3/2, \pi/6)$, $(3/2, 5\pi/6)$, and $(0, 0)$ as seen in the accompanying figure.

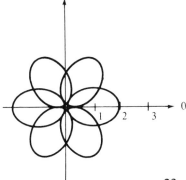

15. Find the area of the region enclosed by one loop of $r = \cos(3\theta/2)$.

Solution

From Exercise 47 (Section 11.4), we see that the area of the loop is given by

$$A = 2\left[\frac{1}{2}\int_0^{\pi/3} \cos^2\left(\frac{3\theta}{2}\right) d\theta\right] = \int_0^{\pi/3} \frac{1 + \cos 3\theta}{2} d\theta$$

$$= \frac{1}{2}\left[\theta + \frac{1}{3}\sin 3\theta\right]_0^{\pi/3} = \frac{\pi}{6}$$

23. Find the area of the region between the loops of $r = 1 + 2\cos\theta$.

Solution

From the symmetry of the graph given in the accompanying figure, we consider the region inside the outer loop and write

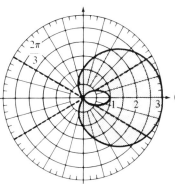

$$(\text{area inside outer loop}) = 2\int_0^{2\pi/3} \frac{1}{2}r^2 \, d\theta = \int_0^{2\pi/3} (1 + 2\cos\theta)^2 \, d\theta$$

$$= \int_0^{2\pi/3} (1 + 4\cos\theta + 4\cos^2\theta) \, d\theta$$

$$= \int_0^{2\pi/3} \left(1 + 4\cos\theta + 4\frac{1 + \cos 2\theta}{2}\right) d\theta$$

$$= \int_0^{2\pi/3} (3 + 4\cos\theta + 2\cos 2\theta) \, d\theta$$

$$= \left[3\theta + 4\sin\theta + \sin 2\theta\right]_0^{2\pi/3}$$

$$= 2\pi + 2\sqrt{3} - \frac{\sqrt{3}}{2} = 2\pi + \frac{3\sqrt{3}}{2}$$

From Exercise 21 (Section 11.6), we see that the area of the inner loop is given by

$$(\text{area inside inner loop}) = 2 \int_{2\pi/3}^{\pi} \frac{1}{2} r^2 \, d\theta$$

$$= \int_{2\pi/3}^{\pi} (1 + 2 \cos \theta)^2 \, d\theta = \pi - \frac{3\sqrt{3}}{2}$$

Finally, the area of the region between the two loops is

$$A = \left(2\pi + \frac{3\sqrt{3}}{2}\right) - \left(\pi - \frac{3\sqrt{3}}{2}\right) = \pi + 3\sqrt{3}$$

25. Find the area of the region common to the interiors of $r = 4 \sin 2\theta$ and $r = 2$.

Solution

From the accompanying sketch we see that we need only consider the region in one petal common to both curves and multiply the result by four. We obtain the points of intersection from the equation

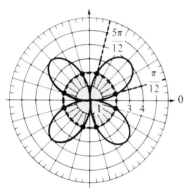

$$4 \sin 2\theta = 2$$

$$\sin 2\theta = \frac{1}{2}$$

$$2\theta = \frac{\pi}{6}, \frac{5\pi}{6}$$

$$\theta = \frac{\pi}{12}, \frac{5\pi}{12}$$

There are three subregions within one petal:

a. for $0 \le \theta \le \pi/12$, $r = 4 \sin 2\theta$
b. for $\pi/12 \le \theta \le 5\pi/12$, $r = 2$
c. for $5\pi/12 \le \theta \le \pi/2$, $r = 4 \sin 2\theta$

Therefore the area within one petal is

$$A = 2 \int_{0}^{\pi/12} \frac{1}{2} (4 \sin 2\theta)^2 \, d\theta + \int_{\pi/12}^{5\pi/12} \frac{1}{2} (2)^2 \, d\theta + \int_{5\pi/12}^{\pi/2} \frac{1}{2} (4 \sin 2\theta)^2 \, d\theta$$

By the symmetry of the petal, the first and third integrals are equal. Thus

$$A = \int_{0}^{\pi/12} \frac{1}{2} (4 \sin 2\theta)^2 \, d\theta + \int_{\pi/12}^{5\pi/12} \frac{1}{2} (2)^2 \, d\theta$$

$$= 16 \int_{0}^{\pi/12} \sin^2 2\theta \, d\theta + 2 \int_{\pi/12}^{5\pi/12} d\theta$$

$$= 8 \int_{0}^{\pi/12} (1 - \cos 4\theta) \, d\theta + 2 \int_{\pi/12}^{5\pi/12} d\theta$$

$$= 8\left[\theta - \frac{1}{4}\sin 4\theta\right]_0^{\pi/12} + [2\theta]_{\pi/12}^{5\pi/12}$$

$$= \frac{2\pi}{3} - \frac{2\sqrt{3}}{2} + \frac{2\pi}{3} = \frac{4\pi}{3} - \sqrt{3}$$

Finally, multiplying by 4, we obtain the total area of $(16\pi/3) - 4\sqrt{3}$.

41. Find the length of the graph of $r = 1/\theta$ over the interval $\pi \le \theta \le 2\pi$.

Solution

$$s = \int_\alpha^\beta \sqrt{[f(\theta)]^2 + [f'(\theta)]^2}\; d\theta$$

$$= \int_\pi^{2\pi} \sqrt{\left(\frac{1}{\theta}\right)^2 + \left(\frac{-1}{\theta^2}\right)^2}\; d\theta$$

$$= \int_\pi^{2\pi} \frac{1}{\theta^2}\sqrt{\theta^2 + 1}\; d\theta$$

$$= \left[-\frac{\sqrt{\theta^2 + 1}}{\theta} + \ln\left|\theta + \sqrt{\theta^2 + 1}\right|\right]_\pi^{2\pi}$$

$$= \frac{2\sqrt{\pi^2 + 1} - \sqrt{4\pi^2 + 1}}{2\pi} + \ln\left|\frac{2\pi + \sqrt{4\pi^2 + 1}}{\pi + \sqrt{\pi^2 + 1}}\right| \approx 0.7112$$

46. Find the area of the surface formed by revolving the curve $r = a(1 + \cos\theta)$ over the interval $0 \le \theta \le \pi$ about the polar axis.

Solution

$$S = 2\pi \int_\alpha^\beta f(\theta)\sin\theta\sqrt{[f(\theta)]^2 + [f'(\theta)]^2}\; d\theta$$

$$= 2\pi \int_0^\pi a(1 + \cos\theta)\sin\theta\sqrt{a^2(1 + \cos\theta)^2 + a^2\sin^2\theta}\; d\theta$$

$$= 2\pi a^2 \int_0^\pi \sin\theta(1 + \cos\theta)\sqrt{2 + 2\cos\theta}\; d\theta$$

$$= -2\sqrt{2}\,\pi a^2 \int_0^\pi (1 + \cos\theta)^{3/2}(-\sin\theta)\; d\theta$$

$$= -\frac{4\sqrt{2}\,\pi a^2}{5}[(1 + \cos\theta)^{5/2}]_0^\pi$$

$$= \frac{32\pi a^2}{5}$$

Review Exercises for Chapter 11

3. (a) Find dy/dx and all points of horizontal tangency, (b) eliminate the parameter if possible, and (c) sketch the curve represented by the parametric equations $x = 3 + 2 \cos \theta$ and
$y = 2 + 5 \sin \theta$.

Solution

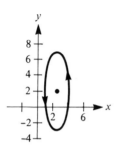

(a) Since $x = 3 + 2 \cos \theta$ and $y = 2 + 5 \sin \theta$, we have

$$\frac{dy}{dx} = \frac{dy/d\theta}{dx/d\theta} = \frac{5 \cos \theta}{-2 \sin \theta} = -\frac{5}{2} \cot \theta$$

The points of horizontal tangency occur at $\theta = \pi/2$ and
$\theta = 3\pi/2$, and the points are $(3, 7)$ and $(3, -3)$.

(b) To eliminate the parameter, we consider the identity $\sin^2 \theta + \cos^2 \theta = 1$, and write

$$\frac{x - 3}{2} = \cos \theta, \qquad \frac{y - 2}{5} = \sin \theta$$

By squaring and adding these equations, we obtain

$$\frac{(x - 3)^2}{4} + \frac{(y - 2)^2}{25} = 1$$

which is an equation for the ellipse centered at $(3, 2)$ with
vertices at $(3, -3)$ and $(3, 7)$.

12. Find a parametric representation of the hyperbola with vertices
at $(0, \pm 4)$ and foci at $(0, \pm 5)$.

Solution

From the vertices and foci we conclude that $h = 0$, $k = 0$,
$a = 4$, $c = 5$, and $b^2 = c^2 - a^2 = 9$. The standard equation
is therefore

$$\frac{y^2}{16} - \frac{x^2}{9} = 1$$

Since $\sec^2 \theta - \tan^2 \theta = 1$, we consider

$$\frac{y}{4} = \sec \theta \qquad \text{and} \qquad \frac{x}{3} = \tan \theta$$

and obtain the parametric equations

$$x = 3 \tan \theta \qquad \text{and} \qquad y = 4 \sec \theta$$

15. Find the length of the involute of Exercise 14, $x = r(\cos\theta + \theta\sin\theta)$, $y = r(\sin\theta - \theta\cos\theta)$, when $\theta = \pi$.

Solution

$$x = r(\cos\theta + \theta\sin\theta) \qquad y = r(\sin\theta - \theta\cos\theta)$$

$$\frac{dx}{d\theta} = r\theta\cos\theta \qquad\qquad \frac{dy}{d\theta} = r\theta\sin\theta$$

$$s = \int_0^\pi \sqrt{\left(\frac{dx}{d\theta}\right)^2 + \left(\frac{dy}{d\theta}\right)^2}\,d\theta = \int_0^\pi \sqrt{(r\theta\cos\theta)^2 + (r\theta\sin\theta)^2}\,d\theta$$

$$= r\int_0^\pi \theta\,d\theta = r\left[\frac{\theta^2}{2}\right]_0^\pi = \frac{1}{2}\pi^2 r$$

32. Sketch the graph of $r = 4(\sec\theta - \cos\theta)$.

Solution

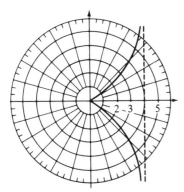

The graph is symmetric with respect to the polar axis, and tangents at the pole are $\theta = 0$ and $\theta = \pi$. Note also that

$$r \longrightarrow \infty \qquad \text{as} \qquad \theta \longrightarrow \frac{\pi^-}{2}$$

and

$$r \longrightarrow -\infty \qquad \text{as} \qquad \theta \longrightarrow \frac{\pi^+}{2}$$

With this information and the following table, we make the accompanying sketch.

θ	0	$\dfrac{\pi}{6}$	$\dfrac{\pi}{3}$	$\dfrac{\pi}{2}$	$\dfrac{2\pi}{3}$	$\dfrac{5\pi}{6}$	π
r	0	1.155	6	undefined	-6	-1.155	0

37. Find a rectangular equation that has the same graph as $r = 4\cos 2\theta\sec\theta$.

Solution

First, we replace $\cos 2\theta$ by $2\cos^2\theta - 1$ and $\sec\theta$ by $1/\cos\theta$ to obtain

$$r = 4\cos 2\theta\sec\theta = 4(2\cos^2\theta - 1)\frac{1}{\cos\theta}$$

$$r\cos\theta = 8\cos^2\theta - 4$$

Now since $x = r \cos \theta$ and $r^2 = x^2 + y^2$, we have

$$x = 8\left(\frac{x^2}{r^2}\right) - 4 = \frac{8x^2 - 4(x^2 + y^2)}{x^2 + y^2}$$

$$x^3 + xy^2 = 4x^2 - 4y^2$$

$$(4 + x)y^2 = (x^2)(4 - x)$$

$$y^2 = x^2\left(\frac{4 - x}{4 + x}\right)$$

41. Find a polar equation for the rectangular equation

$$x^2 + y^2 = a^2\left(\arctan \frac{y}{x}\right)^2$$

Solution

Since $r^2 = x^2 + y^2$ and $\theta = \arctan \dfrac{y}{x}$, we have

$$x^2 + y^2 = a^2\left(\arctan \frac{y}{x}\right)^2$$

$$r^2 = a^2\theta^2$$

$$r = \pm a\theta$$

49. Find the tangent lines at the pole and all points of vertical and horizontal tangency for the graph of $r = 1 - 2 \cos \theta$. Sketch the graph of the equation.

Solution

The graph has polar axis symmetry and the tangents at the pole are $\theta = \pi/3$ and $\theta = -\pi/3$. To find the points of vertical or horizontal tangency, we note that $f'(\theta) = 2 \sin \theta$ and find dy/dx as follows:

$$\frac{dy}{dx} = \frac{f'(\theta) \sin \theta + f(\theta) \cos \theta}{f'(\theta) \cos \theta - f(\theta) \sin \theta}$$

$$= \frac{2 \sin^2 \theta + (1 - 2 \cos \theta) \cos \theta}{2 \sin \theta \cos \theta - (1 - 2 \cos \theta) \sin \theta}$$

$$= \frac{2 \sin^2 \theta + \cos \theta - 2 \cos^2 \theta}{4 \sin \theta \cos \theta - \sin \theta}$$

$$= \frac{2(1 - \cos^2 \theta) + \cos \theta - 2 \cos^2 \theta}{\sin \theta (4 \cos \theta - 1)}$$

$$= \frac{2 + \cos \theta - 4 \cos^2 \theta}{\sin \theta (4 \cos \theta - 1)}$$

The graph has horizontal tangents when $dy/dx = 0$, and this occurs when

$$-4 \cos^2 \theta + \cos \theta + 2 = 0$$

$$\cos \theta = \frac{-1 \pm \sqrt{1 + 32}}{-8} = \frac{1 \mp \sqrt{33}}{8}$$

When $\cos \theta = (1 \mp \sqrt{33})/8$,

$$r = 1 - 2\left(\frac{1 \mp \sqrt{33}}{8}\right) = \frac{3 \pm \sqrt{33}}{4}$$

Therefore the points of horizontal tangency are

$$\left(\frac{3 - \sqrt{33}}{4}, \ \arccos\left[\frac{1 + \sqrt{33}}{8}\right]\right) \approx (-0.686, \ 0.568)$$

$$\left(\frac{3 - \sqrt{33}}{4}, \ -\arccos\left[\frac{1 + \sqrt{33}}{8}\right]\right) \approx (-0.686, \ -0.568)$$

$$\left(\frac{3 + \sqrt{33}}{4}, \ \arccos\left[\frac{1 - \sqrt{33}}{8}\right]\right) \approx (2.186, \ 2.206)$$

$$\left(\frac{3 + \sqrt{33}}{4}, \ -\arccos\left[\frac{1 - \sqrt{33}}{8}\right]\right) \approx (2.186, \ -2.206)$$

The graph has vertical tangents when

$$\sin \theta (4 \cos \theta - 1) = 0$$

$$\sin \theta = 0 \quad \text{or} \quad \cos \theta = \frac{1}{4}$$

$$\theta = 0, \ \pi \quad \text{or} \quad \theta = \pm\arccos \frac{1}{4}$$

When $\cos \theta = \frac{1}{4}$, $r = 1 - 2(\frac{1}{4}) = \frac{1}{2}$. Thus the points of vertical tangency are

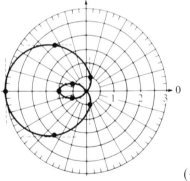

$$(-1, 0), \ (3, \pi), \quad \text{and} \quad \left(\frac{1}{2}, \ \pm\arccos \frac{1}{4}\right) \approx (0.5, \ \pm1.318)$$

as shown in the accompanying figure.

51. Show that the cardioids $r = 1 + \cos \theta$ and $r = 1 - \cos \theta$ are orthogonal at their two points of intersection other than the pole.

Solution

From the accompanying figure, we can see that the points $(1, \pi/2)$ and $(1, 3\pi/2)$ are the two points of intersection (other than the pole). The slope of the graph of $r = 1 + \cos \theta$ is

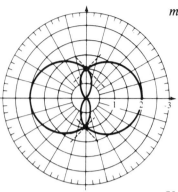

$$m_1 = \frac{dy}{dx} = \frac{r' \sin\theta + r \cos\theta}{r' \cos\theta - r \sin\theta} = \frac{-\sin^2\theta + \cos\theta(1 + \cos\theta)}{-\sin\theta\cos\theta - \sin\theta(1 + \cos\theta)}$$

At $(1, \pi/2)$, $m_1 = -1/(-1) = 1$ and at $(1, 3\pi/2)$, $m_1 = -1/1 = -1$. The slope of the graph of $r = 1 - \cos\theta$ is

$$m_2 = \frac{dy}{dx} = \frac{\sin^2\theta + \cos\theta(1 - \cos\theta)}{\sin\theta\cos\theta - \sin\theta(1 - \cos\theta)},$$

At $(1, \pi/2)$, $m_2 = 1/(-1) = -1$ and at $(1, 3\pi/2)$, $m_2 = 1/1 = 1$. In both cases $m_1 = -1/m_2$ and we conclude that the graphs are orthogonal at $(1, \pi/2)$ and $(1, 3\pi/2)$.

59.   Find the area of the region common to the interiors of $r = 4\cos\theta$ and $r = 2$.

Solution

To find the point of intersection of the two graphs we solve the two equations simultaneously to obtain

$$4\cos\theta = 2$$

$$\cos\theta = \frac{1}{2}$$

$$\theta = \frac{\pi}{3}$$

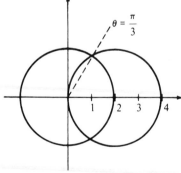

From the accompanying figure we see that the area is given by

$$A = 2\left[\frac{1}{2} \int_0^{\pi/3} 2^2\, d\theta + \frac{1}{2} \int_{\pi/3}^{\pi/2} (4\cos\theta)^2\, d\theta \right]$$

$$= \int_0^{\pi/3} 4\, d\theta + 16 \int_{\pi/3}^{\pi/2} \frac{1 + \cos 2\theta}{2}\, d\theta$$

$$= \left[4\theta\right]_0^{\pi/3} + 8\left[\theta + \frac{1}{2}\sin 2\theta\right]_{\pi/3}^{\pi/2}$$

$$= \frac{8\pi - 6\sqrt{3}}{3}$$

61.   Find the perimeter of a cardioid $r = a(1 - \cos\theta)$.

Solution

$$r = a(1 - \cos\theta) \qquad \text{and} \qquad \frac{dr}{d\theta} = a\sin\theta$$

Due to the symmetry with respect to the polar axis, we have

$$s = 2 \int_0^\pi \sqrt{r^2 + \left(\frac{dr}{d\theta}\right)^2}\, d\theta$$

$$= 2 \int_0^\pi \sqrt{a^2(1 - \cos\theta)^2 + a^2 \sin^2\theta}\, d\theta$$

$$= 2\sqrt{2}\, a \int_0^\pi \sqrt{1 - \cos\theta}\, d\theta$$

$$= 2\sqrt{2}\, a \int_0^\pi \sqrt{1 - \cos\theta}\, \frac{\sqrt{1 + \cos\theta}}{\sqrt{1 + \cos\theta}}\, d\theta$$

$$= 2\sqrt{2}\, a \int_0^\pi \frac{\sin\theta}{\sqrt{1 + \cos\theta}}\, d\theta$$

$$= -4\sqrt{2}\, a[(1 + \cos\theta)^{1/2}]_0^\pi = 8a$$

12 Vectors and the geometry of space

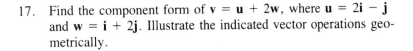

12.1

Vectors in the plane

4. Find the component form and length of the vector **v** in the accompanying figure.

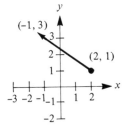

Solution

v is the directed line segment from the point $(x_1, y_1) = (2, 1)$ to the point $(x_2, y_2) = (-1, 3)$. Therefore, $\mathbf{v} = a\mathbf{i} + b\mathbf{j}$, where

$$a = x_2 - x_1 = -1 - 2 = -3$$

and

$$b = y_2 - y_1 = 3 - 1 = 2$$

Thus

$$\mathbf{v} = -3\mathbf{i} + 2\mathbf{j} = \langle -3, 2 \rangle$$

17. Find the component form of $\mathbf{v} = \mathbf{u} + 2\mathbf{w}$, where $\mathbf{u} = 2\mathbf{i} - \mathbf{j}$ and $\mathbf{w} = \mathbf{i} + 2\mathbf{j}$. Illustrate the indicated vector operations geometrically.

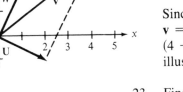

Solution

Since $\mathbf{u} = 2\mathbf{i} - \mathbf{j}$ and $2\mathbf{w} = 2(\mathbf{i} + 2\mathbf{j}) = 2\mathbf{i} + 4\mathbf{j}$, we have $\mathbf{v} = \mathbf{u} + 2\mathbf{w} = (2\mathbf{i} - \mathbf{j}) + (2\mathbf{i} + 4\mathbf{j}) = (2 + 2)\mathbf{i} + (4 - 1)\mathbf{j} = 4\mathbf{i} + 3\mathbf{j} = \langle 4, 3 \rangle$. Geometrically, this sum is illustrated in the accompanying figure.

23. Find a and b such that $\mathbf{v} = a\mathbf{u} + b\mathbf{w}$ where $\mathbf{v} = \langle 3, 0 \rangle$, $\mathbf{u} = \langle 1, 2 \rangle$, and $\mathbf{w} = \langle 1, -1 \rangle$.

Solution

$$\mathbf{v} = a\mathbf{u} + b\mathbf{w}$$
$$3\mathbf{i} = a(\mathbf{i} + 2\mathbf{j}) + b(\mathbf{i} - \mathbf{j}) = (a + b)\mathbf{i} + (2a - b)\mathbf{j}$$

Equating coefficients we have

$$a + b = 3$$

and

$$2a - b = 0$$

Solving the system of equations simultaneously we obtain $a = 1$ and $b = 2$.

37. Demonstrate the triangle inequality using the vectors $\mathbf{u} = \langle 2, 1 \rangle$ and $\mathbf{v} = \langle 5, 4 \rangle$.

Solution

$$\|\mathbf{u}\| = \sqrt{2^2 + 1^2} = \sqrt{5}$$
$$\|\mathbf{v}\| = \sqrt{5^2 + 4^2} = \sqrt{41}$$

Since $\mathbf{u} + \mathbf{v} = \langle 2, 1 \rangle + \langle 5, 4 \rangle = \langle 7, 5 \rangle$ we have

$$\|\mathbf{u} + \mathbf{v}\| = \sqrt{7^2 + 5^2} = \sqrt{74}$$

Therefore,

$$\|\mathbf{u} + \mathbf{v}\| = \sqrt{74} \approx 8.602 \le 2.236 + 6.403$$
$$\approx \sqrt{5} + \sqrt{41} = \|\mathbf{u}\| + \|\mathbf{v}\|$$

43. Find a unit vector (a) parallel to and (b) normal to the graph of $f(x) = x^3$ at $(1, 1)$.

Solution

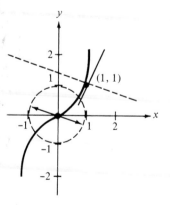

(a) At $(1, 1)$ the slope of the tangent line is

$$f'(1) = 3$$

Therefore a vector $\mathbf{v} = x\mathbf{i} + y\mathbf{j}$ parallel to the tangent line must have a slope of 3 (see the accompanying figure) and $y = 3x$. Since $\mathbf{v}$ is a unit vector, we have $x^2 + y^2 = 1$. Thus we have

$$x^2 + (3x)^2 = 1$$
$$10x^2 = 1$$

$$x = \pm \frac{1}{\sqrt{10}} \quad \text{and} \quad y = \pm \frac{3}{\sqrt{10}}$$

Finally we conclude that

$$\mathbf{v} = \frac{1}{\sqrt{10}}\mathbf{i} + \frac{3}{\sqrt{10}}\mathbf{j} = \left\langle \frac{1}{\sqrt{10}}, \frac{3}{\sqrt{10}} \right\rangle$$

or

$$\mathbf{v} = -\frac{1}{\sqrt{10}}\mathbf{i} - \frac{3}{\sqrt{10}}\mathbf{j} = \left\langle -\frac{1}{\sqrt{10}}, -\frac{3}{\sqrt{10}} \right\rangle$$

(b) Similarly, a vector $\mathbf{v} = x\mathbf{i} + y\mathbf{j}$ normal to the tangent line must have a slope of $-\frac{1}{3}$ (see the accompanying figure), and $x = -3y$. Since $\mathbf{v}$ is a unit vector, we have $x^2 + y^2 = 1$. Thus we have

$$(-3y)^2 + y^2 = 1$$
$$10y^2 = 1$$
$$y = \pm \frac{1}{\sqrt{10}} \qquad \text{and} \qquad x = \mp \frac{3}{\sqrt{10}}$$

Finally, we conclude that

$$\mathbf{v} = \frac{3}{\sqrt{10}}\mathbf{i} - \frac{1}{\sqrt{10}}\mathbf{j} = \left\langle \frac{3}{\sqrt{10}}, -\frac{1}{\sqrt{10}} \right\rangle$$

or

$$\mathbf{v} = \frac{-3}{\sqrt{10}}\mathbf{i} + \frac{1}{\sqrt{10}}\mathbf{j} = \left\langle -\frac{3}{\sqrt{10}}, \frac{1}{\sqrt{10}} \right\rangle$$

49. Find the component form of the vector $\mathbf{v}$ if it makes an angle of $150°$ with the positive x-axis and has magnitude $\|\mathbf{v}\| = 2$.

Solution

We first find a unit vector $\mathbf{u}$ making an angle of $150°$ with the positive x-axis. Since $\mathbf{u}$ is a unit vector, we consider it as the radius of a unit circle as shown in the accompanying figure. Therefore $x = \cos\theta$ and $y = \sin\theta$, and the component form for $\mathbf{u}$ is

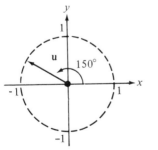

$$\mathbf{u} = x\mathbf{i} + y\mathbf{j} = \cos(150°)\mathbf{i} + \sin(150°)\mathbf{j} = \left\langle -\frac{\sqrt{3}}{2}, \frac{1}{2} \right\rangle$$

Since $\mathbf{v} = 2\mathbf{u}$, we have $\mathbf{v} = \langle -\sqrt{3}, 1 \rangle$

58. To carry a 100-lb cylindrical weight, two men lift on the ends of short ropes that are tied to an eyelet on the top center of the cylinder. If one rope makes a $20°$ angle away from the vertical

and the other a 30° angle, find the following:

(a) the tension in each rope if the resultant force is vertical

(b) the vertical component of each man's force

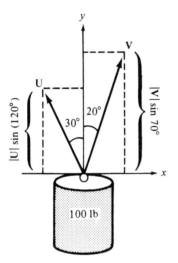

Solution

For the weight to be lifted vertically, the sum of the vertical components of **u** and **v** must be 100 and the sum of the horizontal components must be zero. (See the accompanying sketch.) Thus

$$\|\mathbf{u}\| \sin 120° + \|\mathbf{v}\| \sin 70° = 100$$

and

$$\|\mathbf{u}\| \cos 120° + \|\mathbf{v}\| \cos 70° = 0$$

Or, equivalently,

(1) $$\|\mathbf{u}\| \left(\frac{\sqrt{3}}{2}\right) + \|\mathbf{v}\| \sin 70° = 100$$

(2) $$\|\mathbf{u}\| \left(-\frac{1}{2}\right) + \|\mathbf{v}\| \cos 70° = 0$$

If we multiply the second equation by $\sqrt{3}$ and then add the two equations, we have

$$\|\mathbf{v}\| [\sqrt{3} \cos 70° + \sin 70°] = 100$$

$$\|\mathbf{v}\| = \frac{100}{\sqrt{3} \cos 70° + \sin 70°} \approx 65.27 \text{ lb}$$

Substituting 65.27 for $\|\mathbf{v}\|$ into the second equation, we have

$$65.27(\cos 70°) = \frac{\|\mathbf{u}\|}{2}$$

or

$$\|\mathbf{u}\| = 130.54(\cos 70°) \approx 44.65 \text{ lb}$$

(a) The tension in each rope is

$$\|\mathbf{u}\| = 44.65 \text{ lb}, \qquad \|\mathbf{v}\| = 65.27 \text{ lb}$$

(b) The vertical component of each man's force is

$$\|\mathbf{u}\| \sin 120° \approx (44.65)(0.8660) = 38.67 \text{ lb}$$
$$\|\mathbf{v}\| \sin 70° \approx (65.27)(0.9397) = 61.33 \text{ lb}$$

(Note that the sum of the two vertical components is 100 lb.)

12.2
Space coordinates and vectors in space

7. Find the lengths of the sides of the triangle with vertices
   $(1, -3, -2)$, $(5, -1, 2)$, and $(-1, 1, 2)$ and determine
   whether the triangle is a right triangle, an isosceles triangle, or
   neither of these.

   ### Solution

   Let us denote the three given points by A, B, and C,
   respectively. Then

   $$|AB| = \sqrt{(5 - 1)^2 + (-1 + 3)^2 + (2 + 2)^2} = \sqrt{16 + 4 + 16}$$
   $$= \sqrt{36} = 6$$
   $$|AC| = \sqrt{(-1 - 1)^2 + (1 + 3)^2 + (2 + 2)^2} = \sqrt{4 + 16 + 16}$$
   $$= \sqrt{36} = 6$$
   $$|BC| = \sqrt{(-1 - 5)^2 + (1 + 1)^2 + (2 - 2)^2} = \sqrt{36 + 4}$$
   $$= \sqrt{40} = 2\sqrt{10}$$

   Since two sides have equal lengths, the triangle is isosceles.

15. Find the center and radius of the sphere
    $$x^2 + y^2 + z^2 - 2x + 6y + 8z + 1 = 0$$

    ### Solution

    Writing the equation in standard form, we have

    $$x^2 + y^2 + z^2 - 2x + 6y + 8z + 1 = 0$$
    $$(x^2 - 2x) + (y^2 + 6y) + (z^2 + 8z) = -1$$
    $$(x^2 - 2x + 1) + (y^2 + 6y + 9) + (z^2 + 8z + 16)$$
    $$= -1 + 1 + 9 + 16$$
    $$(x - 1)^2 + (y + 3)^2 + (z + 4)^2 = 25$$

    Thus the sphere is centered at $(1, -3, -4)$ with radius 5.

32. Find $\mathbf{z}$, where $\mathbf{u} = \langle 1, 2, 3 \rangle$, $\mathbf{v} = \langle 2, 2, -1 \rangle$, $\mathbf{w} = \langle 4, 0, -4 \rangle$,
    and $2\mathbf{u} + \mathbf{v} - \mathbf{w} + 3\mathbf{z} = \mathbf{0}$.

    ### Solution

    Since $2\mathbf{u} = 2\mathbf{i} + 4\mathbf{j} + 6\mathbf{k}$, $\mathbf{v} = 2\mathbf{i} + 2\mathbf{j} - \mathbf{k}$, and
    $\mathbf{w} = 4\mathbf{i} - 4\mathbf{k}$, we have

    $$2\mathbf{u} + \mathbf{v} - \mathbf{w} + 3\mathbf{z} = \mathbf{0}$$

$$3z = -2u - v + w$$
$$= (-2 - 2 + 4)i + (-4 - 2 - 0)j + (-6 + 1 - 4)k$$
$$= 0i - 6j - 9k = -6j - 9k$$
$$z = -2j - 3k = \langle 0, -2, -3 \rangle$$

38. Use vectors to determine whether or not the points $(1, -1, 5)$, $(0, -1, 6)$, and $(3, -1, 3)$ lie on a straight line.

Solution

If we denote the three given points by A, B, and C, respectively, then

$$\overrightarrow{AB} = (0 - 1)i + (-1 + 1)j + (6 - 5)k = -i + k$$
$$\overrightarrow{AC} = (3 - 1)i + (-1 + 1)j + (3 - 5)k = 2i - 2k$$
$$= -2(-i + k)$$

Since $\overrightarrow{AC}$ is a scalar multiple of vector $\overrightarrow{AB}$, the three points lie on a straight line.

51. Find a unit vector (a) in the direction of $u = \langle 2, -1, 2 \rangle$ and (b) in the opposite direction of u.

Solution

(a) Since the magnitude of u is

$$\|u\| = \sqrt{2^2 + (-1)^2 + 2^2} = \sqrt{9} = 3$$

the unit vector in the direction of u is

$$\frac{u}{\|u\|} = \frac{1}{3}\langle 2, -1, 2 \rangle = \left\langle \frac{2}{3}, -\frac{1}{3}, \frac{2}{3} \right\rangle$$

(b) Since the unit vector in the opposite direction is obtained by multiplying by the scalar -1, we have

$$(-1)\frac{u}{\|u\|} = \left\langle -\frac{2}{3}, \frac{1}{3}, -\frac{2}{3} \right\rangle$$

59. Use vectors to find the point that lies two-thirds of the way from $P = (4, 3, 0)$ to $Q = (1, -3, 3)$.

Solution

By finding the component form of the vector from P to Q, we have

$$\overrightarrow{PQ} = \langle 1 - 4, -3 - 3, 3 - 0 \rangle = \langle -3, -6, 3 \rangle$$

Let $R = (x, y, z)$ be a point on the line segment two-thirds of the way from P to Q. (See the accompanying figure.) Then

$$\overrightarrow{PR} = \langle x - 4, y - 3, z - 0 \rangle = \langle x - 4, y - 3, z \rangle$$

and

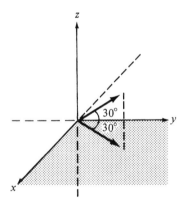

$$\overrightarrow{PR} = \frac{2}{3}\overrightarrow{PQ} = \frac{2}{3}\langle -3, -6, 3 \rangle = \langle -2, -4, 2 \rangle = \langle x - 4, y - 3, z \rangle$$

Therefore

$$\begin{aligned} x - 4 &= -2 & \rightarrow & \quad x = 2 \\ y - 3 &= -4 & \rightarrow & \quad y = -1 \\ z &= 2 & & \end{aligned}$$

and the required point is $(2, -1, 2)$.

63. Write the component form of vector $\mathbf{v}$ that lies in the yz-plane, has magnitude 2, and makes an angle of $30°$ with the positive y-axis. Sketch the vector.

Solution

We consider the form

$$\mathbf{v} = x\mathbf{i} + y\mathbf{j} + z\mathbf{k}$$

where $x = 0$, and

$$y = \|\mathbf{v}\| \cos \theta = 2 \cos 30° = 2\frac{\sqrt{3}}{2} = \sqrt{3}$$

$$z = \|\mathbf{v}\| \sin \theta = 2 \sin 30° = 2(\tfrac{1}{2}) = 1$$

or

$$y = 2 \cos (-30°) = \frac{2\sqrt{3}}{2} = \sqrt{3}$$

$$z = 2 \sin (-30°) = 2(-\tfrac{1}{2}) = -1$$

Therefore

$$\mathbf{v} = \sqrt{3}\,\mathbf{j} + \mathbf{k} \qquad \text{or} \qquad \mathbf{v} = \sqrt{3}\,\mathbf{j} - \mathbf{k}$$
$$= \langle 0, \sqrt{3}, 1 \rangle \qquad\qquad\qquad = \langle 0, \sqrt{3}, -1 \rangle$$

67. Let $\mathbf{u} = \mathbf{i} + \mathbf{j}$, $\mathbf{v} = \mathbf{j} + \mathbf{k}$, and $\mathbf{w} = a\mathbf{u} + b\mathbf{v}$.

(a) Sketch $\mathbf{u}$ and $\mathbf{v}$.

(b) If $\mathbf{w} = \mathbf{0}$, show that a and b are both zero.

(c) Find a and b such that $\mathbf{w} = \mathbf{i} + 2\mathbf{j} + \mathbf{k}$.

(d) Prove that no choice of a and b yields $\mathbf{w} = \mathbf{i} + 2\mathbf{j} + 3\mathbf{k}$.

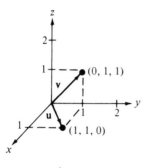

Solution

(a) See accompanying figure.

(b) $\mathbf{w} = a(\mathbf{i} + \mathbf{j}) + b(\mathbf{j} + \mathbf{k}) = \mathbf{0}$
$$a\mathbf{i} + (a + b)\mathbf{j} + b\mathbf{k} = 0\mathbf{i} + 0\mathbf{j} + 0\mathbf{k}$$
Therefore $a = b = 0$.

(c) $\mathbf{w} = a(\mathbf{i} + \mathbf{j}) + b(\mathbf{j} + \mathbf{k}) = \mathbf{i} + 2\mathbf{j} + \mathbf{k}$
$$a\mathbf{i} + (a + b)\mathbf{j} + b\mathbf{k} = \mathbf{i} + 2\mathbf{j} + \mathbf{k}$$
Therefore $a = b = 1$.

(d) $\mathbf{w} = a(\mathbf{i} + \mathbf{j}) + b(\mathbf{j} + \mathbf{k}) = \mathbf{i} + 2\mathbf{j} + 3\mathbf{k}$
$$a\mathbf{i} + (a + b)\mathbf{j} + b\mathbf{k} = \mathbf{i} + 2\mathbf{j} + 3\mathbf{k}$$
Therefore $a = 1$, $b = 3$, and $a + b = 2$, which is a contradiction.

12.3

The dot product

3. Given the vectors $\mathbf{u} = \langle 2, -3, 4 \rangle$ and $\mathbf{v} = \langle 0, 6, 5 \rangle$ find
(a) $\mathbf{u} \cdot \mathbf{v}$, (b) $\mathbf{u} \cdot \mathbf{u}$, (c) $\|\mathbf{u}\|^2$, (d) $(\mathbf{u} \cdot \mathbf{v})\mathbf{v}$, and (e) $\mathbf{u} \cdot (2\mathbf{v})$.

Solution

(a) For the vectors $\mathbf{u} = \langle u_1, u_2, u_3 \rangle$ and $\mathbf{v} = \langle v_1, v_2, v_3 \rangle$

$$\mathbf{u} \cdot \mathbf{v} = u_1 v_1 + u_2 v_2 + u_3 v_3$$
$$= 2(0) + (-3)(6) + 4(5) = 2$$

(b) $\mathbf{u} \cdot \mathbf{u} = 2(2) + (-3)(-3) + 4(4) = 29$

(c) $\|\mathbf{u}\|^2 = 2^2 + (-3)^2 + 4^2 = 29 = \mathbf{u} \cdot \mathbf{u}$

(d) From part (a) we have $\mathbf{u} \cdot \mathbf{v} = 2$. Therefore,

$$(\mathbf{u} \cdot \mathbf{v})\mathbf{v} = 2\mathbf{v} = \langle 0, 12, 10 \rangle$$

(e) $\mathbf{u} \cdot (2\mathbf{v}) = 2\mathbf{u} \cdot \mathbf{v} = 2(2)$ From part (a)
$$= 4$$

10. Find the angle θ between the vectors
$\mathbf{u} = \cos(\pi/6)\mathbf{i} + \sin(\pi/6)\mathbf{j}$ and
$\mathbf{v} = \cos(3\pi/4)\mathbf{i} + \sin(3\pi/4)\mathbf{j}$.

Solution

Let $\theta_\mathbf{u}$ be the angle between the positive x-axis and the unit vector $\mathbf{u}$. Then $\mathbf{u} = (\cos\theta_\mathbf{u})\mathbf{i} + (\sin\theta_\mathbf{u})\mathbf{j}$. Therefore

$$\mathbf{u} = \cos\left(\frac{\pi}{6}\right)\mathbf{i} + \sin\left(\frac{\pi}{6}\right)\mathbf{j} \quad \text{and} \quad \mathbf{v} = \cos\left(\frac{3\pi}{4}\right)\mathbf{i} + \sin\left(\frac{3\pi}{4}\right)\mathbf{j}$$

are unit vectors with $\pi/6$ and $3\pi/4$ the magnitudes of the angles between the positive x-axis and $\mathbf{u}$ and $\mathbf{v}$, respectively. Thus the angle θ between $\mathbf{u}$ and $\mathbf{v}$ is

$$\theta = \frac{3\pi}{4} - \frac{\pi}{6} = \frac{7\pi}{12}$$

13. Find the angle θ between $\mathbf{u} = 3\mathbf{i} + 4\mathbf{j}$ and $\mathbf{v} = -2\mathbf{j} + 3\mathbf{k}$.

    Solution

    Since

    $$\cos \theta = \frac{\mathbf{u} \cdot \mathbf{v}}{\|\mathbf{u}\| \, \|\mathbf{v}\|}$$

    we have

    $$\cos \theta = \frac{0 - 8 + 0}{\sqrt{25} \, \sqrt{13}} = \frac{-8\sqrt{13}}{65}$$

    and

    $$\theta = \arccos \left(\frac{-8\sqrt{13}}{65} \right) \approx 116.344°$$

17. Determine whether $\mathbf{u} = \langle 4, 3 \rangle$ and $\mathbf{v} = \langle \frac{1}{2}, -\frac{2}{3} \rangle$ are orthogonal, parallel, or neither.

    Solution

    $$\mathbf{u} \cdot \mathbf{v} = 4\left(\frac{1}{2}\right) + 3\left(-\frac{2}{3}\right) = 2 - 2 = 0$$

    Therefore $\mathbf{u}$ and $\mathbf{v}$ are orthogonal.

18. Determine whether $\mathbf{u} = -\frac{1}{3}(\mathbf{i} - 2\mathbf{j})$ and $\mathbf{v} = 2\mathbf{i} - 4\mathbf{j}$ are orthogonal, parallel, or neither.

    Solution

    $$\mathbf{u} = -\frac{1}{3}(\mathbf{i} - 2\mathbf{j}) = -\frac{1}{6}(2\mathbf{i} - 4\mathbf{j}) = -\frac{1}{6}\mathbf{v}$$

    Therefore $\mathbf{u}$ and $\mathbf{v}$ are parallel.

23. Find the direction cosines of $\mathbf{u} = \mathbf{i} + 2\mathbf{j} + 2\mathbf{k}$ and demonstrate that the sum of the squares of the direction cosines is 1.

    Solution

    Given a vector $\mathbf{u} = u_1\mathbf{i} + u_2\mathbf{j} + u_3\mathbf{k}$ the direction cosines are

$$\cos \alpha = \frac{u_1}{\|\mathbf{u}\|} \qquad \cos \beta = \frac{u_2}{\|\mathbf{u}\|} \qquad \cos \gamma = \frac{u_3}{\|\mathbf{u}\|}$$

Therefore, for the given vector, we have

$$\cos \alpha = \frac{1}{\sqrt{1^2 + 2^2 + 2^2}} = \frac{1}{\sqrt{9}} = \frac{1}{3}$$

$$\cos \beta = \frac{2}{\sqrt{1^2 + 2^2 + 2^2}} = \frac{2}{\sqrt{9}} = \frac{2}{3}$$

$$\cos \gamma = \frac{2}{\sqrt{1^2 + 2^2 + 2^2}} = \frac{2}{\sqrt{9}} = \frac{2}{3}$$

The sum of the squares of the direction cosines is

$$\cos^2 \alpha + \cos^2 \beta + \cos^2 \gamma = \left(\frac{1}{3}\right)^2 + \left(\frac{2}{3}\right)^2 + \left(\frac{2}{3}\right)^2 = 1$$

31. For the vectors $\mathbf{u} = \langle 1, 1, 1 \rangle$ and $\mathbf{v} = \langle -2, -1, 1 \rangle$ find (a) the vector component of $\mathbf{u}$ along $\mathbf{v}$, and (b) the vector component of $\mathbf{u}$ orthogonal to $\mathbf{v}$.

Solution

(a) The vector component of $\mathbf{u}$ along $\mathbf{v}$ is the projection of $\mathbf{u}$ onto $\mathbf{v}$ and is given by

$$\mathbf{w}_1 = \left(\frac{\mathbf{u} \cdot \mathbf{v}}{\|\mathbf{v}\|^2}\right)\mathbf{v}$$

$$= \left(\frac{-2}{6}\right)\langle -2, -1, 1 \rangle = \left\langle \frac{2}{3}, \frac{1}{3}, -\frac{1}{3} \right\rangle$$

(c) The vector component of $\mathbf{u}$ orthogonal to $\mathbf{v}$ is given by

$$\mathbf{w}_2 = \mathbf{u} - \mathbf{w}_1$$

$$= \langle 1, 1, 1 \rangle - \left\langle \frac{2}{3}, \frac{1}{3}, -\frac{1}{3} \right\rangle = \left\langle \frac{1}{3}, \frac{2}{3}, \frac{4}{3} \right\rangle$$

37. An implement is dragged 10 feet across a floor, using a force of 85 pounds. Find the work done if the direction of the force is $60°$ above the horizontal.

Solution

The work done by a constant force $\mathbf{F}$ as its point of application moves along the vector $\overrightarrow{PQ}$ is given by

$$W = \mathbf{F} \cdot \overrightarrow{PQ}$$

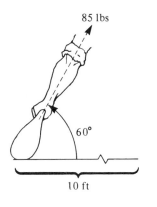

85 lbs

60°

10 ft

From the accompanying figure we see that

$$\mathbf{F} = 85[(\cos 60°)\mathbf{i} + (\sin 60°)\mathbf{j}]$$

so

$$\overrightarrow{PQ} = 10\mathbf{i}$$

Therefore,

$$W = 85[(\cos 60°)(10) + (\sin 60°)(0)] = 425 \text{ ft·lb}$$

43.   What is known about θ, the angle between vectors $\mathbf{u}$ and $\mathbf{v}$, if:
(a) $\mathbf{u} \cdot \mathbf{v} = 0$?   (b) $\mathbf{u} \cdot \mathbf{v} > 0$?   (c) $\mathbf{u} \cdot \mathbf{v} < 0$?

Solution

Assuming that $\mathbf{u}$ and $\mathbf{v}$ are nonzero vectors and
$\mathbf{u} \cdot \mathbf{v} = \|\mathbf{u}\| \|\mathbf{v}\| \cos \theta$, we have the following:
(a) $\mathbf{u} \cdot \mathbf{v} = 0$ implies that $\cos \theta = 0$, or $\theta = \pi/2$.
(b) $\mathbf{u} \cdot \mathbf{v} > 0$ implies that $\cos \theta > 0$, or $0 \le \theta < \pi/2$.
(c) $\mathbf{u} \cdot \mathbf{v} < 0$ implies that $\cos \theta < 0$, or $\pi/2 < \theta \le \pi$.

12.4
The cross product

11.   If $\mathbf{u} = \langle 1, 1, 1 \rangle$ and $\mathbf{v} = \langle 2, 1, -1 \rangle$, find $\mathbf{u} \times \mathbf{v}$ and show that
it is orthogonal to both $\mathbf{u}$ and $\mathbf{v}$.

Solution

First we have

$$\mathbf{u} \times \mathbf{v} = \begin{vmatrix} \mathbf{i} & \mathbf{j} & \mathbf{k} \\ 1 & 1 & 1 \\ 2 & 1 & -1 \end{vmatrix} = \mathbf{i} \begin{vmatrix} 1 & 1 \\ 1 & -1 \end{vmatrix} - \mathbf{j} \begin{vmatrix} 1 & 1 \\ 2 & -1 \end{vmatrix} + \mathbf{k} \begin{vmatrix} 1 & 1 \\ 2 & 1 \end{vmatrix}$$

$$= (-1 - 1)\mathbf{i} - (-1 - 2)\mathbf{j} + (1 - 2)\mathbf{k}$$

$$= -2\mathbf{i} + 3\mathbf{j} - \mathbf{k}$$

Now

$$\mathbf{u} \cdot (\mathbf{u} \times \mathbf{v}) = (\mathbf{i} + \mathbf{j} + \mathbf{k}) \cdot (-2\mathbf{i} + 3\mathbf{j} - \mathbf{k}) = -2 + 3 - 1 = 0$$

and

$$\mathbf{v} \cdot (\mathbf{u} \times \mathbf{v}) = (2\mathbf{i} + \mathbf{j} - \mathbf{k}) \cdot (-2\mathbf{i} + 3\mathbf{j} - \mathbf{k}) = -4 + 3 + 1 = 0$$

Therefore $\mathbf{u} \times \mathbf{v}$ is orthogonal to both $\mathbf{u}$ and $\mathbf{v}$.

16.   Find the area of the parallelogram that has $\mathbf{u} = \mathbf{i} + \mathbf{j} + \mathbf{k}$ and
$\mathbf{v} = \mathbf{j} + \mathbf{k}$ as adjacent sides.

Solution

By Theorem 12.9 the area is given by $\| \mathbf{u} \times \mathbf{v} \|$.

$$\mathbf{u} \times \mathbf{v} = \begin{vmatrix} \mathbf{i} & \mathbf{j} & \mathbf{k} \\ 1 & 1 & 1 \\ 0 & 1 & 1 \end{vmatrix} = \mathbf{i} \begin{vmatrix} 1 & 1 \\ 1 & 1 \end{vmatrix} - \mathbf{j} \begin{vmatrix} 1 & 1 \\ 0 & 1 \end{vmatrix} + \mathbf{k} \begin{vmatrix} 1 & 1 \\ 0 & 1 \end{vmatrix}$$

$$= (1 - 1)\mathbf{i} - (1 - 0)\mathbf{j} + (1 - 0)\mathbf{k} = -\mathbf{j} + \mathbf{k}$$

Therefore

$$\text{area} = \| \mathbf{u} \times \mathbf{v} \| = \sqrt{2}$$

23. Find the area of the triangle with vertices $(1, 3, 5)$, $(3, 3, 0)$ and $(-2, 0, 5)$.

Solution

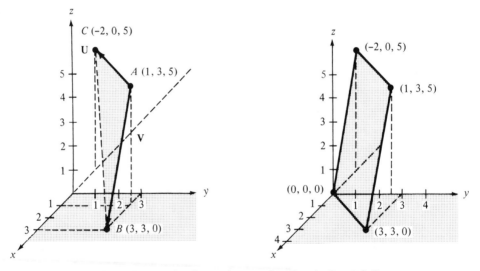

If we denote the three given points by A, B, and C, respectively, then the adjacent sides at vertex A can be denoted by the vectors

$$\overrightarrow{AB} = (3 - 1)\mathbf{i} + (3 - 3)\mathbf{j} + (0 - 5)\mathbf{k} = 2\mathbf{i} - 5\mathbf{k}$$
$$\overrightarrow{AC} = (-2 - 1)\mathbf{i} + (0 - 3)\mathbf{j} + (5 - 5)\mathbf{k} = -3\mathbf{i} - 3\mathbf{j}$$

Theorem 12.9 says that the area of the parallelogram having $\mathbf{v}$ and $\mathbf{u}$ as adjacent sides is $\| \mathbf{v} \times \mathbf{u} \|$. (See accompanying figure.) From this we may conclude that the area of the triangle having these same adjacent sides is $\frac{1}{2}\| \mathbf{v} \times \mathbf{u} \|$. Therefore, since

$$\mathbf{v} \times \mathbf{u} = \begin{vmatrix} \mathbf{i} & \mathbf{j} & \mathbf{k} \\ 2 & 0 & -5 \\ -3 & -3 & 0 \end{vmatrix}$$

$$= \mathbf{i} \begin{vmatrix} 0 & -5 \\ -3 & 0 \end{vmatrix} - \mathbf{j} \begin{vmatrix} 2 & -5 \\ -3 & 0 \end{vmatrix} + \mathbf{k} \begin{vmatrix} 2 & 0 \\ -3 & -3 \end{vmatrix}$$

$$= -15\mathbf{i} + 15\mathbf{j} - 6\mathbf{k} = 3(-5\mathbf{i} + 5\mathbf{j} - 2\mathbf{k})$$

we have

$$(\text{area of triangle}) = \frac{1}{2}\|\mathbf{v} \times \mathbf{u}\| = \frac{1}{2}(3)\sqrt{25 + 25 + 4}$$

$$= \frac{3}{2}\sqrt{54} = \frac{9}{2}\sqrt{6}$$

26. Find the triple scalar product $\mathbf{u} \cdot (\mathbf{v} \times \mathbf{w})$ if $\mathbf{u} = \langle 1, 1, 1 \rangle$, $\mathbf{v} = \langle 2, 1, 0 \rangle$, and $\mathbf{w} = \langle 0, 0, 1 \rangle$.

Solution

$$\mathbf{v} \times \mathbf{w} = \begin{vmatrix} \mathbf{i} & \mathbf{j} & \mathbf{k} \\ 2 & 1 & 0 \\ 0 & 0 & 1 \end{vmatrix} = \mathbf{i} \begin{vmatrix} 1 & 0 \\ 0 & 1 \end{vmatrix} - \mathbf{j} \begin{vmatrix} 2 & 0 \\ 0 & 1 \end{vmatrix} + \mathbf{k} \begin{vmatrix} 2 & 1 \\ 0 & 0 \end{vmatrix} = \mathbf{i} - 2\mathbf{j}$$

Therefore the triple scalar product is

$$\mathbf{u} \cdot (\mathbf{v} \times \mathbf{w}) = 1(1) + 1(-2) + 1(0) = -1$$

29. Find the volume of the parallelepiped with $\mathbf{u} = \mathbf{i} + \mathbf{j}$, $\mathbf{v} = \mathbf{j} + \mathbf{k}$, and $\mathbf{w} = \mathbf{i} + \mathbf{k}$ as adjacent edges.

Solution

From Theorem 12.11 the volume of the parallelepiped is the absolute value of the triple scalar product $\mathbf{u} \cdot (\mathbf{v} \times \mathbf{w})$. Since

$$\mathbf{v} \times \mathbf{w} = \begin{vmatrix} \mathbf{i} & \mathbf{j} & \mathbf{k} \\ 0 & 1 & 1 \\ 1 & 0 & 1 \end{vmatrix} = \mathbf{i} + \mathbf{j} - \mathbf{k}$$

the volume of the parallelepiped is $\mathbf{u} \cdot (\mathbf{v} \times \mathbf{w}) = 1 + 1 - 0 = 2$.

12.5

Lines and planes in space

3. Find a set of (a) parametric equations and (b) symmetric equations of the line through the point $(-2, 0, 3)$ and parallel to $\mathbf{v} = 2\mathbf{i} + 4\mathbf{j} - 2\mathbf{k}$. Express the direction numbers as integers.

Solution

(a) Using the form

$$x = x_1 + at \qquad y = y_1 + bt \qquad z = z_1 + ct$$

where $x_1 = -2$, $y_1 = 0$, and $z_1 = 3$ with direction numbers $a = 2$, $b = 4$, and $c = -2$, a set of parametric equations for the line is

$$x = -2 + 2t \qquad y = 4t \qquad z = 3 - 2t$$

(b) Solving for t in each equation gives us

$$t = \frac{x + 2}{2} = \frac{y}{4} = \frac{z - 3}{-2}$$

Consequently, we have the symmetric form

$$\frac{x + 2}{2} = \frac{y}{4} = \frac{z - 3}{1}$$

8. Find a set of (a) parametric equations and (b) symmetric equations for the line through the point $(-3, 5, 4)$ that is parallel to the line

$$\frac{x - 1}{3} = \frac{y + 1}{-2} = \frac{z - 3}{1}$$

Express the direction numbers as integers.

Solution

(a) Since the new line is parallel to the given line, the direction numbers are $a = 3$, $b = -2$, and $c = 1$. Thus the parametric equations are

$$x = x_1 + at \qquad y = y_1 + bt \qquad z = z_1 + ct$$

or

$$x = -3 + 3t \qquad y = 5 - 2t \qquad z = 4 + t$$

(b) The corresponding symmetric form is

$$\frac{x + 3}{3} = \frac{y - 5}{-2} = \frac{z - 4}{1}$$

13. Determine if the lines

$$\begin{array}{ll} x = 4t + 2 & x = 2s + 2 \\ y = 3 & y = 2s + 3 \\ z = -t + 1 & z = s + 1 \end{array}$$

intersect and, if so, find the point of intersection and the cosine of the angle of intersection.

Solution

At the point of intersection, the coordinates for one line equal the corresponding coordinates for the other line. Thus we have the three equations

(1) $$4t + 2 = 2s + 2$$

(2) $$3 = 2s + 3$$

(3) $$-t + 1 = s + 1$$

From equation (2) we find that $s = 0$ and consequently, from equation (3), $t = 0$. Letting $s = t = 0$, we see that equation (1) is satisfied and we conclude that the two lines intersect. Substituting zero for s or for t, we obtain the point $(2, 3, 1)$.

To find the cosine of the angle of intersection, we consider the vectors

$$\mathbf{u} = 4\mathbf{i} - \mathbf{k} \qquad \text{and} \qquad \mathbf{v} = 2\mathbf{i} + 2\mathbf{j} + \mathbf{k}$$

that have the respective directions of the two given lines. Therefore

$$\cos \theta = \frac{|\mathbf{u} \cdot \mathbf{v}|}{\|\mathbf{u}\|\|\mathbf{v}\|} = \frac{8 - 1}{\sqrt{17}\sqrt{9}} = \frac{7}{3\sqrt{17}} = \frac{7\sqrt{17}}{51}$$

21. Find an equation of the plane through the points $(0, 0, 0)$, $(1, 2, 3)$, and $(-2, 3, 3)$.

Solution

To use the form

$$a(x - x_1) + b(y - y_1) + c(z - z_1) = 0$$

we need to know a point in the plane and a vector $\mathbf{n}$ that is normal to the plane. To obtain a normal vector, we use the cross product of the vectors $\mathbf{v}_1$ and $\mathbf{v}_2$ from the point $(0, 0, 0)$ to $(1, 2, 3)$ and to $(-2, 3, 3)$, respectively. We have

$$\mathbf{v}_1 = \mathbf{i} + 2\mathbf{j} + 3\mathbf{k} \qquad \text{and} \qquad \mathbf{v}_2 = -2\mathbf{i} + 3\mathbf{j} + 3\mathbf{k}$$

Thus the vector

$$\mathbf{n} = \mathbf{v}_1 \times \mathbf{v}_2 = \begin{vmatrix} \mathbf{i} & \mathbf{j} & \mathbf{k} \\ 1 & 2 & 3 \\ -2 & 3 & 3 \end{vmatrix} = -3\mathbf{i} - 9\mathbf{j} + 7\mathbf{k}$$

is normal to the given plane. Using the direction numbers from $\mathbf{n}$ and the point $(0, 0, 0)$, we have

$$-3(x - 0) - 9(y - 0) + 7(z - 0) = 0$$
$$3x + 9y - 7z = 0$$

27. Find an equation of the plane determined by the two intersecting lines

$$\frac{x - 1}{-2} = y - 4 = z \qquad \text{and} \qquad \frac{x - 2}{-3} = \frac{y - 1}{4} = \frac{z - 2}{-1}$$

Solution

Writing the equations of the lines in parametric form, we have

$$x = 1 - 2t \qquad x = 2 - 3t$$
$$y = 4 + t \qquad y = 1 + 4t$$
$$z = t \qquad z = 2 - t$$

To find the point of intersection of the lines, we observe that the z-coordinates are equal when

$$t = 2 - t \qquad \text{or} \qquad t = 1$$

Therefore the point of intersection is $(-1, 5, 1)$ and occurs when $t = 1$. Since the direction vectors of the given lines are

$$\mathbf{v}_1 = -2\mathbf{i} + \mathbf{j} + \mathbf{k} \qquad \text{and} \qquad \mathbf{v}_2 = -3\mathbf{i} + 4\mathbf{j} - \mathbf{k}$$

the vector $\mathbf{n}$ normal to the plane is

$$\mathbf{n} = \mathbf{v}_1 \times \mathbf{v}_2 = \begin{vmatrix} \mathbf{i} & \mathbf{j} & \mathbf{k} \\ -2 & 1 & 1 \\ -3 & 4 & -1 \end{vmatrix} = -5(\mathbf{i} + \mathbf{j} + \mathbf{k})$$

Therefore the equation of the plane is

$$1(x + 1) + 1(y - 5) + 1(z - 1) = 0$$
$$x + y + z = 5$$

29. Find the equation of the plane through the points $(2, 2, 1)$ and $(-1, 1, -1)$ that is perpendicular to the plane $2x - 3y + z = 3$.

Solution

Let $\mathbf{v}$ be the vector from $(-1, 1, -1)$ to $(2, 2, 1)$, and let $\mathbf{n}$ be a vector normal to the plane $2x - 3y + z = 3$. Then $\mathbf{v}$ and $\mathbf{n}$ both lie in the required plane, where

$$\mathbf{v} = 3\mathbf{i} + \mathbf{j} + 2\mathbf{k} \qquad \text{and} \qquad \mathbf{n} = 2\mathbf{i} - 3\mathbf{j} + \mathbf{k}$$

The vector

$$\mathbf{v} \times \mathbf{n} = \begin{vmatrix} \mathbf{i} & \mathbf{j} & \mathbf{k} \\ 3 & 1 & 2 \\ 2 & -3 & 1 \end{vmatrix} = 7\mathbf{i} + \mathbf{j} - 11\mathbf{k}$$

is normal to the required plane, and therefore this plane has direction numbers 7, 1, and -11. Finally, since the point $(2, 2, 1)$ lies in the plane, an equation is

$$7(x - 2) + 1(y - 2) - 11(z - 1) = 0$$
$$7x + y - 11z - 5 = 0$$

49. Find a set of parametric equations for the line of intersection of the planes $3x + 2y - z = 7$ and $x - 4y + 2z = 0$.

Solution

Let $\mathbf{n_1} = 3\mathbf{i} + 2\mathbf{j} - \mathbf{k}$ and $\mathbf{n_2} = \mathbf{i} - 4\mathbf{j} + 2\mathbf{k}$ be the normal vectors to the respective planes. The line of intersection of the two planes will have the same direction as the vector $\mathbf{n_1} \times \mathbf{n_2}$. Since

$$\mathbf{n_1} \times \mathbf{n_2} = \begin{vmatrix} \mathbf{i} & \mathbf{j} & \mathbf{k} \\ 3 & 2 & -1 \\ 1 & -4 & 2 \end{vmatrix} = 0\mathbf{i} - 7\mathbf{j} - 14\mathbf{k}$$

the direction numbers for the line of intersection are 0, -7, -14, or more simply, 0, 1, 2. By solving the equations for the two planes simultaneously, we can find points on the line of intersection.

$$\begin{array}{ll} 3x + 2y - z = 7 \rightarrow & 6x + 4y + 2z = 14 \\ x - 4y + 2z = 0 \rightarrow & \underline{x - 4y + 2z = 0} \\ & 7x = 14 \qquad \text{or} \qquad x = 2 \end{array}$$

By substituting 2 for x, we obtain the equation $2y - z = 1$. If $y = 1$, then $z = 1$. Thus $(2, 1, 1)$ lies on the line of intersection, and we conclude that a set of parametric equations for the line of intersection is

$$x = 2, \qquad y = 1 + t, \qquad z = 1 + 2t$$

51. Find the point of intersection (if any) of the line

$$\frac{x - \frac{1}{2}}{1} = \frac{y + \frac{3}{2}}{-1} = \frac{z + 1}{2}$$

and the plane $2x - 2y + z = 12$.

Solution

The parametric equations for the line are

$$x = \frac{1}{2} + t, \qquad y = \frac{-3}{2} - t, \qquad z = -1 + 2t$$

Now if the line intersects the plane, then the values of x, y, and z must satisfy the equation of the plane. Thus

$$2x - 2y + z = 12$$

$$2\left(\frac{1}{2} + t\right) - 2\left(\frac{-3}{2} - t\right) + (-1 + 2t) = 12$$

$$6t + 3 = 12$$

$$6t = 9$$

$$t = \frac{3}{2}$$

and we conclude that the point of intersection occurs when $t = \frac{3}{2}$, which yields the point $(2, -3, 2)$.

61. Find the distance between the two lines

$$\frac{x}{1} = \frac{y}{2} = \frac{z}{3} \quad \text{and} \quad \frac{x-1}{-1} = \frac{y-4}{1} = \frac{z+1}{1}$$

Solution

Let

$$\mathbf{n}_1 = \mathbf{i} + 2\mathbf{j} + 3\mathbf{k} \quad \text{and} \quad \mathbf{n}_2 = -\mathbf{i} + \mathbf{j} + \mathbf{k}$$

be vectors along the given lines, respectively. Then the vector $\mathbf{n}_1 \times \mathbf{n}_2$ will be orthogonal to both given lines. By choosing an arbitrary vector $\mathbf{v}$ from one line to the other, we can project $\mathbf{v}$ onto $\mathbf{n}_1 \times \mathbf{n}_2$ to find the distance between the two lines. Thus since $(0, 0, 0)$ lies on the first line and $(0, 5, 0)$ lies on the second line, we have $\mathbf{v} = 5\mathbf{j}$. Now the absolute value of the component of $\mathbf{v}$ in the direction of $\mathbf{n}_1 \times \mathbf{n}_2$ will be the actual distance between the lines. Since

$$\mathbf{n}_1 \times \mathbf{n}_2 = \begin{vmatrix} \mathbf{i} & \mathbf{j} & \mathbf{k} \\ 1 & 2 & 3 \\ -1 & 1 & 1 \end{vmatrix} = -\mathbf{i} - 4\mathbf{j} + 3\mathbf{k}$$

the distance between the lines is

$$\left| \begin{matrix} \text{component of } \mathbf{v} \\ \text{in direction of } \mathbf{n}_1 \times \mathbf{n}_2 \end{matrix} \right| = \frac{|\mathbf{v} \cdot (\mathbf{n}_1 \times \mathbf{n}_2)|}{\|\mathbf{n}_1 \times \mathbf{n}_2\|}$$

$$= \left| \frac{-20}{\sqrt{1 + 16 + 9}} \right| = \frac{20}{\sqrt{26}} \approx 3.92$$

12.6
Surfaces in space

13. Describe and sketch the surface defined by the equation $x^2 - y = 0$.

Solution

Since the z-coordinate is missing in the equation, the surface is a cylindrical surface with rulings parallel to the z-axis. The generating curve is the parabola $y = x^2$ and the surface is called a *parabolic cylinder*. (See accompanying figure.)

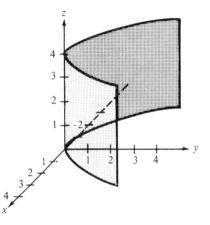

21. Identify and sketch the quadric surface given by
$16x^2 - y^2 + 16z^2 = 4$.

Solution

The given equation has the form

$$\frac{x^2}{\frac{1}{4}} - \frac{y^2}{4} + \frac{z^2}{\frac{1}{4}} = 1$$

which is the form for a **hyperboloid of one sheet.** The axis of the hyperboloid is the y-axis. The xz-trace ($y = 0$) is the circle

$$\frac{x^2}{\frac{1}{4}} + \frac{z^2}{\frac{1}{4}} = 1$$

and the xy and yz traces are the hyperbolas

$$\frac{x^2}{\frac{1}{4}} - \frac{y^2}{4} = 1 \quad \text{and} \quad \frac{z^2}{\frac{1}{4}} - \frac{y^2}{4} = 1$$

(See the accompanying figure.)

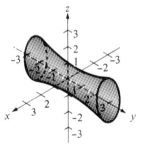

25. Identify and sketch the quadric surface given by
$x^2 - y^2 + z = 0$.

Solution

The given equation can be written as

$$\frac{y^2}{1} - \frac{x^2}{1} = z$$

which has the form for a **hyperbolic paraboloid.** The xy-trace

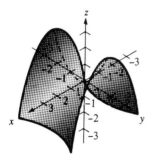

($z = 0$) consists of the intersecting lines $y = \pm x$. The yz-trace, $y^2 = z$, is a parabola opening upward, and the traces parallel to the xz-trace, $x^2 = -z$, are parabolas opening downward. (See the accompanying figure.)

27. Identify and sketch the quadric surface given by

$$4x^2 - y^2 + 4z^2 = -16$$

Solution

Dividing the equation by -16, we obtain the form

$$-\frac{x^2}{4} + \frac{y^2}{16} - \frac{z^2}{4} = 1$$

The equation is the standard form of a **hyperboloid of two sheets.** The axis of the hyperboloid is the y-axis. The xy-trace ($z = 0$) is the hyperbola

$$\frac{y^2}{16} - \frac{x^2}{4} = 1$$

The yz-trace ($x = 0$) is the hyperbola

$$\frac{y^2}{16} - \frac{z^2}{4} = 1$$

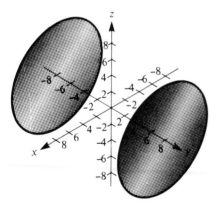

Traces parallel to the xz-plane are circles. For example, when $y = 5$, we have

$$4x^2 - (5)^2 + 4z^2 = -16$$
$$4x^2 + 4z^2 = 9$$
$$x^2 + z^2 = \frac{9}{4}$$

a circle of radius $\frac{3}{2}$.

43. Find an equation for the surface of revolution generated by revolving the graph of $z = 2y$ in the yz-plane about the z-axis.

Solution

Since we are revolving the curve about the z-axis, the equation for the surface of revolution has the form

$$x^2 + y^2 = [a(z)]^2$$

But $y = a(z) = z/2$. Therefore the equation is

$$x^2 + y^2 = \frac{z^2}{4} \qquad \text{or} \qquad 4x^2 + 4y^2 = z^2$$

48. Find an equation of a generating curve if the equation of its
     surface of revolution is $x^2 + z^2 = \sin^2 y$.

Solution

First note that the given equation has the form

$$x^2 + z^2 = [a(y)]^2$$

which means that the rotation is about the y-axis.
Consequently, the generating curve could lie in either the
yz-plane or the xy-plane. If the generating curve lies in the
yz-plane, then

$$z = a(y) \qquad \text{or} \qquad z = \sin y$$

is an equation of the generating curve. If the curve lies in the
xy-plane, then

$$x = a(y) \qquad \text{or} \qquad x = \sin y$$

is an equation of the generating curve.

12.7
Cylindrical and spherical coordinates

5. Convert the point $(2, -2, -4)$ from rectangular to cylindrical
    coordinates.

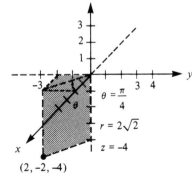

Solution

Since $x = 2$, $y = -2$, and $z = -4$, we have

$$r = \pm\sqrt{x^2 + y^2} = \pm\sqrt{4 + 4} = \pm 2\sqrt{2}$$

$$\theta = \arctan \frac{y}{x} = \arctan \frac{-2}{2} = \arctan (-1) = -\frac{\pi}{4} \qquad \text{or} \qquad \frac{3\pi}{4}$$

$$z = -4$$

Therefore, one set of corresponding cylindrical coordinates for
the given point is $(r, \theta, z) = (2\sqrt{2}, -\pi/4, -4)$. (See the
accompanying sketch.) Another set of cylindrical coordinates
for the given point would be $(-2\sqrt{2}, 3\pi/4, -4)$.

11. Convert the point $(4, 7\pi/6, 3)$ from cylindrical to rectangular
     coordinates.

Solution

Since $(4, 7\pi/6, 3) = (r, \theta, z)$, we have

$$x = r \cos \theta = 4 \cos \frac{7\pi}{6} = -2\sqrt{3}$$

$$y = r \sin \theta = 4 \sin \frac{7\pi}{6} = -2$$

$$z = 3$$

Therefore, in rectangular coordinates the point is $(-2\sqrt{3}, -2, 3)$.

15. Convert the point $(-2, 2\sqrt{3}, 4)$ from rectangular to spherical coordinates.

Solution

Since $(-2, 2\sqrt{3}, 4) = (x, y, z)$, we have

$$\rho = \sqrt{x^2 + y^2 + z^2} = \sqrt{4 + 12 + 16} = 4\sqrt{2}$$

$$\theta = \arctan \frac{y}{x} = \arctan \frac{2\sqrt{3}}{-2} = \frac{2\pi}{3}$$

$$\phi = \arccos \frac{z}{\rho} = \arccos \frac{4}{4\sqrt{2}} = \frac{\pi}{4}$$

Therefore in spherical coordinates the point is $(4\sqrt{2}, 2\pi/3, \pi/4)$.

19. Convert the point $(4, \pi/6, \pi/4)$ from spherical to rectangular coordinates.

Solution

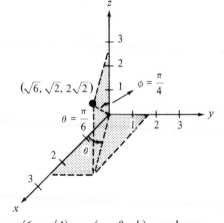

Since $(4, \pi/6, \pi/4) = (\rho, \theta, \phi)$, we have

$$x = \rho \sin \phi \cos \theta = 4 \sin \frac{\pi}{4} \cos \frac{\pi}{6} = 4\left(\frac{\sqrt{2}}{2}\right)\left(\frac{\sqrt{3}}{2}\right) = \sqrt{6}$$

$$y = \rho \sin \phi \sin \theta = 4 \sin \frac{\pi}{4} \sin \frac{\pi}{6} = 4\left(\frac{\sqrt{2}}{2}\right)\left(\frac{1}{2}\right) = \sqrt{2}$$

$$z = \rho \cos \phi = 4 \cos \frac{\pi}{4} = 4\left(\frac{\sqrt{2}}{2}\right) = 2\sqrt{2}$$

Therefore, in rectangular coordinates the given point is $(\sqrt{6}, \sqrt{2}, 2\sqrt{2})$.

27.   Convert the point $(4, -\pi/6, 6)$ from cylindrical to spherical coordinates.

Solution

Since $(4, -\pi/6, 6) = (r, \theta, z)$, we have

$$\rho^2 = x^2 + y^2 + z^2 = r^2 + z^2 = 16 + 36 = 52$$

or

$$\rho = 2\sqrt{13} > 0$$

Furthermore,

$$\cos \phi = \frac{z}{\sqrt{x^2 + y^2 + z^2}} = \frac{z}{\sqrt{r^2 + z^2}} = \frac{6}{\sqrt{52}} = \frac{3}{\sqrt{13}}$$

or

$$\phi = \arccos \frac{3}{\sqrt{13}}$$

Therefore, in spherical coordinates, the given point is $(2\sqrt{13}, -\pi/6, \arccos(3/\sqrt{13}))$.

46.   Find an equation in rectangular coordinates for $r = z/2$.

Solution

Since $r^2 = x^2 + y^2$, we have

$$x^2 + y^2 = \left(\frac{z}{2}\right)^2 \qquad \text{or} \qquad x^2 + y^2 = \frac{z^2}{4}$$

which is the equation for a cone.

55.   Find an equation in rectangular coordinates for $\rho = 4 \cos \phi$.

Solution

Since

$$\cos \phi = \frac{z}{\sqrt{x^2 + y^2 + z^2}}$$

we write

$$\frac{\rho}{4} = \cos \phi = \frac{z}{\sqrt{x^2 + y^2 + z^2}}$$

Furthermore, since

$$\rho = \sqrt{x^2 + y^2 + z^2}$$

we have

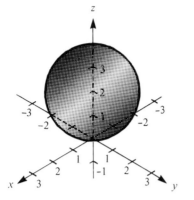

$$\frac{\sqrt{x^2 + y^2 + z^2}}{4} = \frac{z}{\sqrt{x^2 + y^2 + z^2}}$$

$$x^2 + y^2 + z^2 = 4z$$

$$x^2 + y^2 + z^2 - 4z = 0$$

By completing the square on variable z, we have the equation

$$x^2 + y^2 + (z - 2)^2 = 4$$

which represents a sphere of radius 2, centered at $(0, 0, 2)$.

61. Find an equation in (a) cylindrical coordinates and (b) spherical coordinates for the equation $x^2 + y^2 + z^2 - 2z = 0$.

Solution

(a) Since $x^2 + y^2 = r^2$ and $z = z$, the equation in cylindrical coordinates is

$$r^2 + z^2 - 2z = 0 \qquad \text{or} \qquad r^2 + (z - 1)^2 = 1$$

(b) In spherical coordinates, since $x^2 + y^2 + z^2 = \rho^2$ and $z = \rho \cos \phi$, we have

$$\rho^2 - 2\rho \cos \phi = 0$$

$$\rho(\rho - 2 \cos \phi) = 0$$

$$\rho = 2 \cos \phi$$

63. Find an equation in (a) cylindrical coordinates and (b) spherical coordinates for the equation $x^2 + y^2 = 4y$.

Solution

(a) Since $x^2 + y^2 = r^2$ and $y = r \sin \theta$, the equation in cylindrical coordinates is

$$r^2 = 4r \sin \theta \qquad \text{or} \qquad r = 4 \sin \theta$$

(b) In spherical coordinates, since $x^2 + y^2 = r^2 = \rho^2 \sin^2 \phi$ and $y = \rho \sin \phi \sin \theta$, we have

$$x^2 + y^2 = 4y$$

$$\rho^2 \sin^2 \phi = 4\rho \sin \phi \sin \theta$$
$$\rho \sin \phi = 4 \sin \theta$$
$$\rho = 4 \sin \theta \csc \phi$$

69. Sketch the solid that has the following description in spherical coordinates: $0 \le \theta \le 2\pi$, $0 \le \phi \le \pi/6$, and $0 \le \rho \le a \sec \phi$.

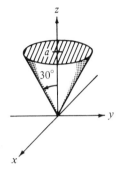

Solution

From the constraint $0 \le \rho \le a \sec \phi$, we have

$$0 \le \rho \le a \left(\frac{1}{\cos \phi} \right)$$
$$0 \le \rho \cos \phi \le a$$
$$0 \le z \le a$$

Review Exercises for Chapter 12

3. Given the points $P = (5, 0, 0)$, $Q = (4, 4, 0)$ and $R = (2, 0, 6)$, let $\mathbf{u} = \overrightarrow{PQ}$ and $\mathbf{v} = \overrightarrow{PR}$. Find (a) the component forms of $\mathbf{u}$ and $\mathbf{v}$, (b) $\mathbf{u} \cdot \mathbf{v}$, (c) $\mathbf{u} \times \mathbf{v}$, (d) an equation of the plane containing P, Q and R, and (e) a set of parametric equations of the line through P and Q.

Solution

(a) $\mathbf{u} = \overrightarrow{PQ} = \langle 4 - 5, 4 - 0, 0 - 0 \rangle = \langle -1, 4, 0 \rangle$
   $\mathbf{v} = \overrightarrow{PR} = \langle 2 - 5, 0 - 0, 6 - 0 \rangle = \langle -3, 0, 6 \rangle$

(b) $\mathbf{u} \cdot \mathbf{v} = (-1)(-3) + (4)(0) + (0)(6) = 3$

(c) $\mathbf{u} \times \mathbf{v} = \begin{vmatrix} \mathbf{i} & \mathbf{j} & \mathbf{k} \\ -1 & 4 & 0 \\ -3 & 0 & 6 \end{vmatrix} = 24\mathbf{i} + 6\mathbf{j} + 12\mathbf{k}$

   $= 6(4\mathbf{i} + \mathbf{j} + 2\mathbf{k})$

(d) A vector normal to the plane is

   $$\frac{1}{6}(\mathbf{u} \times \mathbf{v}) = 4\mathbf{i} + \mathbf{j} + 2\mathbf{k} \qquad \text{See part (c)}$$

   Therefore, using the point $(5, 0, 0)$, an equation of the plane is

   $$4(x - 5) + 1(y - 0) + 2(z - 0) = 0$$
   $$4x + y + 2z - 20 = 0$$

(e) Since the direction of the line is determined by
$\mathbf{u} = \langle -1, 4, 0 \rangle$ [see part (a)], a set of parametric
equations of the line passing through the point $(5, 0, 0)$ is

$$
\begin{array}{ll}
x = 5 - t & x = 5 - t \\
y = 0 + 4t \quad \text{or} & y = 4t \\
z = 0 & z = 0
\end{array}
$$

21. Find (a) a set of parametric equations and (b) a set of symmetric equations for the line perpendicular to the xz-coordinate
plane passing through the point $(1, 2, 3)$.

Solution

(a) Any line perpendicular to the xz-plane must have the
direction of the vector $\mathbf{v} = 0\mathbf{i} + \mathbf{j} + 0\mathbf{k}$. Thus direction
numbers for the required line are $0, 1, 0$. Since the line
passes through the point $(1, 2, 3)$, the parametric
equations are

$$
\begin{aligned}
x &= 1 + (0)t = 1 \\
y &= 2 + t \\
z &= 3 + (0)t = 3
\end{aligned}
$$

(b) Since two of the direction numbers are zero, there is no
symmetric form.

27. Find an equation of the plane containing the lines

$$
\frac{x - 1}{-2} = y = z + 1 \qquad \text{and} \qquad \frac{x + 1}{-2} = y - 1 = z - 2
$$

Solution

We first observe that the lines are parallel since they have the
same direction numbers, $-2, 1, 1$. Therefore, a vector parallel
to the plane is $\mathbf{u} = \langle -2, 1, 1 \rangle$. A point on the first line is
$(1, 0, -1)$ and a point on the second line is $(-1, 1, 2)$. The
vector $\mathbf{v} = \langle 2, -1, -3 \rangle$ connecting these two points is also
parallel to the plane. Thus, a normal to the plane is

$$
\mathbf{u} \times \mathbf{v} = \begin{vmatrix} \mathbf{i} & \mathbf{j} & \mathbf{k} \\ -2 & 1 & 1 \\ 2 & -1 & -3 \end{vmatrix} = -2(\mathbf{i} + 2\mathbf{j})
$$

Therefore, an equation of the plane is

$$
\begin{aligned}
1(x - 1) + 2(y - 0) + 0(z + 1) &= 0 \\
x + 2y &= 1
\end{aligned}
$$

41.   Sketch the graph of the surface $16x^2 + 16y^2 - 9z^2 = 0$.

Solution

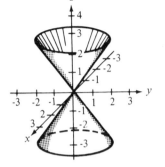

The given equation has the form

$$\frac{x^2}{9} + \frac{y^2}{9} - \frac{z^2}{16} = 0$$

which is the form for a cone. The axis of the cone is the z-axis. The xz-trace ($y = 0$) is given by

$$\frac{x^2}{9} - \frac{z^2}{16} = 0 \qquad \text{or} \qquad z = \pm\frac{4}{3}x$$

The yz-trace ($x = 0$) is given by

$$\frac{y^2}{9} - \frac{z^2}{16} = 0 \qquad \text{or} \qquad z = \pm\frac{4}{3}y$$

Traces parallel to the xy-plane are circles. For example, when $z = 4$ we have

$$16x^2 + 16y^2 - 9(16) = 0$$
$$x^2 + y^2 = 9$$

48.   Sketch the graph of the surface $y = \cos z$.

Solution

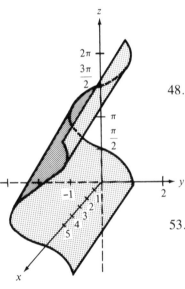

In a three-dimensional system the x-variable is missing from the equation. Therefore the graph is a cylindrical surface with rulings parallel to the x-axis and $y = \cos z$ as its generating curve. The graph is shown in the accompanying figure.

53.   Write the equation $r^2(\cos^2 \theta - \sin^2 \theta) + z^2 = 1$ in rectangular coordinates.

Solution

We rewrite the equation as

$$r^2 \cos^2 \theta - r^2 \sin^2 \theta + z^2 = 1$$

Now since $x = r \cos \theta$ and $y = r \sin \theta$, we have

$$x^2 - y^2 + z^2 = 1$$

as the rectangular equation.

13 Vector-valued functions

13.1
Vector-valued functions

5. Sketch the curve represented by the vector-valued function

$$\mathbf{r}(t) = 2 \cos t\,\mathbf{i} + 2 \sin t\,\mathbf{j} + t\,\mathbf{k}$$

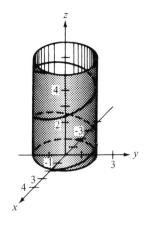

Solution

From the first two parametric equations $x = 2 \cos t$ and $y = 2 \sin t$, we obtain

$$x^2 + y^2 = 4 \cos^2 t + 4 \sin^2 t = 4$$

This means that the curve lies on a right circular cylinder of radius 2 centered about the z-axis. To locate the curve on this cylinder, we use the third parametric equation $z = t$. Thus the graph of the vector-valued function spirals counterclockwise up the cylinder to produce a circular helix.

8. Sketch the curve represented by the vector-valued function

$$\mathbf{r}(t) = t\,\mathbf{i} + t^2\,\mathbf{j} + \frac{3t}{2}\,\mathbf{k}$$

Solution

By eliminating the parameter from the two equations,

$$x = t \qquad \text{and} \qquad y = t^2$$

we obtain the parabolic cylinder $y = x^2$. Since $z = 3t/2 = 3x/2$, the points $(x, y, 3x/2)$ on the space curve will lie $3x/2$ units above or below the parabola $y = x^2$ in the xy-plane. The points lie above the xy-plane for $x > 0$ and below the xy-plane for $x < 0$. (See the accompanying table and sketch.)

t	-2	-1	0	1	2
x	-2	-1	0	1	2
y	4	1	0	1	4
z	-3	$-\dfrac{3}{2}$	0	$\dfrac{3}{2}$	3

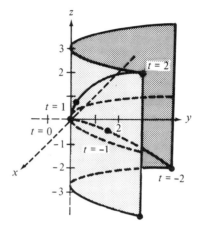

15. Sketch the curve represented by the intersection of the surfaces $x^2 + y^2 + z^2 = 4$ and $x + z = 2$. Find a vector-valued function for the curve using the parameter $x = 1 + \sin t$.

Solution

The equation $x^2 + y^2 + z^2 = 4$ represents a sphere of radius 2 centered at the origin. The equation $x + z = 2$ represents a plane. We let $x = 1 + \sin t$, then

$$z = 2 - x = 1 - \sin t$$

Substituting into the equation of the sphere, we have

$$x^2 + y^2 + z^2 = 4$$
$$(1 + \sin t)^2 + y^2 + (1 - \sin t)^2 = 4$$
$$2 + 2 \sin^2 t + y^2 = 4$$
$$y^2 = 2 - 2 \sin^2 t$$
$$y^2 = 2 \cos^2 t$$
$$y = \pm \sqrt{2} \cos t$$

Therefore $x = 1 + \sin t$, $y = \pm \sqrt{2} \cos t$, $z = 1 - \sin t$ and the two vector-valued functions are

$$\mathbf{r}(t) = (1 + \sin t)\mathbf{i} + (\sqrt{2} \cos t)\mathbf{j} + (1 - \sin t)\mathbf{k}$$

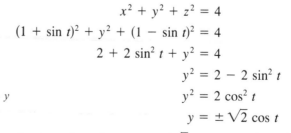

and

$$r(t) = (1 + \sin t)\mathbf{i} - (\sqrt{2} \cos t)\mathbf{j} + (1 - \sin t)\mathbf{k}$$

t	$-\dfrac{\pi}{2}$	$-\dfrac{\pi}{6}$	0	$\dfrac{\pi}{6}$	$\dfrac{\pi}{2}$
x	0	$\dfrac{1}{2}$	1	$\dfrac{3}{2}$	2
y	0	$\pm\dfrac{\sqrt{6}}{2}$	$\pm\sqrt{2}$	$\pm\dfrac{\sqrt{6}}{2}$	0
z	2	$\dfrac{3}{2}$	1	$\dfrac{1}{2}$	0

17. Sketch the curve (first octant portion) represented by the inter-section of the surfaces $x^2 + z^2 = 4$ and $y^2 + z^2 = 4$. Find a vector-valued function for the curve using the parameter $x = t$.

Solution

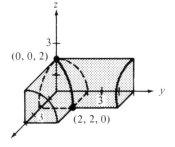

The equation $x^2 + z^2 = 4$ represents a cylinder of radius 2 with its axis along the y-axis, while the equation $y^2 + z^2 = 4$ represents a cylinder of radius 2 with its axis along the x-axis. Thus the curve of intersection lies along both cylinders, as shown in the accompanying sketch. Two points on the curve are $(2, 2, 0)$ and $(0, 0, 2)$. To find a set of parametric equations for the curve of intersection, we subtract the second equation from the first to obtain

$$\begin{aligned} x^2 + z^2 &= 4 \\ -(y^2 + z^2 &= 4) \\ \hline x^2 - y^2 &= 0 \qquad \text{or} \qquad y = \pm x \end{aligned}$$

Therefore, in the first octant, if we let $x = t$ $(t > 0)$, then parametric equations for the curve are

$$x = t, \qquad y = t, \qquad z = \sqrt{4 - t^2}$$

and the vector-valued function is

$$r(t) = t\mathbf{i} + t\mathbf{j} + \sqrt{4 - t^2}\,\mathbf{k}$$

21. Evaluate the limit

$$\lim_{t \to 0} \left(t^2\mathbf{i} + 3t\mathbf{j} + \frac{1 - \cos t}{t}\mathbf{k} \right)$$

Solution

$$\lim_{t \to 0} \left(t^2 \mathbf{i} + 3t\mathbf{j} + \frac{1 - \cos t}{t}\mathbf{k} \right)$$

$$= [\lim_{t \to 0} t^2]\mathbf{i} + [\lim_{t \to 0} 3t]\mathbf{j} + \left[\lim_{t \to 0} \frac{1 - \cos t}{t} \right]\mathbf{k}$$

$$= 0\mathbf{i} + 0\mathbf{j} + 0\mathbf{k} = \mathbf{0}$$

13.2
Differentiation and integration of vector-valued functions

5. Find $\mathbf{r}'(t)$ for $\mathbf{r}(t) = \langle t \sin t, t \cos t, t \rangle$.

Solution

$$\mathbf{r}(t) = \langle t \sin t, t \cos t, t \rangle$$

$$\mathbf{r}'(t) = \left\langle \frac{d}{dt}[t \sin t], \frac{d}{dt}[t \cos t], \frac{d}{dt}[t] \right\rangle$$

$$= \langle t \cos t + \sin t, -t \sin t + \cos t, 1 \rangle$$

15. Find the position vector $\mathbf{r}$ if $\mathbf{r}'(t) = 4e^{2t}\mathbf{i} + 3e^t\mathbf{j}$ and $\mathbf{r}(0) = 2\mathbf{i}$.

Solution

$$\mathbf{r}(t) = \mathbf{i} \int 4e^{2t}\, dt + \mathbf{j} \int 3e^t\, dt = \mathbf{i}[2e^{2t} + c_1] + \mathbf{j}[3e^t + c_2]$$

$$\mathbf{r}(0) = \mathbf{i}[2 + c_1] + \mathbf{j}[3 + c_2] = 2\mathbf{i}$$

Therefore $2 + c_1 = 2$, or $c_1 = 0$. Also, $3 + c_2 = 0$, or $c_2 = -3$. Thus

$$\mathbf{r}(t) = (2e^{2t})\mathbf{i} + (3e^t - 3)\mathbf{j} = (2e^{2t})\mathbf{i} + 3(e^t - 1)\mathbf{j}$$

21. Evaluate the definite integral $\int_0^1 (8t\mathbf{i} + t\mathbf{j} - \mathbf{k})\, dt$.

Solution

$$\int_0^1 (8t\mathbf{i} + t\mathbf{j} - \mathbf{k})\, dt = \mathbf{i} \int_0^1 8t\, dt + \mathbf{j} \int_0^1 t\, dt - \mathbf{k} \int_0^1 dt$$

$$= [4t^2]_0^1\, \mathbf{i} + \left[\frac{1}{2}t^2 \right]_0^1 \mathbf{j} - [t]_0^1\, \mathbf{k}$$

$$= 4\mathbf{i} + \frac{1}{2}\mathbf{j} - \mathbf{k}$$

13.3
Velocity and acceleration

3. The motion of an object in the xy-plane is described by the vector-valued function $\mathbf{r}(t) = t^2\mathbf{i} + t\mathbf{j}$. Sketch a graph of the path and sketch the velocity and acceleration vectors at the point $(4, 2)$.

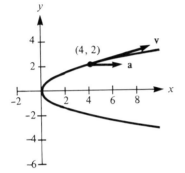

    Solution

    Eliminating the parameter from the parametric equations $x = t^2$ and $y = t$, we obtain the rectangular equation $x = y^2$. Therefore, the object is moving in a parabolic path. (See the accompanying figure.)
    The velocity is given by

    $$\mathbf{v}(t) = \mathbf{r}'(t) = 2t\mathbf{i} + \mathbf{j}$$

    and the acceleration is given by

    $$\mathbf{a}(t) = \mathbf{r}''(t) = 2\mathbf{i}$$

    At the point $(4, 2)$, $t = 2$. Thus

    $$\mathbf{v}(2) = 4\mathbf{i} + \mathbf{j}$$

    and

    $$\mathbf{a}(2) = 2\mathbf{i}$$

15. The position function

    $$\mathbf{r}(t) = \langle 4t, 3\cos t, 3\sin t \rangle$$

    describes the path of an object moving in space. Find the velocity, speed, and acceleration of the object.

    Solution

    Since

    $$\mathbf{r}(t) = \langle 4t, 3\cos t, 3\sin t \rangle$$

    we have

    $$\mathbf{v}(t) = \mathbf{r}'(t) = \langle 4, -3\sin t, 3\cos t \rangle$$
    $$\|\mathbf{v}(t)\| = \text{speed} = \sqrt{16 + 9(\sin^2 t + \cos^2 t)} = \sqrt{25} = 5$$

    and

    $$\mathbf{a}(t) = \mathbf{r}''(t) = \langle 0, -3\cos t, -3\sin t \rangle$$

18. Find the velocity and position functions and the position of the object at $t = 2$ if $\mathbf{a}(t) = \mathbf{i} + \mathbf{k}$, $\mathbf{v}(0) = 5\mathbf{j}$ and $\mathbf{r}(0) = \mathbf{0}$.

Solution

To find the velocity function, we integrate and use the initial condition $\mathbf{v}(0) = 5\mathbf{j}$ to solve for the constant of integration.

$$\mathbf{v}(t) = \int \mathbf{a}(t) \, dt = \int (\mathbf{i} + \mathbf{k}) \, dt = t\mathbf{i} + t\mathbf{k} + \mathbf{C}$$

where $\mathbf{C} = C_1\mathbf{i} + C_2\mathbf{j} + C_3\mathbf{k}$. From the initial condition $\mathbf{v}(0) = 5\mathbf{j}$, we have

$$\mathbf{v}(0) = C_1\mathbf{i} + C_2\mathbf{j} + C_3\mathbf{k} = 5\mathbf{j} \qquad \rightarrow \qquad C_1 = C_3 = 0$$
$$\text{and} \qquad C_2 = 5$$

Thus, the *velocity* at any time t is

$$\mathbf{v}(t) = t\mathbf{i} + 5\mathbf{j} + t\mathbf{k}$$

Now, integrating once more, we have

$$\mathbf{r}(t) = \int \mathbf{v}(t) \, dt = \int (t\mathbf{i} + 5\mathbf{j} + t\mathbf{k}) \, dt = \frac{1}{2}t^2\mathbf{i} + 5t\mathbf{j} + \frac{1}{2}t^2\mathbf{k} + \mathbf{C}$$

where $\mathbf{C} = C_4\mathbf{i} + C_5\mathbf{j} + C_6\mathbf{k}$. Letting $t = 0$ and applying the initial condition $\mathbf{r}(0) = \mathbf{0}$, we have

$$\mathbf{r}(0) = C_4\mathbf{i} + C_5\mathbf{j} + C_6\mathbf{k} = \mathbf{0} \qquad \rightarrow \qquad C_4 = C_5 = C_6 = 0$$

Thus, the *position function* is

$$\mathbf{r}(t) = \frac{1}{2}t^2\mathbf{i} + 5t\mathbf{j} + \frac{1}{2}t^2\mathbf{k}$$

At time $t = 2$, the location of the object is given by

$$\mathbf{r}(2) = 2\mathbf{i} + 10\mathbf{j} + 2\mathbf{k}$$

21.   A batted baseball, hit three feet above the ground, leaves the bat at an angle of 45° and is caught by an outfielder 300 ft from the plate. What was the initial velocity of the ball and how high did it rise if it was caught 3 feet above the ground?

Solution

Using Theorem 13.3 with $h = 3$, the path of the ball is given by

$$\mathbf{r}(t) = (v_0 \cos 45°)t\,\mathbf{i} + [3 + (v_0 \sin 45°)t - 16t^2]\mathbf{j}$$
$$= \left(\frac{tv_0}{\sqrt{2}}\right)\mathbf{i} + \left(3 + \frac{tv_0}{\sqrt{2}} - 16t^2\right)\mathbf{j}$$

We know that the horizontal component is 300 when the vertical component is three. Thus

$$\frac{tv_0}{\sqrt{2}} = 300 \qquad \text{and} \qquad 3 + \frac{tv_0}{\sqrt{2}} - 16t^2 = 3$$

Therefore $t = 300\sqrt{2}/v_0$ and it follows that

$$\frac{300\sqrt{2}}{v_0}\left(\frac{v_0}{\sqrt{2}}\right) - 16\left(\frac{300\sqrt{2}}{v_0}\right)^2 = 0$$

$$300 = \frac{16(300^2)(2)}{v_0{}^2}$$

$$v_0{}^2 = 32(300)$$

and

$$v_0 = \sqrt{9600}$$
$$= 40\sqrt{6} \approx 97.98 \text{ ft/s}$$

The maximum height is reached when the derivative of the vertical component is zero. Thus we have

$$y(t) = 3 + \frac{tv_0}{\sqrt{2}} - 16t^2 = 3 + \frac{t(40\sqrt{6})}{\sqrt{2}} - 16t^2$$
$$= 3 + 40\sqrt{3}t - 16t^2$$

and

$$y'(t) = 40\sqrt{3} - 32t = 0$$
$$t = \frac{40\sqrt{3}}{32} = \frac{5\sqrt{3}}{4}$$

Finally, the maximum height is

$$y\left(\frac{5\sqrt{3}}{4}\right) = 3 + 40\sqrt{3}\left(\frac{5\sqrt{3}}{4}\right) - 16\left(\frac{5\sqrt{3}}{4}\right)^2$$
$$= 3 + 150 - 75 = 78 \text{ ft}$$

29. The path of a ball is given by the rectangular equation $y = x - 0.005x^2$. Use the result of Exercise 28 to find the position function. Then find the speed and direction of the ball at the point when it has traveled a horizontal distance of 60 feet.

Solution

From Exercise 28 we know that the equation $y = -0.005x^2 + x$ is of the form

$$y = -\left[\frac{16(\sec^2 \theta)}{v_0{}^2}\right]x^2 + (\tan \theta)x$$

Thus, $\tan \theta = 1$ and $\theta = 45°$. Furthermore, since $\sec 45° = \sqrt{2}$, we have

$$-0.005 = -\frac{16(\sec 45°)^2}{v_0^2}$$

$$v_0^2 = \frac{-16(2)}{-0.005} = 6400$$

$$v_0 = 80 \text{ ft/s}$$

The vector-valued function for the path of the ball is given by

$$\mathbf{r}(t) = t(v_0 \cos \theta)\mathbf{i} + \left(h + tv_0 \sin \theta - \frac{gt^2}{2}\right)\mathbf{j}$$

(See Theorem 13.3.) Since the path of the ball starts at (0, 0), we let $h = 0$. Therefore

$$\mathbf{r}(t) = t(80 \cos 45°)\mathbf{i} + \left[t(80) \sin 45° - \frac{32t^2}{2}\right]\mathbf{j}$$

$$= t\left(80\frac{\sqrt{2}}{2}\right)\mathbf{i} + \left[t(80)\frac{\sqrt{2}}{2} - 16t^2\right]\mathbf{j}$$

$$= 40\sqrt{2}t\,\mathbf{i} + (40\sqrt{2}t - 16t^2)\mathbf{j}$$

The velocity of the ball is given by

$$\mathbf{v}(t) = \mathbf{r}'(t) = 40\sqrt{2}\,\mathbf{i} + (40\sqrt{2} - 32t)\mathbf{j}$$

Furthermore, when $x = 60$, we have

$$40\sqrt{2}\,t = 60 \quad \text{or} \quad t = \frac{60}{40\sqrt{2}} = \frac{3\sqrt{2}}{4}$$

Thus the velocity when $x = 60$ is given by

$$\mathbf{v} = 40\sqrt{2}\,\mathbf{i} + \left[40\sqrt{2} - 32\left(\frac{3\sqrt{2}}{4}\right)\right]\mathbf{j}$$

$$= 40\sqrt{2}\,\mathbf{i} + (40\sqrt{2} - 24\sqrt{2})\mathbf{j}$$

$$= 40\sqrt{2}\,\mathbf{i} + 16\sqrt{2}\,\mathbf{j} = 8\sqrt{2}(5\mathbf{i} + 2\mathbf{j})$$

which implies that the speed is

$$\|\mathbf{v}\| = 8\sqrt{2}\,\sqrt{5^2 + 2^2} = 8\sqrt{2}\,\sqrt{29} = 8\sqrt{58} \text{ ft/s}$$

32. Find the maximum velocity of a point on the circumference of the tire of an automobile when the automobile is traveling at 55 mi/h and the radius of the wheel is 1 ft. Compare this velocity with the velocity of the automobile.

Solution

The path of the point on the circumference of the wheel is the cycloid

$$\mathbf{r}(t) = b(\omega t - \sin \omega t)\mathbf{i} + b(1 - \cos \omega t)\mathbf{j}$$

$$\mathbf{v}(t) = b\omega[(1 - \cos \omega t)\mathbf{i} + (\sin \omega t)\mathbf{j}]$$
$$\text{speed} = \|\mathbf{v}(t)\| = b\omega\sqrt{(1 - \cos \omega t)^2 + (\sin \omega t)^2}$$
$$= b\omega\sqrt{1 - 2\cos \omega t + \cos^2 \omega t + \sin^2 \omega t}$$
$$= \sqrt{2}\, b\omega\sqrt{1 - \cos \omega t}$$

We observe that the maximum speed is $2b\omega$ when $\cos \omega t = -1$ or $\omega t = \pi,\ 3\pi,\ 5\pi,\ \ldots$. Since $b = 1$, the maximum speed is 2ω. Finally,

$$55 \text{ mi/h} = \frac{(55)(5280)}{3600} = 80\tfrac{2}{3} \text{ ft/s}$$
$$= 80\tfrac{2}{3} \text{ rad/s} \qquad \text{Since } b = 1$$
$$= \omega$$

Therefore the maximum speed of the point is

$$2(80\tfrac{2}{3}) \text{ ft/s} = 110 \text{ mi/h}$$

and occurs when the point is at the top of the tire.

In Exercises 33–36, consider a particle moving on a circular path of radius b described by

$$\mathbf{r}(t) = b \cos \omega t\, \mathbf{i} + b \sin \omega t\, \mathbf{j}$$

where $\omega = d\theta/dt$ is the angular velocity.

33.   Find the velocity vector and show that it is orthogonal to $\mathbf{r}(t)$.

Solution

The velocity vector is

$$\mathbf{v}(t) = \mathbf{r}'(t) = [-b\omega \sin \omega t]\mathbf{i} + [b\omega \cos \omega t]\mathbf{j}$$

and since

$$\mathbf{r}(t) \cdot \mathbf{v}(t) = -b^2\omega \sin \omega t \cos \omega t + b^2\omega \sin \omega t \cos \omega t = 0$$

we conclude that $\mathbf{r}(t)$ and $\mathbf{v}(t)$ are orthogonal.

34.   Show that the speed of the particle is $b\omega$.

Solution

$$\text{speed} = \|\mathbf{v}\| = \sqrt{b^2\omega^2 \sin^2 \omega t + b^2\omega^2 \cos^2 \omega t}$$
$$= \sqrt{b^2\omega^2[\sin^2 \omega t + \cos^2 \omega t]} = b\omega$$

35.   Find the acceleration vector and show that its direction is always toward the center of the circle.

Solution

Since

$$\mathbf{a}(t) = \mathbf{r}''(t) = [-b\omega^2 \cos \omega t]\mathbf{i} - [b\omega^2 \sin \omega t]\mathbf{j}$$
$$= -b\omega^2[\cos \omega t\,\mathbf{i} + \sin \omega t\,\mathbf{j}]$$

we see that $\mathbf{a}(t)$ is a negative multiple of a unit vector from $(0, 0)$ to $(\cos \omega t, \sin \omega t)$, and thus $\mathbf{a}(t)$ is directed toward the origin.

36. Show that the magnitude of the acceleration vector is $\omega^2 b$.

Solution

$$\|\mathbf{a}(t)\| = b\omega^2\|\cos \omega t\,\mathbf{i} + \sin \omega t\,\mathbf{j}\| = b\omega^2$$

13.4
Tangent vectors and normal vectors

3. Find a set of parametric equations for the line tangent to the helix $\mathbf{r}(t) = 2 \cos t\,\mathbf{i} + 2 \sin t\,\mathbf{j} + t\,\mathbf{k}$ at the point $(2, 0, 0)$.

Solution

$$\mathbf{r}(t) = (2 \cos t)\mathbf{i} + (2 \sin t)\mathbf{j} + t\,\mathbf{k}$$
$$\mathbf{r}'(t) = (-2 \sin t)\mathbf{i} + (2 \cos t)\mathbf{j} + \mathbf{k}$$

Since $t = 0$ at the point $(2, 0, 0)$, the direction vector for the line is given by

$$\mathbf{r}'(0) = 2\mathbf{j} + \mathbf{k}$$

and at the point $(2, 0, 0)$ the parametric representation of the line is

$$x = 2, \qquad y = 2s, \qquad z = s$$

13. If $\mathbf{r} = t\mathbf{i} + (1/t)\mathbf{j}$, find $\mathbf{T}(t)$, $\mathbf{N}(t)$, $\mathbf{a}(t) \cdot \mathbf{T}(t)$, and $\mathbf{a}(t) \cdot \mathbf{N}(t)$ when $t = 1$.

Solution

Since

$$\mathbf{r}(t) = t\mathbf{i} + \frac{1}{t}\mathbf{j}$$

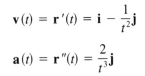

$$\mathbf{v}(t) = \mathbf{r}'(t) = \mathbf{i} - \frac{1}{t^2}\mathbf{j}$$

$$\mathbf{a}(t) = \mathbf{r}''(t) = \frac{2}{t^3}\mathbf{j}$$

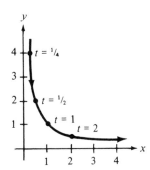

Then at $t = 1$, we have

$$\mathbf{v}(1) = \mathbf{i} - \mathbf{j}, \qquad \|\mathbf{v}(1)\| = \sqrt{2}, \qquad \text{and} \qquad \mathbf{a}(1) = 2\mathbf{j}$$

Therefore when $t = 1$,

$$\mathbf{T}(1) = \frac{\mathbf{v}(1)}{\|\mathbf{v}(1)\|} = \frac{\mathbf{i}}{\sqrt{2}} - \frac{\mathbf{j}}{\sqrt{2}}$$

and since $\mathbf{N}(1)$ points toward the concave side of the curve (see the accompanying figure)

$$\mathbf{N}(1) = \frac{\mathbf{i}}{\sqrt{2}} + \frac{\mathbf{j}}{\sqrt{2}}$$

Furthermore,

$$\mathbf{a}(1) \cdot \mathbf{T}(1) = (2\mathbf{j}) \cdot \left(\frac{\mathbf{i}}{\sqrt{2}} - \frac{\mathbf{j}}{\sqrt{2}} \right) = \frac{-2}{\sqrt{2}} = -\sqrt{2}$$

and

$$\mathbf{a}(1) \cdot \mathbf{N}(1) = (2\mathbf{j}) \cdot \left(\frac{\mathbf{i}}{\sqrt{2}} + \frac{\mathbf{j}}{\sqrt{2}} \right) = \frac{2}{\sqrt{2}} = \sqrt{2}$$

19. If $\mathbf{r}(t) = (e^t \cos t)\mathbf{i} + (e^t \sin t)\mathbf{j}$, find $\mathbf{T}(t)$, $\mathbf{N}(t)$, $\mathbf{a}(t) \cdot \mathbf{T}(t)$, and $\mathbf{a}(t) \cdot \mathbf{N}(t)$, when $t = \pi/2$.

Solution

$$\mathbf{r}(t) = (e^t \cos t)\mathbf{i} + (e^t \sin t)\mathbf{j}$$
$$\mathbf{v}(t) = \mathbf{r}'(t) = e^t(\cos t - \sin t)\mathbf{i} + e^t(\cos t + \sin t)\mathbf{j}$$
$$\mathbf{a}(t) = \mathbf{r}''(t)$$
$$= e^t(-\sin t - \cos t + \cos t - \sin t)\mathbf{i}$$
$$+ e^t(-\sin t + \cos t + \cos t + \sin t)\mathbf{j}$$
$$= e^t(-2 \sin t)\mathbf{i} + e^t(2 \cos t)\mathbf{j}$$

Then at $t = \pi/2$, we have

$$\mathbf{v}(\pi/2) = -e^{\pi/2}\mathbf{i} + e^{\pi/2}\mathbf{j}$$
$$\|\mathbf{v}(\pi/2)\| = e^{\pi/2}\sqrt{2}$$
and $\qquad \mathbf{a}(\pi/2) = -2e^{\pi/2}\mathbf{i}$

Therefore at $t = \pi/2$,

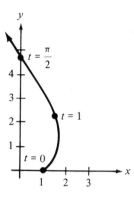

$$T(\pi/2) = \frac{v(\pi/2)}{\|v(\pi/2)\|} = \frac{-i}{\sqrt{2}} + \frac{j}{\sqrt{2}} = \frac{1}{\sqrt{2}}(-i + j)$$

and since $N(\pi/2)$ points toward the concave side of the curve (see the accompanying figure)

$$N(\pi/2) = \frac{1}{\sqrt{2}}(-i - j)$$

$$a(\pi/2) \cdot T(\pi/2) = (-2e^{\pi/2}i) \cdot \frac{(-i + j)}{\sqrt{2}} = \sqrt{2}e^{\pi/2}$$

$$a(\pi/2) \cdot N(\pi/2) = (-2e^{\pi/2}i) \cdot \frac{(-i - j)}{\sqrt{2}} = \sqrt{2}e^{\pi/2}$$

25. If $r(t) = 4t\,i + (3\cos t)j + (3\sin t)k$ find $T(t)$, $N(t)$, $a(t) \cdot T(t)$, and $a(t) \cdot N(t)$ when $t = \pi/2$.

Solution

$$r(t) = 4t\,i + (3\cos t)j + (3\sin t)k$$
$$v(t) = r'(t) = 4i - (3\sin t)j + (3\cos t)k$$
$$\|v(t)\| = \sqrt{16 + 9(\sin^2 t + \cos^2 t)} = \sqrt{25} = 5$$
$$a(t) = r''(t) = -(3\cos t)j - (3\sin t)k$$
$$T(t) = \frac{v(t)}{\|v(t)\|} = \frac{1}{5}[4i - (3\sin t)j + (3\cos t)k]$$
$$N(t) = \frac{T'(t)}{\|T'(t)\|} = \frac{(\frac{1}{5})[-(3\cos t)j - (3\sin t)k]}{(\frac{1}{5})\sqrt{9(\cos^2 t + \sin^2 t)}}$$
$$= \frac{(-\frac{3}{5})[(\cos t)j + (\sin t)k]}{\frac{3}{5}}$$
$$= -(\cos t)j - (\sin t)k$$

Therefore, we have

$$a(\pi/2) = -3k, \qquad T(\pi/2) = \frac{1}{5}[4i - 3j],$$

and $$N(\pi/2) = -k$$

Thus,

$$a\left(\frac{\pi}{2}\right) \cdot T\left(\frac{\pi}{2}\right) = 0 \quad \text{and} \quad a\left(\frac{\pi}{2}\right) \cdot N\left(\frac{\pi}{2}\right) = 3$$

27. Find the tangential and normal components of acceleration for a projectile fired at an angle θ with the horizontal and with an initial speed of v_0. What are the components when the projectile is at its maximum height?

Solution

From Theorem 13.3, we have

$$\mathbf{r}(t) = (v_0 t \cos\theta)\mathbf{i} + (h + v_0 t \sin\theta - 16t^2)\mathbf{j}$$
$$\mathbf{v}(t) = (v_0 \cos\theta)\mathbf{i} + (v_0 \sin\theta - 32t)\mathbf{j}$$
$$\mathbf{a}(t) = -32\mathbf{j}$$
$$\mathbf{T}(t) = \frac{\mathbf{v}(t)}{\|\mathbf{v}(t)\|} = \frac{(v_0 \cos\theta)\mathbf{i} + (v_0 \sin\theta - 32t)\mathbf{j}}{\sqrt{v_0^2 \cos^2\theta + (v_0 \sin\theta - 32t)^2}}$$

Since the path of a projectile is concave downward, we choose

$$\mathbf{N}(t) = \frac{(v_0 \sin\theta - 32t)\mathbf{i} + (-v_0 \cos\theta)\mathbf{j}}{\sqrt{v_0^2 \cos^2\theta + (v_0 \sin\theta - 32t)^2}}$$

Therefore

$$\mathbf{a}(t) \cdot \mathbf{T}(t) = \frac{-32(v_0 \sin\theta - 32t)}{\sqrt{v_0^2 \cos^2\theta + (v_0 \sin\theta - 32t)^2}}$$
$$\mathbf{a}(t) \cdot \mathbf{N}(t) = \frac{32v_0 \cos\theta}{\sqrt{v_0^2 \cos^2\theta + (v_0 \sin\theta - 32t)^2}}$$

The projectile will reach its maximum height when the vertical component of velocity is zero, or

$$v_0 \sin\theta - 32t = 0$$

Therefore at the maximum height $\mathbf{a}(t) \cdot \mathbf{T}(t) = 0$ and $\mathbf{a}(t) \cdot \mathbf{N}(t) = 32$. Thus all the acceleration is normal at the maximum height.

30. An object of mass m moves at a constant speed v in a circular path of radius r. The force required to produce the centripetal component of acceleration is called the **centripetal force**. Show that this force is $F = mv^2/r$. Newton's **Law of Universal Gravitation** is given by $F = GMm/d^2$, where d is the distance between the centers of the two bodies of mass M and m. Use this to show that the speed required for circular motion is $v = \sqrt{GM/r}$.

Solution

Since the motion is circular, the path is described by

$$\mathbf{r}(t) = (r \cos\omega t)\mathbf{i} + (r \sin\omega t)\mathbf{j}$$

Furthermore,

$$\mathbf{v}(t) = (-r\omega \sin\omega t)\mathbf{i} + (r\omega \cos\omega t)\mathbf{j}$$
$$\|\mathbf{v}(t)\| = r\omega\sqrt{1} = r\omega$$

and

$$\mathbf{a}(t) = (-r\omega^2 \cos \omega t)\mathbf{i} - (r\omega^2 \sin \omega t)\mathbf{j}$$
$$\|\mathbf{a}(t)\| = r\omega^2 \sqrt{1} = r\omega^2$$

Since force equals mass times acceleration, we have

$$F = m[\mathbf{a}(t)] = m(r\omega^2) = \frac{m}{r}(r^2\omega^2) = \frac{mv^2}{r}$$

If we consider the moving object to have a point mass m, then in Newton's Law we let $d = r$ and we have

$$\frac{mv^2}{r} = \frac{GMm}{r^2} \quad \text{or} \quad v^2 = \frac{GM}{r} \quad \text{or} \quad v = \sqrt{\frac{GM}{r}}$$

34. Use the result of Exercise 30 to find the speed necessary for the circular orbit of a syncom satellite in a geosynchronous orbit r miles above the surface of the earth. Let $GM = 9.56 \times 10^4 \text{ mi}^3/\text{s}^2$ and assume the radius of the earth is 4000 miles. [The satellite completes one orbit per sidereal day (23 hours, 56 minutes) and thus appears to remain stationary above a point on the earth.]

Solution

We let r be the radius of the orbit measured from the center of the earth. Then the distance the satellite travels in one day is $d = 2\pi r$ and the speed is $v = 2\pi r/t = 2\pi r/(24)(3600)$. However, from Exercise 30 the speed is also given by $v = \sqrt{9.56(10^4)/r}$. Solving these two equations simultaneously, we have

$$\frac{2\pi r}{(24)(3600)} = \sqrt{\frac{9.56(10^4)}{r}}$$

$$\frac{4\pi^2 r^2}{(24^2)(3600^2)} = \frac{9.56(10^4)}{r}$$

$$r^3 = \frac{9.56(10^4)(24^2)(3600^2)}{4\pi^2}$$

$$r = \sqrt[3]{\frac{9.56(10^4)(24^2)(3600^2)}{4\pi^2}} = 26{,}245 \text{ mi}$$

Therefore the altitude above the earth is $26{,}245 - 4{,}000 = 22{,}245$ mi, and the speed is

$$v = \frac{\text{distance}}{\text{time}} = \frac{2\pi(26{,}245)}{(24)(3{,}600)} = 1.91 \text{ mi/s} = 6{,}871 \text{ mi/h}$$

13.5
Arc length and curvature

3. Sketch and find the length of the circular helix

$$\mathbf{r}(t) = a \cos t\mathbf{i} + a \sin t\mathbf{j} + bt\mathbf{k}$$

over the interval $[0, 2\pi]$.

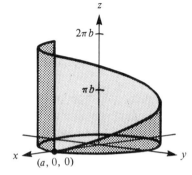

Solution

The graph is a circular helix of radius a. (See the accompanying figure.)
    Since

$$\mathbf{r}'(t) = -a \sin t\mathbf{i} + a \cos t\mathbf{j} + b\mathbf{k}$$

the arc length on the interval $[0, 2\pi]$ is

$$s = \int_0^{2\pi} \|\mathbf{r}'(t)\| \, dt$$

$$= \int_0^{2\pi} \sqrt{(-a \sin t)^2 + (a \cos t)^2 + b^2} \, dt$$

$$= \int_0^{2\pi} \sqrt{a^2(\sin^2 t + \cos^2 t) + b^2} \, dt$$

$$= \int_0^{2\pi} \sqrt{a^2 + b^2} \, dt = [\sqrt{a^2 + b^2} \, t]_0^{2\pi} = 2\pi\sqrt{a^2 + b^2}$$

9. Find the curvature and radius of curvature of the plane curve
    $y = 2x^2 + 3$ at $x = -1$.

Solution

$$y = 2x^2 + 3 \qquad y' = 4x \qquad y'' = 4$$

$$K = \left| \frac{y''}{[1 + (y')^2]^{3/2}} \right| = \frac{4}{[1 + (4x)^2]^{3/2}}$$

When $x = -1$,

$$K = \frac{4}{(1 + 16)^{3/2}} = \frac{4}{17^{3/2}} \approx 0.057$$

The radius of curvature when $x = -1$ is

$$r = \frac{1}{K} = \frac{17^{3/2}}{4} \approx 17.523$$

17. Find the curvature K of the curve $\mathbf{r}(t) = t\mathbf{i} + \frac{1}{t}\mathbf{j}$ at $t = 1$.

Solution

From Exercise 1, Section 13.4, we have

$$\mathbf{a}(1) \cdot \mathbf{N}(1) = \sqrt{2} \quad \text{and} \quad \|\mathbf{v}(1)\|^2 = 2$$

Therefore, the curvature is

$$K = \frac{\mathbf{a}(1) \cdot \mathbf{N}(1)}{\|\mathbf{v}(1)\|} = \frac{\sqrt{2}}{2} \approx 0.707$$

23. Find the curvature K of the curve $\mathbf{r}(t) = e^t \cos t\,\mathbf{i} + e^t \sin t\,\mathbf{j}$.

Solution

$$\mathbf{r}(t) = e^t \cos t\,\mathbf{i} + e^t \sin t\,\mathbf{j}$$

$$\mathbf{r}'(t) = e^t(\cos t - \sin t)\mathbf{i} + e^t(\cos t + \sin t)\mathbf{j}$$

$$\|\mathbf{r}'(t)\| = e^t\sqrt{(\cos t - \sin t)^2 + (\cos t + \sin t)^2} = \sqrt{2}\,e^t$$

$$\mathbf{T}(t) = \frac{\mathbf{r}'(t)}{\|\mathbf{r}'(t)\|} = \frac{1}{\sqrt{2}}[(\cos t - \sin t)\mathbf{i} + (\cos t + \sin t)\mathbf{j}]$$

$$\mathbf{T}'(t) = \frac{1}{\sqrt{2}}[(-\sin t - \cos t)\mathbf{i} + (-\sin t + \cos t)\mathbf{j}]$$

$$\|\mathbf{T}'(t)\| = \frac{1}{\sqrt{2}}\sqrt{(-\sin t - \cos t)^2 + (-\sin t + \cos t)^2} = 1$$

$$K = \frac{\|\mathbf{T}'(t)\|}{\|\mathbf{r}'(t)\|} = \frac{1}{\sqrt{2}\,e^t} = \frac{\sqrt{2}}{2}e^{-t}$$

29. Find the curvature K of the curve $\mathbf{r}(t) = 4t\,\mathbf{i} + 3\cos t\,\mathbf{j} + 3\sin t\,\mathbf{k}$.

Solution

From Exercise 25, Section 13.4, we have

$$\|\mathbf{T}'(t)\| = \frac{3}{5}$$

and

$$\|\mathbf{r}'(t)\| = 5$$

Therefore, the curvature is

$$K = \frac{\|\mathbf{T}'(t)\|}{\|\mathbf{r}'(t)\|} = \frac{\frac{3}{5}}{5} = \frac{3}{25}$$

31. Given the function $y = (x - 1)^2 + 3$, (a) find the point on the curve at which the curvature K is maximum and (b) find the limit of K as $x \to \infty$.

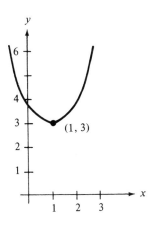

Solution

(a) $y = (x - 1)^2 + 3$

$y' = 2(x - 1)$

$y'' = 2$

$$K = \left| \frac{y''}{[1 + (y')^2]^{3/2}} \right| = \frac{2}{[1 + 4(x - 1)^2]^{3/2}}$$

We can see that K is largest when the denominator $[1 + 4(x - 1)^2]^{3/2}$ is the smallest, and this occurs when $x = 1$. Thus the point on the graph of greatest curvature is $(1, 3)$.

(b) $\displaystyle\lim_{x \to \pm\infty} K = \lim_{x \to \pm\infty} \frac{2}{[1 + 4(x - 1)^2]^{3/2}} = 0$

35.   Find the circle of curvature of the graph of $y = x + (1/x)$ at the point $(1, 2)$.

Solution

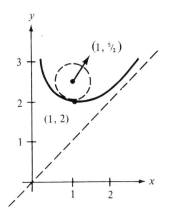

$$f(x) = x + \frac{1}{x} \qquad f'(x) = 1 - \frac{1}{x^2} = \frac{x^2 - 1}{x^2} \qquad f''(x) = \frac{2}{x^3}$$

At the point $(1, 2)$, $f'(1) = 0$ and $f''(1) = 2$. Thus at $(1, 2)$ the curvature is

$$K = \left| \frac{2}{(1 + 0^2)^{3/2}} \right| = 2$$

and the radius of curvature is

$$r = \frac{1}{K} = \frac{1}{2}$$

Since the slope at $(1, 2)$ is 0, the normal line at $(1, 2)$ is vertical and the center of the closest circular approximation is $(1, \frac{5}{2})$. (See the accompanying figure.) Finally, the equation of the closest circular approximation is

$$(x - 1)^2 + \left(y - \frac{5}{2} \right)^2 = \left(\frac{1}{2} \right)^2$$

37.   Show that the curvature is greatest at the endpoints of the major axis and least at the endpoints of the minor axis for the ellipse given by $x^2 + 4y^2 = 4$.

Solution

The endpoints of the major axis are $(\pm 2, 0)$ and the endpoints of the minor axis are $(0, \pm 1)$.

$$x^2 + 4y^2 = 4$$

$$2x + 8yy' = 0$$

$$y' = \frac{-x}{4y}$$

$$y'' = \frac{(4y)(-1) - (-x)(4y')}{16y^2} = \frac{-4y - (x^2/y)}{16y^2}$$

$$= \frac{-(4y^2 + x^2)}{16y^3} = \frac{-4}{16y^3} = \frac{-1}{4y^3}$$

The curvature is given by

$$K = \left| \frac{-1/4y^3}{[1 + (-x/4y)^2]^{3/2}} \right| = \left| \frac{-1}{4y^3[(16y^2 + x^2)/16y^2]^{3/2}} \right|$$

$$= \left| \frac{-16}{(16y^2 + x^2)^{3/2}} \right| = \frac{16}{(12y^2 + 4y^2 + x^2)^{3/2}} = \frac{16}{(12y^2 + 4)^{3/2}}$$

Therefore, since $-1 \leq y \leq 1$, K is largest when $y = 0$ and smallest when $y = \pm 1$.

38. Find all a and b such that the two curves given by $y_1 = ax(b - x)$ and $y_2 = x/(x + 2)$ intersect at only one point and have a common tangent line and equal curvature at the point. Sketch a graph for each of the set of values for a and b.

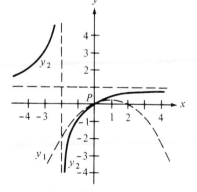

Solution

We observe that $(0, 0)$ is a solution point to both equations, and therefore, the point of intersection is the origin.

$$y_1 = ax(b - x) \qquad y_1' = a(b - 2x) \qquad y_1'' = -2a$$

$$y_2 = \frac{x}{x + 2} \qquad y_2' = \frac{2}{(x + 2)^2} \qquad y_2'' = \frac{-4}{(x + 2)^3}$$

At the origin, $y_1'(0) = ab$ and $y_2'(0) = \frac{1}{2}$. Since the curves have a common tangent at the point of intersection, $y_1'(0) = y_2'(0)$ or $ab = \frac{1}{2}$. Therefore, $y_1'(0) = \frac{1}{2}$. Since the curves have the same curvature at the point of intersection, $K_1(0) = K_2(0)$.

$$K_1(0) = \left| \frac{y_1''(0)}{[1 + (y_1'(0))^2]^{3/2}} \right| = \left| \frac{-2a}{[1 + (\frac{1}{2})^2]^{3/2}} \right|$$

$$K_2(0) = \left| \frac{y_2''(0)}{[1 + (y_2'(0))^2]^{3/2}} \right| = \left| \frac{-1/2}{[1 + (\frac{1}{2})^2]^{3/2}} \right|$$

Therefore, $2a = \pm\frac{1}{2}$ or $a = \pm\frac{1}{4}$. In order that the curves intersect in only one point, the parabola must be concave downward, and thus $a = \frac{1}{4}$ and $b = 1/2a = 2$.

$$y_1 = \frac{1}{4}x(2 - x) \qquad \text{and} \qquad y_2 = \frac{x}{x + 2}$$

Review Exercises for Chapter 13

6. Sketch the space curve represented by the intersection of the surfaces given by $x^2 + z^2 = 4$ and $x - y = 0$. Find a vector-valued function using the parameter $t = x$.

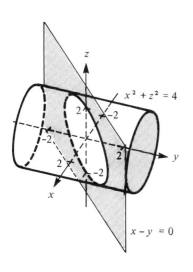

Solution

The equation $x^2 + z^2 = 4$ represents a cylinder of radius 2 with its axis along the y-axis, while $x - y = 0$ is a plane parallel to the z-axis. Since $x = t$, we have

$$y = x = t \qquad\qquad \text{From the equation of the plane}$$

and

$$z = \pm\sqrt{4 - x^2} \qquad \text{From the equation of the cylinder}$$
$$= \pm\sqrt{4 - t^2}$$

Therefore, we obtain the following two vector-valued functions

$$\mathbf{r}(t) = t\mathbf{i} + t\mathbf{j} + \sqrt{4 - t^2}\,\mathbf{k}$$

and

$$\mathbf{r}(t) = t\mathbf{i} + t\mathbf{j} - \sqrt{4 - t^2}\,\mathbf{k}$$

t	-2	-1	0	1	2
x	-2	-1	0	1	2
y	-2	-1	0	1	2
z	0	$\pm\sqrt{3}$	± 2	$\pm\sqrt{3}$	0

11. Given the vector-valued function $\mathbf{r}(t) = \ln t\,\mathbf{i} + t\mathbf{j} + t\mathbf{k}$, (a) find the domain of $\mathbf{r}$ and (b) determine the values of t for which the function is discontinuous.

Solution

(a) *Function*   *Domain*

$x = \ln t$ $(0, \infty)$

$y = t$ $(-\infty, \infty)$

$z = t$ $(-\infty, \infty)$

From observing the domains of the functions in the table, we see that the domain of $\mathbf{r}$ is $(0, \infty)$.

(b) Since each of the functions $x = \ln t$, $y = t$, and $z = t$ is continuous, it follows that $\mathbf{r}$ is continuous in its domain.

17. Find the indefinite integral $\int \| \cos t\mathbf{i} + \sin t\mathbf{j} + t\mathbf{k} \| \, dt$.

Solution

$$\int \| \cos t\mathbf{i} + \sin t\mathbf{j} + t\mathbf{k} \| \, dt$$

$$= \int \sqrt{\cos^2 t + \sin^2 t + t^2} \, dt$$

$$= \int \sqrt{1 + t^2} \, dt = \frac{1}{2}(t\sqrt{1 + t^2} + \ln | t + \sqrt{1 + t^2} |) + C$$

26. Given the vector-valued function

$$\mathbf{r}(t) = t \cos t\mathbf{i} + t \sin t\mathbf{j}$$

find the velocity, speed and acceleration at any time t. Then find $\mathbf{a} \cdot \mathbf{T}$, $\mathbf{a} \cdot \mathbf{N}$, and the curvature at time t.

Solution

Since

$$\mathbf{r}(t) = t \cos t\mathbf{i} + t \sin t\mathbf{j}$$

we have

$$\mathbf{v}(t) = \mathbf{r}'(t) = (-t \sin t + \cos t)\mathbf{i} + (t \cos t + \sin t)\mathbf{j}$$

$$\| \mathbf{v}(t) \| = \text{speed} = \sqrt{(-t \sin t + \cos t)^2 + (t \cos t + \sin t)^2}$$

$$= \sqrt{t^2 + 1}$$

and

$$\mathbf{a}(t) = \mathbf{r}''(t) = (-t \cos t - 2 \sin t)\mathbf{i} + (-t \sin t + 2 \cos t)\mathbf{j}$$

$$\mathbf{T}(t) = \frac{\mathbf{v}(t)}{\| \mathbf{v}(t) \|} = \frac{(-t \sin t + \cos t)\mathbf{i} + (t \cos t + \sin t)\mathbf{j}}{\sqrt{t^2 + 1}}$$

Since the direction of $\mathbf{N}(t)$ is toward the concave side of the curve and orthogonal to $\mathbf{T}(t)$, we have

$$\mathbf{N}(t) = \frac{-(t \cos t + \sin t)\mathbf{i} + (-t \sin t + \cos t)\mathbf{j}}{\sqrt{t^2 + 1}}$$

Therefore

$$\mathbf{a}(t) \cdot \mathbf{T}(t) = \frac{t}{\sqrt{t^2 + 1}}$$

$$\mathbf{a}(t) \cdot \mathbf{N}(t) = \frac{t^2 + 2}{\sqrt{t^2 + 1}}$$

and

$$K = \frac{\mathbf{a}(t) \cdot \mathbf{N}(t)}{\|\mathbf{v}(t)\|} = \frac{t^2 + 2}{(t^2 + 1)^{3/2}}$$

29. Find the length of the space curve given by the vector-valued function

$$\mathbf{r}(t) = \frac{1}{2}t\mathbf{i} + \sin t\mathbf{j} + \cos t\mathbf{k}$$

over the interval $0 \le t \le \pi$.

Solution

$$s = \int_0^\pi \|\mathbf{r}'(t)\| \, dt$$

$$= \int_0^\pi \sqrt{(\tfrac{1}{2})^2 + \cos^2 t + (-\sin t)^2} \, dt$$

$$= \frac{\sqrt{5}}{2} \int_0^\pi dt = \frac{\sqrt{5}}{2} [t]_0^\pi = \frac{\sqrt{5}\pi}{2}$$

14 *Functions of several variables*

Introduction to functions of several variables

15. Find the functional values (a) $f(0, 4)$ and (b) $f(1, 4)$ if

$$f(x, y) = \int_x^y (2t - 3) \, dt$$

Solution

$$f(x, y) = \int_x^y (2t - 3) \, dt = [t^2 - 3t]_x^y = (y^2 - 3y) - (x^2 - 3x)$$

(a) $f(0, 4) = (16 - 12) - (0 - 0) = 4$
(b) $f(1, 4) = (16 - 12) - (1 - 3) = 6$

17. Describe the region R, in the xy-coordinate plane, that corresponds to the domain of $f(x, y) = \sqrt{4 - x^2 - y^2}$. Find the range of $f(x, y)$.

Solution

Since $f(x, y) = \sqrt{4 - x^2 - y^2}$, we have

$$4 - x^2 - y^2 \geq 0$$
$$4 \geq x^2 + y^2$$

Therefore the region R is the set of all points inside and on the boundary of the circle $x^2 + y^2 = 4$. The range of f is the set of all real numbers in the interval $[0, 2]$.

20. Describe the region R, in the xy-coordinate plane, that corresponds to the domain of $f(x, y) = \arcsin (x + y)$. Find the range of $f(x, y)$.

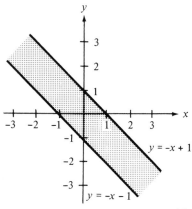

Solution

Since $z = \arcsin(x + y)$ implies that $\sin z = x + y$, we conclude that $|x + y| \leq 1$. Therefore the region R is such that

$$-1 \leq x + y \leq 1$$
$$-1 - x \leq y \leq -x + 1$$

which means that R lies on and between the parallel lines

$$y = -1 - x \quad \text{and} \quad y = -x + 1$$

as shown in the accompanying figure. The range of the arcsine function is the set of all reals in the interval $[-\pi/2, \pi/2]$.

43. Describe the level curves for the function $f(x, y) = x/(x^2 + y^2)$. Sketch the level curves for $c = \pm\frac{1}{2}, \pm 1, \pm\frac{3}{2}, \pm 2$.

Solution

If $f(x, y) = c$, then the level curves are of the form

$$c = \frac{x}{x^2 + y^2}$$

$$x^2 + y^2 = \frac{x}{c}$$

$$x^2 - \frac{x}{c} + y^2 = 0$$

$$\left(x^2 - \frac{x}{c} + \frac{1}{4c^2}\right) + y^2 = \frac{1}{4c^2}$$

$$\left(x - \frac{1}{2c}\right)^2 + y^2 = \left(\frac{1}{2c}\right)^2$$

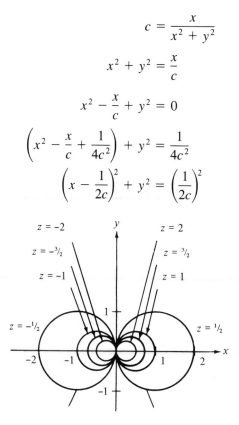

Therefore each level curve is a circle centered at $(1/2c, 0)$ with radius equal to $1/2c$. For example, if $c = 1$, the level curve has the equation

$$\left(x - \frac{1}{2}\right)^2 + y^2 = \frac{1}{4}$$

The required level curves are shown in the accompanying figure.

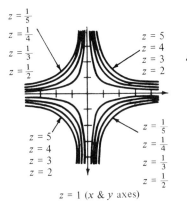

$z = \frac{1}{5}$

$z = \frac{1}{4}$

$z = \frac{1}{3}$

$z = \frac{1}{2}$

$z = 5$

$z = 4$

$z = 3$

$z = 2$

$z = 5$

$z = 4$

$z = 3$

$z = 2$

$z = \frac{1}{5}$

$z = \frac{1}{4}$

$z = \frac{1}{3}$

$z = \frac{1}{2}$

$z = 1$ (x & y axes)

47.   Describe the level curves for the function $f(x, y) = e^{xy}$. Sketch the level curves for $c = 1, 2, 3, 4, \frac{1}{2}, \frac{1}{3}, \frac{1}{4},$ and $\frac{1}{5}$.

Solution

If $f(x, y) = c$, then the level curves are of the form

$$e^{xy} = c \qquad \text{or} \qquad xy = \ln c$$

Therefore, each level curve is a hyperbola centered at the origin with the x- and y-axes as asymptotes.

14.2
Limits and continuity

9.   Find

$$\lim_{(x, y) \to (0, 1)} \frac{\arcsin (x/y)}{1 + xy}$$

and discuss the continuity of the function.

Solution

Since the limit of a quotient is the quotient of the limits, we have

$$\lim_{(x, y) \to (0, 1)} \frac{\arcsin (x/y)}{1 + xy} = \frac{\arcsin 0}{1 + 0} = \frac{0}{1} = 0$$

Since a rational function is continuous at every point in its domain, we conclude that the given function is continuous at each point except those on the hyperbola $1 + xy = 0$ and the line $y = 0$.

17.   Find

$$\lim_{(x, y) \to (1, 2)} f(x, y) = \lim_{(x, y) \to (1, 2)} \frac{2x(y - 1)}{(x + 1)y}$$

by finding the limits

$$\lim_{x \to 1} g(x) \quad \text{and} \quad \lim_{y \to 2} h(y)$$

where $f(x, y) = g(x)h(y)$.

Solution

$$\lim_{(x, y) \to (1, 2)} \frac{2x(y - 1)}{(x + 1)y} = \left[\lim_{x \to 1} \frac{2x}{x + 1}\right]\left[\lim_{y \to 2} \frac{y - 1}{y}\right]$$

$$= (1)\left(\frac{1}{2}\right) = \frac{1}{2}$$

26. Discuss the continuity of the function $f(x, y) = (2x - y^2)/(2x^2 + y)$ and evaluate

$$\lim_{(x, y) \to (0, 0)} \frac{2x - y^2}{2x^2 + y}$$

if it exists.

Solution

Direct substitution yields the indeterminate form

$$\lim_{(x, y) \to (0, 0)} \frac{2x - y^2}{2x^2 + y} = \frac{0}{0}$$

However, along $x = 0$ we have

$$\lim_{(0, y) \to (0, 0)} \frac{-y^2}{y} = \lim_{(0, y) \to (0, 0)} \frac{-y}{1} = 0$$

and along $y = 0$ we have

$$\lim_{(x, 0) \to (0, 0)} \frac{2x}{2x^2} = \lim_{(x, 0) \to (0, 0)} \frac{1}{x} = \infty$$

Therefore

$$\lim_{(x, y) \to (0, 0)} \frac{2x - y^2}{2x^2 + y}$$

does not exist and the function is discontinuous at $(0, 0)$.

30. Use polar coordinates to show that

$$\lim_{(x, y) \to (0, 0)} \frac{xy^2}{x^2 + y^2} = 0$$

Solution

We first observe that direct substitution yields the indeterminate form $0/0$. Letting $x = r \cos \theta$, $y = r \sin \theta$, and $r^2 = x^2 + y^2$, we have

$$\lim_{(x, y) \to (0, 0)} \frac{xy^2}{x^2 + y^2} = \lim_{r \to 0} \frac{(r \cos \theta)(r^2 \sin^2 \theta)}{r^2}$$

$$= \lim_{r \to 0} (r \cos \theta \sin^2 \theta) = 0$$

34.   Discuss the continuity of $f(x, y, z) = z/(x^2 + y^2 - 4)$.

Solution

The function $f(x, y, z) = z/(x^2 + y^2 - 4)$ is discontinuous when

$$x^2 + y^2 - 4 = 0 \qquad \text{or} \qquad x^2 + y^2 = 4$$

Therefore the function is discontinuous at any point on the cylinder $x^2 + y^2 = 4$.

41.   Given the function $f(x, y) = x^2 - 4y$, find

(a) $\displaystyle \lim_{\Delta x \to 0} \frac{f(x + \Delta x, y) - f(x, y)}{\Delta x}$

(b) $\displaystyle \lim_{\Delta y \to 0} \frac{f(x, y + \Delta y) - f(x, y)}{\Delta y}$

Solution

(a) $\displaystyle \lim_{\Delta x \to 0} \frac{f(x + \Delta x, y) - f(x, y)}{\Delta x}$

$$= \lim_{\Delta x \to 0} \frac{[(x + \Delta x)^2 - 4y] - (x^2 - 4y)}{\Delta x}$$

$$= \lim_{\Delta x \to 0} \frac{x^2 + 2x\Delta x + (\Delta x)^2 - 4y - x^2 + 4y}{\Delta x}$$

$$= \lim_{\Delta x \to 0} (2x + \Delta x) = 2x$$

(b) $\displaystyle \lim_{\Delta y \to 0} \frac{f(x, y + \Delta y) - f(x, y)}{\Delta y}$

$$= \lim_{\Delta y \to 0} \frac{x^2 - 4(y + \Delta y) - (x^2 - 4y)}{\Delta y}$$

$$= \lim_{\Delta y \to 0} \frac{x^2 - 4y - 4\Delta y - x^2 + 4y}{\Delta y}$$

$$= \lim_{\Delta y \to 0} (-4) = -4$$

14.3

Partial derivatives

8. Find the first partial derivatives with respect to x and with respect to y for

$$z = xe^{x/y}$$

Solution

First, considering y to be constant and using the Product Rule, we have

$$\frac{\partial z}{\partial x} = x\left[e^{x/y}\left(\frac{1}{y}\right)\right] + e^{x/y}(1) = e^{x/y}\left(\frac{x}{y} + 1\right)$$

Now considering x to be constant, we have

$$\frac{\partial z}{\partial y} = xe^{x/y}\left(\frac{-x}{y^2}\right) = \frac{-x^2}{y^2}e^{x/y}$$

19. Find the first partial derivatives with respect to x and with respect to y for

$$z = e^y \sin xy$$

Solution

First, considering y to be constant, we have

$$\frac{\partial z}{\partial x} = e^y(\cos xy)(y) = ye^y \cos xy$$

Now considering x to be constant and using the Product Rule, we have

$$\frac{\partial z}{\partial y} = e^y(\cos xy)(x) + (\sin xy)e^y(1) = e^y(x \cos xy + \sin xy)$$

23. Given the function $f(x, y) = \arctan(y/x)$, find $f_x(2, -2)$ and $f_y(2, -2)$.

Solution

First, considering y to be constant, we have

$$\frac{\partial z}{\partial x} = \frac{1}{1 + (y^2/x^2)}\left(\frac{-y}{x^2}\right) = \frac{-y}{x^2 + y^2}$$

Now considering x to be constant, we have

$$\frac{\partial z}{\partial y} = \frac{1}{1 + (y^2/x^2)}\left(\frac{1}{x}\right) = \frac{x}{x^2 + y^2}$$

Therefore

$$f_x(2, -2) = \frac{1}{4} \quad \text{and} \quad f_y(2, -2) = \frac{1}{4}$$

29. Find all first partial derivatives for $F(x, y, z) = \ln \sqrt{x^2 + y^2 + z^2}$.

Solution

First we rewrite the function in the form

$$w = \ln \sqrt{x^2 + y^2 + z^2} = \frac{1}{2} \ln (x^2 + y^2 + z^2)$$

Now considering both y and z to be constant, we have

$$\frac{\partial w}{\partial x} = \left(\frac{1}{2}\right)\left(\frac{2x}{x^2 + y^2 + z^2}\right) = \frac{x}{x^2 + y^2 + z^2}$$

Considering x and z to be constant, we have

$$\frac{\partial w}{\partial y} = \left(\frac{1}{2}\right)\left(\frac{2y}{x^2 + y^2 + z^2}\right) = \frac{y}{x^2 + y^2 + z^2}$$

Similarly,

$$\frac{\partial w}{\partial z} = \left(\frac{1}{2}\right)\left(\frac{2z}{x^2 + y^2 + z^2}\right) = \frac{z}{x^2 + y^2 + z^2}$$

33. Find $\partial^2 z/\partial x^2$, $\partial^2 z/\partial y^2$, $\partial^2 z/\partial x\, \partial y$, and $\partial^2 z/\partial y\, \partial x$ for $z = x^2 - 2xy + 3y^2$.

Solution

The first partials are

$$\frac{\partial z}{\partial x} = 2x - 2y \quad \text{and} \quad \frac{\partial z}{\partial y} = -2x + 6y$$

The second partials are

$$\frac{\partial^2 z}{\partial x^2} = \frac{\partial}{\partial x}\left[\frac{\partial z}{\partial x}\right] = 2 \qquad \frac{\partial^2 z}{\partial y\, \partial x} = \frac{\partial}{\partial y}\left[\frac{\partial z}{\partial x}\right] = -2$$

$$\frac{\partial^2 z}{\partial y^2} = \frac{\partial}{\partial y}\left[\frac{\partial z}{\partial y}\right] = 6 \qquad \frac{\partial^2 z}{\partial x\, \partial y} = \frac{\partial}{\partial x}\left[\frac{\partial z}{\partial y}\right] = -2$$

40. Find $\partial^2 z/\partial x^2$, $\partial^2 z/\partial y^2$, $\partial^2 z/\partial x\, \partial y$, and $\partial^2 z/\partial y\, \partial x$ for $z = xy/(x - y)$.

Solution

The first partials are

$$\frac{\partial z}{\partial x} = \frac{(x - y)y - xy(1)}{(x - y)^2} = \frac{-y^2}{(x - y)^2}$$

$$\frac{\partial z}{\partial y} = \frac{(x - y)x - xy(-1)}{(x - y)^2} = \frac{x^2}{(x - y)^2}$$

The second partials are

$$\frac{\partial^2 z}{\partial x^2} = \frac{\partial}{\partial x}\left[\frac{\partial z}{\partial x}\right] = \frac{(x - y)^2(0) + y^2(2)(x - y)}{(x - y)^4} = \frac{2y^2}{(x - y)^3}$$

$$\frac{\partial^2 z}{\partial y^2} = \frac{\partial}{\partial y}\left[\frac{\partial z}{\partial y}\right] = \frac{(x - y)^2(0) - x^2(2)(x - y)(-1)}{(x - y)^4} = \frac{2x^2}{(x - y)^3}$$

$$\frac{\partial^2 z}{\partial y \, \partial x} = \frac{\partial}{\partial y}\left[\frac{\partial z}{\partial x}\right] = \frac{(x - y)^2(-2y) + y^2(2)(x - y)(-1)}{(x - y)^4}$$

$$= \frac{(x - y)[-2y(x - y) - 2y^2]}{(x - y)^4} = \frac{-2xy}{(x - y)^3}$$

$$\frac{\partial^2 z}{\partial x \, \partial y} = \frac{\partial}{\partial x}\left[\frac{\partial z}{\partial y}\right] = \frac{(x - y)^2(2x) - x^2(2)(x - y)}{(x - y)^4}$$

$$= \frac{(x - y)[2x(x - y) - 2x^2]}{(x - y)^4} = \frac{-2xy}{(x - y)^3}$$

55. Show that $z = \sin(x - ct)$ is a solution of the equation $\partial^2 z / \partial t^2 = c^2(\partial^2 z / \partial x^2)$.

Solution

$$\frac{\partial z}{\partial x} = \cos(x - ct) \qquad \text{and} \qquad \frac{\partial^2 z}{\partial x^2} = -\sin(x - ct)$$

$$\frac{\partial z}{\partial t} = -c \cos(x - ct) \qquad \text{and} \qquad \frac{\partial^2 x}{\partial t^2} = -c^2 \sin(x - ct)$$

Therefore

$$\frac{\partial^2 z}{\partial t^2} = -c^2 \sin(x - ct)$$

$$= c^2 \frac{\partial^2 z}{\partial x^2}$$

62. Use the limit definition of partial derivatives to find $f_x(x, y)$ and $f_y(x, y)$ for $f(x, y) = x^2 - 2xy + y^2$.

Solution

By definition,

$$f_x(x, y) = \lim_{\Delta x \to 0} \frac{f(x + \Delta x, y) - f(x, y)}{\Delta x}$$

$$= \lim_{\Delta x \to 0} \frac{[(x + \Delta x)^2 - 2(x + \Delta x)y + y^2] - (x^2 - 2xy + y^2)}{\Delta x}$$

$$= \lim_{\Delta x \to 0} \frac{x^2 + 2x\Delta x + (\Delta x)^2 - 2xy - 2y\Delta x + y^2 - x^2 + 2xy - y^2}{\Delta x}$$

$$= \lim_{\Delta x \to 0} \frac{\Delta x(2x - 2y + \Delta x)}{\Delta x} = 2x - 2y$$

Similarly,

$$f_y(x, y) = \lim_{\Delta y \to 0} \frac{f(x, y + \Delta y) - f(x, y)}{\Delta y}$$

$$= \lim_{\Delta y \to 0} \frac{x^2 - 2xy - 2x\Delta y + y^2 + 2y\Delta y + (\Delta y)^2 - x^2 + 2xy - y^2}{\Delta y}$$

$$= \lim_{\Delta y \to 0} \frac{\Delta y(2y - 2x + \Delta y)}{\Delta y} = 2y - 2x$$

66. Sketch the curve formed by the intersection of the paraboloid
     $z = 9x^2 - y^2$ and the plane $x = 1$. Find the slope of the curve
     at the point $(1, 3, 0)$.

Solution

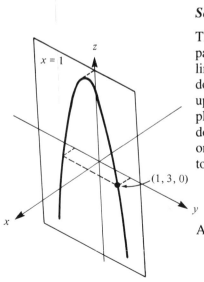

The graph of the equation $z = 9x^2 - y^2$ is a hyperbolic
paraboloid. The xy-trace ($z = 0$) consists of the intersecting
lines $y = \pm 3x$. The yz-trace, $z = -y^2$, is a parabola opening
downward, and the xz-trace, $z = 9x^2$, is a parabola opening
upward. The curve of intersection of the paraboloid and the
plane $x = 1$ is given by $z = 9 - y^2$, is a parabola opening
downward. (See the accompanying figure.) Since x is constant
on the curve of intersection, we differentiate with respect to y
to obtain

$$\frac{\partial z}{\partial y} = -2y$$

At the point $(1, 3, 0)$ the slope is

$$\frac{\partial z}{\partial y} = -2(3) = -6$$

14.4
Differentials

5. Find the total differential for $z = x \cos y - y \cos x$.

   ### Solution

   $$dz = \frac{\partial z}{\partial x} \, dx + \frac{\partial z}{\partial y} \, dy$$

   $$= (\cos y + y \sin x) \, dx + (-x \sin y - \cos x) \, dy$$

   $$= (\cos y + y \sin x) \, dx - (x \sin y + \cos x) \, dy$$

9. Find the total differential for $u = (x + y)/(z - 2y)$.

   ### Solution

   Since $u = (x + y)/(z - 2y)$, we have

   $$du = \frac{\partial u}{\partial x} \, dx + \frac{\partial u}{\partial y} \, dy + \frac{\partial u}{\partial z} \, dz$$

   $$= \frac{1}{z - 2y} \, dx + \frac{(z - 2y)(1) - (x + y)(-2)}{(z - 2y)^2} \, dy + \frac{0 - (x + y)(1)}{(z - 2y)^2} \, dz$$

   $$= \frac{1}{z - 2y} \, dx + \frac{2x + z}{(z - 2y)^2} \, dy - \frac{x + y}{(z - 2y)^2} \, dz$$

12. Given the function $f(x, y) = \sqrt{x^2 + y^2}$, (a) evaluate $f(1, 2)$ and $f(1.05, 2.1)$ and calculate Δz, and (b) use the total differential dz to approximate Δz.

   ### Solution

   (a) $f(1, 2) = \sqrt{1^2 + 2^2} = \sqrt{5} \approx 2.23607$

   $$f(1.05, 2.1) = \sqrt{(1.05)^2 + (2.1)^2} = \sqrt{5.4504}$$

   $$\approx 2.33461$$

   Therefore $\Delta z = f(1.05, 2.1) - f(1, 2) = 0.09854$

   (b) $dz = f_x(x, y) \, dx + f_y(x, y) \, dy$

   $$= \frac{x}{\sqrt{x^2 + y^2}} \, dx + \frac{y}{\sqrt{x^2 + y^2}} \, dy$$

   Letting $x = 1$, $y = 2$, $dx = 0.05$ and $dy = 0.1$, we have

   $$dz = \frac{1}{\sqrt{5}}(0.05) + \frac{2}{\sqrt{5}}(0.1) \approx 0.11180$$

17. The radius r and height h of a right circular cylinder are measured with a possible error of 4% and 2%, respectively. Approximate the maximum possible percentage error in measuring the volume.

Solution

First we consider the percentage errors in r and h as

$$\frac{dr}{r} = 4\% = 0.04 \quad \text{and} \quad \frac{dh}{h} = 2\% = 0.02$$

Now since $V = \pi r^2 h$, we have

$$dV = 2\pi rh\, dr + \pi r^2\, dh$$

or the percentage error in V is

$$\frac{dV}{V} = \frac{2\pi rh\, dr}{\pi r^2 h} + \frac{\pi r^2\, dh}{\pi r^2 h} = 2\frac{dr}{r} + \frac{dh}{h} = 2(0.04) + 0.02$$

$$= 0.10 = 10\%$$

14.5
The Chain Rule

2. If $w = \sqrt{x^2 + y^2}$, $x = \sin t$, and $y = e^t$, find dw/dt by using the Chain Rule.

Solution

$$\frac{dw}{dt} = \frac{\partial w}{\partial x}\frac{dx}{dt} + \frac{\partial w}{\partial y}\frac{dy}{dt}$$

$$= \left[\frac{1}{2}(x^2 + y^2)^{-1/2}(2x)\right]\cos t + \left[\frac{1}{2}(x^2 + y^2)^{-1/2}(2y)\right]e^t$$

$$= \frac{x\cos t}{\sqrt{x^2 + y^2}} + \frac{ye^t}{\sqrt{x^2 + y^2}} = \frac{x\cos t + ye^t}{\sqrt{x^2 + y^2}} = \frac{\sin t\cos t + e^{2t}}{\sqrt{\sin^2 t + e^{2t}}}$$

9. If $w = x^2 - y^2$, $x = s\cos t$, and $y = s\sin t$, find $\partial w/\partial s$ and $\partial w/\partial t$ by using the Chain Rule. Evaluate each partial derivative when $s = 3$ and $t = \pi/4$.

Solution

By the Chain Rule,

$$\frac{\partial w}{\partial s} = \frac{\partial w}{\partial x}\frac{\partial x}{\partial s} + \frac{\partial w}{\partial y}\frac{\partial y}{\partial s} \quad \text{and} \quad \frac{\partial w}{\partial t} = \frac{\partial w}{\partial x}\frac{\partial x}{\partial t} + \frac{\partial w}{\partial y}\frac{\partial y}{\partial t}$$

Therefore

$$\frac{\partial w}{\partial s} = 2x(\cos t) + (-2y)(\sin t)$$

$$= (2s \cos t)(\cos t) - (2s \sin t)(\sin t)$$

$$= 2s(\cos^2 t - \sin^2 t) = 2s \cos 2t$$

when $s = 3$ and $t = \pi/4$, $\partial w/\partial s = 2(3) \cos (\pi/2) = 0$.
Similarly,

$$\frac{\partial w}{\partial t} = 2x(-s \sin t) - 2y(s \cos t)$$

$$= 2s \cos t(-s \sin t) - 2s \sin t(s \cos t)$$

$$= -4s^2 \sin t \cos t = -2s^2 \sin 2t$$

when $s = 3$ and $t = \pi/4$, $\partial w/\partial t = -2(9) \sin (\pi/2) = -18$.

11. If $w = xy$, $x = 2 \sin t$, and $y = \cos t$, find dw/dt (a) by the Chain Rule and (b) by converting w to a function of t before differentiating.

Solution

(a) First we calculate dx/dt and dy/dt.

$$\frac{dx}{dt} = 2 \cos t \qquad \frac{dy}{dt} = -\sin t$$

Now by the Chain Rule

$$\frac{dw}{dt} = \frac{\partial w}{\partial x}\frac{dx}{dt} + \frac{\partial w}{\partial y}\frac{dy}{dt} = y(2 \cos t) + x(-\sin t)$$

$$= (\cos t)(2 \cos t) + (2 \sin t)(-\sin t)$$

$$= 2(\cos^2 t - \sin^2 t) = 2 \cos 2t$$

(b) Now

$$w = xy = (2 \sin t)(\cos t) = \sin 2t$$

$$\frac{dw}{dt} = 2 \cos 2t$$

15. If $w = \arctan (y/x)$, $x = r \cos \theta$, and $y = r \sin \theta$, find $\partial w/\partial r$ and $\partial w/\partial \theta$ (a) by the Chain Rule and (b) by converting w to a function of r and θ before differentiating.

Solution

(a) First we calculate $\partial w/\partial x$ and $\partial w/\partial y$.

$$w = \arctan \frac{y}{x}$$

$$\frac{\partial w}{\partial x} = \frac{-y/x^2}{1 + (y^2/x^2)} = \frac{-y/x^2}{(x^2 + y^2)/x^2} = \frac{-y}{x^2 + y^2}$$

$$\frac{\partial w}{\partial y} = \frac{1/x}{1 + (y^2/x^2)} = \frac{1/x}{(x^2 + y^2)/x^2} = \frac{x}{x^2 + y^2}$$

Now by the Chain Rule,

$$\frac{\partial w}{\partial r} = \frac{\partial w}{\partial x}\frac{\partial x}{\partial r} + \frac{\partial w}{\partial y}\frac{\partial y}{\partial r} = \frac{-y}{x^2 + y^2}(\cos \theta) + \frac{x}{x^2 + y^2}(\sin \theta)$$

$$= \frac{x \sin \theta - y \cos \theta}{x^2 + y^2} = \frac{r \cos \theta \sin \theta - r \sin \theta \cos \theta}{r^2} = 0$$

Furthermore,

$$\frac{\partial w}{\partial \theta} = \frac{\partial w}{\partial x}\frac{\partial x}{\partial \theta} + \frac{\partial w}{\partial y}\frac{\partial y}{\partial \theta} = \frac{-y}{x^2 + y^2}(-r \sin \theta) + \frac{x}{x^2 + y^2}(r \cos \theta)$$

$$= \frac{-r \sin \theta(-r \sin \theta) + r \cos \theta(r \cos \theta)}{r^2} = \frac{r^2}{r^2} = 1$$

(b) Since

$$w = \arctan \frac{y}{x} = \arctan \frac{r \sin \theta}{r \cos \theta} = \arctan (\tan \theta) = \theta + n\pi$$

we have

$$\frac{\partial w}{\partial r} = 0 \qquad \text{and} \qquad \frac{\partial w}{\partial \theta} = 1$$

22. Differentiate implicitly to find $\partial w/\partial x$, $\partial w/\partial y$, and $\partial w/\partial z$ for $x^2 + y^2 + z^2 + 6xw - 8w^2 = 5$.

Solution

Let $F(x, y, z, w) = x^2 + y^2 + z^2 + 6xw - 8w^2 - 5$. Then

$$F_x(x, y, z, w) = 2x + 6w \qquad F_z(x, y, z, w) = 2z$$

$$F_y(x, y, z, w) = 2y \qquad F_w(x, y, z, w) = 6x - 16w$$

Now by an extension of the formulas in Theorem 14.8 we have

$$\frac{\partial w}{\partial x} = \frac{-F_x(x, y, z, w)}{F_w(x, y, z, w)} = \frac{-2x - 6w}{6x - 16w} = -\frac{x + 3w}{3x - 8w}$$

$$\frac{\partial w}{\partial y} = \frac{-F_y(x, y, z, w)}{F_w(x, y, z, w)} = \frac{-2y}{6x - 16w} = \frac{-y}{3x - 8w}$$

$$\frac{\partial w}{\partial z} = \frac{-F_z(x, y, z, w)}{F_w(x, y, z, w)} = \frac{-2z}{6x - 16w} = \frac{-z}{3x - 8w}$$

25. The length, width, and depth of a rectangular chamber are increasing at the rate of 3 ft/min, 2 ft/min, and $\frac{1}{2}$ ft/min, respectively. Find the rate at which the volume and surface area are changing the instant the length, width, and depth are 10 ft, 6 ft, and 4 ft, respectively.

Solution

Let the volume of the chamber be given by $V = xyz$. Then

$$\frac{dV}{dt} = V_x\frac{dx}{dt} + V_y\frac{dy}{dt} + V_z\frac{dz}{dt} = yz\frac{dx}{dt} + xz\frac{dy}{dt} + xy\frac{dz}{dt}$$

$$= (6)(4)(3) + (10)(4)(2) + (10)(6)\left(\frac{1}{2}\right) = 182 \text{ ft}^3/\text{min}$$

Let the surface area of the chamber be given by
$S = 2(xy + yz + xz)$. Then

$$\frac{dS}{dt} = S_x\frac{dx}{dt} + S_y\frac{dy}{dt} + S_z\frac{dz}{dt}$$

$$= 2\left[(y + z)\frac{dx}{dt} + (x + z)\frac{dy}{dt} + (x + y)\frac{dz}{dt}\right]$$

$$= 2\left[(6 + 4)(3) + (10 + 4)(2) + (10 + 6)\left(\frac{1}{2}\right)\right]$$

$$= 132 \text{ ft}^2/\text{min}$$

32. A function $f(x, y)$ is **homogeneous of degree n** if

$$f(tx, ty) = t^n f(x, y)$$

Find the degree of the homogeneous function

$$f(x, y) = \frac{xy}{\sqrt{x^2 + y^2}}$$

Solution
Since

$$f(tx, ty) = \frac{(tx)(ty)}{\sqrt{(tx)^2 + (ty)^2}}$$

$$= \frac{t^2 xy}{t\sqrt{x^2 + y^2}} = \frac{t(xy)}{\sqrt{x^2 + y^2}} = tf(x, y)$$

the function is homogeneous of degree 1.

14.6
Directional derivatives and gradients

3. Find the directional derivative of $f(x, y) = xy$ at $P = (2, 3)$ in the direction of $\mathbf{v} = \mathbf{i} + \mathbf{j}$.

Solution

We begin by finding a unit vector, $\mathbf{u}$, in the direction of $\mathbf{v}$.

$$\mathbf{u} = \frac{\mathbf{v}}{\|\mathbf{v}\|} = \frac{\sqrt{2}}{2}\mathbf{i} + \frac{\sqrt{2}}{2}\mathbf{j} = \cos\theta\,\mathbf{i} + \sin\theta\,\mathbf{j}$$

Thus

$$\cos\theta = \frac{\sqrt{2}}{2} = \sin\theta \qquad \text{and} \qquad \theta = \frac{\pi}{4}$$

Now, using Theorem 14.9, we have

$$D_{\mathbf{u}}f(x, y) = f_x(x, y)\cos\theta + f_y(x, y)\sin\theta$$
$$= y\left(\frac{\sqrt{2}}{2}\right) + x\left(\frac{\sqrt{2}}{2}\right) = \frac{\sqrt{2}}{2}(y + x)$$

Therefore

$$D_{\mathbf{u}}f(2, 3) = \frac{5\sqrt{2}}{2}$$

(Note that $D_{\mathbf{u}}f(x, y) = \nabla f(x, y) \cdot \mathbf{u}$ where $\nabla f(x, y) = y\mathbf{i} + x\mathbf{j}$.)

5. Find the directional derivative of the function $g(x, y) = \sqrt{x^2 + y^2}$ at the point $P = (3, 4)$ in the direction of $\mathbf{v} = 3\mathbf{i} - 4\mathbf{j}$.

Solution

We begin by finding a unit vector, $\mathbf{u}$ in the direction of $\mathbf{v}$

$$\mathbf{u} = \frac{\mathbf{v}}{\|\mathbf{v}\|} = \frac{3}{5}\mathbf{i} - \frac{4}{5}\mathbf{j} = \cos\theta\,\mathbf{i} + \sin\theta\,\mathbf{j}$$

Thus

$$\cos\theta = \frac{3}{5} \qquad \text{and} \qquad \sin\theta = -\frac{4}{5}$$

By Theorem 14.9 we have

$$D_{\mathbf{u}}f(x, y) = f_x(x, y)\cos\theta + f_y(x, y)\sin\theta$$

$$= \frac{x}{\sqrt{x^2 + y^2}}\left(\frac{3}{5}\right) + \frac{y}{\sqrt{x^2 + y^2}}\left(-\frac{4}{5}\right)$$

$$= \frac{1}{5\sqrt{x^2 + y^2}}(3x - 4y)$$

Therefore

$$D_\mathbf{u}f(3, 4) = \frac{-7}{25}$$

(Note that $D_\mathbf{u}f(x, y) = \nabla f(x, y) \cdot \mathbf{u}$ where $\nabla f(x, y) = \frac{x}{\sqrt{x^2 + y^2}}\mathbf{i} + \frac{y}{\sqrt{x^2 + y^2}}\mathbf{j}$.)

9. Find the directional derivative of $f(x, y, z) = xy + yz + xz$ at the point $P = (1, 1, 1)$ in the direction of $\mathbf{v} = 2\mathbf{i} + \mathbf{j} - \mathbf{k}$.

Solution

We begin by finding $\nabla f(x, y, z)$ and a unit vector, $\mathbf{u}$, in the direction of $\mathbf{v}$.

$$\nabla f(x, y, z) = f_x(x, y, z)\mathbf{i} + f_y(x, y, z)\mathbf{j} + f_z(x, y, z)\mathbf{k}$$
$$= (y + z)\mathbf{i} + (x + z)\mathbf{j} + (x + y)\mathbf{k}$$

and

$$\mathbf{u} = \frac{\mathbf{v}}{\|\mathbf{v}\|} = \frac{\sqrt{6}}{6}(2\mathbf{i} + \mathbf{j} - \mathbf{k})$$

Therefore

$$D_\mathbf{u}f(x, y, z) = \nabla f(x, y, z) \cdot \mathbf{u}$$
$$= \frac{\sqrt{6}}{6}[2(y + z) + (x + z) - (x + y)]$$
$$= \frac{\sqrt{6}}{6}(y + 3z)$$

and

$$D_\mathbf{u}f(1, 1, 1) = \frac{4\sqrt{6}}{6} = \frac{2\sqrt{6}}{3}$$

23. Find the gradient of the function $h(x, y) = x \tan y$ and the maximum value of the directional derivative at the point $P = (2, \pi/4)$.

Solution

The gradient vector is given by

$$\nabla f(x, y) = f_x(x, y)\mathbf{i} + f_y(x, y)\mathbf{j} = \tan y\,\mathbf{i} + x\sec^2 y\,\mathbf{j}$$

and at the point $P = (2, \pi/4)$ we have

$$\nabla f(2, \pi/4) = \tan\frac{\pi}{4}\mathbf{i} + 2\sec^2\frac{\pi}{4}\mathbf{j} = \mathbf{i} + 4\mathbf{j}$$

Hence, it follows that the maximum value of the directional derivative at the point $P = (2, \pi/4)$ is

$$\|\nabla f(2, \pi/4)\| = \sqrt{17}$$

28. Find the gradient of the function $f(x, y, z) = xe^{yz}$ and the maximum value of the directional derivative at the point $(2, 0, -4)$.

Solution

The gradient vector is given by

$$\nabla f(x, y, z) = f_x(x, y, z)\mathbf{i} + f_y(x, y, z)\mathbf{j} + f_z(x, y, z)\mathbf{k}$$
$$= e^{yz}\mathbf{i} + xze^{yz}\mathbf{j} + xye^{yz}\mathbf{k} = e^{yz}(\mathbf{i} + xz\mathbf{j} + xy\mathbf{k})$$

and at the point $P = (2, 0, -4)$ we have

$$\nabla f(2, 0, -4) = \mathbf{i} - 8\mathbf{j}$$

Hence, it follows that the maximum value of the directional derivative at the point $P = (2, 0, -4)$ is

$$\|\nabla f(x, y, z)\| = \sqrt{65}$$

32. If $f(x, y) = 3 - (x/3) - (y/2)$, find $D_{\mathbf{u}}f(3, 2)$ if (a) $\theta = \pi/4$ and (b) $\theta = 2\pi/3$.

Solution

(a) By Theorem 14.9 the directional derivative is

$$D_{\mathbf{u}}f(x, y) = f_x(x, y)\cos\theta + f_y(x, y)\sin\theta$$

$$= -\frac{1}{3}\cos\theta - \frac{1}{2}\sin\theta$$

For $\theta = \pi/4$, $x = 3$, and $y = 2$, we have

$$D_{\mathbf{u}}f(2, 3) = -\frac{1}{3}\left(\frac{\sqrt{2}}{2}\right) - \frac{1}{2}\left(\frac{\sqrt{2}}{2}\right) = -\frac{5\sqrt{2}}{12}$$

(b) For $\theta = 2\pi/3$, $x = 3$, and $y = 2$, we have

$$D_{\mathbf{u}}f(3, 2) = -\frac{1}{3}\left(-\frac{1}{2}\right) - \frac{1}{2}\left(\frac{\sqrt{3}}{2}\right) = \frac{1}{6} - \frac{\sqrt{3}}{4} = \frac{1}{12}(2 - 3\sqrt{3})$$

35. If $f(x, y) = 3 - (x/3) - (y/2)$, find $D_v f(3, 2)$ if (a) **v** is the vector from $(1, 2)$ to $(-2, 6)$ and (b) **v** is the vector from $(3, 2)$ to $(4, 5)$.

Solution

(a) Let **u** be a unit vector in the direction of **v**. Then

$$\mathbf{v} = (-2 - 1)\mathbf{i} + (6 - 2)\mathbf{j} = -3\mathbf{i} + 4\mathbf{j}$$

and

$$\mathbf{u} = \frac{\mathbf{v}}{\|\mathbf{v}\|} = \frac{-3\mathbf{i} + 4\mathbf{j}}{\sqrt{25}} = -\frac{3}{5}\mathbf{i} + \frac{4}{5}\mathbf{j}$$

At $(3, 2)$

$$\nabla f(3, 2) = f_x(3, 2)\mathbf{i} + f_y(3, 2)\mathbf{j} = -\frac{1}{3}\mathbf{i} - \frac{1}{2}\mathbf{j}$$

Therefore the directional derivative in the direction of **v** is

$$\nabla f(3, 2) \cdot \mathbf{u} = \left(-\frac{1}{3}\right)\left(-\frac{3}{5}\right) + \left(-\frac{1}{2}\right)\left(\frac{4}{5}\right) = \frac{1}{5} - \frac{2}{5} = -\frac{1}{5}$$

(b) Let **u** be a unit vector in the direction of **v**. Then

$$\mathbf{v} = (4 - 3)\mathbf{i} + (5 - 2)\mathbf{j} = \mathbf{i} + 3\mathbf{j}$$

and

$$\mathbf{u} = \frac{\mathbf{v}}{\|\mathbf{v}\|} = \frac{\mathbf{i} + 3\mathbf{j}}{\sqrt{10}} = \frac{\sqrt{10}}{10}\mathbf{i} + \frac{3\sqrt{10}}{10}\mathbf{j}$$

Therefore, the directional derivative in the direction of **v** is

$$\nabla f(3, 2) \cdot \mathbf{u} = \left(-\frac{1}{3}\right)\left(\frac{\sqrt{10}}{10}\right) + \left(-\frac{1}{2}\right)\left(\frac{3\sqrt{10}}{10}\right) = -\frac{11\sqrt{10}}{60}$$

37. If $f(x, y) = 3 - (x/3) - (y/2)$, find the maximum value of the directional derivative at $(3, 2)$.

Solution

By Theorem 14.10 the maximum value of the directional derivative is $\|\nabla f(3, 2)\|$.

$$f(x, y) = 3 - \frac{x}{3} - \frac{y}{2}$$

$$\nabla f(x, y) = f_x(x, y)\mathbf{i} + f_y(x, y)\mathbf{j} = -\frac{1}{3}\mathbf{i} - \frac{1}{2}\mathbf{j}$$

Therefore the maximum value of the directional derivative at $(3, 2)$ is

$$\|\nabla f(3, 2)\| = \sqrt{\frac{1}{9} + \frac{1}{4}} = \frac{1}{6}\sqrt{13}$$

42.   Find a unit vector **u** orthogonal to ∇f and calculate $D_{\mathbf{u}}f(1, 2)$ if $f(x, y) = 9 - x^2 - y^2$.

Solution

$$f(x, y) = 9 - x^2 - y^2$$
$$\nabla f(x, y) = (-2x)\mathbf{i} + (-2y)\mathbf{j} = -2(x\mathbf{i} + y\mathbf{j})$$

A unit vector in the direction of ∇f is given by

$$\frac{\nabla f(x, y)}{\|\nabla f(x, y)\|} = \frac{-(x\mathbf{i} + y\mathbf{j})}{\sqrt{x^2 + y^2}}$$

Therefore, a unit vector orthogonal to $\nabla f(x, y)$ is

$$\mathbf{u} = \frac{y\mathbf{i} - x\mathbf{j}}{\sqrt{x^2 + y^2}}$$

(Note that the only two unit vectors that are orthogonal to the unit vector $\mathbf{w} = a\mathbf{i} + b\mathbf{j}$ are $\mathbf{w}_1 = b\mathbf{i} - a\mathbf{j}$ and $\mathbf{w}_2 = -b\mathbf{i} + a\mathbf{j}$.)
    At $(1, 2)$

$$\mathbf{u} = \frac{1}{\sqrt{5}}(2\mathbf{i} - \mathbf{j}) \qquad \text{and} \qquad \nabla f(1, 2) = -2(\mathbf{i} + 2\mathbf{j})$$

Thus

$$D_{\mathbf{u}}f(1, 2) = \nabla f(1, 2) \cdot \mathbf{u} = 0$$

We conclude that since ∇f gives the direction of the greatest rate of increase in $f(x, y)$, then a direction orthogonal to ∇f will result in no change in $f(x, y)$.

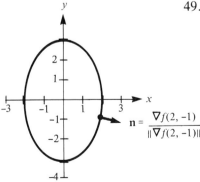

49.   Use the gradient to find a normal vector to the graph of $9x^2 + 4y^2 = 40$ at $P = (2, -1)$. Sketch the results.

Solution

The ellipse given by the equation $9x^2 + 4y^2 = 40$ corresponds to the level curve with $c = 0$ to the function

$$f(x, y) = 9x^2 + 4y^2 - 40$$

By Theorem 14.11, $\nabla f(x_0, y_0)$ yields a normal vector to the level curve at the point (x_0, y_0). Therefore,

$$\nabla f(x, y) = 18x\mathbf{i} + 8y\mathbf{j}$$

and at $(2, -1)$, the normal vector is

$$\nabla f(2, -1) = 36\mathbf{i} - 8\mathbf{j} = 4(9\mathbf{i} - 2\mathbf{j})$$

51. The temperature field at any point in a plate is given by $T = x/(x^2 + y^2)$. Find the direction of greatest increase in heat at the point $(3, 4)$.

Solution

The direction of greatest increase in temperature at $(3, 4)$ will be the direction of the gradient $\nabla T(x, y)$ at that point. Since

$$T_x(x, y) = \frac{(x^2 + y^2)(1) - x(2x)}{(x^2 + y^2)^2} = \frac{y^2 - x^2}{(x^2 + y^2)^2}$$

$$T_y(x, y) = \frac{(x^2 + y^2)(0) - x(2y)}{(x^2 + y^2)^2} = \frac{-2xy}{(x^2 + y^2)^2}$$

the gradient at $(3, 4)$ is

$$\nabla T(3, 4) = T_x(3, 4)\mathbf{i} + T_y(3, 4)\mathbf{j}$$

$$= \frac{7}{(25)^2}\mathbf{i} - \frac{24}{(25)^2}\mathbf{j} = \frac{1}{625}(7\mathbf{i} - 24\mathbf{j})$$

14.7
Tangent planes and normal lines

7. Find a unit normal vector to the surface $z - x \sin y = 4$ at the point $P = (6, \pi/6, 7)$.

Solution

We begin by writing the equation for the surface as a function of three variables obtaining

$$F(x, y, z) = z - x \sin y - 4$$

By Theorem 14.13, a normal vector to the surface, $F(x, y, z) = 0$, at (x_0, y_0, z_0) is given by $\nabla F(x_0, y_0, z_0)$.

$$\nabla F(x, y, z) = F_x(x, y, z)\mathbf{i} + F_y(x, y, z)\mathbf{j} + F_z(x, y, z)\mathbf{k}$$

$$= -\sin y\, \mathbf{i} - x \cos y\, \mathbf{j} + \mathbf{k}$$

$$\nabla F(6, \pi/6, 7) = -\frac{1}{2}\mathbf{i} - 3\sqrt{3}\, \mathbf{j} + \mathbf{k}$$

Now, the unit normal vector to the surface is

$$\frac{\nabla F(6, \pi/6, 7)}{\|\nabla F(6, \pi/6, 7)\|} = \frac{\sqrt{113}}{113}(-\mathbf{i} - 6\sqrt{3}\, \mathbf{j} + 2\mathbf{k})$$

13. Find the equation of the tangent plane for the function
     $f(x, y) = y/x$ at the point $(1, 2, 2)$.

     Solution

     We begin by writing the level surface equation

     $$F(x, y, z) = f(x, y) - z = \frac{y}{x} - z = 0$$

     Since

     $$F_x(x, y, z) = \frac{-y}{x^2}, \qquad F_y(x, y, z) = \frac{1}{x}, \qquad \text{and} \qquad F_z(x, y, z) = -1$$

     we have

     $$F_x(1, 2, 2) = -2, \qquad F_y(1, 2, 2) = 1, \qquad \text{and} \qquad F_z(1, 2, 2) = -1$$

     Therefore, by Theorem 14.12, the equation of the tangent
     plane at $(1, 2, 2)$ is

     $$[F_x(1, 2, 2)](x - 1) + [F_y(1, 2, 2)](y - 2) + [F_z(1, 2, 2)](z - 2) = 0$$
     $$-2(x - 1) + 1(y - 2) - (z - 2) = 0$$
     $$-2x + y - z + 2 = 0$$

23. Find the equation of the tangent plane to $xy^2 + 3x - z^2 = 4$ at
     the point $(2, 1, -2)$.

     Solution

     Let $F(x, y, z) = xy^2 + 3x - z^2 - 4$. Then $\nabla F(2, 1, -2)$ is
     normal to the tangent plane at $(2, 1, -2)$.

     $$\nabla F(x, y, z) = F_x(x, y, z)\mathbf{i} + F_y(x, y, z)\mathbf{j} + F_z(x, y, z)\mathbf{k}$$
     $$= (y^2 + 3)\mathbf{i} + 2xy\mathbf{j} - 2z\mathbf{k}$$
     $$\nabla F(2, 1, -2) = 4\mathbf{i} + 4\mathbf{j} + 4\mathbf{k}$$

     Therefore the equation of the tangent plane is

     $$4(x - 2) + 4(y - 1) + 4(z + 2) = 0$$
     $$x + y + z = 1$$

25. Find an equation for the tangent plane and find symmetric
     equations of the normal line to the surface $x^2 + y^2 + z = 9$ at
     the point $(1, 2, 4)$.

     Solution

     Let $F(x, y, z) = x^2 + y^2 + z - 9$. Then $\nabla F(1, 2, 4)$ is
     normal to the surface at $(1, 2, 4)$.

$$\nabla F(x, y, z) = F_x(x, y, z)\mathbf{i} + F_y(x, y, z)\mathbf{j} + F_z(x, y, z)\mathbf{k}$$
$$= 2x\mathbf{i} + 2y\mathbf{j} + \mathbf{k}$$
$$\nabla F(1, 2, 4) = 2\mathbf{i} + 4\mathbf{j} + \mathbf{k}$$

Therefore the equation of the tangent plane is

$$2(x - 1) + 4(y - 2) + 1(z - 4) = 0$$
$$2x + 4y + z - 14 = 0$$

Now since a normal line at $(1, 2, 4)$ has the same direction as $\nabla F(1, 2, 4)$, the direction numbers for this line are 2, 4, and 1. Therefore symmetric equations for a normal line at $(1, 2, 4)$ are

$$\frac{x - 1}{2} = \frac{y - 2}{4} = \frac{z - 4}{1}$$

29. Find an equation for the tangent plane and find symmetric equations of the normal line to the surface $z = \arctan(y/x)$ at the point $(1, 1, \pi/4)$.

Solution

We let $F(x, y, z) = \arctan(y/x) - z$. Then

$$\nabla F(x, y, z) = F_x(x, y, z)\mathbf{i} + F_y(x, y, z)\mathbf{j} + F_z(x, y, z)\mathbf{k}$$

$$= \frac{-y}{x^2 + y^2}\mathbf{i} + \frac{x}{x^2 + y^2}\mathbf{j} - \mathbf{k}$$

$$\nabla F(1, 1, \pi/4) = -\frac{1}{2}\mathbf{i} + \frac{1}{2}\mathbf{j} - \mathbf{k} = -\frac{1}{2}(\mathbf{i} - \mathbf{j} + 2\mathbf{k})$$

Since $\nabla F(1, 1, \pi/4)$ is normal to the surface at the point $(1, 1, \pi/4)$, an equation of the tangent plane is

$$(x - 1) - (y - 1) + 2\left(z - \frac{\pi}{4}\right) = 0$$

$$x - y + 2z = \frac{\pi}{2}$$

and symmetric equations for the normal line are

$$\frac{x - 1}{1} = \frac{y - 1}{-1} = \frac{z - (\pi/4)}{2}$$

35. For the surfaces described by $x^2 + y^2 = 5$ and $z = x$, (a) find symmetric equations of the tangent line to their curve of intersection at the point $(2, 1, 2)$, and (b) find the cosine of the angle between the gradients of the surfaces at the same point.

Solution

Let $f(x, y, z) = x^2 + y^2 - 5$ and $g(x, y, z) = x - z$. Then at $(2, 1, 2)$

$$\nabla f(x, y, z) = 2x\mathbf{i} + 2y\mathbf{j} = 4\mathbf{i} + 2\mathbf{j} \quad \text{and} \quad \nabla g(x, y, z) = \mathbf{i} - \mathbf{k}$$

(a) Since ∇f and ∇g are each normal to their respective surfaces, then the vector $\nabla f \times \nabla g$ will be tangent to both surfaces at the point $(2, 1, 2)$ on the curve of intersection. Therefore from

$$\nabla f \times \nabla g = \begin{vmatrix} \mathbf{i} & \mathbf{j} & \mathbf{k} \\ 4 & 2 & 0 \\ 1 & 0 & -1 \end{vmatrix} = -2\mathbf{i} + 4\mathbf{j} - 2\mathbf{k}$$

we obtain the direction numbers $-2, 4, -2$ or $1, -2, 1$. Hence symmetric equations for the tangent line at $(2, 1, 2)$ are

$$\frac{x - 2}{1} = \frac{y - 1}{-2} = \frac{z - 2}{1}$$

(b) The angle between ∇f and ∇g at $(2, 1, 2)$ is such that

$$\cos \theta = \frac{\nabla f \cdot \nabla g}{|\nabla f||\nabla g|} = \frac{4 + 0 - 0}{\sqrt{20}\sqrt{2}} = \frac{4}{\sqrt{40}} = \frac{4}{2\sqrt{10}} = \frac{2}{\sqrt{10}} = \frac{\sqrt{10}}{5}$$

Therefore the surfaces are **not** orthogonal at the point of intersection.

46. Show that the tangent plane to the hyperboloid

$$\frac{x^2}{a^2} + \frac{y^2}{b^2} - \frac{z^2}{c^2} = 1$$

at the point (x_0, y_0, z_0) can be written in the form

$$\frac{x_0 x}{a^2} + \frac{y_0 y}{b^2} - \frac{z_0 z}{c^2} = 1$$

Solution

We let $F(x, y, z) = \dfrac{x^2}{a^2} + \dfrac{y^2}{b^2} - \dfrac{z^2}{c^2} - 1$. Then

$$\nabla F(x, y, z) = F_x(x, y, z)\mathbf{i} + F_y(x, y, z)\mathbf{j} + F_z(x, y, z)\mathbf{k}$$

$$= \frac{2x}{a^2}\mathbf{i} + \frac{2y}{b^2}\mathbf{j} - \frac{2z}{c^2}\mathbf{k}$$

$$\nabla F(x_0, y_0, z_0) = 2\left[\frac{x_0}{a^2}\mathbf{i} + \frac{y_0}{b^2}\mathbf{j} - \frac{z_0}{c^2}\mathbf{k}\right]$$

Now since $\nabla F(x_0, y_0, z_0)$ is normal to the surface at the point (x_0, y_0, z_0), an equation of the tangent plane is

$$\frac{x_0}{a^2}(x - x_0) + \frac{y_0}{b^2}(y - y_0) - \frac{z_0}{c^2}(z - z_0) = 0$$

$$\left[\frac{x_0 x}{a^2} + \frac{y_0 y}{b^2} - \frac{z_0 z}{c^2} \right] - \left[\frac{x_0^2}{a^2} + \frac{y_0^2}{b^2} - \frac{z_0^2}{c^2} \right] = 0$$

$$\frac{x_0 x}{a^2} + \frac{y_0 y}{b^2} - \frac{z_0 z}{c^2} = 1$$

14.8
Extrema of functions of two variables

7. Examine the function $h(x, y) = x^2 - y^2 - 2x - 4y - 4$ for relative extrema and saddle points.

Solution

Since

$$h_x(x, y) = 2x - 2 = 2(x - 1) = 0 \qquad \text{when} \qquad x = 1$$

and

$$h_y(x, y) = -2y - 4 = -2(y + 2) = 0 \qquad \text{when} \qquad y = -2$$

we have one critical point, $(1, -2)$. Since

$$h_{xx}(x, y) = 2, \qquad h_{yy}(x, y) = -2, \qquad h_{xy}(x, y) = 0$$

we have

$$d = h_{xx}(1, -2)h_{yy}(1, -2) - [h_{xy}(1, -2)]^2 = -4 - 0 = -4 < 0$$

Therefore by part 3 of Theorem 14.16, the critical point $(1, -2)$ yields the saddle point $(1, -2, -1)$.

11. Examine the function $f(x, y) = x^3 - 3xy + y^3$ for relative extrema and saddle points.

Solution

Since

$$f_x(x, y) = 3x^2 - 3y \qquad \text{and} \qquad f_y(x, y) = -3x + 3y^2$$

we can solve the system of equations

$$3x^2 - 3y = 0 \qquad \text{and} \qquad 3y^2 - 3x = 0$$

by substitution. From the first equation we see that $y = x^2$.

Making this substitution for y in the second equation yields

$$3(x^2)^2 - 3x = 0$$
$$x^4 - x = 0$$
$$x(x^3 - 1) = 0$$

Therefore the critical points are $(0, 0)$ and $(1, 1)$. Since

$$f_{xx}(x, y) = 6x, \qquad f_{yy}(x, y) = 6y, \qquad f_{xy}(x, y) = -3$$

we have $f_{xx}(0, 0) = 0$ and

$$d = f_{xx}(0, 0)f_{yy}(0, 0) - [f_{xy}(0, 0)]^2 = 0 - 9 < 0$$

Therefore, by Theorem 14.16, part 3, we conclude that $(0, 0)$ yields a saddle point of f. At $(1, 1)$ we have $f_{xx}(1, 1) = 6 > 0$ and

$$d = f_{xx}(1, 1)f_{yy}(1, 1) - [f_{xy}(1, 1)]^2 = (6)(6) - 9 > 0$$

and by part 1 of Theorem 14.16, the point $(1, 1)$ yields the relative minimum value $f(1, 1) = -1$. (See the accompanying figure.)

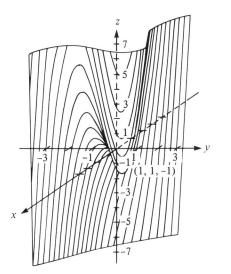

22. Find the absolute extrema of $f(x, y) = x^2 + 2xy + y^2$ over the region $R = \{(x, y) : |x| \le 2, |y| \le 1\}$.

Solution

We first observe that f is a perfect square trinomial and can be written as

$$f(x, y) = (x + y)^2$$

Therefore, the minimum value of f is zero and occurs at all points in R such that $y = -x$. From the partial derivatives, we have

$$f_x(x, y) = 2(x + y) = f_y(x, y)$$

Each point lying on the line $y = -x$ in the xy-coordinate plane is a critical point. However, from the discussion above, we know these points always yield the minimum value of zero. Thus, by Theorem 14.14, the maximum must occur at point(s) on the boundary of R. There the function f will be maximum when the absolute value of $x + y$ is maximum. Thus, we have

$$f(-2, -1) = f(2, 1) = 9$$

as the absolute maximum of f over the region R.

30. For the function $f(x, y) = (x^2 + y^2)^{2/3}$, find the critical points and test for relative extrema. List the critical points for which the Second-Partials Test fails.

Solution

Since $x^2 + y^2 \geq 0$ for all x and y, we observe that the minimum of f is $f(0, 0) = 0$. From the first partial derivatives we have

$$f_x(x, y) = \frac{2}{3}(x^2 + y^2)^{-1/3}(2x) = \frac{4x}{3\sqrt[3]{x^2 + y^2}}$$

and

$$f_y(x, y) = \frac{2}{3}(x^2 + y^2)^{-1/3}(2y) = \frac{4y}{3\sqrt[3]{x^2 + y^2}}$$

Since neither first partial derivative exists at $(0, 0)$, it is a critical point. Moreover,

$$f_{xx}(x, y) = \frac{4(x^2 + 3y^2)}{9(x^2 + y^2)^{4/3}}$$

also does not exist at $(0, 0)$ and thus the Second-Partials Test fails.

14.9

Applications of extrema of functions of two variables

7. Find three positive numbers whose sum is 30 and the sum of whose squares is minimum.

Solution

Let x, y, and z be the numbers and let $s(x, y, z) = x^2 + y^2 + z^2$. Since $x + y + z = 30$, we wish to minimize

$$s(x, y, z) = x^2 + y^2 + (30 - x - y)^2$$

We solve the system of equations

$$s_x(x, y, z) = 2x - 2(30 - x - y) = 0 \qquad 2x + y = 30$$
$$s_y(x, y, z) = 2y - 2(30 - x - y) = 0 \qquad \underline{2x + 4y = 60}$$
$$-3y = -30$$

to obtain the critical values $x = y = z = 10$. These values give us the desired minimum, since $s_{xx}(10, 10) = 4 > 0$ and

$$s_{xx}(10, 10)s_{yy}(10, 10) - [s_{xy}(10, 10)]^2 = (4)(4) - [2]^2 > 0$$

9. Find the dimensions of a rectangular package of largest volume that may be sent by parcel post if the sum of the length and girth (perimeter of a cross section) cannot exceed 108 inches.

Solution

Let x, y, and z be the length, width, and height, respectively, of the box. Then the sum of its length and its girth is given by

$$x + (2y + 2z) = 108 \quad \text{or} \quad x = 108 - 2y - 2z$$

Since we are to maximize the volume of the box, we have

$$V(x, y, z) = xyz = (108 - 2y - 2z)yz = 108yz - 2zy^2 - 2yz^2$$

To find the critical dimensions, we solve the following equations simultaneously:

$$V_y(x, y, z) = 108z - 4yz - 2z^2 = 0$$

or

$$z(108 - 4y - 2x) = 0$$

and

$$V_z(x, y, z) = 108y - 2y^2 - 4yz = 0$$

or

$$y(108 - 2y - 4z) = 0$$

Ignoring the solutions $z = y = 0$, we solve the system

$$4y + 2x = 108 \qquad 8y + 4z = 216$$
$$2y + 4z = 108 \qquad \underline{2y + 4z = 108}$$
$$6y = 108$$

Therefore $y = 18$, $z = 18$, and $x = 108 - 72 = 36$ are the dimensions of the rectangular package of largest volume that can be sent.

15. An eaves trough with trapezoidal cross sections is formed by turning up the edges of a sheet of aluminum which is w inches wide. Find the cross section of maximum area.

Solution

From the accompanying figure we observe that the area of a trapezoidal cross section is given by

$$A = h\left[\frac{(w - 2r) + [(w - 2r) + 2x]}{2}\right]$$

$$= (w - 2r + x)h$$

where $x = r \cos \theta$ and $h = r \sin \theta$. Substituting these expressions for x and h, we have

$$A(r, \theta) = (w - 2r + r \cos \theta)(r \sin \theta)$$

$$= wr \sin \theta - 2r^2 \sin \theta + r^2 \sin \theta \cos \theta$$

Now $A_r(r, \theta) = w \sin \theta - 4r \sin \theta + 2r \sin \theta \cos \theta$

$$= \sin \theta (w - 4r + 2r \cos \theta) = 0$$

Therefore $\rightarrow w = r(4 - 2 \cos \theta)$

$$A_\theta(r, \theta) = wr \cos \theta - 2r^2 \cos \theta + r^2 \cos 2\theta = 0$$

Substituting the expression for w from $A_r(r, \theta) = 0$ into the equation $A_\theta(r, \theta) = 0$, we have

$$r^2(4 - 2 \cos \theta) \cos \theta - 2r^2 \cos \theta + r^2(2 \cos^2 \theta - 1) = 0$$

$$r^2(2 \cos \theta - 1) = 0 \text{ or } \cos = \frac{1}{2}$$

The first partial derivatives are zero when $\theta = \pi/3$ and $r = w/3$. (Ignore the solution $r = \theta = 0$.) Thus, the trapezoid of maximum area occurs when each edge of width $w/3$ is turned up 60° from the horizontal.

23. For the points in the accompanying figure, (a) find the least squares regression line, and (b) calculate S, the sum of squared errors.

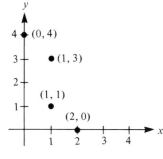

Solution

(a)

x	y	xy	x^2
0	4	0	0
1	3	3	1
1	1	1	1
2	0	0	4
$\sum x_i = 4$	$\sum y_i = 8$	$\sum x_i y_i = 4$	$\sum x_i^2 = 6$

By Theorem 14.17, we have

$$a = \frac{n \sum x_i y_i - \sum x_i \sum y_i}{n \sum x_i^2 - \left(\sum x_i \right)^2} = \frac{4(4) - 4(8)}{4(6) - 4^2} = -2$$

$$b = \frac{1}{n}\left(\sum y_i - a \sum x_i \right) = \frac{1}{4}[8 + 2(4)] = 4$$

Therefore the least squares regression line is $f(x) = -2x + 4$.

(b) $S = \sum [f(x_i) - y_i]^2 = (4 - 4)^2 + (2 - 3)^2 +$

$$(2 - 1)^2 + (0 - 0)^2 = 2$$

26. Use the result of Exercise 19 to find the least squares regression quadratic for the points $(0, 10)$, $(1, 9)$, $(2, 6)$, and $(3, 0)$.

Solution

From Exercise 19, we have that the least squares regression quadratic for the points (x_1, y_1), (x_2, y_2), . . . , (x_n, y_n) is

$$y = ax^2 + bx + c$$

where a, b, and c are the solutions to the system

$$a \sum_{i=1}^{n} x_i^4 + b \sum_{i=1}^{n} x_i^3 + c \sum_{i=1}^{n} x_i^2 = \sum_{i=1}^{n} x_i^2 y_i$$

$$a \sum_{i=1}^{n} x_i^3 + b \sum_{i=1}^{n} x_i^2 + c \sum_{i=1}^{n} x_i = \sum_{i=1}^{n} x_i y_i$$

$$a \sum_{i=1}^{n} x_i^2 + b \sum_{i=1}^{n} x_i + cn = \sum_{i=1}^{n} y_i$$

For the given points, we have

$$\sum x_i = 6, \qquad \sum x_i^2 = 14, \qquad \sum x_i^3 = 36, \qquad \sum x_i^4 = 98$$

$$\sum y_i = 25, \qquad \sum x_i y_i = 21, \qquad \sum x_i^2 y_i = 33$$

The resulting system of equations is

$$98a + 36b + 14c = 33$$
$$36a + 14b + 6c = 21$$
$$14a + 6b + 4c = 25$$

Solving this system we have $a = -1.25$, $b = 0.45$, and $c = 9.95$. Therefore, the least squares regression quadratic is

$$y = -1.25x^2 + 0.45x + 9.95$$

14.10

Lagrange multipliers

5. Use Lagrange multipliers to find the maximum of $f(x, y) = x^2 - y^2$ subject to the constraint $y - x^2 = 0$. Assume that x and y are positive.

Solution

We define the function $F(x, y, \lambda)$ by

$$F(x, y, \lambda) = x^2 - y^2 + \lambda(y - x^2)$$

By setting the first partial derivatives of F equal to zero, we obtain the system of equations

$$F_x(x, y, \lambda) = 2x - 2x\lambda = 0 \quad \rightarrow \quad 2x(1 - \lambda) = 0$$

$$F_y(x, y, \lambda) = -2y + \lambda = 0 \quad \rightarrow \quad y = \frac{1}{2}\lambda$$

$$F_\lambda(x, y, \lambda) = \quad y - x^2 = 0 \quad \rightarrow \quad x = \sqrt{y}$$

From the first equation we have $\lambda = 1$. Using this result in the second and third equations we have $x = \sqrt{2}/2$ and $y = 1/2$. Therefore, the maximum value of f, subject to the given constraint is

$$f\left(\frac{\sqrt{2}}{2}, \frac{1}{2}\right) = \frac{1}{4}$$

15. Use Lagrange multipliers to maximize $f(x, y, z) = xyz$, subject to the constraints $x + y + z = 32$ and $x - y + z = 0$.

Solution

Let $g(x, y) = x + y + z - 32$ and $h(x, y) = x - y + z$ and let

$$F(x, y, z, \lambda, \mu) = xyz + \lambda(x + y + z - 32) + \mu(x - y + z)$$

Then we have the system

$$F_x(x, y, \lambda, \mu) = yz + \lambda + \mu = 0$$
$$F_y(x, y, \lambda, \mu) = xz + \lambda - \mu = 0$$
$$F_z(x, y, \lambda, \mu) = xy + \lambda + \mu = 0$$
$$F_\lambda(x, y, \lambda, \mu) = x + y + z - 32 = 0$$
$$F_\mu(x, y, \lambda, \mu) = x - y + z = 0$$

From the first and third equations we have

$$yz = xy \quad \text{or} \quad z = x$$

Substituting this result into the equation for F_μ, we have

$$y = 2x$$

Finally, substituting these two results into the equation for F_λ, we obtain

$$4x = 32 \quad \text{or} \quad x = 8$$
$$y = 16 \qquad\qquad z = 8$$

Therefore the maximum value of f, subject to the given constraints is

$$f(8, 16, 8) = 8(16)(8) = 1024$$

19. Use Lagrange multipliers to find any extrema of $f(x, y, z) = xyz$ subject to the constraints $x^2 + z^2 = 5$ and $x - 2y = 0$.

Solution

Let $g(x, z) = x^2 + z^2 - 5$ and $h(x, y) = x - 2y$ and let

$$F(x, y, z, \lambda, \mu) = f(x, y, z) + \lambda g(x, z) + \mu h(x, y)$$
$$= xyz + \lambda(x^2 + z^2 - 5) + \mu(x - 2y)$$

Then we have the system

$$F_x(x, y, z, \lambda, \mu) = yz + 2\lambda x + \mu = 0$$

$$F_y(x, y, z, \lambda, \mu) = xz - 2\mu = 0 \qquad\qquad \rightarrow \qquad \frac{xz}{2} = \mu$$

$$F_z(x, y, z, \lambda, \mu) = xy + 2\lambda z = 0 \qquad\qquad \rightarrow \qquad -\frac{xy}{2z} = \lambda$$

$$F_\lambda(x, y, z, \lambda, \mu) = x^2 + z^2 - 5 = 0 \qquad\qquad \rightarrow \qquad z^2 = 5 - x^2$$

$$F_\mu(x, y, z, \lambda, \mu) = x - 2y = 0 \qquad\qquad \rightarrow \qquad x = 2y$$

From these equations we observe that since $x = 2y$,

$$\mu = \frac{xz}{2} = yz, \qquad \lambda = -\frac{y^2}{z}, \qquad z^2 = -4y^2$$

Substituting these values for μ and λ into the equation for F_x, we obtain

$$yz + 4y\left(\frac{y^2}{z}\right) + yz = 0$$
$$2yz^2 - 4y^3 = 0$$
$$2y(z^2 - 2y^2) = 0$$
$$2y(5 - 4y^2 - 2y^2) = 0$$
$$2y(5 - 6y^2) = 0$$
$$y = 0 \qquad \text{or} \qquad y = \pm\sqrt{\frac{5}{6}}$$

For $y = 0$, we have $x = 0$ and $z = \pm\sqrt{5}$. For $y = \pm\sqrt{\frac{5}{6}}$, we have $x = \pm 2\sqrt{\frac{5}{6}}$ and $z = \pm\sqrt{\frac{10}{6}} = \pm\sqrt{\frac{5}{3}}$. Finally, since we require that x, y, and z be positive we have

$$f\left(2\sqrt{\frac{5}{6}}, \sqrt{\frac{5}{6}}, \sqrt{\frac{5}{3}}\right) = \left(2\sqrt{\frac{5}{6}}\right)\left(\sqrt{\frac{5}{6}}\right)\sqrt{\frac{5}{3}} = \frac{5}{3}\sqrt{\frac{5}{3}}$$

as the maximum value of f. Note that if we did not require x, y, and z to be be positive and if we wanted all the extrema, then we would have

$$f\left(-2\sqrt{\frac{5}{6}}, -\sqrt{\frac{5}{6}}, \sqrt{\frac{5}{3}}\right) = \left(-2\sqrt{\frac{5}{6}}\right)\left(-\sqrt{\frac{5}{6}}\right)\sqrt{\frac{5}{3}} = \frac{5}{3}\sqrt{\frac{5}{3}}$$

and

$$f\left(\pm 2\sqrt{\frac{5}{6}}, \pm\sqrt{\frac{5}{6}}, -\sqrt{\frac{5}{3}}\right) = -\frac{5}{3}\sqrt{\frac{5}{3}}$$

as the maximum and minimum values, respectively, of f.

27. Use Lagrange multipliers to find the minimum distance from the point $(2, 1, 1)$ to the plane $x + y + z = 1$.

Solution

Let (x, y, z) be an arbitrary point in the given plane. Then

$$s = \sqrt{(x - 2)^2 + (y - 1)^2 + (z - 1)^2}$$

represents the distance between $(2, 1, 1)$ and a point in the plane. To simplify our calculations, we will minimize s^2 rather than s. With $g(x, y, z) = x + y + z - 1$ as the constraint, we let

$$F(x, y, z, \lambda) = s^2 + \lambda g$$
$$= (x - 2)^2 + (y - 1)^2 + (z - 1)^2 + \lambda(x + y + z - 1)$$

Then we have the system

$$F_x(x, y, z, \lambda) = 2(x - 2) + \lambda = 0$$
$$F_y(x, y, z, \lambda) = 2(y - 1) + \lambda = 0$$
$$F_z(x, y, z, \lambda) = 2(z - 1) + \lambda = 0$$
$$F_\lambda(x, y, z, \lambda) = x + y + z - 1 = 0$$

From the first three equations, we conclude that

$$-\lambda = 2(x - 2) = 2(y - 1) = 2(z - 1)$$

or that $x = y + 1$ and $z = y$. Therefore, from the equation for F_λ, we have

$$(y + 1) + y + (y - 1) = 0 \qquad \text{or} \qquad y = 0$$

which means $x = 1$ and $z = 0$. Thus in the plane $x + y + z = 1$, the point $(1, 0, 0)$ is closest to the given point $(2, 1, 1)$ and this minimum distance is

$$s = \sqrt{(1 - 2)^2 + (0 - 1)^2 + (0 - 1)^2} = \sqrt{3}$$

35. Use Lagrange multipliers to find the dimensions of the rectangular box (edges parallel to the coordinate axes) of maximum volume that can be inscribed in the ellipsoid $(x^2/a^2) + (y^2/b^2) + (z^2/c^2) = 1$.

Solution

Let (x, y, z) be a point on the ellipsoid and a vertex of the rectangular solid. Then the dimensions of the box are $2x$ by $2y$ by $2z$. Now we must maximize $f(x, y, z) = (2x)(2y)(2z)$ subject to the constraint $(x^2/a^2) + (y^2/b^2) + (z^2/c^2) = 1$. Let

$$F(x, y, z, \lambda) = 8xyz + \lambda\left(\frac{x^2}{a^2} + \frac{y^2}{b^2} + \frac{z^2}{c^2} - 1\right)$$

Then

$$F_x = 8yz + \frac{2\lambda x}{a^2} = 0 \quad \rightarrow \quad 8xyz + \frac{2\lambda x^2}{a^2} = 0$$

$$F_y = 8xz + \frac{2\lambda y}{b^2} = 0 \quad \rightarrow \quad 8xyz + \frac{2\lambda y^2}{b^2} = 0$$

$$F_z = 8xy + \frac{2\lambda z}{c^2} = 0 \quad \rightarrow \quad 8xyz + \frac{2\lambda z^2}{c^2} = 0$$

$$F_\lambda = \frac{x^2}{a^2} + \frac{y^2}{b^2} + \frac{z^2}{c^2} - 1 = 0$$

From the first three equations, we have $(x^2/a^2) = (y^2/b^2) = (z^2/c^2)$. Substituting into $F_\lambda = 0$ yields

$$\frac{3x^2}{a^2} = 1 \quad \text{or} \quad x = \frac{a}{\sqrt{3}}$$

$$\frac{3y^2}{b^2} = 1 \quad \text{or} \quad y = \frac{b}{\sqrt{3}}$$

$$\frac{3z^2}{c^2} = 1 \quad \text{or} \quad z = \frac{c}{\sqrt{3}}$$

Therefore the dimensions of the box are $\dfrac{2a}{\sqrt{3}}$ by $\dfrac{2b}{\sqrt{3}}$ by $\dfrac{2c}{\sqrt{3}}$.

Review Exercises for Chapter 14

16. Find the first partial derivatives for $f(x, y, z) = 1/\sqrt{1 - x^2 - y^2 - z^2}$.

Solution

First we rewrite f as

$$f(x, y, z) = (1 - x^2 - y^2 - z^2)^{-1/2}$$

Then

$$f_x(x, y, z) = -\frac{1}{2}(1 - x^2 - y^2 - z^2)^{-3/2}(-2x)$$

$$= \frac{x}{(1 - x^2 - y^2 - z^2)^{3/2}}$$

Similarly,

$$f_y(x, y, z) = \frac{y}{(1 - x^2 - y^2 - z^2)^{3/2}}$$

and

$$f_z(x, y, z) = \frac{z}{(1 - x^2 - y^2 - z^2)^{3/2}}$$

23. Find all second partial derivatives for $f(x, y) = x \sin y + y \cos x$ and verify that the second mixed partials are equal.

Solution

Since $\qquad f(x, y) = x \sin y + y \cos x$

we have

$$f_x(x, y) = \sin y - y \sin x \qquad f_{xx}(x, y) = -y \cos x$$
$$f_y(x, y) = x \cos y + \cos x \qquad f_{yy}(x, y) = -x \sin y$$

Furthermore,

$$f_{xy}(x, y) = \cos y - \sin x \qquad \text{and} \qquad f_{yx}(x, y) = \cos y - \sin x$$

27. Show that the function $z = y/(x^2 + y^2)$ satisfies the **Laplace Equation**

$$\frac{\partial^2 z}{\partial x^2} + \frac{\partial^2 z}{\partial y^2} = 0$$

Solution

$$\frac{\partial z}{\partial x} = \frac{-2xy}{(x^2 + y^2)^2} = -2y\left[\frac{x}{(x^2 + y^2)^2}\right]$$

$$\frac{\partial^2 z}{\partial x^2} = -2y\left[\frac{(x^2 + y^2)^2 - x(2)(x^2 + y^2)(2x)}{(x^2 + y^2)^4}\right] = \frac{2y(3x^2 - y^2)}{(x^2 + y^2)^3}$$

$$\frac{\partial z}{\partial y} = \frac{(x^2 + y^2) - y(2y)}{(x^2 + y^2)^2} = \frac{x^2 - y^2}{(x^2 + y^2)^2}$$

$$\frac{\partial^2 z}{\partial y^2} = \frac{(x^2 + y^2)^2(-2y) - (x^2 - y^2)(2)(x^2 + y^2)(2y)}{(x^2 + y^2)^4}$$

$$= \frac{-2y(3x^2 - y^2)}{(x^2 + y^2)^3}$$

Therefore

$$\frac{\partial^2 z}{\partial x^2} + \frac{\partial^2 z}{\partial y^2} = \frac{2y(3x^2 - y^2)}{(x^2 + y^2)^3} + \frac{(-2y)(3x^2 - y^2)}{(x^2 + y^2)^3} = 0$$

31. Find $\partial u / \partial r$ and $\partial u / \partial t$ if $u = x^2 + y^2 + z^2$, $x = r \cos t$, $y = r \sin t$, and $z = t$. (a) Use the Chain Rule and (b) check the result by substituting for x, y, and z before differentiating.

Solution

By the Chain Rule,

$$\frac{\partial u}{\partial r} = \frac{\partial u}{\partial x}\frac{\partial x}{\partial r} + \frac{\partial u}{\partial y}\frac{\partial y}{\partial r} + \frac{\partial u}{\partial z}\frac{\partial z}{\partial r} = 2x(\cos t) + 2y(\sin t) + 2z(0)$$

$$= 2r \cos t(\cos t) + 2r \sin t(\sin t) + 2r = 2r(\cos^2 t + \sin^2 t)$$

$$= 2r$$

and

$$\frac{\partial u}{\partial t} = \frac{\partial u}{\partial x}\frac{\partial x}{\partial t} + \frac{\partial u}{\partial y}\frac{\partial y}{\partial t} + \frac{\partial u}{\partial z}\frac{\partial z}{\partial t} = 2x(-r \sin t) + 2y(r \cos t) + 2z(1)$$

$$= 2r \cos t(-r \sin t) + 2r \sin t(r \cos t) + 2t$$

$$= -2r^2 \sin t \cos t + 2r^2 \sin t \cos t + 2t = 2t$$

By first substituting for x, y, and z, we have

$$u = r^2 \cos^2 t + r^2 \sin^2 t + t^2 = r^2(\cos^2 t + \sin^2 t) + t^2$$

$$= r^2 + t^2$$

Therefore

$$\frac{\partial u}{\partial r} = 2r \quad \text{and} \quad \frac{\partial u}{\partial t} = 2t$$

38. Find the gradient and the maximum value of the directional derivative of $f(x, y) = x^2/(x - y)$ at the point $(2, 1)$.

Solution

Since

$$\nabla f(x, y) = f_x(x, y)\mathbf{i} + f_y(x, y)\mathbf{j}$$

$$= \frac{x(x - 2y)}{(x - y)^2}\mathbf{i} + \frac{x^2}{(x - y)^2}\mathbf{j}$$

the gradient at (2, 1) is

$$\nabla f(2, 1) = 4\mathbf{j}$$

Consequently, the maximum value of the directional derivative at (2, 1) is

$$\|\nabla f(2, 1)\| = 4$$

43. For the surface defined by $f(x, y) = 9 + 4x - 6y - x^2 - y^2$, find an equation of the tangent plane and parametric equations for the normal line at the point $(2, -3, 4)$.

Solution

Consider $z = f(x, y)$ and let

$$F(x, y, z) = z - 9 - 4x + 6y + x^2 + y^2$$

Then at $(2, -3, 4)$

$$F_x(x, y, z) = -4 + 2x = 0$$
$$F_y(x, y, z) = 6 + 2y = 6 - 6 = 0$$
$$F_z(x, y, z) = 1$$

Therefore, by Theorem 14.12 an equation of the tangent plane at $(2, -3, 4)$ is

$$0(x - 2) + 0(y + 3) + 1(z - 4) = 0 \quad \text{or} \quad z = 4$$

Furthermore, a normal line at $(2, -3, 4)$ has direction numbers 0, 0, 1 and its parametric equations are

$$x = 2, \quad y = -3, \quad z = 4 + t$$

48. Locate and classify any extrema of the function

$$f(x, y) = 2x^2 + 6xy + 9y^2 + 8x + 14$$

Solution

Since

$$f_x(x, y) = 4x + 6y + 8 \quad \text{and} \quad f_y(x, y) = 6x + 18y$$

we can solve the system of equations

$$2x + 3y + 4 = 0 \quad \text{and} \quad x + 3y = 0$$

by substitution. From the second equation we see that

$x = -3y$. Making this substitution for x in the first equation yields

$$2(-3y) + 3y + 4 = 0$$

$$-6y + 8 = 0 \quad \text{or} \quad y = \frac{4}{3}$$

Therefore the critical point is $(-4, \frac{4}{3})$. Since

$$f_{xx}(x, y) = 4, \qquad f_{yy}(x, y) = 18, \qquad f_{xy}(x, y) = 6$$

we have

$$d = (f_{xx})(f_{yy}) - (f_{xy})^2 = 72 - 36 > 0$$

Therefore, by part 1 of Theorem 14.16, $(-4, \frac{4}{3}, -2)$ is a minimum.

54. To determine the height of a tower, the angle of elevation to the top of the tower was measured from a point $100 \pm \frac{1}{2}$ ft from the base. The angle is measured at $33°$ with a possible error of $1°$. Assuming the ground to be horizontal, approximate the maximum error in determining the height of the tower.

Solution

Letting θ, h, and x be the angle of elevation, the height of the tower, and the distance from the base, respectively, we have

$$\tan \theta = \frac{h}{x}$$

or

$$h = x \tan \theta$$

and

$$dh = \frac{\partial h}{\partial x} dx + \frac{\partial h}{\partial \theta} d\theta$$

$$= \tan \theta \, dx + x \sec^2 \theta \, d\theta$$

Letting $x = 100$, $dx = \pm\frac{1}{2}$, $\theta = 11\pi/60$, and $d\theta = \pi/180$ (note that we express the measurement of the angle in radians), the maximum error is approximately

$$dh = \tan \left(\frac{11\pi}{60}\right)\left(\pm\frac{1}{2}\right) + 100 \sec^2 \left(\frac{11\pi}{60}\right)\left(\frac{\pi}{180}\right)$$

$$\approx \pm 2.8 \text{ ft}$$

15 *Multiple integration*

Iterated integrals and area in the plane

7. Evaluate

$$\int_{e^y}^y \frac{y \ln x}{x} \, dx$$

Solution

Consider the integral in the form

$$\int_{e^y}^y y (\ln x) \left(\frac{1}{x}\right) dx$$

Then integration with respect to x yields

$$\left[\frac{y (\ln x)^2}{2}\right]_{e^y}^y = \frac{y}{2}[(\ln y)^2 - (\ln e^y)^2] = \frac{y}{2}[\ln^2 y - y^2]$$

13. Evaluate $\int_1^2 \int_0^4 (x^2 - 2y^2 + 1) \, dx \, dy$.

Solution

$$\int_1^2 \int_0^4 (x^2 - 2y^2 + 1) \, dx \, dy = \int_1^2 \left[\frac{x^3}{3} - 2xy^2 + x\right]_0^4 dy$$

$$= \int_1^2 \left(\frac{64}{3} - 8y^2 + 4\right) dy$$

$$= \int_1^2 \left(\frac{76}{3} - 8y^2\right) dy = \frac{1}{3}[76y - 8y^3]_1^2$$

$$= \frac{1}{3}[152 - 64 - 76 + 8] = \frac{20}{3}$$

19. Evaluate $\int_0^{\pi/2} \int_0^{\sin\theta} \theta r \, dr \, d\theta$.

Solution

$$\int_0^{\pi/2} \int_0^{\sin\theta} \theta r \, dr \, d\theta = \int_0^{\pi/2} \left[\frac{\theta r^2}{2} \right]_0^{\sin\theta} d\theta = \frac{1}{2} \int_0^{\pi/2} \theta \sin^2\theta \, d\theta$$

$$= \frac{1}{2} \int_0^{\pi/2} \theta \left(\frac{1 - \cos 2\theta}{2} \right) d\theta$$

$$= \frac{1}{4} \int_0^{\pi/2} (\theta - \theta \cos 2\theta) \, d\theta$$

Using integration by parts or Integration Formula 53 at the back of the text, we have

$$\int_0^{\pi/2} \int_0^{\sin\theta} \theta r \, dr \, d\theta = \frac{1}{4} \left[\frac{\theta^2}{2} - \left(\frac{1}{4} \cos 2\theta + \frac{\theta}{2} \sin 2\theta \right) \right]_0^{\pi/2}$$

$$= \frac{1}{4} \left(\frac{\pi^2}{8} + \frac{1}{4} - 0 - 0 + \frac{1}{4} + 0 \right)$$

$$= \frac{1}{4} \left(\frac{\pi^2}{8} + \frac{1}{2} \right) = \frac{\pi^2}{32} + \frac{1}{8}$$

24. Sketch the region R whose area is given by the iterated integral

$$\int_0^2 \int_0^x dy \, dx + \int_2^4 \int_0^{4-x} dy \, dx$$

Switch the order of integration, and show that both orders yield the same area.

Solution

From the limits of integration, we know that when $0 \le x \le 2$, $0 \le y \le x$, which means that part of the region R is bounded by the lines $y = 0$, $y = x$, and $x = 2$. From the limits on the second iterated integral, we know that when $2 \le x \le 4$, $0 \le y \le 4 - x$, which means that the remaining part of the region R is bounded by the lines $x = 2$, $y = 0$, and $y = 4 - x$. Therefore the region R can be sketched as in the accompanying figure.

If we interchange the order of integration so that y is the outer variable, we see that $0 \le y \le 2$. Solving for x in the equations $y = x$ and $y = 4 - x$, we obtain $x = y$ and $x = 4 - y$. Thus $y \le x \le 4 - y$, and the area of the region R can be obtained by the one iterated integral

$$\int_0^2 \int_y^{4-y} dx \, dy$$

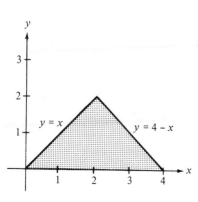

Evaluating the iterated integrals for each order of integration, we have

$$\int_0^2 \int_0^x dy \, dx + \int_2^4 \int_0^{4-x} dy \, dx = \int_0^2 x \, dx + \int_2^4 (4 - x) \, dx$$

$$= \left[\frac{1}{2}x^2\right]_0^2 + \left[4x - \frac{1}{2}x^2\right]_2^4 = 4$$

$$\int_0^2 \int_y^{4-y} dx \, dy = \int_0^2 (4 - 2y) \, dy = \left[4y - y^2\right]_0^2 = 4$$

27.   Sketch the region R whose area is given by the iterated integral

$$\int_0^1 \int_{y^2}^{\sqrt[3]{y}} dx \, dy$$

Switch the order of integration, and show that both orders yield the same area.

Solution

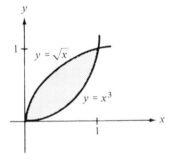

From the limits of integration, we know that when $0 \le y \le 1$, $y^2 \le x \le \sqrt[3]{y}$, which means that the region R is bounded by the curves $x = y^2$ and $x = \sqrt[3]{y}$. Therefore the region R can be sketched as in the accompanying figure.

If we interchange the order of integration so that x is the outer variable, we see that $0 \le x \le 1$. Solving for y in the equations $x = y^2$ and $x = \sqrt[3]{y}$, we have $y = \sqrt{x}$ and $y = x^3$. Thus $x^3 \le y \le \sqrt{x}$, and the area of R is given by the iterated integral

$$\int_0^1 \int_{x^3}^{\sqrt{x}} dy \, dx$$

Evaluating each iterated integral, we have

$$\int_0^1 \int_{y^2}^{\sqrt[3]{y}} dx \, dy = \int_0^1 (\sqrt[3]{y} - y^2) \, dy = \left[\frac{3}{4}y^{4/3} - \frac{1}{3}y^3\right]_0^1 = \frac{5}{12}$$

$$\int_0^1 \int_{x^3}^{\sqrt{x}} dy \, dx = \int_0^1 (\sqrt{x} - x^3) \, dx = \left[\frac{2}{3}x^{3/2} - \frac{1}{4}x^4\right]_0^1 = \frac{5}{12}$$

34.   Use a double integral to find the area of the region R bounded by $x^2 + y^2 = 4$, where $0 \le x$ and $0 \le y$. Evaluate the double integral in both orders.

Solution

Since $x^2 = 4 - y^2$ and $0 \le x$, we have

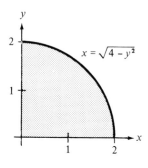

$$\iint\limits_R dx\, dy = \int_0^2 \int_0^{\sqrt{4-y^2}} dx\, dy = \int_0^2 [x]_0^{\sqrt{4-y^2}}\, dy$$

$$= \int_0^2 \sqrt{4 - y^2}\, dy$$

Using trigonometric substitution or Integration Formula 37 at the back of the text, we have

$$\iint\limits_R dx\, dy = \frac{1}{2}\left[y\sqrt{4 - y^2} + 4 \arcsin \frac{y}{2} \right]_0^2$$

$$= \frac{1}{2}\left[0 + 4\left(\frac{\pi}{2}\right) - 0 \right] = \pi$$

Since $0 \le y$ and $y^2 = 4 - x^2$, we also have

$$\iint\limits_R dy\, dx = \int_0^2 \int_0^{\sqrt{4-x^2}} dy\, dx = \int_0^2 \sqrt{4 - x^2}\, dx$$

$$= \frac{1}{2}\left[x\sqrt{4 - x^2} + 4 \arcsin \frac{x}{2} \right]_0^2$$

$$= \frac{1}{2}\left[0 + 4\left(\frac{\pi}{2}\right) - 0 \right] = \pi$$

Note that in this problem the integration steps for both orders are identical.

15.2
Double integrals and volume

4. Sketch the region R and evaluate the integral

$$\int_0^1 \int_y^{\sqrt{y}} x^2 y^2\, dx\, dy$$

Solution

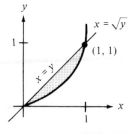

$$\int_0^1 \int_y^{\sqrt{y}} x^2 y^2\, dx\, dy = \int_0^1 \left[y^2\left(\frac{x^3}{3}\right) \right]_y^{\sqrt{y}} dy = \frac{1}{3}\int_0^1 (y^{7/2} - y^5)\, dy$$

$$= \frac{1}{3}\left[\frac{2}{9}y^{9/2} - \frac{y^6}{6} \right]_0^1 = \frac{1}{3}\left[\frac{2}{9} - \frac{1}{6} \right] = \frac{1}{54}$$

The corresponding region R is shown in the accompanying figure, where $y \le x \le \sqrt{y}$ and $0 \le y \le 1$.

9. Given

$$\iint\limits_R \frac{y}{x^2 + y^2}\, dA$$

where R is the triangle bounded by $y = x$, $y = 2x$, and $x = 2$, set up the integrals for $dx\,dy$ and for $dy\,dx$ and use the most convenient order to evaluate the integral over the region R.

Solution

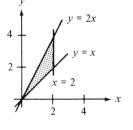

$$\iint_R \frac{y}{x^2 + y^2}\,dx\,dy = \int_0^2 \int_{y/2}^y \frac{y}{x^2 + y^2}\,dx\,dy + \int_2^4 \int_{y/2}^2 \frac{y}{x^2 + y^2}\,dx\,dy$$

and

$$\iint_R \frac{y}{x^2 + y^2}\,dy\,dx = \int_0^2 \int_x^{2x} \frac{y}{x^2 + y^2}\,dy\,dx$$

$$= \frac{1}{2} \int_0^2 [\ln\,(x^2 + y^2)]_x^{2x}\,dx$$

$$= \frac{1}{2} \int_0^2 [\ln\,(5x^2) - \ln\,(2x^2)]\,dx$$

$$= \frac{1}{2} \int_0^2 \ln\left(\frac{5}{2}\right) dx$$

$$= \frac{1}{2} \left[x\,\ln\left(\frac{5}{2}\right) \right]_0^2 = \ln\left(\frac{5}{2}\right)$$

17. Use a double integral to find the volume of the solid

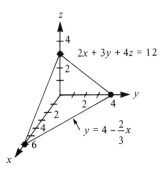

Solution

By letting $z = 0$, we see that the base of the solid is the triangle in the xy-plane bounded by the graphs of $2x + 3y = 12$, $x = 0$, and $y = 0$. This plane region is both vertically and horizontally simple and we choose the order $dy\,dx$.

variable bounds for y: $0 \le y \le 4 - \dfrac{2}{3}x$

constant bounds of x: $0 \le x \le 6$

Therefore, the volume is given by

$$V = \int_0^6 \int_0^{4-(2/3)x} \left(3 - \frac{1}{2}x - \frac{3}{4}y\right) dy\, dx$$

$$= \int_0^6 \left(3y - \frac{1}{2}xy - \frac{3}{8}y^2\right)\Bigg]_0^{4-(2/3)x} dx$$

$$= \int_0^6 \left(6 - 2x + \frac{1}{6}x^2\right) dx$$

$$= \left[6x - x^2 + \frac{1}{18}x^3\right]_0^6 = 12$$

29. Use a double integral to find the volume of the solid of intersection of the cylinders $x^2 + z^2 = 1$ and $y^2 + z^2 = 1$.

Solution

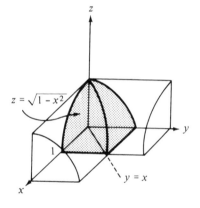

In the accompanying figure we show the solid in the first
octant. We divide this solid in two equal parts by the plane
$y = x$ and thus find $\frac{1}{16}$ of the total volume. Therefore we
integrate the function $z = \sqrt{1 - x^2}$ over the triangle bounded
by $y = 0$, $y = x$, and $x = 1$.

Constant bounds for x: $0 \le x \le 1$

Variable bounds for y: $0 \le y \le x$

$$V = 16 \int_0^1 \int_0^x \sqrt{1 - x^2}\, dy\, dx = 16 \int_0^1 x\sqrt{1 - x^2}\, dx$$

$$= \left[-\frac{16}{3}(1 - x^2)^{3/2}\right]_0^1 = \frac{16}{3}$$

Thus the volume is 16 times that of Exercise 22.

33. Use Wallis's Formula as an aid in finding the volume of the
solid bounded by the paraboloid $z = 4 - x^2 - y^2$ and the
xy-plane.

Solution

Because of the symmetry of the paraboloid, we find the
volume of the solid in the first octant (one-fourth the total
volume). Thus we integrate over the first-quadrant portion of
the circle in the accompanying figure.

Constant bounds for x: $0 \le x \le 2$

Variable bounds for y: $0 \le y \le \sqrt{4 - x^2}$

Therefore

$$V = 4 \int_0^2 \int_0^{\sqrt{4-x^2}} (4 - x^2 - y^2)\, dy\, dx$$

$$= 4 \int_0^2 \left[4y - x^2 y - \frac{1}{3} y^3 \right]_0^{\sqrt{4-x^2}} dx$$

$$= 4 \int_0^2 \left[4\sqrt{4 - x^2} - x^2 \sqrt{4 - x^2} - \frac{1}{3}(4 - x^2)^{3/2} \right] dx$$

$$= \frac{8}{3} \int_0^2 (4 - x^2)^{3/2}\, dx$$

Now we let $x = 2 \sin \theta$. Then $\sqrt{4 - x^2} = 2 \cos \theta$, $dx = 2 \cos \theta\, d\theta$, and

$$V = \frac{8}{3} \int_0^{\pi/2} (2 \cos \theta)^3 (2 \cos \theta)\, d\theta = \frac{128}{3} \int_0^{\pi/2} \cos^4 \theta\, d\theta$$

$$= \frac{128}{3} \left(\frac{1 \cdot 3}{2 \cdot 4} \cdot \frac{\pi}{2} \right) \qquad \text{by Wallis's Formula}$$

$$= 8\pi$$

43. Approximate the double integral $\int_0^4 \int_0^2 (x + y)\, dy\, dx$ by dividing the region of integration R into 8 equal squares and finding the sum $\sum_{i=1}^{8} f(x_i, y_i)\, \Delta x_i\, \Delta y_i$. Let (x_i, y_i) be the center of the ith square. Evaluate the double integral and compare the result to the approximation.

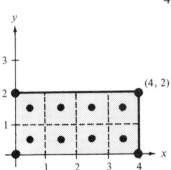

Solution

Region R is the rectangle with vertices $(0, 0)$, $(4, 0)$, $(4, 2)$, and $(0, 2)$, as shown in the accompanying figure. By dividing the rectangle into 8 equal squares of area $\Delta A = \Delta x_i \Delta y_i = 1$, we obtain the following table:

centers (x_i, y_i)	$\left(\frac{1}{2}, \frac{1}{2}\right)$	$\left(\frac{3}{2}, \frac{1}{2}\right)$	$\left(\frac{5}{2}, \frac{1}{2}\right)$	$\left(\frac{7}{2}, \frac{1}{2}\right)$	$\left(\frac{1}{2}, \frac{3}{2}\right)$	$\left(\frac{3}{2}, \frac{3}{2}\right)$	$\left(\frac{5}{2}, \frac{3}{2}\right)$	$\left(\frac{7}{2}, \frac{3}{2}\right)$
$f(x_i, y_i) = x_i + y_i$	1	2	3	4	2	3	4	5
$\Delta A_i = \Delta x_i\, \Delta y_i$	1	1	1	1	1	1	1	1

Therefore

$$\sum_{i=1}^{8} f(x_i, y_i)\, \Delta x_i\, \Delta y_i = 1 + 2 + 3 + 4 + 2 + 3 + 4 + 5 = 24$$

By integration,

$$\int_0^4 \int_0^2 (x + y)\, dy\, dx = \int_0^4 \left[xy + \frac{y^2}{2} \right]_0^2 dx = \int_0^4 (2x + 2)\, dx$$
$$= [x^2 + 2x]_0^4 = 16 + 8 = 24$$

15.3
Change of variables: Polar coordinates

6.   Evaluate the double integral

$$\int_0^{\pi/2} \int_0^{1-\cos\theta} \sin\theta\, dr\, d\theta$$

and sketch the region R.

Solution

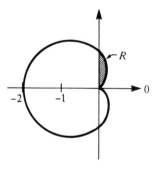

$$\int_0^{\pi/2} \int_0^{1-\cos\theta} \sin\theta\, dr\, d\theta = \int_0^{\pi/2} [r\,\sin\theta]_0^{1-\cos\theta}\, d\theta$$

$$= \int_0^{\pi/2} (\sin\theta - \sin\theta \cos\theta)\, d\theta$$

$$= \left[-\cos\theta + \frac{1}{2}\cos^2\theta \right]_0^{\pi/2} = \frac{1}{2}$$

15.   Evaluate the double integral

$$\int_0^2 \int_0^x \sqrt{x^2 + y^2}\, dy\, dx + \int_0^{2\sqrt{2}} \int_0^{\sqrt{8-x^2}} \sqrt{x^2 + y^2}\, dy\, dx$$

by changing to polar coordinates.

Solution

From the accompanying figure we can see that R has the bounds

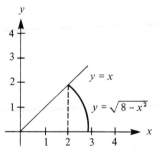

$$0 \le y \le x \qquad\qquad 0 \le x \le 2$$
$$0 \le y \le \sqrt{8 - x^2} \qquad 2 \le x \le 2\sqrt{2}$$

and these bounds form a sector of a circle. In polar coordinates the bounds are

$$0 \le r \le 2\sqrt{2} \qquad 0 \le \theta \le \frac{\pi}{4}$$

with height $z = \sqrt{x^2 + y^2} = r$. Consequently, the double integral in polar coordinates is

$$\int_0^{\pi/4}\int_0^{2\sqrt{2}}(r)r\,dr\,d\theta = \int_0^{\pi/4}\left[\frac{1}{3}r^3\right]_0^{2\sqrt{2}}d\theta \qquad dA = r\,dr\,d\theta$$

$$= \frac{16\sqrt{2}}{3}\int_0^{\pi/4}d\theta = \frac{4\sqrt{2}\pi}{3}$$

21. Use polar coordinates to evaluate $\iint_R f(x, y)\,dA$, where $f(x, y) = \arctan(y/x)$ and R is such that $x^2 + y^2 \le 1$, $0 \le x$, and $0 \le y$.

Solution

Since $x^2 + y^2 = r^2 \le 1$, $\tan\theta = y/x$, and $dA = r\,dr\,d\theta$, we have

$$\int_0^1\int_0^{\sqrt{1-x^2}}\arctan\left(\frac{y}{x}\right)dy\,dx = \int_0^{\pi/2}\int_0^1 \theta r\,dr\,d\theta$$

$$= \frac{1}{2}\int_0^{\pi/2}\left[\theta r^2\right]_0^1 d\theta$$

$$= \frac{1}{2}\int_0^{\pi/2}\theta\,d\theta = \frac{1}{4}\left[\theta^2\right]_0^{\pi/2} = \frac{\pi^2}{16}$$

27. Use a double integral in polar coordinates to find the volume of the solid inside the hemisphere $z = \sqrt{16 - x^2 - y^2}$ and the cylinder $x^2 + y^2 - 4x = 0$.

Solution

Writing the equation for the cylinder in polar form we have $r = 4\cos\theta$. We can see from the accompanying figure that in polar coordinates R has the bounds

$$0 \le r \le 4\cos\theta \qquad \text{and} \qquad -\frac{\pi}{2} \le \theta \le \frac{\pi}{2}$$

and $z = \sqrt{16 - x^2 - y^2} = \sqrt{16 - r^2}$. Since the solid is symmetric with respect to the xz-plane, the volume V is given by

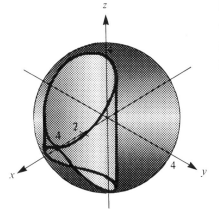

$$V = 2\int_0^{\pi/2}\int_0^{4\cos\theta}\sqrt{16 - r^2}\,r\,dr\,d\theta$$

$$= -\frac{2}{3}\int_0^{\pi/2}\left[(16 - 16\cos^2\theta)^{3/2} - 64\right]d\theta$$

$$= \frac{128}{3}\int_0^{\pi/2}(1 - \sin^3\theta)\,d\theta = \frac{64}{9}(3\pi - 4)$$

15.4

Change of variables: Jacobians

7. Find the Jacobian for the change of variables

$$x = e^u \sin v \qquad y = e^u \cos v$$

Solution

From the definition we evaluate the determinant

$$\frac{\partial(x, y)}{\partial(u, v)} = \begin{vmatrix} \dfrac{\partial x}{\partial u} & \dfrac{\partial y}{\partial u} \\ \dfrac{\partial x}{\partial v} & \dfrac{\partial y}{\partial v} \end{vmatrix} = \begin{vmatrix} e^u \sin v & e^u \cos v \\ e^u \cos v & -e^u \sin v \end{vmatrix}$$

$$= -e^{2u} \sin^2 v - e^{2u} \cos^2 v = -e^{2u}$$

11. Evaluate

$$\iint_R 4(x + y)e^{x-y} \, dy \, dx$$

using the change of variables $x = \frac{1}{2}(u + v)$ and $y = \frac{1}{2}(u - v)$ and letting R be the following region

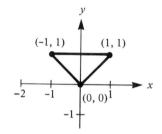

Solution

We begin by solving for u and v in terms of x and y and obtain

$$u = x + y \qquad \text{and} \qquad v = x - y$$

Since the region R is bounded by the lines $x + y = 0$, $x - y = 0$ and $y = 1$, the bounds for the region S in the uv-plane are $u = 0$, $v = 0$, and $v = u - 2$. The Jacobian for the change of variables is

$$\frac{\partial(x, y)}{\partial(u, v)} = \begin{vmatrix} \dfrac{\partial x}{\partial u} & \dfrac{\partial y}{\partial u} \\ \dfrac{\partial x}{\partial v} & \dfrac{\partial y}{\partial v} \end{vmatrix} = \begin{vmatrix} \dfrac{1}{2} & \dfrac{1}{2} \\ \dfrac{1}{2} & -\dfrac{1}{2} \end{vmatrix} = -\frac{1}{2}$$

Therefore, by Theorem 15.5 we obtain

$$\iint_R 4(x + y)e^{x-y} \, dy \, dx = \iint_S 4ue^v \left| \frac{\partial(x, y)}{\partial(u, v)} \right| \, dv \, du$$

$$= 2 \int_0^2 \int_{u-2}^0 ue^v \, dv \, du$$

$$= 2 \int_0^2 (u - ue^{u-2}) \, du$$

$$= 2 \left[\frac{1}{2}u^2 - e^{u-2}(u - 1) \right]_0^2 = 2(1 - e^{-2})$$

17.   Use a change of variables to evaluate

$$\iint_R \sqrt{(x - y)(x + 4y)} \, dy \, dx$$

where R is the region bounded by the parallelogram with vertices $(0, 0)$, $(1, 1)$, $(5, 0)$, and $(4, -1)$.

Solution

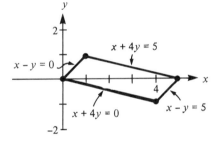

The region R is bounded by the graphs of $x - y = 0$, $x - y = 5$, $x + 4y = 0$ and $x + 4y = 5$. (See the accompanying figure.) By letting $u = x - y$ and $v = x + 4y$, we have

$$x = \frac{1}{5}(4u + v) \qquad \text{and} \qquad y = -\frac{1}{5}(u - v)$$

Thus, the Jacobian is

$$\frac{\partial(x, y)}{\partial(u, v)} = \begin{vmatrix} \dfrac{\partial x}{\partial u} & \dfrac{\partial y}{\partial u} \\[2mm] \dfrac{\partial x}{\partial v} & \dfrac{\partial y}{\partial v} \end{vmatrix} = \begin{vmatrix} \dfrac{4}{5} & -\dfrac{1}{5} \\[2mm] \dfrac{1}{5} & \dfrac{1}{5} \end{vmatrix} = \frac{1}{5}$$

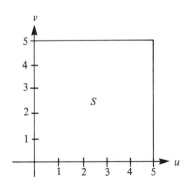

Therefore, by Theorem 15.5 we obtain

$$\iint_R \sqrt{(x - y)(x + 4y)} \, dy \, dx = \iint_S \sqrt{uv} \left| \frac{\partial(x, y)}{\partial(u, v)} \right| \, dv \, du$$

$$= \frac{1}{5} \int_0^5 \int_0^5 \sqrt{uv} \, dv \, du$$

$$= \frac{2\sqrt{5}}{3} \int_0^5 \sqrt{u} \, du = \frac{100}{9}$$

15.5

Center of mass

1. Find the mass and center of mass for the rectangular lamina with vertices $(0, 0)$, $(a, 0)$, $(0, b)$, and (a, b) if (a) $\rho = k$ and (b) $\rho = ky$.

Solution

(a) $m = \int_0^a \int_0^b k \, dy \, dx = kab$

$M_x = \int_0^a \int_0^b ky \, dy \, dx = \dfrac{kab^2}{2}$

$M_y = \int_0^a \int_0^b kx \, dy \, dx = \dfrac{ka^2 b}{2}$

$\bar{x} = \dfrac{M_y}{m} = \dfrac{ka^2 b/2}{kab} = \dfrac{a}{2}, \qquad \bar{y} = \dfrac{M_x}{m} = \dfrac{kab^2/2}{kab} = \dfrac{b}{2}$

(b) $m = \int_0^a \int_0^b ky \, dy \, dx = \dfrac{kab^2}{2}$

$M_x = \int_0^a \int_0^b ky^2 \, dy \, dx = \dfrac{kab^3}{3}$

$M_y = \int_0^a \int_0^b kxy \, dy \, dx = \dfrac{ka^2 b^2}{4}$

$\bar{x} = \dfrac{M_y}{m} = \dfrac{ka^2 b^2/4}{kab^2/2} = \dfrac{a}{2}, \qquad \bar{y} = \dfrac{M_y}{m} = \dfrac{kab^3/3}{kab^2/2} = \dfrac{2}{3} b$

2. Find the mass and center of mass for the rectangular lamina with vertices $(0, 0)$, $(a, 0)$, $(0, b)$, and (a, b) if (a) $\rho = kxy$ and (b) $\rho = k(x^2 + y^2)$.

Solution

(a) $m = \int_0^a \int_0^b kxy \, dy \, dx = \dfrac{ka^2 b^2}{4}$

$M_x = \int_0^a \int_0^b kxy^2 \, dy \, dx = \dfrac{ka^2 b^3}{6}$

$M_y = \int_0^a \int_0^b kx^2 y \, dy \, dx = \dfrac{ka^3 b^2}{6}$

$\bar{x} = \dfrac{M_y}{m} = \dfrac{ka^3 b^2/6}{ka^2 b^2/4} = \dfrac{2}{3} a$

$\bar{y} = \dfrac{M_x}{m} = \dfrac{ka^2 b^3/6}{ka^2 b^2/4} = \dfrac{2}{3} b$

(b) $m = \int_0^a \int_0^b k(x^2 + y^2)\, dy\, dx = \dfrac{kab}{3}(a^2 + b^2)$

$M_x = \int_0^a \int_0^b k(x^2 y + y^3)\, dy\, dx = \dfrac{kab^2}{12}(2a^2 + 3b^2)$

$M_y = \int_0^a \int_0^b k(x^3 + xy^2)\, dy\, dx = \dfrac{ka^2 b}{12}(3a^2 + 2b^2)$

$\bar{x} = \dfrac{M_y}{m} = \dfrac{(ka^2 b/12)(3a^2 + 2b^2)}{(kab/3)(a^2 + b^2)} = \dfrac{a}{4}\left(\dfrac{3a^2 + 2b^2}{a^2 + b^2}\right)$

$\bar{y} = \dfrac{M_x}{m} = \dfrac{(kab^2/12)(2a^2 + 3b^2)}{(kab/3)(a^2 + b^2)} = \dfrac{b}{4}\left(\dfrac{2a^2 + 3b^2}{a^2 + b^2}\right)$

15. Find the mass and center of mass of the lamina of density
$\rho = ky$ and bounded by $y = \sin(\pi x/L)$, $y = 0$, $x = 0$, and
$x = L$.

Solution

$m = \int_0^L \int_0^{\sin(\pi x/L)} ky\, dy\, dx = \dfrac{k}{2}\int_0^L \sin^2\left(\dfrac{\pi x}{L}\right) dx$

$\quad = \dfrac{k}{4}\int_0^L\left[1 - \cos\left(\dfrac{2\pi x}{L}\right)\right] dx$

$\quad = \dfrac{k}{4}\left[x - \dfrac{L}{2\pi}\sin\left(\dfrac{2\pi x}{L}\right)\right]_0^L = \dfrac{kL}{4}$

$M_x = \int_0^L \int_0^{\sin(\pi x/L)} ky^2\, dy\, dx = \dfrac{k}{3}\int_0^L \sin^3\left(\dfrac{\pi x}{L}\right) dx$

$\quad = \dfrac{k}{3}\int_0^L\left[\sin\left(\dfrac{\pi x}{L}\right) - \cos^2\left(\dfrac{\pi x}{L}\right)\sin\left(\dfrac{\pi x}{L}\right)\right] dx$

$\quad = \dfrac{kL}{3\pi}\left[-\cos\left(\dfrac{\pi x}{L}\right) + \dfrac{1}{3}\cos^3\left(\dfrac{\pi x}{L}\right)\right]_0^L = \dfrac{4kL}{9\pi}$

Now by symmetry we have $\bar{x} = \dfrac{L}{2}$ and $\bar{y} = \dfrac{M_x}{m} = \dfrac{4kL/9\pi}{kL/4} = \dfrac{16}{9\pi}$.

28. Find I_x, I_y, I_0, $\bar{\bar{x}}$, and $\bar{\bar{y}}$ for the lamina with density $\rho = ky$ and
bounded by the graphs of $y = \sqrt{a^2 - x^2}$ and $y = 0$.

Solution

Since $\rho = ky$, we have

$m = \int_{-a}^a \int_0^{\sqrt{a^2 - x^2}} ky\, dy\, dx = 2k\int_0^a \left[\dfrac{y^2}{2}\right]_0^{\sqrt{a^2 - x^2}} dx$

$\quad = k\int_0^a (a^2 - x^2)\, dx = k\left[a^2 x - \dfrac{x^3}{3}\right]_0^a = \dfrac{2ka^3}{3}$

Furthermore,

$$I_x = \iint\limits_R y^2 \rho \, dA = 2 \int_0^a \int_0^{\sqrt{a^2-x^2}} ky^3 \, dy \, dx = \frac{2k}{4} \int_0^a (a^2 - x^2)^2 \, dx$$

$$= \frac{k}{2} \int_0^a (a^4 - 2a^2x^2 + x^4) \, dx = \frac{k}{2} \left[a^4 x - \frac{2a^2 x^3}{3} + \frac{x^5}{5} \right]_0^a$$

$$= \frac{k}{2} \left(a^5 - \frac{2a^5}{3} + \frac{a^5}{5} \right) = \frac{8ka^5}{30} = \frac{4ka^5}{15}$$

and

$$I_y = \iint\limits_R x^2 \rho \, dA = 2 \int_0^a \int_0^{\sqrt{a^2-x^2}} kx^2 y \, dy \, dx$$

$$= k \int_0^a x^2 (a^2 - x^2) \, dx$$

$$= k \left[\frac{a^2 x^3}{3} - \frac{x^5}{5} \right]_0^a = \frac{2ka^5}{15}$$

Therefore

$$I_0 = I_x + I_y = \frac{4ka^5}{15} + \frac{2ka^5}{15} = \frac{2ka^5}{5}$$

Finally,

$$\bar{\bar{x}} = \sqrt{\frac{I_y}{m}} = \sqrt{\frac{2ka^5/15}{2ka^3/3}} = \sqrt{\frac{a^2}{5}} = \frac{\sqrt{5}a}{5}$$

$$\bar{\bar{y}} = \sqrt{\frac{I_x}{m}} = \sqrt{\frac{4ka^5/15}{2ka^3/3}} = \sqrt{\frac{2a^2}{5}} = \frac{\sqrt{10}a}{5}$$

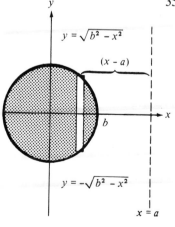

35. Find the moment of inertia of a uniform circular disk $x^2 + y^2 = b^2$ about the line $x = a$ $(a > b)$.

Solution

$$I = \iint\limits_R (\text{distance})^2 \text{mass}$$

$$= \int_{-r}^r \int_{-\sqrt{b^2-x^2}}^{\sqrt{b^2-x^2}} (x - a)^2 (k) \, dy \, dx$$

$$= 2k \int_0^\pi \int_0^b (r \cos \theta - a)^2 \, r \, dr \, d\theta \qquad \text{Double integral in polar coordinates}$$

$$= 2k \int_0^\pi \int_0^b (r^3 \cos^2 \theta - 2r^2 \cos \theta + ra^2) \, dr \, d\theta$$

$$= 2k \int_0^\pi \left[\frac{b^4}{4} \cos^2 \theta - \frac{2}{3} b^3 \cos \theta + \frac{a^2 b^2}{2} \right] d\theta$$

$$= 2k \left[\frac{b^4}{8} \left(\theta + \frac{1}{2} \sin 2\theta \right) - \frac{2b^3}{3} \sin \theta + \frac{a^2 b^2}{2} \theta \right]_0^\pi$$

$$= 2k \left[\frac{b^4 \pi}{8} + \frac{a^2 b^2 \pi}{2} \right] = \frac{kb^2 \pi}{4} (b^2 + 4a^2)$$

15.6
Surface area

3. Find the area of the surface $f(x, y) = 8 + 2x + 2y$ over the region $R = \{(x, y): x^2 + y^2 \le 4\}$.

    ### Solution

    The first partial derivatives of f are

    $$f_x(x, y) = 2 \qquad \text{and} \qquad f_y(x, y) = 2$$

    and from the formula for surface area we have

    $$\sqrt{1 + [f_x(x, y)]^2 + [f_y(x, y)]^2} = 3$$

    Therefore, the surface area is given by

    $$S = \iint_R 3 \, dA = 3 \iint_R dA = 3(\text{area of } R) = 3(4\pi) = 12\pi$$

15. Find the area of the surface $f(x, y) = \sqrt{x^2 + y^2}$ over the region $R = \{(x, y): 0 \le f(x, y) \le 1\}$.

    ### Solution

    We first observe that we are to find the surface area of that part of the cone $f(x, y) = \sqrt{x^2 + y^2}$ inside the cylinder $x^2 + y^2 = 1$. Therefore, the region R in the xy-plane is a circle of radius 1 centered at the origin. Furthermore, the first partial derivatives of f are

    $$f_x(x, y) = \frac{x}{\sqrt{x^2 + y^2}} \qquad \text{and} \qquad f_y(x, y) = \frac{y}{\sqrt{x^2 + y^2}}$$

    and from the formula for surface area, we have

    $$\sqrt{1 + [f_x(x, y)]^2 + [f_y(x, y)]^2} = \sqrt{1 + \frac{x^2 + y^2}{x^2 + y^2}} = \sqrt{2}$$

    Therefore, the surface area is given by

    $$S = \iint_R \sqrt{2} \, dA = \sqrt{2} \iint_R dA = \sqrt{2}(\text{area of } R) = \sqrt{2}\pi$$

19. Find the surface area of the solid of intersection of the cylinders $x^2 + z^2 = a^2$ and $y^2 + z^2 = a^2$.

Solution

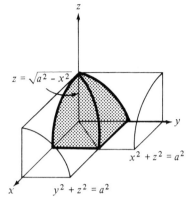

$z = \sqrt{a^2 - x^2}$

$x^2 + z^2 = a^2$

$y^2 + z^2 = a^2$

In the accompanying figure we show the surface in the first octant. We divide this surface into two equal parts by the plane $y = x$ and thus find $\frac{1}{16}$ of the total surface area. Therefore, we find the area of the surface $z = \sqrt{a^2 - x^2}$ over the triangle bounded by $y = 0$, $y = x$, and $x = a$.

$$\frac{\partial z}{\partial x} = \frac{-x}{\sqrt{a^2 - x^2}} \qquad \frac{\partial z}{\partial y} = 0$$

Therefore

$$S = 16 \int_0^a \int_0^x \sqrt{1 + \left(\frac{\partial z}{\partial x}\right)^2 + \left(\frac{\partial z}{\partial y}\right)^2}\ dy\ dx$$

$$= 16 \int_0^a \int_0^x \sqrt{1 + \frac{x^2}{a^2 - x^2}}\ dy\ dx$$

$$= 16 \int_0^a \int_0^x \frac{a}{\sqrt{a^2 - x^2}}\ dy\ dx$$

$$= 16a \int_0^a \frac{x}{\sqrt{a^2 - x^2}}\ dx = \left[-16a\sqrt{a^2 - x^2}\right]_0^a$$

$$= 16a^2$$

23. Set up the double integral which gives the area of the surface on the graph of $f(x, y) = e^{-x} \sin y$ over the region $R = \{(x, y): x^2 + y^2 \leq 4\}$.

Solution

The first partial derivatives of f are

$$f_x(x, y) = -e^{-x} \sin y \qquad \text{and} \qquad f_y(x, y) = e^{-x} \cos y$$

and from the formula for surface area, we have

$$\sqrt{1 + [f_x(x, y)]^2 + [f_y(x, y)]^2} = \sqrt{1 + e^{-2x} \sin^2 y + e^{-2x} \cos^2 y}$$
$$= \sqrt{1 + e^{-2x}}$$

We integrate over the circle $x^2 + y^2 = 4$.

Constant bounds for x: $\qquad\qquad -2 \leq x \leq 2$

Variable bounds for y: $\qquad -\sqrt{4 - x^2} \leq y \leq \sqrt{4 - x^2}$

Therefore

$$S = \int_{-2}^2 \int_{-\sqrt{4-x^2}}^{\sqrt{4-x^2}} \sqrt{1 + e^{-2x}}\ dy\ dx$$

15.7
Triple integrals and applications

9.   Evaluate $\int_0^2 \int_{-\sqrt{4-x^2}}^{\sqrt{4-x^2}} \int_0^{x^2} x \, dz \, dy \, dx$.

Solution

$$\int_0^2 \int_{-\sqrt{4-x^2}}^{\sqrt{4-x^2}} \int_0^{x^2} x \, dz \, dy \, dx$$

$$= \int_0^2 \int_{-\sqrt{4-x^2}}^{\sqrt{4-x^2}} [xz]_0^{x^2} \, dy \, dx$$

$$= \int_0^2 \int_{-\sqrt{4-x^2}}^{\sqrt{4-x^2}} x^3 \, dy \, dx$$

$$= \int_{-\pi/2}^{\pi/2} \int_0^2 (r \cos \theta)^3 \, r \, dr \, d\theta \qquad \text{Double integral in polar coordinates}$$

$$= \frac{32}{5} \int_{-\pi/2}^{\pi/2} \cos^3 \theta \, d\theta$$

$$= \frac{32}{5} \left[\sin \theta - \frac{1}{3} \sin^3 \theta \right]_{-\pi/2}^{\pi/2} = \frac{128}{15}$$

13.   Sketch the solid region whose volume is given by the triple integral

$$\int_0^1 \int_y^1 \int_0^{\sqrt{1-y^2}} dz \, dx \, dy$$

and rewrite the integral using $dz \, dy \, dx$ as the order of integration.

Solution

We have

Constant bounds on y: $0 \le y \le 1$

Variable bounds on x: $y \le x \le 1$

Variable bounds on z: $0 \le z \le \sqrt{1 - y^2}$

From the upper bound on z, we have

$$z = \sqrt{1 - y^2} \qquad \text{or} \qquad y^2 + z^2 = 1$$

a cylinder of radius 1 with the x-axis as its axis. Therefore, the triple integral gives the volume of the solid in the first octant bounded by the graphs of $z = \sqrt{1 - y^2}$, $z = 0$, $x = y$, and $x = 1$. From the accompanying sketch of the solid we observe the following bounds when we change the order of integration to $dz \, dy \, dx$.

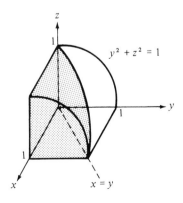

Constant bounds on x: $0 \le x \le 1$

Variable bounds on y: $0 \le y \le x$

Variable bounds on z: $0 \le z \le \sqrt{1 - y^2}$

Therefore the integral is $\int_0^1 \int_0^x \int_0^{\sqrt{1-y^2}} dz \, dy \, dx$.

19. Use a triple integral to find the volume of the solid bounded by the cylinders $z = 4 - x^2$ and $y = 4 - x^2$ in the first octant.

Solution

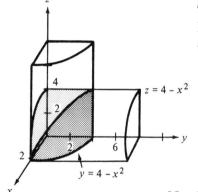

In the first octant we have $0 \le z \le 4 - x^2$, $0 \le y \le 4 - x^2$, and $0 \le x \le 2$. Therefore the volume is

$$V = \int_0^2 \int_0^{4-x^2} \int_0^{4-x^2} dz \, dy \, dx = \int_0^2 \int_0^{4-x^2} (4 - x^2) \, dy \, dx$$

$$= \int_0^2 (4 - x^2)^2 \, dx = \int_0^2 (16 - 8x^2 + x^4) \, dx$$

$$= \left[16x - \frac{8x^3}{3} + \frac{x^5}{5} \right]_0^2 = 32 - \frac{64}{3} + \frac{32}{5} = \frac{256}{15}$$

25. Find the mass and center of mass of the solid with density $\rho(x, y, z) = kxy$ and bounded by the graphs of $x = 0$, $x = b$, $y = 0$, $y = b$, $z = 0$, and $z = b$.

Solution

The mass of the cube is

$$m = \int_0^b \int_0^b \int_0^b kxy \, dz \, dy \, dx = k \int_0^b \int_0^b bxy \, dy \, dx$$

$$= kb \int_0^b \left[\frac{xy^2}{2} \right]_0^b dx = \frac{kb^3}{2} \int_0^b x \, dx = \frac{kb^3}{4} [x^2]_0^b = \frac{kb^5}{4}$$

Furthermore,

$$M_{yz} = \int_0^b \int_0^b \int_0^b x(kxy) \, dz \, dy \, dx = k \int_0^b \int_0^b x^2 y (b) \, dy \, dx$$

$$= kb \int_0^b \left[\frac{x^2 y^2}{2} \right]_0^b dx = \frac{kb^3}{2} \int_0^b x^2 \, dx = \frac{kb^3}{6} [x^3]_0^b = \frac{kb^6}{6}$$

By the symmetry of the cube and of $\rho = kxy$, we have $M_{xz} = M_{yz} = kb^6/6$. Moreover,

$$M_{xy} = \int_0^b \int_0^b \int_0^b z(kxy) \, dz \, dy \, dx = k \int_0^b \int_0^b \left[\frac{xyz^2}{2} \right]_0^b dy \, dx$$

$$= \frac{kb^2}{2} \int_0^b \int_0^b xy \, dy \, dx = \frac{kb^2}{2} \int_0^b \left[\frac{xy^2}{2} \right]_0^b dx = \frac{kb^4}{4} \int_0^b x \, dx$$

$$= \frac{kb^4}{4} \left[\frac{x^2}{2} \right]_0^b = \frac{kb^6}{8}$$

Finally

$$\bar{x} = \frac{M_{yz}}{m} = \frac{kb^6/6}{kb^5/4} = \frac{2b}{3}$$

$$\bar{y} = \bar{x} = \frac{2b}{3}$$

$$\bar{z} = \frac{M_{xy}}{m} = \frac{kb^6/8}{kb^5/4} = \frac{b}{2}$$

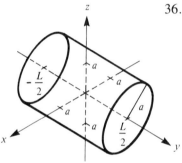

Right circular cylinder

36. Verify the moments of inertia

$$I_x = I_z = \frac{1}{12} m (3a^2 + L^2) \qquad \text{and} \qquad I_y = \frac{1}{2} ma^2$$

for the cylinder of uniform density in the accompanying figure.

Solution

The solid is a right circular cylinder of radius a, length L, and uniform density $\rho(x, y, z) = k$. Thus the mass of the cylinder is

$$m = k(\text{volume}) = k\pi a^2 L$$

Now

$$I_x = \int_{-a}^a \int_{-\sqrt{a^2-x^2}}^{\sqrt{a^2-x^2}} \int_{-L/2}^{L/2} (y^2 + z^2) k \, dy \, dz \, dx$$

$$= 8k \int_0^a \int_0^{\sqrt{a^2-x^2}} \int_0^{L/2} (y^2 + z^2) \, dy \, dz \, dx \qquad \text{By symmetry}$$

$$= \frac{kL}{3} \int_0^a \int_0^{\sqrt{a^2-x^2}} (L^2 + 12z^2) \, dz \, dx$$

$$= \frac{kL}{3} \int_0^a [L^2\sqrt{a^2 - x^2} + 4(a^2 - x^2)^{3/2}] \, dx$$

Now let $x = a \sin \theta$. Then $\sqrt{a^2 - x^2} = a \cos \theta$, $dx = a \cos \theta \, d\theta$, and

$$I_x = \frac{ka^2L^3}{3} \int_0^{\pi/2} \cos^2 \theta \, d\theta + \frac{4ka^4L}{3} \int_0^{\pi/2} \cos^4 \theta \, d\theta$$

By Wallis's Formula (Section 8.3) we have

$$I_x = \frac{ka^2L^3}{3}\left(\frac{1}{2}\right)\left(\frac{\pi}{2}\right) + \frac{4ka^4L}{3}\left(\frac{1}{2}\right)\left(\frac{3}{4}\right)\left(\frac{\pi}{2}\right)$$

$$= \frac{k\pi a^2 L}{12}(L^2 + 3a^2) = \frac{1}{12}m(3a^2 + L^2)$$

By symmetry $I_x = I_z$. To find I_y, we change only the integrand in the triple integral given above and obtain

$$I_y = 8k\int_0^a\int_0^{\sqrt{a^2-x^2}}\int_0^{L/2}(x^2 + z^2)\,dy\,dz\,dx$$

Proceeding with integrations similar to those given above yields

$$I_y = \frac{k\pi a^4 L}{2} = k\pi a^2 L\left(\frac{a^2}{2}\right) = \frac{1}{2}ma^2$$

15.8
Triple integrals in cylindrical and spherical coordinates

6. Evaluate the triple integral $\int_0^{\pi/2}\int_0^\pi\int_0^{\sin\theta}(2\cos\phi)\rho^2\,d\rho\,d\theta\,d\phi$.

Solution

$$\int_0^{\pi/2}\int_0^\pi\int_0^{\sin\theta}(2\cos\phi)\rho^2\,d\rho\,d\theta\,d\phi$$

$$= 2\int_0^{\pi/2}\int_0^\pi\left[\frac{1}{3}\rho^3\cos\phi\right]_0^{\sin\theta}d\theta\,d\phi$$

$$= \frac{2}{3}\int_0^{\pi/2}\int_0^\pi\sin^3\theta\cos\phi\,d\theta\,d\phi$$

$$= \frac{2}{3}\int_0^{\pi/2}\left[\left(-\cos\theta + \frac{1}{3}\cos^3\theta\right)\cos\phi\right]_0^\pi d\phi$$

$$= \frac{8}{9}\int_0^{\pi/2}\cos\phi\,d\phi = \frac{8}{9}[\sin\phi]_0^{\pi/2} = \frac{8}{9}$$

9. Sketch the solid region whose volume is given by the integral

$$\int_0^{\pi/2}\int_0^3\int_0^{e^{-r^2}}r\,dz\,dr\,d\theta$$

and evaluate the integral.

Solution

We first observe that the triple integral is written in terms of cylindrical coordinates. From the limits of integration we have

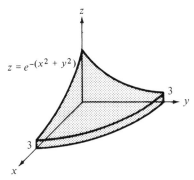

Constant bounds on θ: $0 \le \theta \le \pi/2$

Constant bounds on r: $0 \le r \le 3$

Variable bounds on z: $0 \le z \le e^{-r^2}$

The limits on r determine a circular cylinder of radius 3 having the z-axis as its axis. The limits on θ and z restrict us to the first-octant portion of the cylinder bounded by the surface

$$z = e^{-r^2} = e^{-(x^2+y^2)}$$

The solid is shown in the accompanying figure. The value of the integral is given by

$$\int_0^{\pi/2}\int_0^3\int_0^{e^{-r^2}} r\; dz\; dr\; d\theta = \int_0^{\pi/2}\int_0^3 re^{-r^2}\; dr\; d\theta$$

$$= -\frac{1}{2}(e^{-9} - 1)\int_0^{\pi/2} d\theta = \frac{\pi}{4}(1 - e^{-9})$$

15.   Convert the integral

$$\int_{-a}^{a}\int_{-\sqrt{a^2-x^2}}^{\sqrt{a^2-x^2}}\int_{a}^{a+\sqrt{a^2-x^2-y^2}} x\; dz\; dy\; dx$$

from rectangular coordinates to both cylindrical and spherical coordinates and evaluate the simplest integral.

Solution

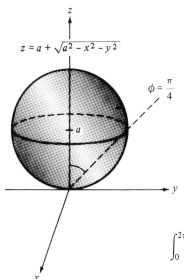

We first observe that the solid S over which we are integrating is a hemisphere, the top half of the sphere of radius a and center $(0, 0, a)$. Projecting the solid onto the xy-plane forms a circle of radius a. Thus we have

Constant bounds on θ: $0 \le \theta \le 2\pi$

Constant bounds on r: $0 \le r \le a$

Variable bounds on z: $a \le z \le a + \sqrt{a^2 - x^2 - y^2}$

$$= a + \sqrt{a^2 - r^2}$$

Finally, since $x = r\cos\theta$, we obtain the integral in cylindrical coordinates:

$$\int_0^{2\pi}\int_0^a\int_a^{a+\sqrt{a^2-r^2}} (r\cos\theta)r\; dz\; dr\; d\theta = \int_0^{2\pi}\int_0^a\int_a^{a+\sqrt{a^2-r^2}} r^2\cos\theta\; dz\; dr\; d\theta$$

To write the integral in spherical coordinates, we first write the equation of the sphere in spherical coordinates:

$$z = a + \sqrt{a^2 - x^2 - y^2}$$

$$x^2 + y^2 + (z - a)^2 = a^2$$

$$x^2 + y^2 + z^2 - 2az + a^2 = a^2$$

$$\rho - 2a(\rho \cos \phi) = 0 \qquad \text{Since } z = \rho \cos \phi$$

$$\rho = 2a \cos \phi$$

We next write the equation of the plane $z = a$ in spherical coordinates and obtain

$$z = a$$

$$\rho \cos \phi = a$$

$$\rho = a \sec \phi$$

From the accompanying figure we see that the bounds on ϕ are $0 \leq \phi \leq \pi/4$. Now since $x = \rho \sin \phi \cos \theta$, we write the integral in spherical coordinates:

$$\int_0^{\pi/4} \int_0^{2\pi} \int_{a \sec \phi}^{2a \cos \phi} (\rho \sin \phi \cos \theta)(\rho^2 \sin \phi) \, d\rho \, d\theta \, d\phi$$

$$= \frac{1}{4} \int_0^{\pi/4} \int_0^{2\pi} \sin^2 \phi \cos \theta [(2a \cos \phi)^4 - (a \sec \phi)^4] \, d\theta \, d\phi$$

$$= \frac{a^4}{4} \int_0^{\pi/4} [(16 \sin^2 \phi \cos^4 \phi - \sin^2 \phi \sec^4 \phi) \sin \theta]_0^{2\pi} \, d\phi$$

$$= \frac{a^4}{4} \int_0^{\pi/4} 0 \, d\phi = 0$$

17. Use a triple integral in cylindrical coordinates to find the volume of a right circular cone with altitude h and a base of radius r_0.

Solution

In the accompanying figure the line $z = -(h/r_0)y + h$ can be considered to be the generating curve for the cone whose equation is of the form

$$x^2 + y^2 = [r(z)]^2$$

where $r(z) = y(z) = -(r_0/h)(z - h)$. Thus $[y(z)]^2 = x^2 + y^2$, and we have

$$z = -\frac{hy}{r_0} + h = h - \frac{h}{r_0}\sqrt{x^2 + y^2}$$

Therefore, in cylindrical coordinates we obtain

$$V = 4 \int_0^{\pi/2} \int_0^{r_0} \int_0^{h - (hr/r_0)} r \, dz \, dr \, d\theta$$

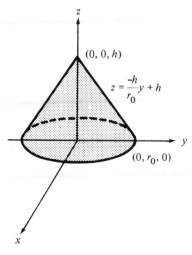

$$= 4 \int_0^{\pi/2} \int_0^{r_0} \left(rh - \frac{r^2 h}{r_0} \right) dr \, d\theta = 4h \int_0^{\pi/2} \left[\frac{r^2}{2} - \frac{r^3}{3r_0} \right]_0^{r_0} d\theta$$

$$= 4h \int_0^{\pi/2} \left[\frac{r_0^2}{2} - \frac{r_0^2}{3} \right] d\theta = \frac{4hr_0^2}{6} \int_0^{\pi/2} d\theta$$

$$= \frac{2hr_0^2}{3} \left(\frac{\pi}{2} \right) = \frac{\pi r_0^2 h}{3}$$

25. Use a triple integral in cylindrical coordinates to find the volume of the solid inside both the sphere $x^2 + y^2 + z^2 = a^2$ and the cylinder $[x - (a/2)]^2 + y^2 = (a/2)^2$.

Solution

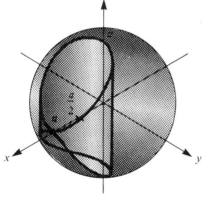

The accompanying figure shows the sphere and cylinder. We will find the volume of the solid by using the cylindrical coordinate system. Thus the equation of the cylinder is given by

$$r = a \cos \theta \qquad (-\pi/2 \le \theta \le \pi/2)$$

and the equation of the sphere is given by

$$r^2 + z^2 = a^2 \qquad \text{or} \qquad z = \pm\sqrt{a^2 - r^2}$$

Therefore the volume is

$$V = \int_{-\pi/2}^{\pi/2} \int_0^{a \cos \theta} \int_{-\sqrt{a^2 - r^2}}^{\sqrt{a^2 - r^2}} r \, dz \, dr \, d\theta$$

$$= 4 \int_0^{\pi/2} \int_0^{a \cos \theta} r\sqrt{a^2 - r^2} \, dr \, d\theta$$

$$= -\frac{4}{3} \int_0^{\pi/2} [(a^2 - a^2 \cos^2 \theta)^{3/2} - a^3] \, d\theta$$

$$= \frac{4a^3}{3} \int_0^{\pi/2} (1 - \sin^3 \theta) \, d\theta = \frac{4a^3}{3} \left(\frac{\pi}{2} - \frac{2}{3} \right)$$

27. Use a triple integral in spherical coordinates to find the volume of the solid inside the sphere $x^2 + y^2 + z^2 = 1$ and outside the cone $z^2 = x^2 + y^2$.

Solution

From the accompanying figure, we can determine that

$$0 \le \rho \le 1, \qquad 0 \le \theta \le 2\pi, \qquad \frac{\pi}{4} \le \phi \le \frac{\pi}{2}$$

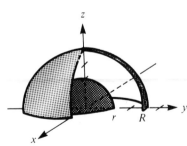

Therefore the volume is

$$V = 8 \int_{\pi/4}^{\pi/2} \int_0^{\pi/2} \int_0^1 \rho^2 \sin\phi \, d\rho \, d\theta \, d\phi = 8 \int_{\pi/4}^{\pi/2} \int_0^{\pi/2} \frac{1}{3} \sin\phi \, d\theta \, d\phi$$

$$= \frac{8}{3} \int_{\pi/4}^{\pi/2} [\theta \sin\phi]_0^{\pi/2} \, d\phi = \frac{4\pi}{3} \int_{\pi/4}^{\pi/2} \sin\phi \, d\phi = -\frac{4\pi}{3} [\cos\phi]_{\pi/4}^{\pi/2}$$

$$= -\frac{4\pi}{3} \left(0 - \frac{\sqrt{2}}{2} \right) = \frac{2\sqrt{2}\pi}{3}$$

32. Use spherical coordinates to find the center of mass of the solid of uniform density lying between two concentric hemispheres of radii r and R, where $r < R$.

Solution

First, without loss of generality, we can position the hemispheres with their centers at the origin and bases on the xy-coordinate plane. Then, by symmetry, $\bar{x} = \bar{y} = 0$. Since the solid has uniform density k, the mass is given by

$$m = k(\text{volume}) = k\left(\frac{2}{3}\pi R^3 - \frac{2}{3}\pi r^3 \right) = \frac{2}{3}k\pi(R^3 - r^3)$$

Now, by symmetry, we have

$$M_{xy} = \iiint_Q z(\text{density}) \, dV$$

$$= 4k \int_0^{\pi/2} \int_0^{\pi/2} \int_r^R \rho^3 \cos\phi \sin\phi \, d\rho \, d\theta \, d\phi$$

$$= \frac{1}{2}k(R^4 - r^4) \int_0^{\pi/2} \int_0^{\pi/2} \sin 2\phi \, d\theta \, d\phi$$

$$= \frac{1}{4}k\pi(R^4 - r^4) \int_0^{\pi/2} \sin 2\phi \, d\phi$$

$$= -\frac{1}{8}k\pi(R^4 - r^4)[\cos 2\phi]_0^{\pi/2} = \frac{1}{4}k\pi(R^4 - r^4)$$

Therefore

$$\bar{z} = \frac{M_{xy}}{m} = \frac{k\pi(R^4 - r^4)/4}{2k\pi(R^3 - r^3)/3} = \frac{3(R^4 - r^4)}{8(R^3 - r^3)}$$

Review Exercises for Chapter 15

1. Evaluate $\int_0^1 \int_0^{1+x} (3x + 2y) \, dy \, dx$.

Solution

$$\int_0^1 \int_0^{1+x} (3x + 2y) \, dy \, dx = \int_0^1 [3xy + y^2]_0^{1+x} \, dx$$

$$= \int_0^1 (3x + 3x^2 + 1 + 2x + x^2) \, dx$$

$$= \int_0^1 (4x^2 + 5x + 1) \, dx$$

$$= \left[\frac{4x^3}{3} + \frac{5x^2}{2} + x \right]_0^1 = \frac{4}{3} + \frac{5}{2} + 1 = \frac{29}{6}$$

7. Evaluate

$$\int_0^h \int_0^x \sqrt{x^2 + y^2} \, dy \, dx$$

by using the coordinate system that makes the integration easiest.

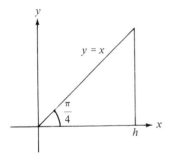

Solution

We choose to use polar coordinates since $r = \sqrt{x^2 + y^2}$. To rewrite the limits, we first find the equation of the line $x = h$ in polar coordinates.

$$x = h$$

$$r \cos \theta = h$$

$$r = h\left(\frac{1}{\cos \theta} \right) = h \sec \theta$$

Therefore

$$\int_0^h \int_0^x \sqrt{x^2 + y^2} \, dy \, dx = \int_0^{\pi/4} \int_0^{h \sec \theta} (r)r \, dr \, d\theta = \frac{h^3}{3} \int_0^{\pi/4} \sec^3 \theta \, d\theta$$

$$= \frac{h^3}{3} \left[\frac{\sec \theta \tan \theta}{2} + \frac{1}{2} \ln |\sec \theta + \tan \theta| \right]_0^{\pi/4}$$

$$= \frac{h^3}{6} [\sqrt{2} + \ln (\sqrt{2} + 1)]$$

17. If R is the larger region between the circle $x^2 + y^2 = 25$ and the line $x = 3$, write the limits for the double integral $\iint_R f(x, y) \, dA$ for both orders of integration. Compute the area by letting $f(x, y) = 1$.

Solution

$$\iint_R f(x, y) \, dA = \int_{-5}^{3} \int_{-\sqrt{25 - x^2}}^{\sqrt{25 - x^2}} f(x, y) \, dy \, dx$$

$$= \int_{-5}^{-4} \int_{-\sqrt{25 - y^2}}^{\sqrt{25 - y^2}} f(x, y) \, dx \, dy$$

$$+ \int_{-4}^{4} \int_{-\sqrt{25 - y^2}}^{3} f(x, y) \, dx \, dy$$

$$+ \int_{4}^{5} \int_{-\sqrt{25 - y^2}}^{\sqrt{25 - y^2}} f(x, y) \, dx \, dy$$

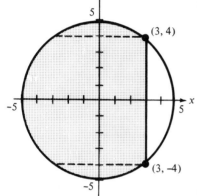

The area of R is

$$A = 2 \int_{-5}^{3} \int_{0}^{\sqrt{25 - x^2}} dy \, dx = 2 \int_{-5}^{3} \sqrt{25 - x^2} \, dx$$

$$= 2 \left(\frac{1}{2} \right) \left[x\sqrt{25 - x^2} + 25 \arcsin \left(\frac{x}{5} \right) \right]_{-5}^{3}$$

$$= 3(4) + 25 \arcsin \left(\frac{3}{5} \right) - 0 - 25 \left(-\frac{\pi}{2} \right)$$

$$= 12 + \frac{25\pi}{2} + 25 \arcsin \left(\frac{3}{5} \right) \approx 66.36$$

(Note: The area of the entire circle is $25\pi \approx 78.54$.)

33. Show that the center of mass of the portion of the solid of constant density in the first octant bounded by $x^2 + y^2 + z^2 = a^2$ is $(3a/8, 3a/8, 3a/8)$.

Solution

The solid S is the first-octant portion of a sphere of radius a and therefore, because of its symmetry, the coordinates of the center of mass are equal. Let k be the constant density; then the mass is

$$m = k(\text{volume}) = k \left(\frac{1}{8} \right) \left(\frac{4}{3} \pi a^3 \right) = \frac{k}{6} \pi a^3$$

Now

$$M_{xy} = \iiint_S z(\text{density}) \, dS \qquad \text{Rectangular coordinates}$$

$$= k \int_{0}^{\pi/2} \int_{0}^{\pi/2} \int_{0}^{a} (\rho \cos \phi)(\rho^2 \sin \phi) \, d\rho \, d\theta \, d\phi \qquad \text{Spherical coordinates}$$

$$= \frac{ka^4}{4} \int_0^{\pi/2} \int_0^{\pi/2} \cos\phi \, \sin\phi \, d\theta \, d\phi = \frac{k\pi a^4}{8} \int_0^{\pi/2} \cos\phi \, \sin\phi \, d\phi$$

$$= \frac{k\pi a^4}{8} \left[\frac{1}{2} \sin^2\phi \right]_0^{\pi/2} = \frac{k\pi a^4}{16}$$

Therefore

$$\bar{x} = \bar{y} = \bar{z} = \frac{M_{xy}}{m} = \frac{k\pi a^4/16}{k\pi a^3/6} = \frac{3a}{8}$$

35. Find the surface area of the solid bounded by $z = 16 - x^2 - y^2$ and $z = 0$. (Use polar coordinates.)

Solution

$$z = 16 - x^2 - y^2 \qquad \frac{\partial z}{\partial x} = -2x \qquad \frac{\partial z}{\partial y} = -2y$$

$$S = \iint_R \sqrt{1 + \left(\frac{\partial z}{\partial x}\right)^2 + \left(\frac{\partial z}{\partial y}\right)^2} \, dy \, dx$$

$$= \int_{-4}^{4} \int_{-\sqrt{16-x^2}}^{\sqrt{16-x^2}} \sqrt{1 + 4x^2 + 4y^2} \, dy \, dx$$

$$= 4 \int_0^4 \int_0^{\sqrt{16-x^2}} \sqrt{1 + 4(x^2 + y^2)} \, dy \, dx$$

$$= \frac{1}{2} \int_0^{\pi/2} \int_0^4 \sqrt{1 + 4r^2}(8r) \, dr \, d\theta$$

$$= \frac{1}{3} \int_0^{\pi/2} (65^{3/2} - 1) \, d\theta = \frac{\pi}{6}(65^{3/2} - 1)$$

16 *Vector analysis*

Vector fields

7. Sketch several representative vectors in the vector field

$$\mathbf{F}(x, y) = -x\mathbf{i} - y\mathbf{j}$$

Solution

We will plot vectors of equal magnitude and, in this case, they lie along circles given by

$$\|\mathbf{F}(x, y)\| = \sqrt{(-x)^2 + (-y)^2} = c \quad \rightarrow \quad x^2 + y^2 = c^2$$

For $c = 1$, we sketch several vectors $-x\mathbf{i} - y\mathbf{j}$ of magnitude 1 on the circle given by $x^2 + y^2 = 1$. For $c = 4$, we sketch several vectors $-x\mathbf{i} - y\mathbf{j}$ of magnitude 2 on the circle given by $x^2 + y^2 = 4$. (See the accompanying figure.)

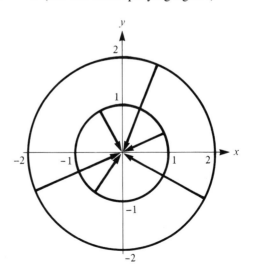

27. Determine if $\mathbf{F}(x, y) = xe^{x^2y}(2y\mathbf{i} + x\mathbf{j})$ is conservative and, if so, find the potential function $f(x, y)$.

Solution

Since $\mathbf{F}(x, y) = 2xye^{x^2y}\mathbf{i} + x^2e^{x^2y}\mathbf{j}$, it follows from Theorem 16.1, that $\mathbf{F}$ is conservative since

$$\frac{\partial}{\partial y}[2xye^{x^2y}] = 2x^3ye^{x^2y} + 2xe^{x^2y} = \frac{\partial}{\partial x}[x^2e^{x^2y}]$$

Now, if f is a function such that $\nabla f(x, y) = f_x(x, y)\mathbf{i} + f_y(x, y)\mathbf{j}$, then we have

$$f_x(x, y) = 2xye^{x^2y} \qquad \text{and} \qquad f_y(x, y) = x^2e^{x^2y}$$

To reconstruct the function f from these two partial derivatives, we integrate $f_x(x, y)$ with respect to x and $f_y(x, y)$ with respect to y as follows

$$f(x, y) = \int f_x(x, y)\, dx = \int 2xye^{x^2y}\, dx = e^{x^2y} + g(y)$$

$$f(x, y) = \int f_y(x, y)\, dy = \int x^2e^{x^2y}\, dy = e^{x^2y} + h(x)$$

Now, we reconcile the differences between these two expressions for $f(x, y)$ by determining that $g(y) = h(x) = C$. Therefore, we have

$$f(x, y) = e^{x^2y} + C$$

35. Determine if $\mathbf{F}(x, y) = (\sin y)\mathbf{i} - (x\cos y)\mathbf{j}$ is conservative and, if so, find the potential function $f(x, y)$.

Solution

Using Theorem 16.1 it follows that $\mathbf{F}$ is **not** conservative since

$$\frac{\partial}{\partial y}[\sin y] = \cos y \neq -\cos y = \frac{\partial}{\partial x}[-x\cos y]$$

39. Determine if $\mathbf{F}(x, y, z) = (1/y)\mathbf{i} - (x/y^2)\mathbf{j} + (2z - 1)\mathbf{k}$ is conservative and if so find the potential function $f(x, y, z)$.

Solution

Using Theorem 16.2, we let

$$\mathbf{F}(x, y, z) = \frac{1}{y}\mathbf{i} - \frac{x}{y^2}\mathbf{j} + (2z - 1)\mathbf{k} = M\mathbf{i} + N\mathbf{j} + P\mathbf{k}$$

Since

$$\frac{\partial P}{\partial y} = 0 = \frac{\partial N}{\partial z}, \qquad \frac{\partial P}{\partial x} = 0 = \frac{\partial M}{\partial z}, \qquad \frac{\partial N}{\partial x} = -\frac{1}{y^2} = \frac{\partial M}{\partial y}$$

it follows that $\mathbf{F}$ is conservative. Now, if f is a function such that $\mathbf{F}(x, y, z) = \nabla f(x, y, z)$, then

$$f_x(x, y, z) = \frac{1}{y}, \qquad f_y(x, y, z) = -\frac{x}{y^2}, \qquad f_z(x, y, z) = 2z - 1$$

and by integrating with respect to x, y, and z separately, we obtain

$$f(x, y, z) = \int M \, dx = \int \frac{1}{y} \, dx = \frac{x}{y} + g(y, z)$$

$$f(x, y, z) = \int N \, dy = \int -\frac{x}{y^2} \, dy = \frac{x}{y} + h(x, z)$$

$$f(x, y, z) = \int P \, dz = \int (2z - 1) \, dz = z^2 - z + k(x, y)$$

By comparing these three versions of f, we conclude that

$$f(x, y, z) = \frac{x}{y} + z^2 - z + C$$

41.   Find the curl of the vector field $\mathbf{F}(x, y, z) = xyz\mathbf{i} + y\mathbf{j} + z\mathbf{k}$ at the point $(1, 2, 1)$.

Solution

By definition of curl, we have

$$\mathbf{curl} \, \mathbf{F}(x, y, z) = \begin{vmatrix} \mathbf{i} & \mathbf{j} & \mathbf{k} \\ \dfrac{\partial}{\partial x} & \dfrac{\partial}{\partial y} & \dfrac{\partial}{\partial z} \\ xyz & y & z \end{vmatrix}$$

$$= \begin{vmatrix} \dfrac{\partial}{\partial y} & \dfrac{\partial}{\partial z} \\ y & z \end{vmatrix} \mathbf{i} - \begin{vmatrix} \dfrac{\partial}{\partial x} & \dfrac{\partial}{\partial z} \\ xyz & z \end{vmatrix} \mathbf{j} + \begin{vmatrix} \dfrac{\partial}{\partial x} & \dfrac{\partial}{\partial y} \\ xyz & y \end{vmatrix} \mathbf{k}$$

$$= (0 - 0)\mathbf{i} - (0 - xy)\mathbf{j} + (0 - xz)\mathbf{k} = xy\mathbf{j} - xz\mathbf{k}$$

Therefore, $\mathbf{curl} \, \mathbf{F}(1, 2, 1) = 2\mathbf{j} - \mathbf{k}$.

45.   Find the curl of the vector field $\mathbf{F}(x, y, z) = \left(\arctan \dfrac{x}{y}\right)\mathbf{i} + (\ln \sqrt{x^2 + y^2}) \, \mathbf{j} + \mathbf{k}$.

Solution

$$\text{curl } \mathbf{F}(x, y, z) = \begin{vmatrix} \mathbf{i} & \mathbf{j} & \mathbf{k} \\ \dfrac{\partial}{\partial x} & \dfrac{\partial}{\partial y} & \dfrac{\partial}{\partial z} \\ \arctan \dfrac{x}{y} & \ln \sqrt{x^2 + y^2} & 1 \end{vmatrix}$$

$$= (0 - 0)\mathbf{i} - (0 - 0)\mathbf{j}$$

$$+ \left[\frac{x}{x^2 + y^2} - \frac{-x/y^2}{1 + (x^2/y^2)} \right]\mathbf{k} = \frac{2x}{x^2 + y^2}\mathbf{k}$$

51. Find **curl(curl F)** if $\mathbf{F}(x, y, z) = xyz\mathbf{i} + y\mathbf{j} + z\mathbf{k}$.

Solution

From Exercise 41 we have **curl** $\mathbf{F}(x, y, z) = xy\mathbf{j} - xz\mathbf{k}$. Thus

$$\text{curl}[\text{curl } \mathbf{F}(x, y, z)] = \begin{vmatrix} \mathbf{i} & \mathbf{j} & \mathbf{k} \\ \dfrac{\partial}{\partial x} & \dfrac{\partial}{\partial y} & \dfrac{\partial}{\partial z} \\ 0 & xy & -xz \end{vmatrix}$$

$$= (0 - 0)\mathbf{i} - (-z - 0)\mathbf{j} + (y - 0)\mathbf{k}$$

$$= z\mathbf{j} + y\mathbf{k}$$

16.2
Line integrals

9. Evaluate $\displaystyle\int_C (x^2 + y^2)\, ds$ along the line from $(0, 0)$ to $(1, 1)$.

Solution

Path C is given by $x = t$ and $y = t$ with $0 \leq t \leq 1$. Therefore $ds = \sqrt{1^2 + 1^2}\, dt = \sqrt{2}\, dt$, and we have

$$\int_C (x^2 + y^2)\, ds = \int_0^1 2t^2\sqrt{2}\, dt = \left[2\sqrt{2}\,\frac{t^3}{3} \right]_0^1 = \frac{2\sqrt{2}}{3}$$

11. Evaluate $\displaystyle\int_C (x^2 + y^2)\, ds$ counterclockwise around the triangle with vertices $(0, 0)$, $(1, 0)$, and $(0, 1)$.

Solution

Path C has parts as shown in the accompanying figure.

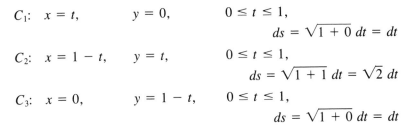

$$C_1: \quad x = t, \qquad\qquad y = 0, \qquad\qquad 0 \le t \le 1,$$
$$ds = \sqrt{1 + 0} \; dt = dt$$

$$C_2: \quad x = 1 - t, \qquad y = t, \qquad\qquad 0 \le t \le 1,$$
$$ds = \sqrt{1 + 1} \; dt = \sqrt{2} \; dt$$

$$C_3: \quad x = 0, \qquad\qquad y = 1 - t, \qquad 0 \le t \le 1,$$
$$ds = \sqrt{1 + 0} \; dt = dt$$

Therefore

$$\int_C (x^2 + y^2) \, ds$$

$$= \int_{C_1} t^2 \, dt + \int_{C_2} [(1 - t)^2 + t^2] \sqrt{2} \, dt + \int_{C_3} (1 - t)^2 \, dt$$

$$= \int_0^1 t^2 \, dt + \sqrt{2} \int_0^1 (1 - 2t + 2t^2) \, dt + \int_0^1 (1 - 2t + t^2) \, dt$$

$$= \left[\frac{t^3}{3} \right]_0^1 + \sqrt{2} \left[t - t^2 + \frac{2t^3}{3} \right]_0^1 + \left[t - t^2 + \frac{t^3}{3} \right]_0^1$$

$$= \frac{1}{3} + \sqrt{2} \left(\frac{2}{3} \right) + \frac{1}{3} = \frac{2}{3} (1 + \sqrt{2})$$

19.   Evaluate $\displaystyle\int_C (2x - y) \, dx + (x + 3y) \, dy$ along the parabolic path $x = t$ and $y = 2t^2$ from $(0, 0)$ to $(2, 8)$.

Solution

Path C is given by $x = t$, $y = 2t^2$, with $0 \le t \le 2$, $dx = dt$, and $dy = 4t \, dt$. Therefore

$$\int_C (2x - y) \, dx + (x + 3y) \, dy = \int_0^2 (2t - 2t^2) \, dt + (t + 6t^2) \, 4t \, dt$$

$$= \int_0^2 (2t + 2t^2 + 24t^3) \, dt$$

$$= \left[t^2 + \frac{2t^3}{3} + 6t^4 \right]_0^2 = 4 + \frac{16}{3} + 96$$

$$= \frac{316}{3}$$

23.   Evaluate

$$\int_C \mathbf{F} \cdot d\mathbf{r}$$

for $\mathbf{F}(x, y) = 3x\mathbf{i} + 4y\mathbf{j}$ and $\mathbf{r}(t) = (2 \cos t)\mathbf{i} + (2 \sin t)\mathbf{j}$ for $0 \le t \le \pi/2$.

Solution

Since $x = 2 \cos t$ and $y = 2 \sin t$, we have $d\mathbf{r} = (-2 \sin t\mathbf{i} + 2 \cos t\mathbf{j})\, dt$ and $\mathbf{F}(x, y) = 6 \cos t\mathbf{i} + 8 \sin t\mathbf{j}$. Therefore, with $0 \le t \le \dfrac{\pi}{2}$, we have

$$\int_C \mathbf{F} \cdot d\mathbf{r} = \int_0^{\pi/2}(-12 \sin t \cos t + 16 \sin t \cos t)\, dt$$

$$= 4 \int_0^{\pi/2} \sin t \cos t \, dt = 4\left[\frac{\sin^2 t}{2}\right]_0^{\pi/2} = 2$$

31. Find the area of the lateral surface under the function $f(x, y) = xy$ and over the circle $x^2 + y^2 = 1$ from $(1, 0)$ to $(0, 1)$.

Solution

We represent C parametrically as $x = \cos t$ and $y = \sin t$ for $0 \le t \le \pi/2$. Then we have

$$f(x, y) = xy = \cos t \sin t$$

and

$$ds = \sqrt{[x'(t)]^2 + [y'(t)]^2}\, dt = \sqrt{(-\sin t)^2 + (\cos t)^2}\, dt = dt$$

Therefore

$$\text{area} = \int_C f(x, y)\, ds$$

$$= \int_0^{\pi/2} \cos t \sin t \, dt = \left[\frac{1}{2}\sin^2 t\right]_0^{\pi/2} = \frac{1}{2}$$

37. Find the work done by the force field $\mathbf{F}(x, y) = -x\mathbf{i} - 2y\mathbf{j}$ on an object moving along the path given by $y = x^3$ from $(0, 0)$ to $(2, 8)$.

Solution

Since the path C is given by $x = t$ and $y = t^3$, with $0 \le t \le 2$, we have $\mathbf{F}(t) = -t\mathbf{i} - 2t^3\mathbf{j}$ and $\mathbf{r}'(t) = \mathbf{i} + 3t^2\mathbf{j}$. Therefore

$$W = \int_C \mathbf{F}(t) \cdot \mathbf{r}'(t)\, dt = \int_0^2 (-t - 6t^5)\, dt = \left[-\frac{t^2}{2} - t^6\right]_0^2 = -66$$

45. If the tangent vector $\mathbf{r}'(t)$ is orthogonal to the force vector $\mathbf{F}$ then

$$\int_C \mathbf{F} \cdot d\mathbf{r} = 0$$

regardless of the initial and terminal points of C. Demonstrate this for

$$\mathbf{F}(x, y) = (x^3 - 2x^2)\mathbf{i} + \left(x - \frac{y}{2}\right)\mathbf{j} \qquad \text{and} \qquad \mathbf{r}(t) = t\mathbf{i} + t^2\mathbf{j}$$

Solution

Since the path C is given by $x = t$ and $y = t^2$ we have

$$\mathbf{F}(x, y) = (t^3 - 2t^2)\mathbf{i} + \left(t - \frac{t^2}{2}\right)\mathbf{j}$$

Also

$$d\mathbf{r} = (\mathbf{i} + 2t\mathbf{j})\, dt$$

$$\int_C \mathbf{F} \cdot d\mathbf{r} = \int_C (t^3 - 2t^2 + 2t^2 - t^3)\, dt = \int_C 0\, dt = 0$$

16.3
Conservative vector fields and independence of path

3. Show that the value of $\int_C \mathbf{F} \cdot d\mathbf{r}$ is the same for each parametric representation of C if $\mathbf{F}(x, y) = y\mathbf{i} - x\mathbf{j}$.

(a) $\mathbf{r}_1(\theta) = \sec\theta\,\mathbf{i} + \tan\theta\,\mathbf{j}, \qquad 0 \le \theta \le \pi/3$

(b) $\mathbf{r}_2(t) = \sqrt{t + 1}\,\mathbf{i} + \sqrt{t}\,\mathbf{j}, \qquad 0 \le t \le 3$

Solution

(a) For $\mathbf{r}_1(\theta) = \sec\theta\,\mathbf{i} + \tan\theta\,\mathbf{j}$, we have $\mathbf{F}(x, y) = \tan\theta\,\mathbf{i} - \sec\theta\,\mathbf{j}$ and $\mathbf{r}_1'(\theta) = \sec\theta\tan\theta\,\mathbf{i} + \sec^2\theta\,\mathbf{j}$. Therefore

$$\int_C \mathbf{F} \cdot d\mathbf{r}_1 = \int_0^{\pi/3} \mathbf{F} \cdot \mathbf{r}_1'\, d\theta = \int_0^{\pi/3} (\sec\theta\tan^2\theta - \sec^3\theta)\, d\theta$$

$$= -\int_0^{\pi/3} \sec\theta\, d\theta = [-\ln|\sec\theta + \tan\theta|]_0^{\pi/3}$$

$$= -\ln(2 + \sqrt{3})$$

(b) For $\mathbf{r}_2(t) = \sqrt{t + 1}\,\mathbf{i} + \sqrt{t}\,\mathbf{j}$, we have $\mathbf{F}(t) = \sqrt{t}\,\mathbf{i} - \sqrt{t + 1}\,\mathbf{j}$ and $\mathbf{r}_2'(t) = \dfrac{1}{2\sqrt{t + 1}}\mathbf{i} + \dfrac{1}{2\sqrt{t}}\mathbf{j}$. Therefore

$$\int_C \mathbf{F} \cdot d\mathbf{r}_2 = \int_0^3 \mathbf{F} \cdot \mathbf{r}_2'\, dt = \int_0^3 \left(\frac{\sqrt{t}}{2\sqrt{t + 1}} - \frac{\sqrt{t + 1}}{2\sqrt{t}}\right) dt$$

$$= \frac{1}{2}\int_0^3 \frac{-1}{\sqrt{t + 1}\,\sqrt{t}}\, dt$$

By letting $u = \sqrt{t+1}$, we have $u^2 = t + 1$, $u^2 - 1 = t$, $2u\,du = dt$, and $\sqrt{u^2 - 1} = \sqrt{t}$. Thus

$$\frac{1}{2} \int_0^3 \frac{-1}{\sqrt{t+1}\sqrt{t}}\, dt = -\int_1^2 \frac{du}{\sqrt{u^2 - 1}}$$

$$= [-\ln |u + \sqrt{u^2 + 1}\,|]_1^2 = -\ln(2 + \sqrt{3})$$

11. Find the value of the line integral

$$\int_C 2xy\, dx + (x^2 + y^2)\, dy$$

along the paths

(a) C: ellipse $(x^2/25) + (y^2/16) = 1$ from $(5, 0)$ to $(0, 4)$.

(b) C: parabola $y = 4 - x^2$ from $(2, 0)$ to $(0, 4)$.

Solution

(a)   We first observe that the vector field
$\mathbf{F}(x, y) = 2xy\mathbf{i} + (x^2 + y^2)\mathbf{j}$ is conservative, since

$$\frac{\partial}{\partial y}[2xy] = 2x = \frac{\partial}{\partial x}[x^2 + y^2]$$

Therefore, the line integral is independent of path and we can replace the path along the ellipse from $(5, 0)$ to $(0, 4)$ with a path which will simplify the integration. We select the path along the coordinate axes from $(5, 0)$ to $(0, 0)$ and then from $(0, 0)$ to $(0, 4)$. Along the path from $(5, 0)$ to $(0, 0)$ we have $y = 0$ and $dy = 0$. On the path from $(0, 0)$ to $(0, 4)$ we have $x = 0$ and $dx = 0$. Hence

$$\int_C 2xy\, dx + (x^2 + y^2)\, dy = \int_5^0 0\, dx + (x^2)0 + \int_0^4 (0)(0)$$

$$+ (0 + y^2)\, dy$$

$$= \frac{y^3}{3}\bigg]_0^4 = \frac{64}{3}$$

(b)   Using the same method as in part (a) we replace the path along the parabola by the path along the axes from $(2, 0)$ to $(0, 0)$ and then from $(0, 0)$ to $(0, 4)$. Thus, we have

$$\int_C 2xy\, dx + (x^2 + y^2)\, dy = \int_2^0 0\, dx + (x^2 + 0)(0)$$

$$+ \int_0^4 (0)(0) + (0 + y^2)\, dy$$

$$= \frac{y^3}{3}\bigg]_0^4 = \frac{64}{3}$$

[Since the line integral is path-independent, we could have used the Fundamental Theorem.]

15. Find the value of the line integral $\int_C \mathbf{F} \cdot d\mathbf{r}$ where

$$\mathbf{F}(x, y, z) = (2y + x)\mathbf{i} + (x^2 - z)\mathbf{j} + (2y - 4z)\mathbf{k}$$

and

(a) $\mathbf{r}_1(t) = t\mathbf{i} + t^2\mathbf{j} + \mathbf{k}$ $0 \le t \le 1$

(b) $\mathbf{r}_2(t) = t\mathbf{i} + t\mathbf{j} + (2t - 1)^2\mathbf{k}$ $0 \le t \le 1$

Solution

(a) Along the path $\mathbf{r}_1(t)$, we have

$$\mathbf{F}(x, y, z) = (2t^2 + t)\mathbf{i} + (t^2 - 1)\mathbf{j} + (2t^2 - 4)\mathbf{k}$$

and

$$d\mathbf{r}_1 = \mathbf{r}_1'(t)\, dt = \mathbf{i} + 2t\mathbf{j}$$

Therefore

$$\int_C \mathbf{F} \cdot d\mathbf{r}_1 = \int_0^1 (2t^3 + 2t^2 - t)\, dt = \frac{2}{3}$$

(b) Along the path $\mathbf{r}_2(t)$, we have

$$\mathbf{F}(x, y, z) = (2t + t)\mathbf{i} + [t^2 - (2t - 1)^2]\mathbf{j} + [2t - 4(2t - 1)^2]\mathbf{k}$$
$$= 3t\mathbf{i} + (-3t^2 + 4t - 1)\mathbf{j} + (-16t^2 + 18t - 4)\mathbf{k}$$

and

$$d\mathbf{r}_2(t) = \mathbf{i} + \mathbf{j} + 4(2t - 1)\mathbf{k}$$

Therefore

$$\int_C \mathbf{F} \cdot d\mathbf{r}_2 = \int_0^1 (-128t^3 + 205t^2 - 97t + 15)\, dt = \frac{17}{6}$$

23. Use the Fundamental Theorem to evaluate

$$\int_C e^x \sin y\, dx + e^x \cos y\, dy$$

where C is one arch of the cycloid $x = \theta - \sin \theta$, $y = 1 - \cos \theta$ from $(0, 0)$ to $(2\pi, 0)$.

Solution

Since

$$\frac{\partial}{\partial y}[e^x \sin y] = e^x \cos y = \frac{\partial}{\partial x}[e^x \cos y]$$

the integral is path-independent. Therefore, we can evaluate the line integral by using the Fundamental Theorem. We begin by finding the potential function for the vector field $\mathbf{F}(x, y) = e^x \sin y\mathbf{i} + e^x \cos y\mathbf{j}$. If f is a potential function of $\mathbf{F}$, then

$$f_x(x, y) = e^x \sin y \qquad \text{and} \qquad f_y(x, y) = e^x \cos y$$

and we have

$$f(x, y) = \int f_x(x, y)\, dx = \int e^x \sin y\, dx = e^x \sin y + g(y)$$

$$f(x, y) = \int f_y(x, y)\, dy = \int e^x \cos y\, dy = e^x \sin y + h(x)$$

Now, we reconcile the differences between these two expressions for $f(x, y)$ by determining that $g(y) = h(x) = C$. Therefore, we have

$$f(x, y) = e^x \sin y + C$$

and

$$\int_C e^x \sin y\, dx + e^x \cos y\, dy = f(2\pi, 0) - f(0, 0)$$

$$= e^{2\pi}(0) - e^0(0) = 0$$

[Since this line integral is path-independent we could have integrated along the x-axis from $(0, 0)$ to $(2\pi, 0)$ with $y = 0$ and $dy = 0$. This would have given the result of zero immediately.]

16.4
Green's Theorem

3. Verify Green's Theorem by evaluating both integrals

$$\int_C y^2\, dx + x^2\, dy = \iint_R \left(\frac{\partial N}{\partial x} - \frac{\partial M}{\partial y} \right) dA$$

where C is the boundary of the region lying between $y = x$ and $y = x^2/4$.

Solution

(a) As a line integral we define C_1 and C_2 as shown in the accompanying figure.

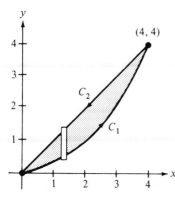

$$C_1 : x = t, \qquad y = \frac{t^2}{4}, \qquad dx = dt,\, dy = \frac{t}{2}\, dt, \qquad 0 \le t \le 4$$

$$C_2: x = 4 - t, \ y = 4 - t, \ dx = dy = -dt, \qquad 0 \le t \le 4$$

Thus

$$
\begin{aligned}
\int_C y^2 \, dx + x^2 \, dy &= \int_{C_1}\left(\frac{t^4}{16} + \frac{t^3}{2}\right) dt + \int_{C_2} 2(4 - t)^2(-dt) \\
&= \int_0^4 \left(\frac{t^4}{16} + \frac{t^3}{2}\right) dt + 2 \int_0^4 (4 - t)^2(-1) \, dt \\
&= \left[\frac{t^5}{5(16)} + \frac{t^4}{2(4)} + \frac{2(4 - t)^3}{3}\right]_0^4 \\
&= \frac{64}{5} + 32 - \frac{128}{3} = \frac{32}{15}
\end{aligned}
$$

(b)   By Green's Theorem we have

$$
\begin{aligned}
\iint_R \left(\frac{\partial N}{\partial x} - \frac{\partial M}{\partial y}\right) dA &= \iint_R (2x - 2y) \, dy \, dx = \int_0^4 \int_{x^2/4}^x (2x - 2y) \, dy \, dx \\
&= \int_0^4 [2xy - y^2]_{x^2/4}^x \, dx = \int_0^4 \left(x^2 - \frac{x^3}{2} + \frac{x^4}{16}\right) dx \\
&= \left[\frac{x^3}{3} - \frac{x^4}{8} + \frac{x^5}{80}\right]_0^4 = \frac{63}{3} - 32 + \frac{64}{5} = \frac{32}{15}
\end{aligned}
$$

7.   Use Green's Theorem to evaluate the integral

$$\int_C (y - x) \, dx + (2x - y) \, dy$$

where C is the boundary of the region lying inside the rectangle with vertices $(5, 3)$, $(-5, 3)$, $(-5, -3)$, and $(5, -3)$ and outside the square with vertices $(1, 1)$, $(-1, 1)$, $(-1, -1)$, and $(1, -1)$.

Solution

Using $M(x, y) = y - x$, $N(x, y) = 2x - y$, and the accompanying figure, we have

$$
\begin{aligned}
\int_C (y - x) \, dx + (2x - y) \, dy &= \iint_R \left(\frac{\partial N}{\partial x} - \frac{\partial M}{\partial y}\right) dA \\
&= \iint_R (2 - 1) \, dA \\
&= \iint_R dA \\
&= \text{area of region}
\end{aligned}
$$

$$= \text{(area of rectangle)} - \text{(area of square)}$$
$$= 6(10) - 2(2) = 56$$

13. Use Green's Theorem to evaluate

$$\int_C 2 \arctan \frac{y}{x} \, dx + \ln (x^2 + y^2) \, dy$$

where C is the boundary of the ellipse $x = 4 + 2 \cos \theta$, $y = 4 + \sin \theta$.

Solution

By Green's Theorem we have

$$\int_C M \, dx + N \, dy = \int_C \left(2 \arctan \frac{y}{x} \right) dx + \ln (x^2 + y^2) \, dy$$

$$= \iint_R \left(\frac{\partial N}{\partial x} - \frac{\partial M}{\partial y} \right) dA$$

$$= \iint_R \left[\frac{2x}{x^2 + y^2} - \frac{2(1/x)}{1 + (y/x)^2} \right] dA$$

$$= \iint_R \left[\frac{2x}{x^2 + y^2} - \frac{2x}{x^2 + y^2} \right] dA = 0$$

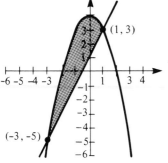

25. Use a line integral to find the area of the region R bounded by $y = 2x + 1$ and $y = 4 - x^2$. (See accompanying figure.)

Solution

The region R is enclosed by the path C given by

$$C_1: x = t, \ y = 2t + 1, \ dx = dt, \ dy = 2 \, dt, \ -3 \le t \le 1$$
$$C_2: x = 1 - t, \ y = 4 - (1 - t)^2 = 3 + 2t - t^2, \ dx = -dt,$$
$$dy = (2 - 2t) \, dt, \ 0 \le t \le 4$$

Therefore by Theorem 16.7 the area of R is

$$A = \frac{1}{2} \int_C x \, dy - y \, dx$$

$$= \frac{1}{2} \int_{C_1} t(2 \, dt) - (2t + 1) \, dt + \frac{1}{2} \int_{C_2} (1 - t)(2 - 2t) \, dt$$
$$- (3 + 2t - t^2)(-dt)$$

$$= \frac{1}{2} \int_{-3}^{1} (-1) \, dt + \frac{1}{2} \int_{0}^{4} (5 - 2t + t^2) \, dt$$

$$= \left[-\frac{t}{2} \right]_{-3}^{1} + \frac{1}{2} \left[5t - t^2 + \frac{t^3}{3} \right]_{0}^{4} = -2 + \frac{1}{2} \left(20 - 16 + \frac{64}{3} \right) = \frac{32}{3}$$

30. The centroid of the region having area A and bounded by the simple closed path C is

$$\bar{x} = \frac{1}{2A} \int_C x^2 \, dy \qquad \text{and} \qquad \bar{y} = \frac{-1}{2A} \int_C y^2 \, dx$$

Find the centroid of the region R bounded by $y = \sqrt{a^2 - x^2}$ and $y = 0$.

Solution

Since the region is a semicircle of radius a, its area is $A = \pi a^2/2$. Now using $y^2 = a^2 - x^2$ and $x^2 = a^2 - y^2$, we have

$$\bar{x} = \frac{1}{2A} \int_C x^2 \, dy = \frac{1}{\pi a^2} \int_0^0 (a^2 - y^2) \, dy = 0$$

$$\bar{y} = \frac{-1}{2A} \int_C y^2 \, dx = \frac{-1}{\pi a^2} \int_a^{-a} (a^2 - x^2) \, dx$$

$$= \frac{-1}{\pi a^2} \left[a^2 x - \frac{x^3}{3} \right]_a^{-a} = \frac{-1}{\pi a^2} \left[-a^3 + \frac{a^3}{3} - \left(a^3 - \frac{a^3}{3} \right) \right] = \frac{4a}{3\pi}$$

35. The area of a plane region in polar coordinates is

$$A = \frac{1}{2} \int_C r^2 \, d\theta$$

Find the area of the region R bounded by the inner loop of the limaçon $r = 1 + 2 \cos \theta$. (See accompanying figure.)

Solution

The inner loop of $r = 1 + 2 \cos \theta$ starts at $\theta = 2\pi/3$ and ends at $\theta = 4\pi/3$. Hence the area enclosed by this inner loop is

$$A = \frac{1}{2} \int_{2\pi/3}^{4\pi/3} (1 + 2 \cos \theta)^2 \, d\theta$$

$$= \frac{1}{2} \int_{2\pi/3}^{4\pi/3} \left[1 + 4 \cos \theta + 4 \left(\frac{1 + \cos 2\theta}{2} \right) \right] d\theta$$

$$= \frac{1}{2} \int_{2\pi/3}^{4\pi/3} (3 + 4 \cos \theta + 2 \cos 2\theta) \, d\theta$$

$$= \frac{1}{2} [3\theta + 4 \sin \theta + \sin 2\theta]_{2\pi/3}^{4\pi/3}$$

$$= \pi - \frac{3\sqrt{3}}{2}$$

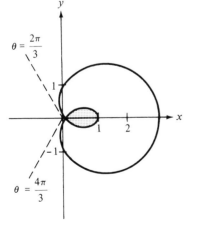

16.5
Surface integrals

7. Evaluate

$$\iint_S xy \, dS$$

over the surface given by $z = 9 - x^2$ for $0 \le x \le 2$, and $0 \le y \le x$.

Solution

We write the equation for the surface S as $z = 9 - x^2 = g(x, y)$ so that $g_x(x, y) = -2x$ and $g_y(x, y) = 0$, and obtain

$$\sqrt{1 + [g_x(x, y)]^2 + [g_y(x, y)]^2} = \sqrt{1 + 4x^2}$$

Using Theorem 16.8 and the accompanying figure, we have

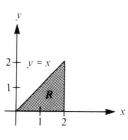

$$\iint_S xy \, dS = \iint_R f(x, y, g(x, y)) \sqrt{1 + [g_x(x, y)]^2 + [g_y(x, y)]^2} \, dA$$

$$= \int_0^2 \int_0^x xy \sqrt{1 + 4x^2} \, dy \, dx$$

$$= \frac{1}{2} \int_0^2 x^3 \sqrt{1 + 4x^2} \, dx$$

Now, we use trigonometric substitution letting $2x = \tan \theta$ and $2 \, dx = \sec^2 \theta \, d\theta$ to obtain

$$\iint_S xy \, dS = \frac{1}{32} \int_0^{\arctan 4} \tan^3 \theta \sec^3 \theta \, d\theta$$

$$= \frac{1}{32} \left[\frac{1}{5} \sec^5 \theta - \frac{1}{3} \sec^3 \theta \right]_0^{\arctan 4}$$

$$= \frac{391 \sqrt{17} + 1}{240}$$

17. Find $\iint_S \mathbf{F} \cdot \mathbf{N} \, dS$ (the flux of $\mathbf{F}$ through S where $\mathbf{N}$ is the unit upper normal to S) if $\mathbf{F}(x, y, z) = x\mathbf{i} + y\mathbf{j} + z\mathbf{k}$ and S is the surface given by $z = 9 - x^2 - y^2$ for $0 \le z$.

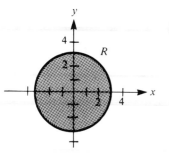

Solution

The vector field $\mathbf{F}$, over the surface S, is given by

$$\mathbf{F}(x, y, z) = x\mathbf{i} + y\mathbf{j} + z\mathbf{k} = x\mathbf{i} + y\mathbf{j} + (9 - x^2 - y^2)\mathbf{k}$$

We write the equation for the surface S as $z = 9 - x^2 - y^2 = g(x, y)$ so that $g_x(x, y) = -2x$ and $g_y(x, y) = -2y$. Then by Theorem 16.9, we have

$$\iint\limits_{S} \mathbf{F} \cdot \mathbf{N} \, dS$$

$$= \iint\limits_{R} \mathbf{F} \cdot [-g_x(x, y)\mathbf{i} - g_y(x, y)\mathbf{j} + \mathbf{k}] \, dA$$

$$= \iint\limits_{R} [x\mathbf{i} + y\mathbf{j} + (9 - x^2 - y^2)\mathbf{k}] \cdot (2x\mathbf{i} + 2y\mathbf{j} + \mathbf{k}) \, dA$$

$$= \iint\limits_{R} (9 + x^2 + y^2) \, dA$$

$$= 4 \int_0^{\pi/2} \int_0^3 (9 + r^2) r \, dr \, d\theta \qquad \text{Double integral in polar coordinates}$$

$$= \frac{243\pi}{2}$$

27. Find I_z for the lamina $x^2 + y^2 = a^2$ $(0 \le z \le h)$ of uniform density 1.

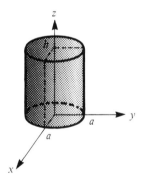

Solution

Note that S does not define z as a function of x and y. Hence, we project onto the xz-plane, so that $y = \sqrt{a^2 - x^2} = g(x, z)$ and obtain

$$\sqrt{1 + [g_x(x, z)]^2 + [g_z(x, z)]^2} = \sqrt{1 + \frac{x^2}{a^2 - x^2}} = \frac{a}{\sqrt{a^2 - x^2}}$$

Therefore

$$I_z = \iint\limits_{S} (x^2 + y^2)(1) \, dS$$

$$= \iint\limits_{R} a^2 \sqrt{1 + [g_x(x, z)]^2 + [g_z(x, z)]^2} \, dA$$

$$= 4a^2 \int_0^a \int_0^h \frac{a}{\sqrt{a^2 - x^2}} \, dz \, dx$$

$$= 4a^3 h \int_0^a \frac{1}{\sqrt{a^2 - x^2}} \, dx$$

$$= 4a^3 h \left[\arcsin \frac{x}{a} \right]_0^a = 2\pi a^3 h$$

16.6

Divergence Theorem

2. Find the divergence of the vector field $F(x, y, z) = xe^x i + ye^y j + z^2 k$.

Solution

By definition we have

$$\text{div } F(x, y, z) = \frac{\partial}{\partial x}[xe^x] + \frac{\partial}{\partial y}[ye^y] + \frac{\partial}{\partial z}[z^2]$$

$$= e^x(x + 1) + e^y(y + 1) + 2z$$

6. Verify the Divergence Theorem by evaluating $\int_S \int F \cdot N \, dS$ as a surface integral and as a triple integral where $F(x, y, z) = 2x i - 2y j + z^2 k$ and S is the cylinder given by $x^2 + y^2 = 1$ for $0 \leq z \leq h$.

Solution

As a *surface integral*, we have

$$S_1: \quad x^2 + y^2 \leq 1, \quad z = h$$
$$S_2: \quad x^2 + y^2 = 1, \quad 0 \leq z \leq h$$
$$S_3: \quad x^2 + y^2 \leq 1, \quad z = 0$$

For the surface S_1 we have $N_1(x, y, z) = k$, $F(x, y, z) \cdot N_1(x, y, z) = z^2$, and $dS_1 = dA$. Therefore

$$\iint\limits_{S_1} F \cdot N \, dS = \iint\limits_{S_1} z \, dS = \iint\limits_{S_1} h^2 \, dS = \pi h^2$$

For the surface S_2 we have

$$N_2(x, y, z) = \frac{2x i + 2y j}{\sqrt{4x^2 + 4y^2}} = x i + y j$$

and

$$F(x, y, z) \cdot N_2(x, y, z) = 2x^2 - 2y^2 = 2x^2 - 2(1 - x^2) = 4x^2 - 2$$

Note that S_2 does **not** define z as a function of x and y. Hence, we project onto the xz-plane so that $y = \sqrt{1 - x^2} = g(x, z)$ and obtain

$$dS_2 = \sqrt{1 + [g_x(x, z)]^2 + [g_z(x, z)]^2} \, dx \, dz = \frac{1}{\sqrt{1 - x^2}} \, dx \, dz$$

Therefore

$$\iint\limits_{S_2} \mathbf{F} \cdot \mathbf{N}\, dS = \iint\limits_{S_2} \frac{4x^2 - 2}{\sqrt{1-x^2}}\, dx\, dz$$

$$= \int_0^h \int_{-1}^1 \left[\frac{4x^2}{\sqrt{1-x^2}} - \frac{2}{\sqrt{1-x^2}} \right] dx\, dz$$

$$= \int [2 \arcsin x - 2x\sqrt{1-x^2} - 2 \arcsin x]_{-1}^1\, dz$$

$$= \int_0^h 0\, dz = 0$$

For the surface S_3 we have $\mathbf{N}_3(x, y, z) = -\mathbf{k}$ and
$\mathbf{F}(x, y, z) \cdot \mathbf{N}(x, y, z) = -z^2$. Since $z = 0$ on S_3, we have

$$\iint\limits_{S_3} \mathbf{F} \cdot \mathbf{N}\, dS = 0$$

Thus, the surface integral for the cylinder S is

$$\iint\limits_{S} \mathbf{F} \cdot \mathbf{N}\, dS = \pi h^2 + 0 + 0 = \pi h^2$$

Using the *Divergence Theorem*, we have

$$\text{div } \mathbf{F}(x, y, z) = 2 - 2 + 2z = 2z$$

Thus

$$\iiint\limits_{Q} 2z\, dV = 4 \int_0^1 \int_0^{\sqrt{1-x^2}} \int_0^h 2z\, dz\, dy\, dx$$

$$= 4 \int_0^1 \int_0^{\sqrt{1-x^2}} [z^2]_0^h\, dy\, dx$$

$$= 4 \int_0^1 \int_0^{\sqrt{1-x^2}} h^2\, dy\, dx$$

$$= 4h^2 \int_0^1 [y]_0^{\sqrt{1-x^2}}\, dx = 4h^2 \int_0^1 \sqrt{1-x^2}\, dx$$

$$= 4h^2(\text{area of quarter circle of radius } 1) = 4h^2\left(\frac{\pi}{4}\right)$$

$$= \pi h^2$$

13. Use the Divergence Theorem to evaluate $\iint \mathbf{F} \cdot \mathbf{N}\, dS$, where
$\mathbf{F}(x, y, z) = x\mathbf{i} + y\mathbf{j} + z\mathbf{k}$ and S is the sphere given by
$x^2 + y^2 + z^2 = 4$.

Solution

Since div $\mathbf{F}(x, y, z) = 1 + 1 + 1 = 3$, we have

$$\iiint\limits_Q \text{div } \mathbf{F} \, dV = 3 \iiint\limits_Q dV = 3(\text{volume of sphere of radius 2})$$

$$= 3\left[\frac{4\pi 2^3}{3}\right] = 32\pi$$

22. Evaluate $\int_S \int \text{curl } \mathbf{F} \cdot \mathbf{N} \, dS$ where $\mathbf{F}(x, y, z) = xy \cos z\mathbf{i} + yz \sin x\mathbf{j} + xyz\mathbf{k}$ and S is the closed surface of the solid bounded by the graphs of $x = 4$, $z = 9 - y^2$, and the coordinate planes.

Solution

Using the Divergence Theorem, we have

$$\iint\limits_S \text{curl } \mathbf{F} \cdot \mathbf{N} \, dS = \iiint\limits_Q \text{div (curl F)} \, dV$$

$$\text{curl } \mathbf{F}(x, y, z) = \begin{vmatrix} \mathbf{i} & \mathbf{j} & \mathbf{k} \\ \dfrac{\partial}{\partial x} & \dfrac{\partial}{\partial y} & \dfrac{\partial}{\partial z} \\ xy \cos z & yz \sin x & xyz \end{vmatrix}$$

$$= (xz - y \sin x)\mathbf{i} - (yz + xy \sin z)\mathbf{j} + (yz \cos x - x \cos z)\mathbf{k}$$

Now

$$\text{div curl } \mathbf{F}(x, y, z) = (z - y \cos x) - (z + x \sin z) + (y \cos x + x \sin z) = 0$$

Therefore $\int_S \int \text{curl } \mathbf{F} \cdot \mathbf{N} \, dS = \iiint_Q \text{div (curl F)} \, dV = 0$

16.7
Stokes' Theorem

9. Verify Stokes' Theorem by evaluating $\int_C \mathbf{F} \cdot \mathbf{T} \, ds$ as a line integral and as a double integral, where $\mathbf{F}(x, y, z) = xyz\mathbf{i} + y\mathbf{j} + z\mathbf{k}$ and S is the portion of the plane $3x + 4y + 2z = 12$ lying in the first octant.

Solution

As a *line integral*, we integrate along the three paths shown in the accompanying figure and obtain

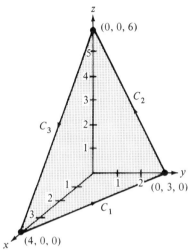

$$\int_C \mathbf{F} \cdot \mathbf{T} \, ds = \int_C xyz \, dx + y \, dy + z \, dz$$

$$= \int_{C_1} 0 \, dx + y \, dy + 0 \, dz + \int_{C_2} 0 \, dx + y \, dy + z \, dz$$

$$+ \int_{C_3} 0 \, dx + 0 \, dy + z \, dz$$

$$= \int_0^3 y \, dy + \int_3^0 y \, dy + \int_0^6 z \, dz + \int_6^0 z \, dz = 0$$

Now, to evaluate the *double integral*, we begin by finding the curl of **F**.

$$\mathbf{curl}\ \mathbf{F}(x, y, z) = \begin{vmatrix} \mathbf{i} & \mathbf{j} & \mathbf{k} \\ \dfrac{\partial}{\partial x} & \dfrac{\partial}{\partial y} & \dfrac{\partial}{\partial z} \\ xyz & y & z \end{vmatrix} = xy\mathbf{j} - xz\mathbf{k}$$

By Theorem 16.9 for an upward normal **N**, we obtain

$$\iint_S (\mathbf{curl}\ \mathbf{F}) \cdot \mathbf{N} \, dS$$

$$= \iint_R (xy\mathbf{j} - xz\mathbf{k}) \cdot (3\mathbf{i} + 4\mathbf{j} + 2\mathbf{k}) \, dA$$

$$= \iint_R (4xy - 2xz) \, dA$$

$$= \int_0^4 \int_0^{3(4-x)/4} \left[4xy - 2x\left(6 - 2y - \frac{3x}{2} \right) \right] dy \, dx$$

$$= \int_0^4 \int_0^{3(4-x)/4} (8xy + 3x^2 - 12x) \, dy \, dx$$

$$= \int_0^4 \left(36x - 18x^2 + \frac{9x^3}{4} + 9x^2 - \frac{9x^3}{4} - 36x + 9x^2 \right) dx$$

$$= \int_0^4 (0) \, dx = 0$$

11.   Use Stokes' Theorem to evaluate $\int_C \mathbf{F} \cdot d\mathbf{r}$ for $\mathbf{F}(x, y, z) = 2y\mathbf{i} + 3z\mathbf{j} - x\mathbf{k}$, where C is the triangle whose vertices are $(0, 0, 0)$, $(0, 2, 0)$, and $(1, 1, 1)$. (See the accompanying figure.)

Solution

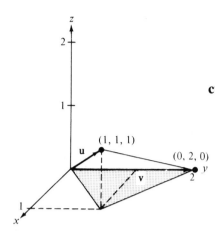

$$\text{curl } \mathbf{F} = \begin{vmatrix} \mathbf{i} & \mathbf{j} & \mathbf{k} \\ \dfrac{\partial}{\partial x} & \dfrac{\partial}{\partial y} & \dfrac{\partial}{\partial z} \\ 2y & 3z & -x \end{vmatrix} = (0 - 3)\mathbf{i} - (-1 - 0)\mathbf{j} + (0 - 2)\mathbf{k}$$

$$= -3\mathbf{i} + \mathbf{j} - 2\mathbf{k}$$

Consider the vectors $\mathbf{u} = \mathbf{i} + \mathbf{j} + \mathbf{k}$ and $\mathbf{v} = 0\mathbf{i} + 2\mathbf{j} + 0\mathbf{k} = 2\mathbf{j}$. Then

$$\mathbf{u} \times \mathbf{v} = \begin{vmatrix} \mathbf{i} & \mathbf{j} & \mathbf{k} \\ 1 & 1 & 1 \\ 0 & 2 & 0 \end{vmatrix} = (0 - 2)\mathbf{i} - (0 - 0)\mathbf{j} + (2 - 0)\mathbf{k}$$

$$= -2\mathbf{i} + 2\mathbf{k}$$

Thus

$$\frac{\mathbf{u} \times \mathbf{v}}{\|\mathbf{u} \times \mathbf{v}\|} = \frac{-2\mathbf{i} + 2\mathbf{k}}{2\sqrt{2}} = \frac{-\mathbf{i} + \mathbf{k}}{\sqrt{2}}$$

and the surface S has direction numbers $-1, 0, 1$. Thus we have $f(x, y, z) = -x + z$, $dS = \sqrt{1 + 1}\, dA = \sqrt{2}\, dA$, and we conclude that

$$\iint_S (\text{curl } \mathbf{F}) \cdot \mathbf{N} \ dS = \iint_R \frac{(3 - 2)}{\sqrt{2}} \sqrt{2}\, dA = \iint_R dA$$

$$= \text{area of triangle} = \left(\frac{1}{2}\right)(2)(1) = 1$$

Review Exercises for Chapter 16

9. Determine if $\mathbf{F}(x, y, z) = (yz\mathbf{i} - xz\mathbf{j} - xy\mathbf{k})/y^2z^2$ is conservative and if it is, find the potential function.

Solution

Since

$$\text{curl } \mathbf{F}(x, y, z) = \begin{vmatrix} \mathbf{i} & \mathbf{j} & \mathbf{k} \\ \dfrac{\partial}{\partial x} & \dfrac{\partial}{\partial y} & \dfrac{\partial}{\partial z} \\ \dfrac{1}{yz} & \dfrac{-x}{y^2z} & \dfrac{-x}{yz^2} \end{vmatrix}$$

$$= \left(\frac{x}{y^2 z^2} - \frac{x}{y^2 z^2}\right)\mathbf{i} - \left(\frac{-1}{yz^2} - \frac{-1}{yz^2}\right)\mathbf{j}$$

$$+ \left(\frac{-1}{y^2 z} - \frac{-1}{y^2 z}\right)\mathbf{k} = 0$$

F is conservative. Now, if f is a function such that
$\mathbf{F}(x, y, z) = \nabla f(x, y, z)$, then

$$f_x(x, y, z) = \frac{1}{yz}, \quad f_y(x, y, z) = -\frac{x}{y^2 z}, \quad f_z(x, y, z) = -\frac{x}{yz^2}$$

and by integrating with respect to x, y, and z separately, we obtain

$$f(x, y, z) = \int \frac{1}{yz}\, dx = \frac{x}{yz} + g(y, z)$$

$$f(x, y, z) = \int -\frac{x}{y^2 z}\, dy = \frac{x}{yz} + h(x, z)$$

$$f(x, y, z) = \int -\frac{x}{yz^2}\, dz = \frac{x}{yz} + k(x, y)$$

By comparing these three versions of $f(x, y, z)$, we conclude that

$$f(x, y, z) = \frac{x}{yz} + C$$

13. Find (a) the divergence and (b) the curl of the vector field
$$\mathbf{F}(x, y, z) = (\cos y + y \cos x)\mathbf{i} + (\sin x - x \sin y)\mathbf{j} + xyz\,\mathbf{k}$$

Solution

(a) $\quad$ div $\mathbf{F}(x, y, z) = \dfrac{\partial}{\partial x}[\cos y + y \cos x]$

$$+ \frac{\partial}{\partial y}[\sin x - x \sin y] + \frac{\partial}{\partial z}[xyz]$$

$$= -y \sin x - x \cos y + xy$$

(b)

$$\text{curl } \mathbf{F}(x, y, z) = \begin{vmatrix} \mathbf{i} & \mathbf{j} & \mathbf{k} \\ \dfrac{\partial}{\partial x} & \dfrac{\partial}{\partial y} & \dfrac{\partial}{\partial z} \\ \cos y + y \cos x & \sin x - x \sin y & xyz \end{vmatrix}$$

$$= (xz - 0)\mathbf{i} - (yz - 0)\mathbf{j}$$

$$+ (\cos x - \sin y + \sin y - \cos x)\mathbf{k}$$

$$= xz\mathbf{i} - yz\mathbf{j}$$

21.   Evaluate $\int_C (2x - y)\, dx + (x + 3y)\, dy$:

    (a) C is the line segment from $(0, 0)$ to $(2, -3)$.

    (b) C is one counterclockwise revolution on the circle
        $x = 3 \cos t$ and $y = 3 \sin t$.

Solution

(a) $C: x = t,\ y = -3t/2,\ 0 \le t \le 2$. Then

$$\int_C (2x - y)\, dx + (x + 3y)\, dy = \int_0^2 \left[\frac{7t}{2}\, dt + \left(-\frac{7t}{2} \right)\left(-\frac{3}{2}\, dt \right) \right]$$

$$= \int_0^2 \frac{35}{4} t\, dt = \left[\frac{35}{8} t^2 \right]_0^2 = \frac{35}{2}$$

(b) $C: x = 3 \cos t,\ y = 3 \sin t,\ 0 \le t \le 2\pi$. Then

$$\int_C (2x - y)\, dx + (x + 3y)\, dy$$

$$= \int_0^{2\pi} [(6 \cos t - 3 \sin t)(-3 \sin t)$$

$$+ (3 \cos t + 9 \sin t)(3 \cos t)]\, dt$$

$$= \int_0^{2\pi} (9 \sin t \cos t + 9)\, dt = \left[\frac{9 \sin^2 t}{2} + 9t \right]_0^{2\pi} = 18\pi$$

29.   Evaluate $\int_C \mathbf{F} \cdot d\mathbf{r}$, where $\mathbf{F}(x, y, z) = (y - z)\mathbf{i} + (z - x)\mathbf{j} + (x - y)\mathbf{k}$ and C is the curve of intersection of the paraboloid $z = x^2 + y^2$ and the plane $x + y = 0$ from the point $(-2, 2, 8)$ to $(2, -2, 8)$.

Solution

On the curve of intersection $z = x^2 + (-x)^2 = 2x^2$. Hence C is given by $x = t,\ y = -t,\ z = 2t^2,\ -2 \le t \le 2$, and we have

$$\mathbf{r}(t) = t\mathbf{i} - t\mathbf{j} + 2t^2\mathbf{k} \qquad \text{and} \qquad d\mathbf{r} = (\mathbf{i} - \mathbf{j} + 4t\mathbf{k})\, dt$$

Therefore with

$$\mathbf{F}(x, y, z) = (y - z)\mathbf{i} + (z - x)\mathbf{j} + (x - y)\mathbf{k}$$
$$= (-t - 2t^2)\mathbf{i} + (2t^2 - t)\mathbf{j} + 2t\mathbf{k}$$

we have

$$\int_C \mathbf{F} \cdot d\mathbf{r} = \int_{-2}^2 (-2t^2 - t - 2t^2 + t + 8t^2)\, dt$$

$$= \int_{-2}^2 4t^2\, dt = \left[\frac{4t^3}{3} \right]_{-2}^2 = \frac{64}{3}$$

35. Use Green's Theorem to evaluate $\int_C xy \, dx + x^2 \, dy$ where C is the boundary of the region between the graphs of $y = x^2$ and $y = x$.

Solution

By Green's Theorem and the accompanying figure, we have

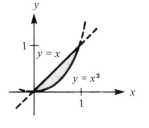

$$\int_C M(x, y) \, dx + N(x, y) \, dy = \int_C xy \, dx + x^2 \, dy$$

$$= \iint_R \left[\frac{\partial N}{\partial x} - \frac{\partial M}{\partial y} \right] dA$$

$$= \int_0^1 \int_{x^2}^x x \, dy \, dx = \int_0^1 (x^2 - x^3) \, dx = \frac{1}{12}$$

38. Use the Fundamental Theorem to evaluate

$$\int_{(0,0,1)}^{(4,4,4)} y \, dx + x \, dy + \frac{1}{z} \, dz.$$

Solution

Since

$$\frac{\partial}{\partial y} \left[\frac{1}{z} \right] = 0 = \frac{\partial}{\partial z} [x], \quad \frac{\partial}{\partial x} \left[\frac{1}{z} \right] = 0 = \frac{\partial}{\partial z} [y], \quad \frac{\partial}{\partial x} [x] = 1 = \frac{\partial}{\partial y} [y]$$

the vector field $\mathbf{F}(x, y, z) = y\mathbf{i} + x\mathbf{j} + (1/z)\mathbf{k}$ is conservative. By integrating y, x, and $1/z$ with respect to x, y, and z respectively we obtain the potential function $f(x, y, z) = xy + \ln z + C$.

Therefore, by the Fundamental Theorem (Theorem 16.4), we have

$$\int_{(0,0,1)}^{(4,4,4)} y \, dx + x \, dy + \frac{1}{z} \, dz = f(4, 4, 4) - f(0, 0, 1) = 16 + \ln 4$$

17 Differential equations

17.1

Definitions and basic concepts

7. Classify the differential equation $(y'')^2 + 3y' - 4y = 0$ according to type, order, and degree (when applicable).

   Solution

   This is an ordinary differential equation and the term $(y'')^2$ indicates that the order is 2 and the degree is 2.

16. Verify that $y = C_1 e^{-x} \cos x + C_2 e^{-x} \sin x$ is a solution of the differential equation $y'' + 2y' + 2y = 0$.

    Solution

    Since $y = C_1 e^{-x} \cos x + C_2 e^{-x} \sin x$, we have

    $$y' = C_1(-e^{-x} \sin x - e^{-x} \cos x) + C_2(e^{-x} \cos x - e^{-x} \sin x)$$
    $$= (C_2 - C_1)e^{-x} \cos x - (C_2 + C_1)e^{-x} \sin x$$
    $$y'' = (C_2 - C_1)(-e^{-x} \sin x - e^{-x} \cos x)$$
    $$- (C_2 + C_1)(e^{-x} \cos x - e^{-x} \sin x)$$
    $$= 2C_1 e^{-x} \sin x - 2C_2 e^{-x} \cos x$$

    Therefore

    $$y'' + 2y' + 2y = 2C_1 e^{-x} \sin x - 2C_2 e^{-x} \cos x + 2C_2 e^{-x} \cos x$$
    $$- 2C_1 e^{-x} \cos x - 2C_2 e^{-x} \sin x$$
    $$- 2C_1 e^{-x} \sin x + 2C_1 e^{-x} \cos x$$
    $$+ 2C_2 e^{-x} \sin x = 0$$

31. Verify that $y = Ce^{-2x}$ is the general solution of the differential

equation $y' + 2y = 0$. Then find the particular solution satisfying the initial condition $y = 3$ when $x = 0$.

Solution

$y = Ce^{-2x}$ is the general solution since $y' = -2Ce^{-2x}$ and $y' + 2y = -2Ce^{-2x} + 2Ce^{-2x} = 0$. Furthermore, $y = 3$ when $x = 0$ implies that $3 = Ce^0 = C$. Therefore the particular solution is $y = 3e^{-2x}$.

38. The general solution to the differential equation $yy' + x = 0$ is given by $x^2 + y^2 = C$. Sketch the solution curves given by $C = 0$, $C = 1$, and $C = 4$.

Solution

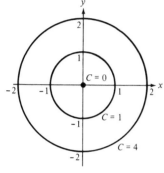

The solution curves $x^2 + y^2 = C$ are circles centered at the origin. The accompanying figure shows the solution curves for $C = 0$, 1, and 4.

$\quad\quad C = 0$: point $(0, 0)$

$\quad\quad C = 1$: circle of radius 1

$\quad\quad C = 4$: circle of radius 2

17.2
Separation of variables

5. Solve the differential equation $yy' = \sin x$ by separation of variables.

Solution

First, we have

$$y \frac{dy}{dx} = \sin x \quad\quad \text{or} \quad\quad y \, dy = \sin x \, dx$$

Then by integration we obtain

$$\int y \, dy = \int \sin x \, dx$$

$$\frac{y^2}{2} = -\cos x + C_1$$

$$y^2 = C - 2 \cos x$$

as the general solution.

9. Solve the differential equation $y(x + 1) + y' = 0$, $y(-2) = 1$, by separation of variables.

Solution

First we have

$$y(x + 1) + y' = 0$$

$$\frac{dy}{dx} = -y(x + 1)$$

$$\frac{dy}{y} = -(x + 1) \, dx$$

Integration yields

$$\int \frac{dy}{y} = -\int (x + 1) \, dx$$

$$\ln |y| = -\frac{(x + 1)^2}{2} + C_1$$

$$y = e^{C_1 - (x+1)^2/2} = Ce^{-(x+1)^2/2}$$

Since $y = 1$ when $x = 2$, it follows that

$$1 = Ce^{-(-2+1)^2/2} = Ce^{-1/2} \qquad \text{or} \qquad C = e^{1/2}$$

Therefore the particular solution is

$$y = e^{1/2} e^{-(x+1)^2/2} = e^{-(x^2+2x)/2}$$

16. Determine whether or not the function $f(x, y) = xy/\sqrt{x^2 + y^2}$ is homogeneous and, if so, determine its degree.

Solution

Since

$$f(tx, ty) = \frac{(tx)(ty)}{\sqrt{(tx)^2 + (ty)^2}} = \frac{t^2 xy}{t\sqrt{x^2 + y^2}} = \frac{t(xy)}{\sqrt{x^2 + y^2}} = t f(x, y)$$

the function is homogeneous of degree 1.

25. Solve the homogeneous differential equation $y' = xy/(x^2 - y^2)$.

Solution

Letting $y = vx$, we have

$$y' = \frac{xy}{x^2 - y^2}$$

$$v + x\frac{dv}{dx} = \frac{x(vx)}{x^2 - v^2x^2} = \frac{v}{1 - v^2}$$

$$x\frac{dv}{dx} = \frac{v}{1 - v^2} - v = \frac{v^3}{1 - v^2}$$

$$x\,dv = \frac{v^3}{1 - v^2}\,dx$$

Separating variables, we obtain

$$\frac{1 - v^2}{v^3}\,dv = \frac{dx}{x}$$

$$\int\left(v^{-3} - \frac{1}{v}\right)dv = \int\frac{dx}{x}$$

$$\frac{v^{-2}}{-2} - \ln|v| = \ln|x| + \ln|C_1|$$

$$\frac{1}{-2v^2} = \ln|v| + \ln|x| + \ln|C_1| = \ln|C_1 vx|$$

$$\frac{x^2}{-2y^2} = \ln|C_1 y|$$

$$e^{-x^2/2y^2} = C_1 y$$

Finally, the general solution can be written as

$$y = Ce^{-x^2/2y^2}$$

29. Solve the homogeneous differential equation

$$\left[x\sec\left(\frac{y}{x}\right) + y\right]dx - x\,dy = 0, \qquad y(1) = 0$$

Solution

Letting $y = vx$, we have

$$\left[x\sec\left(\frac{vx}{x}\right) + vx\right]dx - x(v\,dx + x\,dv) = 0$$

$$x\sec v\,dx + vx\,dx - vx\,dx - x^2\,dx = 0$$

$$x\sec v\,dx = x^2\,dv$$

$$\frac{x}{x^2}\,dx = \frac{dv}{\sec v}$$

Integration yields

$$\int\frac{dx}{x} = \int\cos v\,dv$$

$$\ln|x| = \sin v + C_1 = \sin\frac{y}{x} + C_1$$

Since $y = 0$ when $x = 1$, we have

$$0 = \sin(0) + C_1 = C_1$$

and it follows that

$$\ln|x| = \sin\frac{y}{x} \qquad \text{or} \qquad x = e^{\sin(y/x)}$$

35.   Find the orthogonal trajectories of the family of curves $y^2 = Cx^3$, and sketch several members of each family.

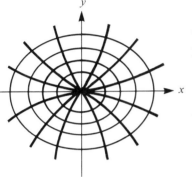

Solution

First, we solve for C in the given equation and obtain

$$C = \frac{y^2}{x^3}$$

Then, differentiating implicitly with respect to x and substituting the expression for C given above, we have

$$2yy' = 3Cx^2 = 3\left(\frac{y^2}{x^3}\right)x^2 = \frac{3y^2}{x}$$

Therefore

$$\frac{dy}{dx} = \frac{3y}{2x} \qquad \text{Slope of given family}$$

Since dy/dx represents the slope of the given family of curves at (x, y), it follows that the orthogonal family has the negative reciprocal slope, and we write

$$\frac{dy}{dx} = -\frac{2x}{3y} \qquad \text{Slope of orthogonal family}$$

Now, we find the orthogonal family by separating variables and integrating

$$3\int y\,dy = -2\int x\,dx$$

$$\frac{3}{2}y^2 = -x^2 + C_1$$

$$2x^2 + 3y^2 = C_2$$

41.   The rate of decomposition of radioactive radium is proportional to the amount present at a given instant. Find the percentage that remains of a present amount after 25 years if the half-life of radioactive radium is 1600 years.

Solution

Letting y represent the percentage amount of radioactive radium present at time t, we have the differential equation

$$y' = ky$$

and the general solution has the form

$$y = Ce^{kt} = C(e^k)^t$$

Since $y = 100\% = 1$ when $t = 0$, we have

$$1 = Ce^0 = C$$

Furthermore, $y = \frac{1}{2}$ when $t = 1600$. Hence

$$\frac{1}{2} = e^{k(1600)} = (e^k)^{1600} \quad \text{or} \quad e^k = \left(\frac{1}{2}\right)^{1/1600}$$

and we have

$$y = \left(\frac{1}{2}\right)^{t/1600}$$

Finally, when $t = 25$, we have

$$y = \left(\frac{1}{2}\right)^{25/1600} = \left(\frac{1}{2}\right)^{1/64} = 0.989 = 98.9\%$$

17.3
First-order exact equations

3. Test the differential equation

$$(3y^2 + 10xy^2)\,dx + (6xy - 2 + 10x^2y)\,dy = 0$$

for exactness and solve the equation if it is exact.

Solution

Since

$$M(x, y)\,dx + N(x, y)\,dy = (3y^2 + 10xy^2)\,dx$$
$$+ (6xy - 2 + 10x^2y)\,dy = 0$$

we have

$$\frac{\partial M}{\partial y} = 6y + 20xy = \frac{\partial N}{\partial x}$$

and the equation is exact. We next find a function f such that $f_x(x, y) = M(x, y)$ and $f_y(x, y) = N(x, y)$. Partial integration yields

$$f(x, y) = \int (3y^2 + 10xy^2) \, dx = 3xy^2 + 5x^2y^2 + g(y)$$

Then

$$\overbrace{\qquad\qquad}^{N(x, y)}$$

$$f_y(x, y) = 6xy + 10x^2y + g'(y) = 6xy - 2 + 10x^2y$$

Therefore

$$g'(y) = -2$$

and $$\qquad g(y) = \int -2 \, dy = -2y + C_1$$

Thus the general solution is

$$f(x, y) = C \qquad \text{or} \qquad 3xy^2 + 5x^2y^2 - 2y = C$$

7. Test the differential equation

$$\frac{1}{x^2 + y^2} (x \, dy - y \, dx) = 0$$

for exactness and solve the equation if it is exact.

Solution

Since

$$M(x, y) \, dx + N(x, y) \, dy = \frac{-y}{x^2 + y^2} \, dx + \frac{x}{x^2 + y^2} \, dy = 0$$

we have

$$\frac{\partial M}{\partial y} = \frac{y^2 - x^2}{x^2 + y^2} = \frac{\partial N}{\partial x}$$

and the equation is exact. We next find a function f such that $f_x(x, y) = M(x, y)$ and $f_y(x, y) = N(x, y)$.

$$f(x, y) = \int \frac{-y}{x^2 + y^2} \, dx = -\arctan \frac{x}{y} + g(y)$$

and

$$\overbrace{\qquad\qquad}^{N(x, y)}$$

$$f_y(x, y) = -\frac{-x/y^2}{1 + (x^2/y^2)} + g'(y) = \frac{x}{x^2 + y^2} + g'(y)$$

Therefore $g'(y) = 0$ and $g(y) = C_1$. Thus, the general solution is

$$f(x, y) = C \qquad \text{or} \qquad \arctan \frac{x}{y} = C$$

13. Find the particular solution of the differential equation

$$\frac{1}{x^2 + y^2}(x \, dx + y \, dy) = 0$$

that satisfies the boundary condition $y(0) = 4$.

Solution

Since $M(x, y) = \dfrac{x}{x^2 + y^2}$ and $N(x, y) = \dfrac{y}{x^2 + y^2}$, we have

$$\frac{\partial M}{\partial y} = \frac{-2xy}{(x^2 + y^2)^2} = \frac{\partial N}{\partial x}$$

and the equation is exact. We next find a function f such that $f_x(x, y) = M(x, y)$ and $f_y(x, y) = N(x, y)$.

$$f(x, y) = \int \frac{x}{x^2 + y^2} \, dx = \ln \sqrt{x^2 + y^2} + g(y)$$

$$f(x, y) = \int \frac{y}{x^2 + y^2} \, dy = \ln \sqrt{x^2 + y^2} + h(x)$$

Therefore, $g(y) = h(x) = C_1$ and the general solution to the differential equation is

$$f(x, y) = C \qquad \text{or} \qquad \ln \sqrt{x^2 + y^2} = C$$

Finally, since $y = 4$ when $x = 0$, we have $\ln 4 = C$, and the particular solution is $\ln \sqrt{x^2 + y^2} = \ln 4$ or $x^2 + y^2 = 16$.

21. Solve the differential equation

$$(x + y) \, dx + \tan x \, dy = 0$$

by using an integrating factor that is a function of x or y alone.

Solution

Since

$$\frac{(\partial M / \partial y) - (\partial N / \partial x)}{N} = \frac{1 - \sec^2 x}{\tan x} = h(x)$$

and

$$\int \frac{1 - \sec^2 x}{\tan x} \, dx = -\int \frac{\tan^2 x}{\tan x} \, dx = -\int \tan x \, dx = \ln |\cos x|$$

it follows that

$$e^{\int h(x)\,dx} = e^{\ln|\cos x|} = \cos x$$

is an integrating factor. Thus

$$(x + y) \cos x \, dx + \cos x \tan x \, dy = 0$$
$$(x + y) \cos x \, dx + \sin x \, dy = 0$$

is exact and we have

$$f(x, y) = \int \sin x \, dy = y \sin x + g(x)$$

and

$$f_x(x, y) = y \cos x + g'(x) = \overbrace{(x + y) \cos x}^{M(x,\ y)}$$

Therefore,

$$g'(x) = x \cos x, \qquad g(x) = \int x \cos x \, dx = x \sin x + \cos x + C_1$$

and

$$f(x, y) = y \sin x + x \sin x + \cos x + C_1$$
$$= (x + y) \sin x + \cos x + C_1$$

Finally the general solution is

$$f(x, y) = C \qquad \text{or} \qquad (x + y) \sin x + \cos x = C$$

29. Find an integrating factor of the form $x^m y^n$ for the differential equation

$$(-y^5 + x^2 y) \, dx + (2xy^4 - 2x^3) \, dy = 0$$

and use this factor to find the general solution of the equation.

Solution

By regrouping terms of equal degrees, we have

$$(-y^5 \, dx + 2xy^4 \, dy) + (x^2 y \, dx - 2x^3 \, dy) = 0$$
$$y^4(-y \, dx + 2x \, dy) + x^2(y \, dx - 2x \, dy) = 0$$

For the left grouping, $m = -1$, $n = 2$, and $x^{m-1}y^{n-1} = x^{-2}y$, whereas for the right grouping, $m = 1$, $n = -2$, and $x^{m-1}y^{n-1} = y^{-3}$. Since the left grouping already has y^4 as a factor and the right grouping has x^2, the integrating factor is

$$\frac{x^{-2}y}{y^4} = \frac{1}{x^2 y^3} = \frac{y^{-3}}{x^2}$$

Multiplication by this factor yields

$$\frac{y}{x^2}(-y\,dx + 2x\,dy) + \frac{1}{y^3}(y\,dx - 2x\,dy) = 0$$

$$d[x^{-1}y^2] + d[xy^{-2}] = 0$$

Integrating, we obtain the general solution

$$\frac{y^2}{x} + \frac{x}{y^2} = C$$

17.4
First-order linear equations

5. Solve the first-order linear differential equation $y' - y = \cos x$, $y(0) = \frac{1}{2}$.

Solution

The equation $y' - y = \cos x$ is linear with $P(x) = -1$. Thus

$$\int P(x)\,dx = -x \qquad \text{and} \qquad e^{\int P(x)\,dx} = e^{-x}$$

Therefore by Theorem 17.4 the general solution has the form

$$ye^{-x} = \int e^{-x}\cos x\,dx + C = \frac{1}{2}(e^{-x}\sin x - e^{-x}\cos x) + C$$

$$y = \frac{1}{2}(\sin x - \cos x) + Ce^x$$

Since $y(0) = \frac{1}{2}$, it follows that $C = 1$, and the particular solution is

$$y = \frac{1}{2}(\sin x - \cos x) + e^x$$

9. Solve the first-order linear differential equation

$$(x - 1)y' + y = x^2 - 1$$

Solution

In standard form the equation is

$$y' + \left(\frac{1}{x - 1}\right)y = x + 1$$

Thus

$$\int P(x)\, dx = \int \frac{1}{x-1}\, dx = \ln |x-1|$$

$$e^{\int P(x)\, dx} = e^{\ln |x-1|} = x - 1$$

Therefore by Theorem 17.4 the general solution has the form

$$y(x-1) = \int (x-1)(x+1)\, dx = \int (x^2-1)\, dx = \frac{x^3}{3} - x + C_1$$

$$y = \frac{x^3 - 3x + C}{3(x-1)}$$

19. Solve the Bernoulli differential equation

$$y' + \left(\frac{1}{x}\right) y = xy^2$$

Solution

For the given equation we have $n = 2$, $1 - n = -1$, and $P(x) = 1/x$. Thus we have

$$\int (1-n)P(x)\, dx = -\int \frac{1}{x}\, dx = -\ln |x|$$

$$e^{-\ln |x|} = \frac{1}{x}$$

Since $Q(x) = x$, the general solution is

$$y^{-1}e^{-\ln |x|} = \int (-1)xe^{-\ln |x|}\, dx$$

$$\frac{1}{xy} = \int (-1)\, dx = -x + C$$

$$\frac{1}{y} = -x^2 + Cx$$

$$y = \frac{1}{Cx - x^2}$$

26. A 200-gal tank is full of a solution containing 25 lb of concentrate. Starting at time $t = 0$, distilled water is admitted to the tank at the rate of 10 gal/min, and the well-stirred solution is withdrawn at the same rate.

(a) Find the amount of the concentrate in the solution as a function of t.

(b)   Find the time at which concentrate must be added if it is done when the amount of concentrate in the tank reaches 15 lb.

Solution

(a)   Using Exercise 25, we have $r_1 = 10$, $r_2 = 10$, $q_1 = 0$, and $v_0 = 200$. Thus we have

$$\frac{dQ}{dt} + \frac{10Q}{200 + (0)t} = 0$$

$$\frac{dQ}{dt} = \frac{-10Q}{200} = -\frac{Q}{20}$$

Separating variables, we have

$$\frac{dQ}{Q} = -\frac{dt}{20}$$

$$\ln |Q| = -\frac{t}{20} + C_1$$

$$Q = e^{C_1 - (t/20)} = Ce^{-t/20}$$

(b)   When $Q = 15$, we have

$$15 = 25e^{-t/20}$$

$$\frac{3}{5} = e^{-t/20}$$

$$\ln 3 - \ln 5 = -\frac{t}{20}$$

$$t = 20(\ln 5 - \ln 3) \approx 10.2 \text{ min}$$

29.   Solve the differential equation $e^{2x+y} \, dx - e^{x-y} \, dy = 0$ by any appropriate method.

Solution

Separating variables yields

$$e^{2x}e^y \, dx = e^x e^{-y} \, dy$$

$$e^x \, dx = e^{-2y} \, dy$$

$$e^x = -\frac{1}{2}e^{-2y} + C_1$$

$$2e^x + e^{-2y} = C$$

35.   Solve the differential equation $2xy \, dx + (x^2 + \cos y) \, dy = 0$ by any appropriate method.

Solution

Since

$$\frac{\partial M}{\partial y} = 2x = \frac{\partial N}{\partial x}$$

the equation is exact. Thus

$$f(x, y) = \int 2xy \, dx + g(y) = x^2y + g(y)$$

$$f_y(x, y) = x^2 + g'(y) = \overbrace{x^2 + \cos y}^{N(x, y)}$$

Therefore

$$g(y) = \int \cos y \, dy = \sin y + C_1$$

and

$$f(x, y) = x^2y + \sin y + C_1$$

Hence, the general solution is

$$x^2y + \sin y = C$$

41. Solve the differential equation $(x^2y^4 - 1) \, dx + x^3y^3 \, dy = 0$ by any appropriate method.

Solution

In linear form we have

$$x^3y^3 \frac{dy}{dx} + x^2y^4 = 1$$

$$\frac{dy}{dx} + \left(\frac{1}{x}\right)y = x^{-3}y^{-3}$$

which is a Bernoulli equation with $P(x) = 1/x$, $Q(x) = x^{-3}$, and $n = -3$. Thus

$$\int (1 - n)P(x) \, dx = \int 4\left(\frac{1}{x}\right) dx = 4 \ln |x|$$

and

$$e^{4 \ln |x|} = x^4$$

Thus the general solution is

$$y^4x^4 = \int 4(x^{-3})(x^4) \, dx = 4 \int x \, dx = 2x^2 + C$$

$$x^4y^4 - 2x^2 = C$$

17.5
Second-order homogeneous linear equations

3. Find the general solution of the linear differential equation
    $y'' - y' - 6y = 0$.

Solution

The characteristic equation

$$m^2 - m - 6 = 0 \quad \text{or} \quad (m - 3)(m + 2) = 0$$

has two distinct real roots, $m_1 = 3$ and $m_2 = -2$ (Case I).
Thus the general solution is

$$y = C_1 e^{m_1 x} + C_2 e^{m_2 x} = C_1 e^{3x} + C_2 e^{-2x}$$

7. Find the general solution of the linear differential equation
    $y'' + 6y' + 9y = 0$.

Solution

The characteristic equation

$$m^2 + 6m + 9 = 0 \quad \text{or} \quad (m + 3)^2 = 0$$

has equal roots, $m_1 = m_2 = -3$ (Case II). Thus the general
solution is

$$y = C_1 e^{m_1 x} + C_2 x e^{m_1 x} = C_1 e^{-3x} + C_2 x e^{-3x} = (C_1 + C_2 x)e^{-3x}$$

17. Find the general solution of the linear differential equation
    $y'' - 3y' + y = 0$.

Solution

The characteristic equation

$$m^2 - 3m + 1 = 0$$

has the distinct real roots (using the quadratic formula)

$$m = \frac{3 \pm \sqrt{9 - 4}}{2} = \frac{3 \pm \sqrt{5}}{2}$$

Thus the general solution is

$$y = C_1 e^{(3+\sqrt{5})x/2} + C_2 e^{(3-\sqrt{5})x/2}$$

19. Find the general solution of the linear differential equation
    $9y'' - 12y' + 11y = 0$.

Solution

The characteristic equation

$$9m^2 - 12m + 11 = 0$$

has complex roots (Case III).

$$m = \frac{12 \pm \sqrt{144 - 4(9)(11)}}{2(9)} = \frac{12 \pm \sqrt{-252}}{18}$$

$$= \frac{12 \pm 6i\sqrt{7}}{18} = \frac{2}{3} \pm \frac{\sqrt{7}}{3}i$$

Thus $\alpha = 2/3$ and $\beta = \sqrt{7}/3$ and the general solution is

$$y = C_1 e^{2x/3} \cos \frac{\sqrt{7}x}{3} + C_2 e^{2x/3} \sin \frac{\sqrt{7}x}{3}$$

$$= e^{2x/3}\left[C_1 \cos \frac{\sqrt{7}x}{3} + C_2 \sin \frac{\sqrt{7}x}{3}\right]$$

33. Describe the motion of a 32-pound weight suspended on a spring. Assume that the weight stretches the spring $\frac{2}{3}$ foot from its natural position, it is pulled $\frac{1}{2}$ foot below the equilibrium and released, and the motion takes place in a medium which furnishes a damping force of magnitude $\frac{1}{8}$ its speed at all times.

Solution

By Hooke's Law, $32 = k(2/3)$ so that $k = 48$. Moreover, since the weight w is given by mg, it follows that $m = w/g = \frac{32}{32} = 1$. Also the damping force is given by $-\frac{1}{8}(dy/dt)$. Thus the differential equation for the oscillations of the weight is

$$m\left(\frac{d^2y}{dt^2}\right) = -\frac{1}{8}\left(\frac{dy}{dt}\right) - 48y$$

$$m\left(\frac{d^2y}{dt^2}\right) + \frac{1}{8}\left(\frac{dy}{dt}\right) + 48y = 0$$

In this case the characteristic equation is

$$8m^2 + m + 384 = 0$$

with complex roots

$$m = -\frac{1}{16} \pm \frac{\sqrt{12287}\,i}{16}$$

Therefore, the general solution is

$$y(t) = e^{-t/16}\left(C_1 \cos \frac{\sqrt{12287}\,t}{16} + C_2 \sin \frac{\sqrt{12287}\,t}{16}\right)$$

Using the initial conditions, we have

$$y(0) = C_1 = \frac{1}{2}$$

$$y'(t) = e^{-t/16}\left[\left(-\frac{\sqrt{12287}}{16}C_1 - \frac{C_2}{16}\right)\sin\frac{\sqrt{12287}\ t}{16}\right.$$

$$\left.+ \left(\frac{\sqrt{12287}}{16}C_2 - \frac{C_1}{16}\right)\cos\frac{\sqrt{12287}\ t}{16}\right]$$

$$y'(0) = \frac{\sqrt{12287}}{16}C_2 - \frac{C_1}{16} = 0 \qquad \longrightarrow \qquad C_2 = \frac{\sqrt{12287}}{24{,}574}$$

and the particular solution

$$y(t) = \frac{e^{-t/16}}{2}\left[\cos\frac{\sqrt{12287}\ t}{16} + \frac{\sqrt{12287}}{12287}\sin\frac{\sqrt{12287}\ t}{16}\right]$$

17.6
Second-order nonhomogeneous linear equations

3. Solve the differential equation $y'' + y = x^3$, $y(0) = 1$, $y'(0) = 0$ by the method of undetermined coefficients.

Solution

The characteristic equation $m^2 + 1 = 0$ has roots $m = \pm i$. Thus we have

$$y_h = C_1 \cos x + C_2 \sin x$$

Since $F(x) = x^3$, we choose y_p to be

$$y_p = A + Bx + Cx^2 + Dx^3$$

Thus $y_p' = B + 2Cx + 3Dx^2$ and $y_p'' = 2C + 6Dx$. Substitution into the given differential equation yields

$$2C + 6Dx + A + Bx + Cx^2 + Dx^3 = x^3$$

Therefore

$$2C + A = 0, \qquad 6D + B = 0, \qquad C = 0, \qquad D = 1$$

from which it follows that $A = 0$ and $B = -6$. Thus the general solution is

$$y = y_h + y_p = C_1 \cos x + C_2 \sin x - 6x + x^3$$

Now since $y(0) = 1$, $y'(0) = 0$, and $y' = -C_1 \sin x + C_2 \cos x - 6 + 3x^2$, we have

$$1 = C_1(1) + C_2(0) = C_1$$
$$0 = -C_1(0) + C_2(1) - 6 \quad \text{or} \quad C_2 = 6$$

Finally, the particular solution is

$$y = \cos x + 6 \sin x - 6x + x^3$$

11. Solve the differential equation $y'' + 9y = \sin 3x$ by the method of undetermined coefficients.

Solution

The characteristic equation $m^2 + 9 = 0$ has roots $m = \pm 3i$ and we have

$$y_h = C_1 \cos 3x + C_2 \sin 3x$$

Since $F(x) = \sin 3x$, we consider

$$y_p = A \cos 3x + B \sin 3x$$

However, both terms of y_p are similar to those of y_h; thus we multiply both terms by x and write

$$y_p = Ax \cos 3x + Bx \sin 3x$$
$$y_p{'} = -3Ax \sin 3x + A \cos 3x + 3Bx \cos 3x + B \sin 3x$$
$$y_p{''} = -6A \sin 3x + 6B \cos 3x - 9Ax \cos 3x - 9Bx \sin 3x$$

Therefore substitution into the given differential equation yields

$$y_p{''} + 9y_p = -6A \sin 3x + 6B \cos 3x = \sin 3x$$

which means that $A = -\frac{1}{6}$ and $B = 0$. Therefore the general solution is

$$y = y_h + y_p = C_1 \cos 3x + C_2 \sin 3x - \frac{x}{6} \cos 3x$$

19. Solve the differential equation $y'' + 4y = \csc 2x$ by the method of variation of parameters.

Solution

The characteristic equation $m^2 + 4 = 0$ has solutions $m = \pm 2i$, and hence

$$y_h = C_1 \cos 2x + C_2 \sin 2x$$

Replacing C_1 and C_2 by u_1 and u_2, respectively, we write

$$y_p = u_1 \cos 2x + u_2 \sin 2x$$

By the method of variation of parameters we obtain the system

$$u_1' \cos 2x + u_2' \sin 2x = 0$$
$$u_1'(-2 \sin 2x) + u_2'(2 \cos 2x) = \csc 2x$$

Multiplying the first equation by $2 \sin 2x$ and the second by $\cos 2x$, and then adding the equations yields $u_2' = \frac{1}{2} \cot 2x$. Then, by substitution in the first equation, we have $u_1' = -\frac{1}{2}$. Integration yields

$$u_1 = \int -\frac{1}{2} \, dx = -\frac{x}{2}$$

$$u_2 = \int \frac{1}{2} \cot 2x \, dx = \frac{1}{4} \ln |\sin 2x|$$

and it follows that

$$y = y_h + y_p$$

$$= C_1 \cos 2x + C_2 \sin 2x - \frac{x}{2} \cos 2x + \frac{1}{4} \sin 2x \ln |\sin 2x|$$

$$= \left(C_1 - \frac{x}{2} \right) \cos 2x + \left(C_2 + \frac{1}{4} \ln |\sin 2x| \right) \sin 2x$$

23. Use the differential equation

$$5y'' + 80y' + 140y = 5 \sin 2t$$

to describe the motion of a 5-kg mass attached to a spring with a spring constant of 140 N/m and an external driving force of $F(t) = 5 \sin 2t$. Assume the mass starts from the equilibrium position with an imparted upward velocity of 2 m/sec and that the motion takes place in a medium which furnishes a damping force of $-80y'$.

Solution

We first divide the equation by 5. Then the characteristic equation $m^2 + 16m + 28 = 0$, has solutions $m = -2$ and $m = -14$. Thus

$$y_h = C_1 e^{-2t} + C_2 e^{-14t}$$

and we obtain the following system:

$$u_1' e^{-2t} + u_2' e^{-14t} = 0$$
$$u_1'(-2e^{-2t}) + u_2'(-14e^{-14t}) = \sin 2t$$

Multiplying the first equation by 14 and then adding the two resulting equations yields

$$u_1' = \frac{1}{12} (\sin 2t) e^{2t}$$

Now, we multiply the first equation by 2 and add the resulting equations to obtain

$$u_2' = -\frac{1}{12}(\sin 2t)e^{14t}$$

Therefore

$$u_1 = \frac{1}{12}\int e^{2t}\sin 2t\, dt = \frac{e^{2t}}{48}(\sin 2t - \cos 2t)$$

$$u_2 = -\frac{1}{12}\int e^{14t}\sin 2t\, dt = \frac{-7e^{14t}}{600}\left(\frac{1}{2}\sin 2t - \frac{1}{14}\cos 2t\right)$$

from which it follows that

$$y_p = \frac{\sin 2t}{48} - \frac{\cos 2t}{48} - \frac{7}{1200}\sin 2t + \frac{1}{1200}\cos 2t$$

$$= \frac{3}{200}\sin 2t - \frac{1}{50}\cos 2t$$

Thus we have

$$y = y_h + y_p = C_1 e^{-2t} + C_2 e^{-14t} + \frac{3}{200}\sin 2t - \frac{1}{50}\cos 2t$$

$$y' = -2C_1 e^{-2t} - 14C_2 e^{-14t} + \frac{3}{100}\cos 2t + \frac{1}{25}\sin 2t$$

Furthermore, if $y(0) = 0$ and $y'(0) = 2$, we have

$$0 = C_1 + C_2 - \frac{1}{50} \quad\text{and}\quad 2 = -2C_1 - 14C_2 + \frac{3}{100}$$

which yields

$$C_1 = \frac{225}{1200} \quad\text{and}\quad C_2 = \frac{-201}{1200}$$

Finally, we obtain the solution

$$y = \frac{225}{1200}e^{-2t} - \frac{201}{1200}e^{-14t} + \frac{3}{200}\sin 2t - \frac{1}{50}\cos 2t$$

17.7
Series solutions of differential equations

3. Use power series to solve the differential equation
    $y'' - 9y = 0$.

Solution

a. Assume $y = \sum_{n=0}^{\infty} a_n x^n$ is a solution; then

$$y' = \sum_{n=0}^{\infty} n a_n x^{n-1} \qquad \text{and} \qquad y'' = \sum_{n=0}^{\infty} n(n-1) a_n x^{n-2}$$

Thus the equation $y'' - 9y = 0$ is written as

$$\sum_{n=0}^{\infty} n(n-1) a_n x^{n-2} - 9 \sum_{n=0}^{\infty} a_n x^n = 0$$

or

$$\sum_{n=0}^{\infty} n(n-1) a_n x^{n-2} = \sum_{n=0}^{\infty} 9 a_n x^n$$

b. Adjusting the indices by replacing n by $n + 2$ in the left sum, we obtain

$$\sum_{n=-2}^{\infty} (n+2)(n+1) a_{n+2} x^n = \sum_{n=0}^{\infty} 9 a_n x^n$$

c. Now by equating coefficients, we have

$$(n+2)(n+1) a_{n+2} = 9 a_n$$

from which we obtain the recursion formula

$$a_{n+2} = \frac{9 a_n}{(n+2)(n+1)} \qquad (n \geq 0)$$

Thus the coefficients of the solution series are

$$a_2 = \frac{9}{2} a_0 = \frac{3^2}{2} a_0 \qquad\qquad a_3 = \frac{9}{6} a_1 = \frac{3^2}{3 \cdot 2} a_1$$

$$a_4 = \frac{9}{12} a_2 = \frac{3^4}{4 \cdot 3 \cdot 2} a_0 \qquad\qquad a_5 = \frac{9}{20} a_3 = \frac{3^4}{5 \cdot 4 \cdot 3 \cdot 2} a_1$$

$$\vdots \qquad\qquad\qquad\qquad \vdots$$

$$a_{2n} = \frac{(3)^{2n}}{(2n)!} a_0 \qquad\qquad a_{2n+1} = \frac{(3)^{2n}}{(2n+1)!} a_1$$

d. In this case we can write the general solution as the sum of two power series: one for the even-powered terms and one for the odd-powered terms. Thus we have

$$y = \left(a_0 x^0 + \frac{3^2}{2!} a_0 x^2 + \frac{3^4}{4!} a_0 x^4 + \cdots \right)$$

$$+ \left(a_1 x + \frac{3^2}{3!} a_1 x^3 + \frac{3^4}{5!} a_1 x^5 + \cdots \right)$$

$$= a_0 \left[1 + \frac{(3x)^2}{2!} + \frac{(3x)^4}{4!} + \cdots \right] + \frac{a_1}{3} \left[3x + \frac{(3x)^3}{3!} + \frac{(3x)^5}{5!} + \cdots \right]$$

$$= a_0 \sum_{n=0}^{\infty} \frac{(3x)^{2n}}{(2n)!} + \frac{a_1}{3} \sum_{n=0}^{\infty} \frac{(3x)^{2n+1}}{(2n+1)!}$$

We observe that $y'' - 9y = 0$ is a second-order homogeneous differential equation with characteristic equation $m^2 - 9 = 0$. Thus, the general solution is

$$y = C_1 e^{3x} + C_2 e^{-3x}$$

We now reconcile this form of the solution with the series solution given above.

$$y = C_1 e^{3x} + C_2 e^{-3x}$$

$$= C_1 \sum_{n=0}^{\infty} \frac{(3x)^n}{n!} + C_2 \sum_{n=0}^{\infty} \frac{(-3x)^n}{n!}$$

$$= C_1 \left[1 + (3x) + \frac{(3x)^2}{2!} + \cdots \right]$$

$$+ C_2 \left[1 + (-3x) + \frac{(-3x)^2}{2!} + \cdots \right]$$

$$= C_1 \left[1 + \frac{(3x)^2}{2!} + \frac{(3x)^4}{4!} + \cdots \right]$$

$$+ C_1 \left[(3x) + \frac{(3x)^3}{3!} + \frac{(3x)^5}{5!} + \cdots \right]$$

$$+ C_2 \left[1 + \frac{(-3x)^2}{2!} + \frac{(-3x)^4}{4!} + \cdots \right]$$

$$+ C_2 \left[(-3x) + \frac{(-3x)^3}{3!} + \frac{(-3x)^5}{5!} + \cdots \right]$$

$$= (C_1 + C_2) \left[1 + \frac{(3x)^2}{2!} + \frac{(3x)^4}{4!} + \cdots \right]$$

$$+ (C_1 - C_2) \left[(3x) + \frac{(3x)^3}{3!} + \cdots \right]$$

$$= a_0 \sum_{n=0}^{\infty} \frac{(3x)^{2n}}{(2n)!} + \frac{a_1}{3} \sum_{n=0}^{\infty} \frac{(3x)^{2n+1}}{(2n+1)!}$$

where

$$C_1 + C_2 = a_0 \qquad \text{and} \qquad C_1 - C_2 = \frac{a_1}{3}$$

9. Use power series to solve the differential equation
    $xy'' - y = 0$.

Solution

a. Assume $y = \sum_{n=0}^{\infty} a_n x^n$ is a solution; then

$$y' = \sum_{n=0}^{\infty} n a_n x^{n-1} \qquad \text{and} \qquad y'' = \sum_{n=0}^{\infty} n(n-1)a_n x^{n-2}$$

Thus the equation $xy'' - y = 0$ becomes

$$x \sum_{n=0}^{\infty} n(n-1)a_n x^{n-2} - \sum_{n=0}^{\infty} a_n x^n = 0$$

or

$$\sum_{n=0}^{\infty} n(n-1)a_n x^{n-1} = \sum_{n=0}^{\infty} a_n x^n$$

b. Adjusting the indices to get equal powers, we replace n by $n + 1$ in the left sum. Thus

$$\sum_{n=-1}^{\infty} (n+1)(n)a_{n+1}x^n = \sum_{n=0}^{\infty} a_n x^n$$

c. Equating coefficients, we obtain

$$(n+1)(n)a_{n+1} = a_n \qquad \text{or} \qquad a_{n+1} = \frac{a_n}{(n+1)n}$$

Thus the coefficients for the solution series are

$$a_2 = \frac{1}{2}a_1$$

$$a_3 = \frac{1}{3 \cdot 2}a_2 = \frac{1}{(3 \cdot 2) \cdot 2}a_1 = \frac{1}{3!\,2!}a_1$$

$$a_4 = \frac{1}{4 \cdot 3}a_3 = \frac{1}{(4 \cdot 3 \cdot 2)(3 \cdot 2)}a_1 = \frac{1}{4!\,3!}a_1$$

$$\vdots$$

$$a_n = \frac{a_1}{n!(n-1)!}$$

d. In this case we have the series solution

$$y = a_0 x^0 + a_1 x^1 + a_2 x^2 + \cdots$$

$$= a_0 + a_1 x + \frac{a_1}{2!}x^2 + \frac{a_1}{3!\,2!}x^3 + \frac{a_1}{4!\,3!}x^4 + \cdots$$

$$= a_0 + a_1 \sum_{n=1}^{\infty} \frac{x^n}{n!(n-1)!}$$

15. Use Taylor's Theorem to find the series solution of

$y'' - 2xy = 0$ given that $y(0) = 1$ and $y'(0) = -3$. Use the first six terms of the series to approximate y when $x = \frac{1}{4}$.

Solution

Taylor's Theorem for $c = 0$ is

$$y = y(0) + y'(0)x + \frac{y''(0)}{2!}x^2 + \frac{y'''(0)}{3!}x^3 + \cdots$$

Since $y'' = 2xy$, $y(0) = 1$ and $y'(0) = -3$, we have

$$y(0) = 1$$
$$y'(0) = -3$$
$$y'' = 2xy \qquad\qquad y''(0) = 0$$
$$y''' = 2xy' + 2y \qquad\qquad y'''(0) = 2$$
$$y^{(4)} = 2xy'' + 4y' \qquad\qquad y^{(4)}(0) = -12$$
$$y^{(5)} = 2xy''' + 6y'' \qquad\qquad y^{(5)}(0) = 0$$
$$y^{(6)} = 2xy^{(4)} + 8y''' \qquad\qquad y^{(6)}(0) = 16$$
$$y^{(7)} = 2xy^{(5)} + 10y^{(4)} \qquad\qquad y^{(7)}(0) = -120$$

Therefore the first six terms of the Taylor series are

$$y = 1 - 3x + 0x^2 + \frac{2}{3!}x^3 - \frac{12}{4!}x^4 + 0x^5 + \frac{16}{6!}x^6 - \frac{120}{7!}x^7 + \cdots$$

$$= 1 - 3x + \frac{1}{3}x^3 - \frac{1}{2}x^4 + \frac{1}{45}x^6 - \frac{1}{42}x^7 + \cdots$$

Finally, at $x = \frac{1}{4}$ we have

$$y = 1 - \frac{3}{4} + \frac{1}{3 \cdot 4^3} - \frac{1}{2 \cdot 4^4} + \frac{1}{45 \cdot 4^6} + \frac{1}{42 \cdot 4^7} \approx 0.253$$

Review Exercises for Chapter 17

5. Find the general solution of the first-order differential equation

$$\frac{dy}{dx} - \frac{y}{x} = \frac{x}{y}$$

Solution

The given equation

$$\frac{dy}{dx} - \left(\frac{1}{x}\right)y = xy^{-1}$$

is a Bernoulli equation with $P(x) = -1/x$, $n = -1$, and

$1 - n = 2$. Thus we have

$$\int (1 - n)P(x)\, dx = \int -\frac{2}{x}\, dx = -2 \ln |x|$$

$$e^{\int (1-n)P(x)\, dx} = e^{-2 \ln |x|} = \frac{1}{x^2}$$

Since $Q(x) = x$, the general solution is

$$y^2\left(\frac{1}{x^2}\right) = \int (2)(x)\left(\frac{1}{x^2}\right)\, dx = \int \frac{2}{x}\, dx = 2 \ln |x| + C$$

or

$$y^2 = 2x^2 \ln |x| + Cx^2 = x^2 \ln x^2 + Cx^2$$

11. Find the general solution of the first-order differential equation

$$(2x - 2y^3 + y)\, dx + (x - 6xy^2)\, dy = 0$$

Solution

Since

$$\frac{\partial M}{\partial y} = 1 - 6y^2 = \frac{\partial N}{\partial x}$$

the equation is exact. Therefore, there exists a function f such that

$$f_x(x, y) = 2x - 2y^3 + y \qquad \text{and} \qquad f_y(x, y) = x - 6xy^2$$

Integration yields

$$f(x, y) = \int (2x - 2y^3 + y)\, dx = x^2 - 2xy^3 + xy + g(y)$$

and

$$f(x, y) = \int (x - 6xy^2)\, dy = xy - 2xy^3 + h(x)$$

Reconciling these two versions of f, we have $h(x) = x^2$, $g(y) = C_1$, and

$$f(x, y) = x^2 - 2xy^3 + xy + C_1$$

Therefore the general solution is

$$x^2 - 2xy^3 + xy = C$$

27. Find the general solution of the first-order differential equation
$(1 + x^2)\, dy = (1 + y^2)\, dx$.

Solution

We can separate the variables and obtain

$$\frac{dy}{1 + y^2} = \frac{dx}{1 + x^2}$$

$$\int \frac{dy}{1 + y^2} = \int \frac{dx}{1 + x^2}$$

$$\arctan y = \arctan x + C_1$$

Now taking the tangent of this equation and using the identity for $\tan (A - B)$, we have

$$\tan (\arctan y - \arctan x) = \tan C_1$$

$$\frac{\tan \arctan y - \tan \arctan x}{1 + \tan (\arctan y) \tan (\arctan x)} = C$$

or

$$\frac{y - x}{1 + xy} = C$$

as the general solution.

35. Find the general solution of the second-order differential equation $y'' - 2y' + y = 2xe^x$.

Solution

Using the method of undetermined coefficients, we have

$$m^2 - 2m + 1 = 0, \qquad m_1 = 1 = m_2$$

and

$$y_h = C_1 e^x + C_2 x e^x$$

Since $F(x) = 2xe^x$, we consider

$$y_p = (A + Bx)e^x = Ae^x + Bxe^x$$

However, since y_h already contains both of these terms, we multiply by x^2 to obtain

$$y_p = Ax^2 e^x + Bx^3 e^x = e^x(Ax^2 + Bx^3)$$
$$y_p' = 2Axe^x + (3B + A)x^2 e^x + Bx^3 e^x$$
$$y_p'' = 2Ae^x + (4A + 6B)xe^x + (A + 6B)x^2 e^x + Bx^3 e^x$$

Substituting into the given differential equation, we have

$$2Ae^x + (4A + 6B)xe^x + (A + 6B)x^2 e^x + Bx^3 e^x - 4Axe^x$$
$$- 2(3B + A)x^2 e^x - 2Bx^3 e^x + Ax^2 e^x + Bx^3 e^x = 2xe^x$$

or

$$2Ae^x + 6Bxe^x = 2xe^x$$

Equating coefficients, we obtain $A = 0$ and $6B = 2$, or $B = \frac{1}{3}$. Therefore $y_p = \frac{1}{3}x^3e^x$, and the general solution is

$$y = y_h + y_p$$

$$= C_1e^x + C_2xe^x + \frac{1}{3}x^3e^x$$

$$= \left(C_1 + C_2x + \frac{1}{3}x^3\right)e^x$$